ORGANIC CHEMISTRY

PROBLEMS AND SOLUTIONS

ORGANIC CHEMISTRY
PROBLEMS AND SOLUTIONS

K.V. RAMAN

Professor of Chemistry (Retired)
St. Joseph's College
Trichy

G.R. VIJAYAGOPAL

Senior Research Scientist
DuPont Knowledge Center
Hyderabad

MJP Publishers

Cataloguing-in-Publication Data

Raman, K.V (1944-).
Organic Chemistry: Problems and Solutions / by
K.V. Raman and G.R. Vijaya Gopal. -
Chennai : MJP Publishers, 2010
xviii, 506 p.; 24 cm.
Includes References and Index.
ISBN 978-81-8094-087-3 (pb.)
1. Chemistry, Organic I. Vijaya Gopal, G.R
II. Title.
547 dc22 RAM MJP 079

ISBN 978-81-8094-087-3

Printed and bound in India

MJP PUBLISHERS
47, Nallathambi Street
Triplicane
Chennai 600 005

Publisher : J.C. Pillai
Managing Editor : C. Sajeesh Kumar
Marketing Manager : S.Y. Sekar
Project Editor : P. Parvath Radha
Acquisitions Editor : C. Janarthanan
Editorial Team : M. Gnanasoundari, Lissy John, N. Yamuna Devi, N. Thilagavathi
CIP Data : Prof. K. Hariharan, Librarian RKM Vivekananda College, Chennai.

Printed at TechShresta Solutions Private Limited, India

To

Our Family

Preface

This book entitled *Organic Chemistry: Problems and Solutions* has been written mainly to help students face competitive examinations like IIT-JEE, AIEEE, JAM, CSIR, NET and GATE examinations. Every chapter has been designed in such a way that the first part consists of a brief discussion of important matter relating to the topic and the concluding part of every chapter has the Problems section which consists of "Explain why" questions, Multiple choice questions, Assertion reason questions, Comprehension passages, Matrix matching questions and Structure-based problems. Solutions are provided for problems. Several solved problems are also given within the text.

The book consists of 13 chapters and an appendix. Chapter 1 gives a brief account of theoretical principles of organic chemistry such as stereoisomerism, tautomerism, structural isomerism, nomenclature, empirical and molecular formula determination, hydrogen bonds, resonance, inductive, mesomeric and steric effects, hyperconjugation, reactive intermediates like carbocations, carbanions, free radicals and carbenes.

Chapter 2 discusses briefly about the preparative methods of alkanes, their physical properties and chemical reactions with special emphasis on halogenation. Chapter 3 deals with the methods of preparation of alkenes, isomerism in alkenes and reactions of alkenes. The preparative methods and reactions are discussed from the point of view of mechanism and stereochemistry. Chapter 4 is on alkynes and dienes. It briefs about the acidity of alkynes, the methods of preparation of alkynes and dienes, the reactions like ozonolysis, hydration, addition of halogens and Diels–Alder reaction.

Chapter 5 is on aromatic hydrocarbons. The Huckel rule, electrophilic aromatic substitution, aromatic nucleophilic substitution, benzyne intermediate and orientation of substitution in substituted benzenes in reactions have been discussed. Chapter 6 gives a brief account of alkyl halides, their methods of preparation, the S_N1, S_N2 reactions, and the E1 and E2 reactions. Phase transfer catalysis and use of crown ethers in S_N2 reactions are also discussed.

Chapter 7 discusses the methods of preparation of alcohols and their reactions, with detailed mechanisms, the different ways of oxidation of alcohols. A discussion about Grignard reagent is also presented. Chapter 8 gives an account of ethers and phenols. The methods of preparation of ethers, the nomenclature of cyclic ethers, the reactions, Williamson synthesis including intramolecular Williamson synthesis, the methods of preparation of phenols, the reactions of phenols such as Kolbe–Schmidt reaction, Reimer–Teimann reaction, and Claisen rearrangement are presented.

Chapter 9 is on aldehydes and ketones. The methods of preparation, reactions such as haloform reaction, aldol condensation, Robinson annulation, Cannizarro reaction, etc. are discussed in detail. Chapter 10 is on acids and derivatives of acids. Acidity of acids, methods of preparation and reactions of acids are discussed. HVZ reaction, Hunsdiecker reaction are some important reactions discussed in this chapter. Chapter 11 is on amines and diazonium compounds. Methods of preparation, reactions like Hoffmann elimination and Hoffmann rearrangement are discussed. The synthetic use of aromatic diazonium salts is given in detail. Chapter 12 gives a brief account of carbohydrates, polymers and amino acids. The thirteenth chapter gives a brief account of elemental analysis and detection of functional groups present and also discusses methods of separation of binary mixtures like acids and phenols or phenols and alcohols, nitro and amino compounds, etc. More structure-based problems are given in the Appendix.

A book such as this could not have appeared without the encouragement of a number of people.

I (K.V. Raman) wish to thank my Professors, Dr. P.T. Narasimhan and Dr. J.C. Kuriacose, who were my teachers at IIT-Kanpur and IIT-Chennai during my M.Sc. and Ph.D. programmes.

We wish to thank our family members Usha Raman, Lavenya Vijayagopal, Arvind, Indu and Shreyas Vijayagopal for their constant encouragement.

We are grateful to MJP Publishers for bringing out this book in the best format within a short period.

K.V. RAMAN

G.R. VIJAYAGOPAL

Contents

1. THEORETICAL ASPECTS OF ORGANIC CHEMISTRY 1

1

THEORETICAL ASPECTS OF ORGANIC CHEMISTRY

INTRODUCTION

This chapter discusses the theoretical principles of organic chemistry. The concepts of sp, sp^2 and sp^3 hybridization are explained. The shapes of molecules like ethylene, acetylene, CO_2, NH_3 are presented. Concepts of isomerism like structural isomerism, stereoisomerism, geometrical isomerism, keto-enol tautomerism are explained with examples. The theory about chirality, prochirality, the elements of symmetry to be absent for chirality are discussed. The calculation of empirical and molecular formulae for organic compounds is presented with examples. Concepts of intermolecular and intramolecular hydrogen bonding and their consequences are explained. The acidity of acids, substituted acids, the basicity of amines are discussed. The inductive, resonance and steric effects are explained.

The reactive intermediates (carbocations, carbanions, carbenes and free radicals) are explained. The concept of resonance and hyperconjugation are outlined with examples.

HYBRIDIZATION

The concept of hybridization was originally introduced by Linus Pauling in the 1930s. There are three important hybridizations from the point of view of carbon compounds namely sp^3, sp^2 and sp.

sp^3 Hybridization

Carbon in its ground state has the configuration $1s^2 2s^2 2p^2$. One of the electrons is promoted to the 2p orbital and the 2s orbital is hybridized with three 2p orbitals to give 4 equivalent hybrids which point towards the corners of a tetrahedron. Geometry decides the type of hybrid. If it is tetrahedral, the hybridization may be sp^3 or d^3s. In CH_4 since no d orbitals are present we have sp^3 hybridization. In MnO_4^- the d orbitals are present and hybridization is d^3s. The tetrahedral angle in sp^3 hybrid is 109° 28′. In methane the 4 C—H bonds are sigma bonds and the structure of methane is given in Figure 1.1.

H
|
C
H H H

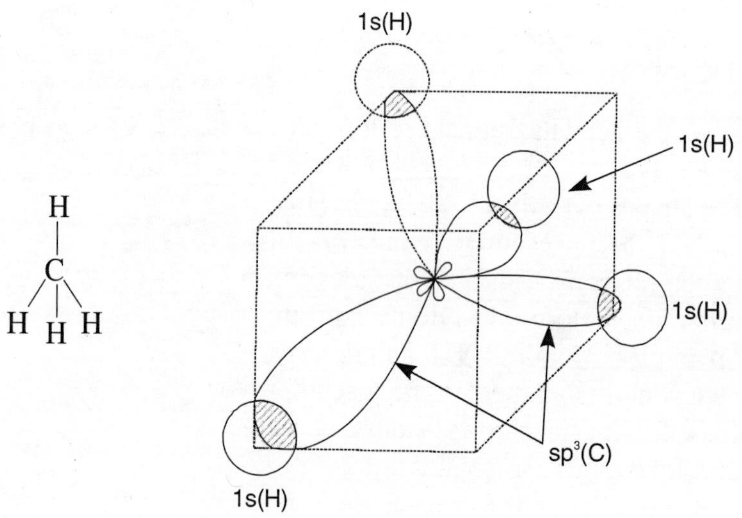

Figure 1.1 Structure of methane

Methane has four sp^3 hybrids.

sp^2 Hybridization

In ethylene each carbon gives 3 sp^2 hybrids and we have 4 C—H sigma bonds. Here the 2s hybridizes with two p orbitals to give 3 equivalent planar triangular hybrids (sp^2). The angle is 120°. There is a C—C sigma bond with third sp^2 hybrid. There are two p orbitals remaining free from each carbon and a lateral overlap between them gives a pi bond (Figure 1.2).

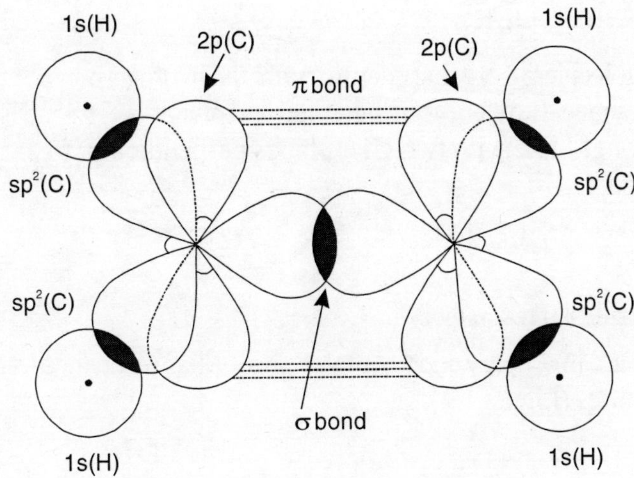

Figure 1.2 Structure of ethylene

sp Hybridization

In acetylene one s orbital and one p orbital of carbon hybridize to give two equivalent hybrids with angle 180°. The two hydrogens form two sigma bonds with two carbons in acetylene. There are two more p orbitals free on each carbon atom that overlap laterally to give two pi bonds (Figure 1.3).

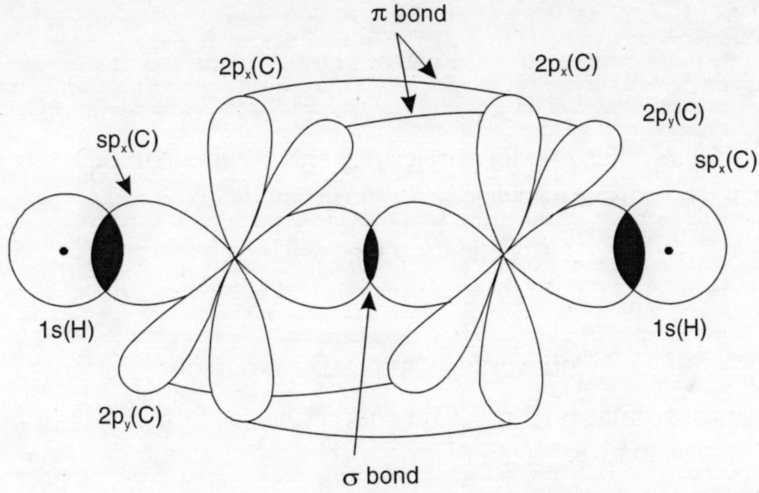

Figure 1.3 Structure of acetylene

The relative sizes of hybrid orbitals is $sp^3 > sp^2 > sp$. The electronegativity increases with increasing s character. Hence the electronegativity order is $sp > sp^2 > sp^3$. That is why we say that acetylene is more acidic than ethylene which is more acidic than ethane. This is because sp is more electronegative than sp^2 which is more electronegative than sp^3.

SHAPES OF MOLECULES

The most important molecules in organic chemistry are the hydrocarbons. Saturated hydrocarbons are tetrahedral like methane. Ethylene is unsaturated and planar. Acetylene is linear. Allene has all three carbons in a straight line with two CH_2 planes perpendicular to each other. The central carbon is sp-hybridized.

RESONANCE

The following are the rules of resonance.

1. Resonance forms are imaginary and not real. Benzene has resonance hybrid consisting of the two Kekule structures (Figure 1.4).

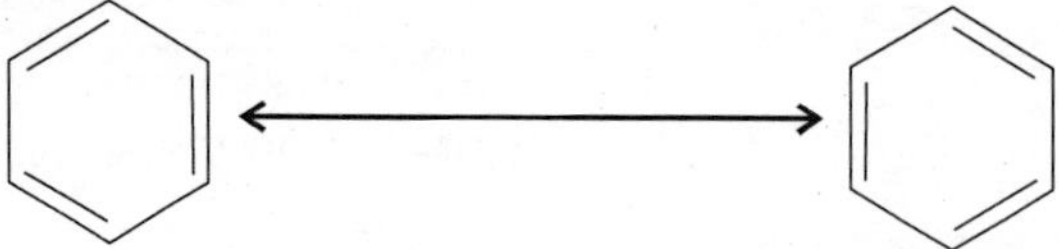

Figure 1.4 Kekule structures of benzene

Dewar structure is shown in Figure1.5.

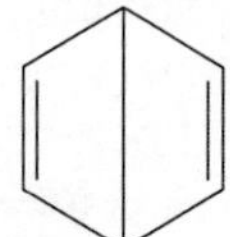

Figure 1.5 Dewar structures of benzene

2. Resonance structures differ in position of electrons only. Neither the position nor the hybridization of atoms changes from one resonance structure to another.

$$CH_3\underset{\underset{\displaystyle O}{\|}}{C}CH_3 \text{ and } CH_2{=}\underset{\underset{\displaystyle OH}{|}}{C}{-}CH_3$$

Figure 1.6 Tautomers of acetone

The tautomers of acetone (Figure 1.6) are not resonance forms since atoms have changed places. They are tautomers.

3. Different resonance forms need not be equivalent.
4. The greater the number of resonance structures, the more stable the molecule.
5. Electronegativity plays a role when we write resonance structures. For example we can consider the enolate anion of acetaldehyde as shown below.

   $—CH_2CHO$ can be represented in the following resonance forms. Figure 1.7 shows the resonance forms of enolate anion of acetaldehyde.

$$\underset{\substack{|\\ H}}{H_2\bar{C}-C{=}O} \longleftrightarrow \underset{\substack{|\\ H}}{H_2C{=}C-\bar{O}}$$

(a) (b)

Figure 1.7 Resonance forms of enolate anion of acetaldehyde

(b) contributes more than (a) since oxygen bears the negative charge.

HYPERCONJUGATION

Let us consider the ethyl cation CH_3—CH_2^+. The carbon with positive charge has an empty p orbital and electron density from the neighbouring hydrogens of the CH_3 can be delocalized on to this empty p orbital (Figure 1.8). This sigma delocalization is called hyperconjugation (no bond resonance or Baker–Nathan effect). This effect is used to explain hyperconjugation in ethyl cation.

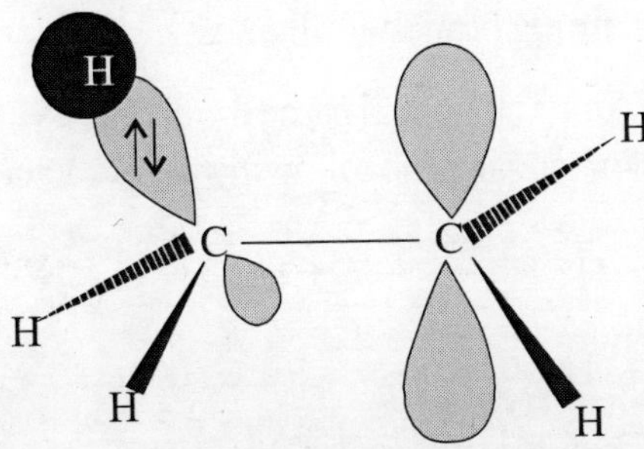

Figure 1.8 Hyperconjugation in ethyl cation

If we replace one of the hydrogens of CH_2^+ attached to CH_3 by a methyl, we get isopropyl cation $(CH_3)_2CH^+$ and now we have six alpha hydrogens, and more electron delocalization is possible. If one more hydrogen is removed from $(CH_3)_2CH^+$ and replaced by another methyl, we have nine alpha hydrogens, and delocalization is further increased. Thus a tertiary carbocation $(CH_3)_3C^+$ is more stable than $(CH_3)_2CH^+$ which is more stable than $CH_3CH_2^+$ due to hyperconjugation. The same explanation holds good for free radicals. The stability of radicals is $(CH_3)_3C. > (CH_3)_2CH. > CH_3CH_2$.

If we compare $CH_3CH_2^+$ and $^+NH_3$ - CH_3, the first carbocation is stabilized by hyperconjugation, an empty p orbital is available, and sigma delocalization happens. But in nitrogen there is no empty p orbital and there cannot be overlap with adjacent filled orbital. Resonance forms can be written but are less stable.

IUPAC NOMENCLATURE

In the nineteenth century many organic compounds were known and their names originated from their properties. Formic acid was obtained from distillation of red ants (formic in Latin means ants).

Ethane is quite inflammable and its name was derived from *aithen* (kindle or blaze). These are called common names. In 1892 a band of chemists of different countries met in Geneva, Switzerland to devise a rational system of naming organic and inorganic compounds. This was called IUPAC system of Nomenclature. IUPAC is the acronym for International Union of Pure and Applied Chemistry.

Naming of Alkanes

The saturated compounds end with the letters *ane*. The first four compounds are the common names (methane, ethane, propane and butane). For alkanes having carbons equal to or greater than five the prefixes penta (five), hexa (six), hepta (seven) etc., were included. Thus the seven-membered hydrocarbon was called heptane. When hydrogen is removed we get alkyl groups and they are named as methyl (CH_3), ethyl (C_2H_5), isopropyl [$(CH_3)_2CH$], tertiary butyl [$(CH_3)_3C$ $(CH_3)_4C$], neopentyl and so on.

The following are rules for naming branched alkanes:

Rule 1 Select the longest continuous chain and name the alkane as a derivative of the parent alkane. The longest chain may be straight or zigzag. The following two structures have the same name.

$$\underset{\displaystyle C_2H_5}{CH_3\overset{}{\underset{|}{C}}HCH_3} \qquad\qquad \underset{\displaystyle CH_3}{CH_3\underset{|}{C}HCH_2CHCH_3}$$

(a) (b)

(a and b refer to the same compound)

Rule 2 The parent chain carbons are numbered 1,2,3,4, etc. beginning with the end nearest to branch. The carbon bearing alkyl must have the lowest number.

Thus a or b has the name 2-methylbutane and not 3-methylbutane.

Rule 3 The complete name of the alkane is written as one word by prefixing the name of the alkyl group in the name of the parent alkane.

Solved Example 1

Write IUPAC name of $CH_3CH_2\underset{\displaystyle CH_2CH_3}{\underset{|}{C}}HCH_2CH_2CH_3$

Answer

There are 6 carbons and ethyl group is at position 3. It is 3-ethylhexane.

Rule 4 When 2 or more different alkyl groups are attached to the parent alkane, assign a locater number to each group and list them in alphabetical order with hyphen in between. If one group is ethyl, another is methyl, methyl must precede ethyl.

Thus the compound

$$\begin{array}{l}CH_3\underset{\displaystyle CH_2CH_3}{\underset{|}{C}}HCH_2\underset{\displaystyle CH_3}{\underset{|}{C}}HCH_3\end{array}$$

is called 2-ethyl 4-methylpentane.

Rule 5 When two or more identical groups are present, indicate by use of prefixes (di, tri, tetra), etc. immediately before the alkyl group. Commas separate position numbers. For example, the compound

$$CH_3\underset{\displaystyle C_2H_5}{\underset{|}{C}}HCH_2\underset{\displaystyle C_2H_5}{\underset{|}{C}}HCH_3$$

is called 2,4-diethylpentane. While determining the alphabetical order, di, tri are ignored.

Rule 6 When there are two long chains with same number of carbon choose the one with greatest number of branches as parent compound. For example, the following compound,

$$\begin{array}{ccccccccccccc} & & & & CH_3 & & & & CH_3 & & & & \\ & & & & | & & & & | & & & & \\ CH_3 & - & CH & - & CH & - & CH & - & CH & - & CH_2 & - & CH_3 \\ & & | & & & & | & & & & & & \\ & & CH_3 & & & & CH_2 & & & & & & \\ & & & & & & | & & & & & & \\ & & & & & & CH_2 & - & CH_3 & & & & \end{array}$$

is named as 2,3,5-trimethyl 4-propylheptane.

Rule 7 When branching occurs at equal distance from both sides of the longest chain, choose the name that gives lower number at the first point of difference. Thus the structure given below is named as

$$\begin{array}{ccccccccccc} & & CH_3 & & & & & & CH_3 & & \\ & & | & & & & & & | & & \\ CH_3 & - & CH & - & CH & - & CH_2 & - & CH & - & CH_3 \\ & & & & | & & & & & & \\ & & & & CH_3 & & & & & & \end{array}$$

2,3,5-trimethylpentane and not 2,4,5-trimethylpentane.

Rule 8 In case the substituent in the parent chain is complex (having more than 4 carbons) name it as substituent alkyl group whose carbon chain is numbered from carbon attached to main chain. Thus the compound

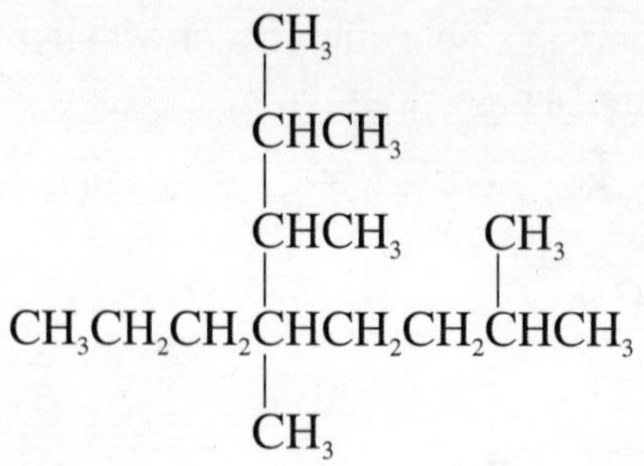

$$
\begin{array}{l}
\qquad\qquad\quad CH_3 \\
\qquad\qquad\quad | \\
\qquad\qquad\quad CHCH_3 \\
\qquad\qquad\quad | \\
\qquad\qquad\quad CHCH_3 \quad\; CH_3 \\
\qquad\qquad\quad | \qquad\qquad\quad | \\
CH_3CH_2CH_2CHCH_2CH_2CHCH_3 \\
\qquad\qquad\quad | \\
\qquad\qquad\quad CH_3
\end{array}
$$

is called 5-(1,2-dimethylpropyl)nonane.

Bicycloalkanes Compounds containing two fused or bridged rings are called bicycloalkanes. The name is based on the total number of carbons, and carbons common to both rings are bridgehead carbons. The following compounds are named as 8-methyl bicyclo [4.3.0] nonane (Figure1.9) and bicyclo [2.2.1] heptane (Figure1.10).

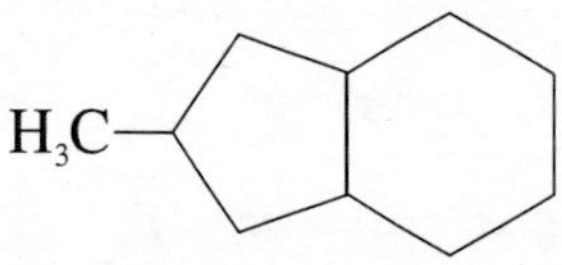

H_3C–

Figure 1.9 8-methyl bicyclo [4.3.0] nonane

H, C, H_2C, CH_2, CH_2, H_2C, CH_2, C, H =

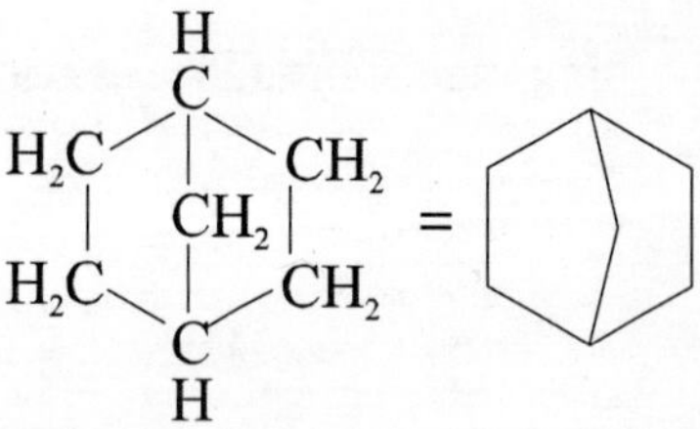

Figure 1.10 Bicyclo [2.2.1] heptane.

Naming of Alkenes

1. Select longest continuous chain containing both carbons of double bond and name with ending letters '*ene*' Thus the following compound

$$
\begin{array}{l}
\qquad\qquad\qquad\quad CH{=}CHCH_3 \\
\qquad\qquad\qquad\quad | \\
CH_3{-}CH_2{-}CH_2
\end{array}
$$

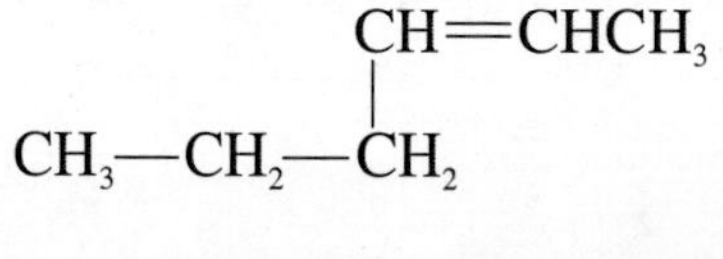

is called 2-hexene.

2. In a branched alkene with several double bonds, the parent chain is the longest continuous chain containing maximum number of double bonds. Thus

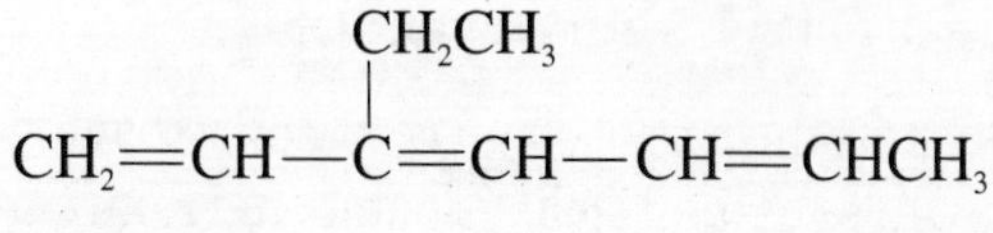

$$
\begin{array}{l}
\qquad\qquad\qquad CH_2CH_3 \\
\qquad\qquad\qquad | \\
CH_2{=}CH{-}C{=}CH{-}CH{=}CHCH_3
\end{array}
$$

is called 3-ethyl 1,3,5-heptatriene.

Naming of Alkynes

When there are two or more triple bonds, it is named as diyne, triyne, and chain containing maximum number of double bonds is selected and numbered so that lowest number is for triple bond carbons. Thus the following compound

$$CH_3—C\equiv C—C\equiv C—C\equiv C—CH_3$$

is called 2,4,6 octatriyne.

Alcohols Select longest chain with OH group and number chain with lowest number for carbon with OH group. The following compound

```
                  CH3
                  |
CH3CH2CH2CHCHCH3
               |
               OH
```

is called 2-methyl 3-hexanol.

More examples on nomenclature will be discussed under each chapter in the text.

STRUCTURAL AND STEREOISOMERISM

Berzelius and Wohler were surprised to find that two compounds with different properties (ammonium cyanate and urea) had the same formula CH_4N_2O. Berzelius coined the word isomer to describe different compounds with same molecular formula. Isomerism can be broadly divided into the following types. Structural isomerism and stereoisomerism. Stereoisomerism is again divided into a) optical isomerism and b) geometrical isomerism.

Structural Isomerism

Structural isomers are also called constitutional isomers and they differ in the order in which atoms are connected. There are three types of structural isomerism.

Chain or nuclear isomerism n-pentane is $CH_3CH_2CH_2\ CH_2CH_3$ and neopentane is $(CH_3)_4C$. The two differ in the arrangement of carbons.

Position isomerism The two molecules differ in the position occupied by a substituent group.

```
CH3CH2CH2OH and CH3CHCH3
                     |
                     OH
    (a)             (b)
```

(a) is n-propyl alcohol and (b) is isopropyl alcohol.

Ortho-nitrophenol, *meta*-nitrophenol and *para*-nitrophenol are aromatic compounds with same formula but with nitro group attached at different positions of the ring.

Functional group isomerism CH_3CH_2CHO and CH_3COCH_3 refer to an aldehyde and ketone. $CH_3CH_2OCH_2CH_3$ (ether) and $CH_3CH_2CH_2CH_2OH$ (alcohol) are also functional isomers. Tautomerism can be considered as a special kind of functional isomerism. $CH_2{=}CHOH$ (vinyl alcohol) and CH_3CHO (acetaldehyde) are also functional isomers.

Conformational isomers Saturated compounds like ethane can be drawn in eclipsed form or staggered form and this aspect is discussed in chapter 2.

Stereoisomerism

This phenomenon is exhibited by isomers having same formula but differing in spatial arrangement. They have different configurations. The two kinds of stereo isomerism are 1) optical isomerism 2) geometrical isomerism. These two types of stereoisomerism are discussed in detail.

OPTICAL ISOMERISM

The compounds which rotate the angle of plane polarized light are called chiral molecules. Chirality may be arising when a compound has a stereogenic centre (carbon attached to 4 different groups) or if the molecule is dissymmetric. Lactic acid is an example of a molecule with one stereogenic centre. *Trans* 1, 2-cyclopropane dicarboxylic acid is an example of molecule with dissymmetry. (presence of axis of symmetry without plane of symmetry or inversion centre or alternating axis of symmetry is called dissymmetry). Molecules can be classified under three important categories:

i. *Asymmetric molecules* An example of asymmetric molecule (stereogenic centre) is lactic acid. It has no plane of symmetry or centre of symmetry or rotational axis of symmetry or alternating axis of symmetry.

ii. *Dissymmetric molecules* 1,3 dibromoallene has C_2 symmetry only and is dissymmetric. It has no stereogenic centre.

iii. *Symmetric molecules* Symmetric molecules are those, that have plane of symmetry or centre of symmetry or alternating axis of symmetry. One of these elements may be present or all of them may be present. They may have rotational axis of symmetry also. They are not optically active (achiral).

Tetrachloroethylene has plane of symmetry and inversion centre in addition to rotational axes and hence is symmetric and achiral.

Optical isomers have the same physical and chemical properties but one form rotates plane polarized light to right (dextrorotatory) and another rotates it top left (laevo rotatory). They are called enantiomers. They correspond to object and a nonsuperimposable mirror image. This is what is present in bromochloroiodomethane. If one chlorine is replaced by another iodine the molecule loses stereogenic centre and is achiral.

The following are some examples of molecules which are asymmetric.

1) lactic acid 2) $CH_3CHCOOH$ (with NH_2 on the CH) 3) 2-butanol 4) 3-methylhexane

The following are dissymmetric molecules.

1. *trans*-1,2-dichlorocyclopentane (two-fold axis of symmetry—C_2 axis is present)
2. 1,3-dibromoallene (C_2 axis of symmetry)
3. Fiest's acid 1,2-*trans*-cyclopropene dicarboxylic acid (C_2 axis) (2 chiral centres)

   The *cis* form of this compound has plane of symmetry and is achiral.
4. $[Co(en)_3]^+$ (3-fold axis)

OPTICAL ROTATION

A beam of ordinary light consists of electromagnetic waves oscillating in infinite number of planes at right angles to the direction of light travel. When a beam of ordinary light is allowed to go through a polarizer, however only light waves oscillating in a single plane pass through. This light is referred to as plane polarized light. This is similar to the observation that if we cut a rectangular opening in a match box which consists of match sticks arranged randomly only those sticks aligned with the opening come out of it. These sticks are restricted to the specific plane.

It was observed by J.B.Biot, a French scientist, that when plane polarized light is passed through molecules like sucrose, or camphor, the plane of polarization is rotated. These compounds are optically active or chiral.

The amount of rotation is measured by the instrument called polarimeter.

POLARIMETER

A solution of chiral molecules is placed in a sample tube and introduced into the polarimeter (Figure 1.11) (polarizer nicol prism) and rotation occurs. Another nicol prism called analyzer is rotated until light passes through it and we can find the angle of rotation. The amount of rotation is expressed by the angle of rotation. The specific rotation $[\alpha]_\lambda$ is given by

$$[\alpha]_\lambda = \frac{\alpha}{xc}$$

where

x is called path length in dm

c is concentration in g/ml

$[\alpha]_\lambda$ is called specific rotation

α is called angle of rotation

λ corresponds to the symbol for wavelength of light

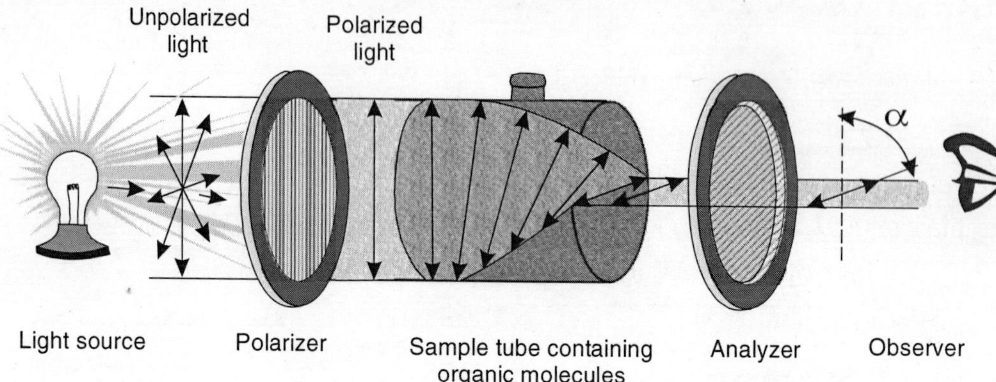

Figure 1.11 Diagrammatic representation of polarimeter

The following are the specific rotations of some chiral molecules:

Compound	Specific rotation $[\alpha]_\lambda$
Camphor	+44.26
2-methyl-1-butanol	+1.151
Morphine	–132
Sucrose	+66.47
Cholesterol	–31.5
Penicillin V	+233
Monosodium glutamate	+25.5
Acetic acid	0

Calculation of specific rotation

SOLVED EXAMPLE 2

A 1500 mg sample of conine was dissolved in 10 ml of ethanol and placed in a sample cell with 50 mm path length. The observed rotation of sodium D line was +1.21 degrees. Find the specific rotation.

Answer

$$[\alpha]_\lambda = \alpha/(xc) = \frac{+1.21}{(5/10) \times 0.15} = +16.1$$

A sample of optically active substance consisting of single enantiomer is enantiomerically pure or is said to have 100% enantiomeric excess. Enantiomeric excess is defined as follows.

$$\text{Enantiomeric excess} = \frac{(\text{Moles of enantiomer A} - \text{Moles of enantiomer B}) \times 100}{(\text{Total moles of the enantiomers})}$$

Optical rotations can be used to calculate enantiomeric excess.

SOLVED EXAMPLE 3

Pure sample of (S)-2-butanol has a specific rotation of 13.52 degrees. If the mixture of 2-butanol isomers showed specific rotation of +6.76°, calculate the enantiomeric excess.

Answer

Enantiomeric excess is (6.76/13.52) × 100=50%.

PROCHIRALITY

If a molecule is achiral and if its reaction with another reagent by substitution or addition leads to a chiral centre then the achiral molecule is called prochiral. CH_2FBr when converted to CHFClBr becomes chiral.CH_2FBr is prochiral. CH_3CHO is achiral but reaction with HCN gives cyanohydrin ($CH_3CHOHCN$), which is chiral. The aldehyde is prochiral. Addition of a group to trigonal carbon or replacement of two identical groups attached to tetrahedral carbon leads to a stereogenic centre. This is prochirality.

Butene-2 on addition of HCl gives the major product $CH_3CH(Cl)\ CH_2CH_3$ which is chiral. Hence 2-butene is prochiral to addition of HCl.

Hell–Volhard–Zelinsky reaction In this reaction the starting acid is not chiral but the bromocompound RCHBrCOOH is chiral. RCH_2COOH is converted to RCHBrCOOH by adding Br_2 and PBr_3. RCH_2COOH is prochiral to bromine addition.

Absolute Configurations of Chiral Molecules

A method for indicating three-dimensional arrangement of the optical isomers is necessary and these absolute configurations can be obtained using Cahn Prelog Ingold rules.

Cahn Prelog Ingold rules

1. We can look at the four atoms directly attached to the stereogenic centre and assign priorities in the order of decreasing atomic number. The atom with highest atomic number is ranked first. Thus in CClBrFH, Br is ranked 1, Cl is ranked 2, F is ranked 3 and H is ranked 4.

2. When a priority cannot be assigned based on atomic number of atoms directly attached to stereo centre then the next set of atoms in the unassigned group are examined. The process is continued until decision is made. We assign a priority on the first point of difference. For example let us consider $CH_3CH_2CHOHCH_3$. Between CH_3 and C_2H_5, in CH_3 we have (H, H, H) but in ethyl we have (C, H, H), so ethyl has higher priority than methyl. The group with lowest priority is directed away from us by rotation.

   Now we trace the path $1 \rightarrow 2 \rightarrow 3 \rightarrow 4$ and if we trace in clockwise direction we designate it as R and it is (R)—2-butanol.

3. If compounds contain multiple bonds the priorities are assigned as if the central atom is connected two times to atom Y if there is double bond and three times to Y if there is a triple bond. Thus CH=CH has priority than isopropyl. Vinyl and isopropyl are diagrammatically represented as in Figure 1.12.

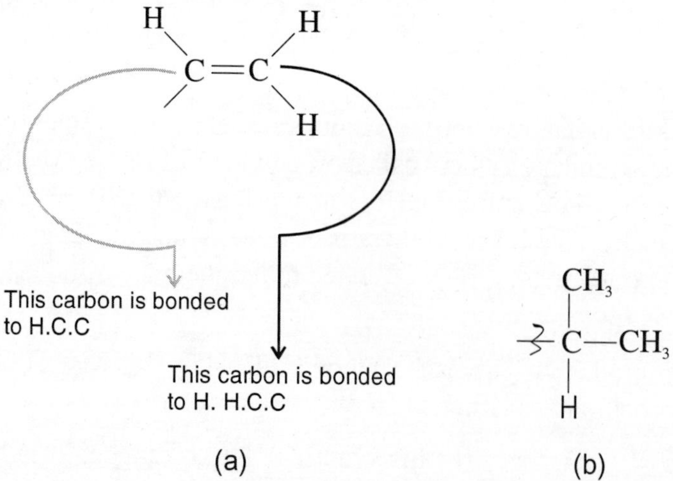

Figure 1.12 Diagrammatic representation of a) vinyl group b) isopropyl group

Let us now consider $CH_3CHOHCOOH$. OH is assigned 1, COOH is assigned 2, CH_3 is assigned 3 and H, 4. If we trace in clockwise direction from OH to methyl we denote the configuration as R (Figure 1.13). It is (R)-Lactic acid.

1 HO, H, C, COOH 2, CH_3 3

Stereo view

R configuration

(a) (–)-Lactic acid

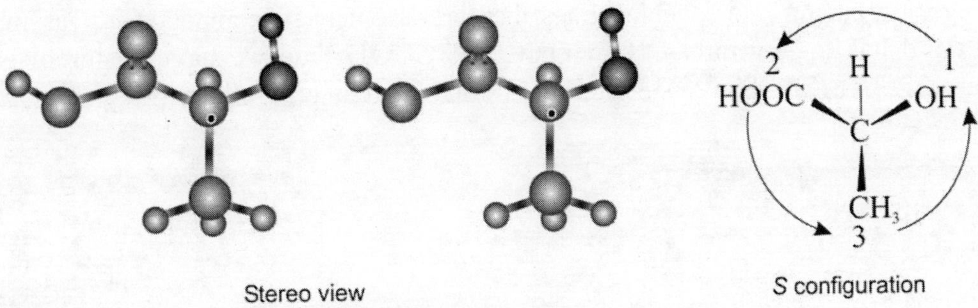

(b) (+)-Lactic acid

Figure 1.13 Configuration R and S for lactic acid

Diastereoisomers

If a molecule has more than one stereogenic centre, the number of isomers increase. Let us consider 2,3-dibromobutane (Figure 1.14). It has 2 stereogenic centres and there are $2^2 - 1$ isomers. (2R,3R), (2S,3S) and (2R,3S) are the stereoisomers. The (2R,3R) form of dibromobutane for carbon 2 and carbon 3 can be illustrated as shown in Figure 1.15.

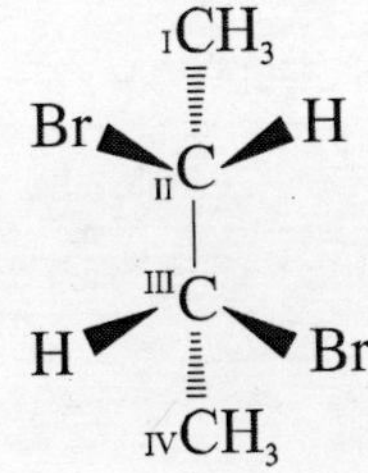

Figure 1.14 2,3-Dibromobutane

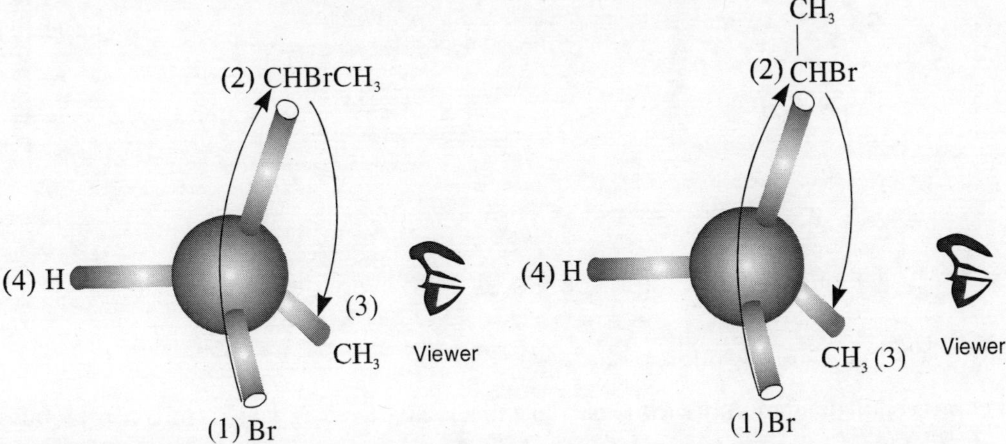

Figure 1.15 R configuration carbon II and carbon III of 2,3-dibromobutane

For glucose having 3 stereogenic centres, there are 8 isomers. If a compound has 2 asymmetric centres and has the formula $CH_3CHOHCHOHCOOH$ then we have 4 stereoisomers. They are (2R,3R), (2R,3S), (2S,3S) and (2S,3R) as shown in Figure 1.16.

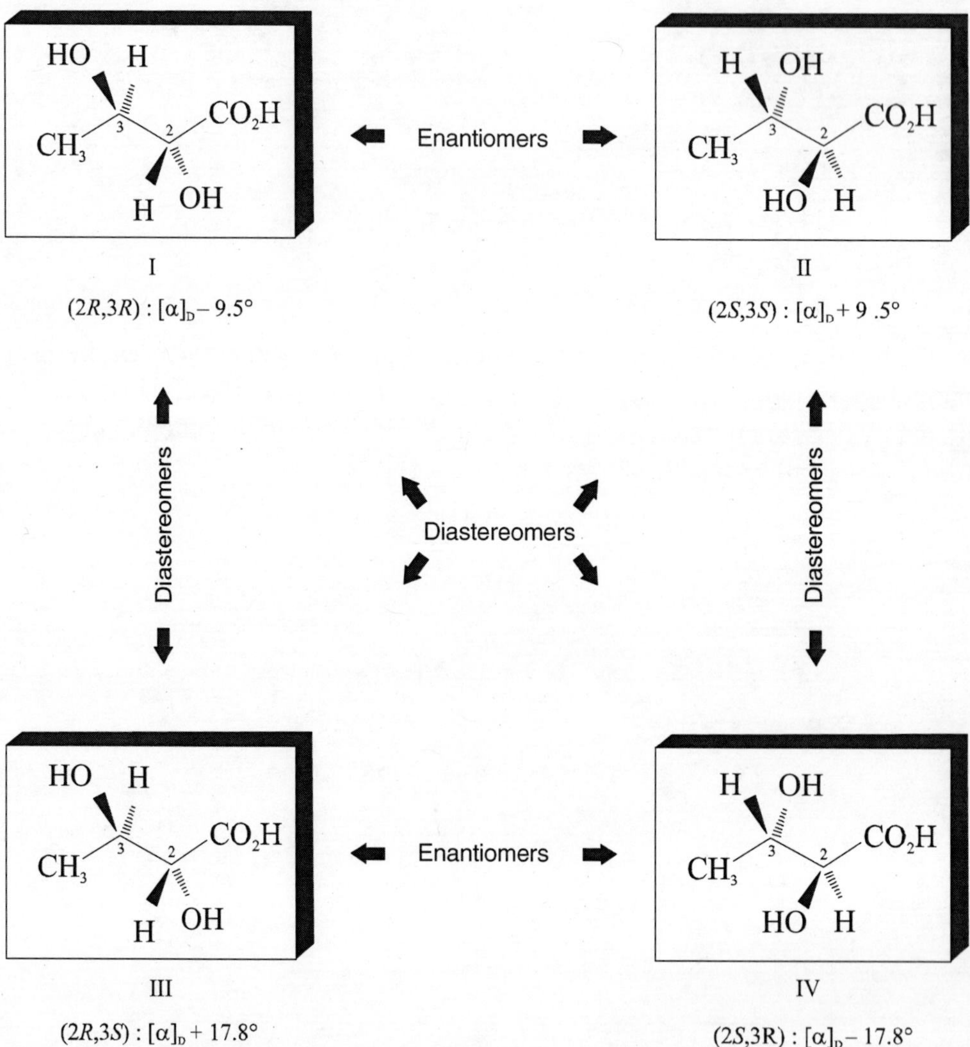

Figure 1.16 Stereoisomers of 2,3-dihydroxybutanoic acid

Fischer Projection Formulae

Fischer projection diagrams are 2 dimensional and it is easy to designate R and S configurations by this procedure.

The three-dimensional model and Fischer projection diagram for R-lactic acid is given below.

$$\begin{array}{c} COOH \\ | \\ H-C-OH \\ | \\ CH_3 \end{array}$$

It is clockwise rotation and represents R form.

A Fischer projection can be rotated by 180 degrees but not by 90 or 270 degrees.

The form

$$\begin{array}{ccc} & COOH & \\ H & + & OH \\ & CH_3 & \end{array}$$

is same as

$$\begin{array}{ccc} & COOH & \\ HO & + & CH_3 \\ & H & \end{array}$$

We can assign R, S by Fischer projection technique.

Lets us consider alanine with the following structure.

$$\begin{array}{c} H \\ | \\ CH_3-C-COOH \\ | \\ NH_2 \end{array}$$

(alanine)

Assign priorities 1 for NH_2, 2 for COOH, and 3 for CH_3 and 4 for hydrogen. Rotate so that hydrogen is on top vertical. Keep CH_3 unmoved and rotate others counterclockwise.

$$\begin{array}{ccc} & \overset{4}{H} & \\ \overset{2}{COOH} & + & \overset{1}{NH_2} \\ & {}^{3}CH_3 & \end{array}$$

Now go from 1, 2, 3 priorities surpassing hydrogen. The rotation is anticlockwise and configuration is S.

Figure 1.17 shows the erythro and threo forms of 2,3-dihydroxybutanoic acid.

I	II	III	IV
CO_2H	CO_2H	CO_2H	CO_2H
H—|—OH	OH—|—H	OH—|—H	OH—|—H
H—|—OH	OH—|—H	OH—|—H	OH—|—H
CH_3	CH_3	CH_3	CH_3
erythro	erythro	threo	threo

Figure 1.17 The erythro and threoforms of 2,3-dihydroxybutanoic acid in Fisher projection

Erythro, threo nomenclature vanishes if a compound has two asymmetric carbons with identical groups as in tartaric acid (COOHCHOHCHOHCOOH).

In tartaric acid we have only (2R,3R), (2S,3S) and (2R,3S).

(2S,3R) is same as (2R,3S) and is called meso form. The Fisher projection forms of tartaric acid are shown below.

```
      COOH
       |
   H—C—OH
       |
  HO—C—H
       |
      COOH
```

This is 2R, 3R form. The other forms are:

```
      COOH            COOH
       |               |
  HO—C—H          H—C—OH
       |               |
   H—C—OH         H—C—OH
       |               |
      COOH            COOH
    (2S, 3S)        (2R, 3S)
```

Since the compound contains three pairs of the same substituents, there are only three isomers (2R, 3R), (2S, 3S) and (2R, 3S). (2S, 3R) is same as (2R, 3S) and is called meso compound. The meso form can be in eclipsed or staggered conformation. There is a plane of symmetry in eclipsed form and centre of symmetry in staggered form, and the meso form is achiral.

(2R, 3R) and (2S, 3S) are enantiomers but (2R, 3R) and (2R, 3S) are diastereoisomers. These are not mirror images of each other. Diastereoisomers have opposite configurations at one or more stereogenic centres and the same configuration at other centres, but enantiomers have opposite configuration at each stereogenic centre.

The following table shows enantiomers and diastereoisomers:

Stereoisomer	**Enantiomeric with**	**Diasteroisomeric with**
2R,3R	2S,3S	2R,3S and 2S,3R
2S,3S	2R,3R	2R,3S and 2S,3R
2R,3S	2S,3R	2R,3R and 2S,3S
2S,3R	2R,3S	2R,3R and 2S,3S

One-to-one relationship does not exist between R and S configurations and *d* and *l* isomers. *d* form rotates the plane polarized light to the right and *l* form to the left.

When we consider enantiomers and diastereoisomers, enantiomers differ in optical rotation but other physical properties are the same. Diastereoisomers have no angle of rotation and their physical properties differ. These are illustrated in the table given below.

Stereoisomer	**Melting point °C**	**Opt. rotation in degrees**	**Density g/cc**	**Solubility G/100 ml of water at 293K**
[+]	168–170	+12	1.7598	139
[–]	168–170	–12	1.7598	139
Meso	146–148	0	1.686	125
[±]	206	0	1.789	20.6

Diastereoisomers have different physical properties but enantiomers have same physical properties except optical rotation (sign is changed).

GEOMETRICAL ISOMERISM

Geometrical isomerism is exhibited by

i. alkenes with two different substituents on each carbon bearing double bond,
ii. cycloalkanes and
iii. dienes.

It is known that free rotation is not possible when there is a carbon–carbon double bond. If the compound is di-substituted (two substituents other than hydrogen are bonded to the 2 carbons then there are two configurations possible for the compound. Consider 2-butene.

$$CH_3CH{=}CHCH_3$$

The two methyl groups can be on the same side of the double bond (*cis*) or on the opposite sides of the double bond (*trans*).

$$\begin{array}{c} CH_3 \quad\quad CH_3 \\ \diagdown \quad\; \diagup \\ C{=}C \\ \diagup \quad\; \diagdown \\ H \quad\quad\quad H \end{array} \quad (I) \qquad\qquad \begin{array}{c} CH_3 \quad\quad H \\ \diagdown \quad\; \diagup \\ C{=}C \\ \diagup \quad\; \diagdown \\ H \quad\quad\quad CH_3 \end{array} \quad (II)$$

(*cis*) (*trans*)

If one of the double bonds is attached to the same substituent, this isomerism is not possible. Compounds like $CH_3CH{=}CH_2$, $(CH_3)_2C{=}CHCH_3$ etc., cannot exhibit geometrical isomerism.

Solved Example 4

Which of the following exist as *cis*, *trans* isomers:

a) $(CH_3)_2C{=}C(CH_3)CH_2CH_3$

b) $ClCH{=}CHCl$

c) $BrCH{=}CHCl$

Answer

b and c.

This approach of naming the two isomers works only when the substituents are two in number. It fails with tri- or tetra-substituted compounds and we have to use Z and E designation. These are based on the rules proposed by Cahn, Ingold and Prelog.

Rule 1 Considering each double-bonded carbon separately, look at the two atoms connected and rank them according to the atomic number. The higher atomic number system has higher priority. The atoms are assigned in the order Br > Cl > O > N > C > H.

Consider the following two compounds

$$\begin{array}{c} H \quad\quad\quad Cl \\ \diagdown \quad\; \diagup \\ C{=}C \\ \diagup \quad\; \diagdown \\ CH_3 \quad\quad CH_3 \end{array} \qquad\qquad \begin{array}{c} H \quad\quad\quad CH_3 \\ \diagdown \quad\; \diagup \\ C{=}C \\ \diagup \quad\; \diagdown \\ CH_3 \quad\quad Cl \end{array}$$

(I) (II)

In the first compound, H has less priority than methyl and in the second carbon, Cl has high priority relative to CH_3. Since atoms with high priority are on opposite sides, it is the E (or *trans*) isomer. E means entgegen (opposite). The second compound is the Z (zussumman) isomer or *cis* isomer.

Rule 2 If a decision cannot be reached by looking at the first atoms, look at the second, third, fourth atoms, until the first difference is found. According to rule 1, ethyl and methyl have equal priority but by rule 2, ethyl has higher priority because second atoms are C, H, H rather than H, H, H. Thus the compound

$$\begin{array}{ccc} (CH_3)_2\,CH & & CH_2Cl \\ & C{=}C & \\ CH_2CH_3 & & CHCH_3NH_2 \end{array}$$

is the Z isomer since both the high priority groups are on the same side.

SOLVED EXAMPLE 5

Give Z, E designation to each of these compounds:

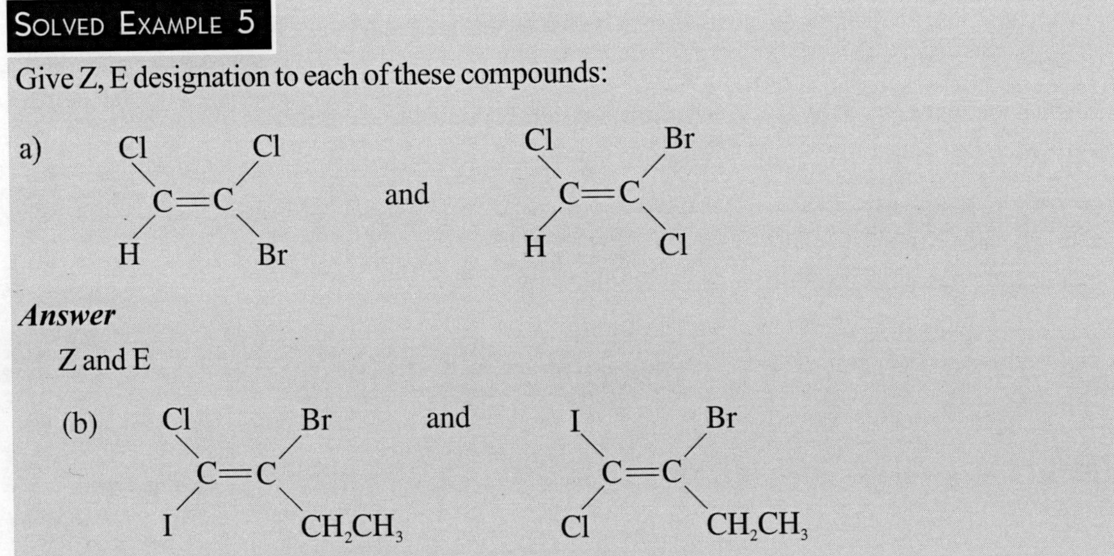

Answer

(a) Z and E

Answer

(b) E and Z

Rule 3 Multiple-bonded atoms are equivalent to the same number of single-bonded atoms. For example, an aldehyde substituent $-\underset{\displaystyle H}{\underset{|}{C}}{=}O$ is equivalent to carbon singly bonded to two oxygens.

For example, the compound $\begin{array}{ccc} H & & CH{=}CH_2 \\ & C{=}C & \\ CH_3 & & CH_3 \end{array}$ belongs to E configuration and is named as E-3-methyl 1,3-pentadiene. The higher priority groups are on the opposite side.

The compound

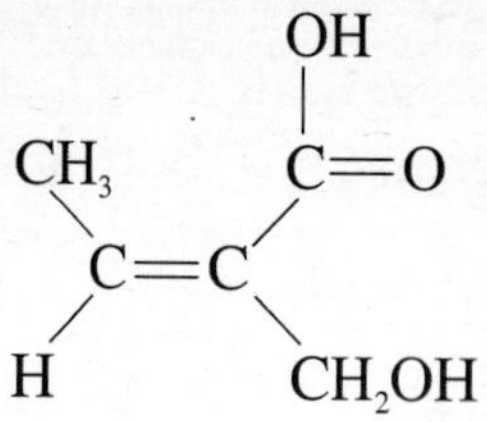

$$\begin{array}{ccc} CH_3 & & C(=O)OH \\ & C=C & \\ H & & CH_2OH \end{array}$$

has the Z designation since both high priority groups are on same side. It is named as Z-2-*n*-hydroxymethyl 2-butenoic acid.

KETO-ENOL TAUTOMERISM

Carbonyl compounds with one or more hydrogens on the alpha carbon rapidly interconvert with the corresponding enols. This is known as tautomerism (in greek, *tauto* means "the same," and *meros* means "part"). Tautomers are different compounds and can be individually isolated whereas resonance forms are imaginary. Unlike isomeric alkenes, ketone and enol are easily interconvertible.

The concept of acidic hydrogens (hydrogens attached to alpha carbon) is important when we discuss keto-enol tautomerism. Acetone has 6 acidic hydrogens.

SOLVED EXAMPLE 6

Find the number of acidic hydrogens in the following compounds

1. cyclopentanone
2. acetylchloride
3. $CH_3COOCH_2CH_3$
4. CH_3CH_2CHO
5. $C_6H_5CH_2COCH_3$
6. $C_6H_5COCH_3$
7. $(CH_3)_3C—COC(CH_3)_3$

Answer

1. 4   2. 3   3. 3   4. 2   5. 5   6. 3   7. 0

7 cannot exist in *enol* form since there is no acidic hydrogen.

Simple aldehydes and ketones are predominantly in keto form. For example, the equilibrium constant for keto-enol equilibrium for acetaldehyde is 6×10^{-7}.

$$\underset{\text{(keto)}}{CH_3\overset{}{\underset{H}{C}}=O} \overset{K}{\rightleftarrows} \underset{\text{(enol)}}{CH_2=\underset{H}{C}—OH} \qquad K=6\times10^{-7}$$

Similarly for acetone, K is 5×10^{-9}.

Simple ketones and aldehydes do not exist much in *enol* form due to the relative stability of C=O compared to C=C. If we see the sum of bond strengths of *keto* forms and *enol* forms, we obtain the following data.

Form	Bond to H	Pi bond	Sum(kJ/mole)
Keto	C—H(440)	C=O (720)	1160
Enol	O—H(500)	C=C (620)	1120

Hence the bonds are stronger in *keto* form.

Compared to aldehydes, ketones are more stable in *keto* form since the second methyl group stabilizes C=O bond.

β-dicarbonyl compounds such as 1,3-diketones are more stable in *enol* form due to conjugation of C=C with C=O.

$$CH_3COCH_2COCH_3 \longleftrightarrow H_3C-\underset{}{\overset{OH}{\overset{|}{C}}}=\underset{\underset{H}{|}}{C}-COCH_3$$

Equilibrium constant K is 3.2.

In such enols, intramolecular hydrogen bonding takes place.

Ethyl acetoacetate exists in *enol* form to a considerable percentage due to intramolecuar hydrogen bonding and conjugation of C=C with C=O. It is more stable in nonpolar solvents compared to protic solvents such as water. In protic solvent intermolecular hydrogen bonding becomes important.

One compound which exists as 100% enol is phenol due to stability of aromatic ring and the corresponding *keto* form is 1,3-cyclohexadienone. The equilibrium constant for *enol* formation is very high equal to 10^{13}.

Sometimes steric factor also favours enol form. This aspect is discussed under the steric effects later in this chapter.

When keto form is converted to *enol* form, an optically active aldehyde or ketone loses chirality since the *enol* is planar and product is racemic.

The reactions of alkynes with $HgSO_4$ (catalyst) in dil. H_2SO_4 produces carbonyl compounds and the reaction proceeds through *enol* formation. Acetylene forms vinyl alcohol which converts to acetaldehyde.

If a compound exists in *enol* form, deuterium can replace hydrogen and deuterium substitution can take place.

EMPIRICAL AND MOLECULAR FORMULAE

Organic compounds contain C, H and other elements such as N, Cl, Br, I, S, P, etc. It is very important to know the correct formula of the organic compound (molecular formula). This can be obtained from the information on 1) molecular weights and 2) empirical formula.

Empirical Formula

The elemental composition of a compound may be used to calculate a formula that denotes the relative number of atoms of each type in the molecule.

There are three important steps necessary to determine the empirical formula of a compound from data on percentage composition. % composition of elements is determined by the following methods:

Combustion method For carbon and hydrogen, combustion method is used. In the combustion experiment, a known weight of the compound (w) is burnt in oxygen. The combustion products are CO_2 and H_2O. H_2O is collected by a drying agent ($CaCl_2$). CO_2 is collected by reaction with NaOH. The weights of H_2O and CO_2 are found out (w_1 and w_2).

% of carbon $= (12/44) \times (w_2/w) \times 100 = x$

% of hydrogen $= (2/18) \times (w_1/w) \times 100 = y$

If no other element is present except oxygen then

% of oxygen $= 100 - (x + y)$

Dumas or Kjeldahl method Percentage composition of nitrogen is determined by Dumas or Kjeldahl method. (*See* question 19, Section VII in Problems.)

Carius method The substance containing halogen is oxidized with HNO_3 in the presence of $AgNO_3$. Silver halides are precipitated. Knowing the weight of the substance and that of silver halide, the % of halogen is computed.

The substance containing sulphur is heated with fuming H_2SO_4 and $BaCl_2$, $BaSO_4$ is precipitated and from this weight and the weight of compound, amount of sulphur is computed.

After finding the % of elements, the following steps are used to determine empirical formula.

Computation of empirical formula

1. Divide the percentage composition of each element by its atomic mass. This is called atomic ratio and gives the number of moles of each element in 100 grams of compound.
2. Take the smallest number obtained in step one and divide all other percentages by this number. This gives ratios of elements in 100 g of compound.
3. In step 2 if we get a fraction for any element, multiply all the numbers obtained in step 2 by the lowest common denominator to clear all fractions.

This gives empirical formula.

The following example illustrates the computation of empirical formula.

SOLVED EXAMPLE 7

A compound contains 60% carbon, 13.3 % hydrogen and remaining oxygen. Find the empirical formula.

Answer

Step 1 Atomic ratio

C = 60/12 = 5

Hydrogen = 13.3/1 = 13.3

Oxygen = 26.7/16 = 1.67

Step 2

C = 5/1.67 = 3

H = 13.3/1.67 = 8

O = 1.67/1.67 = 1

Since these numbers are whole numbers, step 3 is not necessary. The empirical formula is C_3H_8O.

CalculatioOn of molecular formula Using molecular weight and empirical formula we get ratio as number *n*. This number when multiplied with the coefficient for each atom in the empirical formula gives the molecular formula.

In case of the above example molecular weight was found from depression of freezing point method as 60 grams.

Molecular weight/empirical weight = 60/60 = 1. Hence molecular formula is also C_3H_8O.

Determination of molecular weight Molecular weights are determined by various methods, which are given below:

1. *Victor Mayer method* The formula used in this method is

$$V\ (\text{NTP}) = \frac{(P - p) \times V_1 \times 273}{(273 + t) \times 76}$$

where

P is pressure,

p is aqueous tension,

V_1 is volume at temperature t.

V. density = $w/V(\text{NTP}) \times 0.00009$

MWT = 2 × V. density

2. *Ebullioscopic and cryoscopic methods* (*Refer* a physical chemistry book for these methods)
3. *Platinichloride method for bases*

$$M = [(\text{acidity}) \times (w_1/w_2) \times 195 - 410]/2$$

where w_1 and w_2 are the weights of chloroplatinate and platinum respectively.

4. *Silver salt method for acids*

$M = (w_1/w_2) \times 108n - 108n + n$

where n is the number of atoms of silver replaced by n atoms of hydrogen. n is 1 for monobasic acid and 2 for dibasic acid. w_1 and w_2 are the weights of silver salt and silver.

SOLVED EXAMPLE 8

1. 0.4281 g of a monoacid base gave 0.8191 g of CO_2 and 0.1675 g of water, 0.1207 g gave 21 ml of moist nitrogen collected over NaOH at 284 K and 754 torr. 0.1982 g of platinichloride gave 0.0563 g of platinum and aqueous tension is 11.2 torr. Find the empirical and molecular formula of the base.

Answer

$$\%C = \frac{12 \times 0.8191 \times 100}{(44 \times 0.4281)} = 52.18$$

$$\%H = \frac{2 \times 0.0.1675 \times 100}{(18 \times 0.4281)} = 4.35$$

$$VNTP = \frac{742.8 \times 21 \times 273}{(286 \times 760)} = 19.59$$

$$\%N = \frac{19.59 \times 28 \times 100}{22400 \times 0.1207} = 20.28$$

$\%\,O = 100 - (52.18 + 4.35 + 20.28) = 23.19$

The following table gives the name of the element, %, atomic ratio and whole number ratio

Element	%	At. wt	Atomic ratio	Whole number ratio
C	52.18	12	4.35	4.35/1.45=3
H	4,55	1	4.35	3
N	20.28	14	1.45	1
O	23.19	16	1.45	1

The empirical formula is C_3H_3NO.

$$\text{M.wt. of base} = \frac{1 \times 0.1982 \times 195}{(2 \times 0.0563)} - 205 = 138.24$$

$$n = \frac{138.24}{69} = 2$$

Hence molecular formula is $C_6H_6N_2O_2$.

Eudiometry

Eudiometry is a process which deals with gas analysis and determining molecular formula without determining % composition.

Principle A known volume of the hydrocarbon is exploded with excess oxygen in an eudiometer tube with mercury. An electric spark is passed between Pt electrodes and as a result carbon and hydrogen get converted to CO_2 and water. The mixture left in the eudiometer tube after the apparatus is cooled contains CO_2 and unused oxygen. Water vapour condenses and the volume decreases. Caustic potash (KOH) is introduced into the eudiometer tube which absorbs CO_2 to form K_2CO_3. There is an additional contraction in volume due to CO_2 absorption. Finally the tube contains unused oxygen and nitrogen if nitrogen is present initially.

The following problem illustrates the use of this method.

Solved Example 9

0.005 litres of a hydrocarbon was mixed with excess oxygen (30 cc) and the mixture exploded by electric arc. After explosion the volume of mixed gases was found to be 25 cc after cooling. On adding KOH, the volume was further reduced to 15 ml. The remaining gas was oxygen. All volumes were measured at NTP. The compound is obtained by dehydration of ethanol. Find the molecular formula and the name of the compound (IIT 1979)

Answer

Volume of hydrocarbon = 5 ml

Volume of oxygen = 30 ml

Volume of $CO_2 + O_2 = 25$ ml

Volume of unused oxygen = 15 ml

CO_2 absorbed = 10 ml

Oxygen used = 30 – 15 = 15 ml.

$CxHy + [x + (y/4)]O2 \longrightarrow xCO_2 + (y/2)H_2O$

5 ml of hydrocarbon mixes with $5(x+(y/4))O_2$ to give $5x\ CO_2$

$5x = 10$

$x = 2$

$5(x + (y/4)) = 15$

$x + (y/4) = 3$

$y/4 = 3 - 2 = 1$

$y = 4$

Formula is C_2H_4. It is ethylene.

SOLVED EXAMPLE 10

20 ml of hydrocarbon was exploded with excess oxygen in a eudiometer tube. On cooling, the volume reduced to 50 ml. On treatment with KOH, the volume contracted to 10 ml. The compound gives only one monohalogen derivative and is produced by Wurtz reaction of CH_3I with Na. Find the molecular formula and the name of the compound.

Answer

Volume of $CO_2 + O_2$(unused) = 50 ml

Volume of unused oxygen = 10 ml

CO_2 absorbed = 40 ml

$20x = 40$; Thus $x = 2$

Total contraction per ml

$1 + x + (y/4) - x = 1 + y/4$

$20(1 + y/4) = 50$

$1 + y/4 = 2.5$

$y/4 = 1.5$

$y = 6$

The formula is C_2H_6 and the compound is ethane.

Eudiometry can also be used to find % composition of gases in a mixture and this is illustrated below.

SOLVED EXAMPLE 11

A 0.020 litre mixture contains gases A, B and C. A contains carbon and oxygen and is isoelectronic with N_2. B is obtained by hydrolysis of aluminium carbide. C is a monatomic gas with 2 protons and 2 neutrons. This mixture was exploded with oxygen, and the volume contraction was 13 cc. Further contraction of 14 ml occurred when KOH was passed. Find % composition of A, B and C after identifying them.

Answer

A is CO, B is CH_4 and C is He.

$v_1 + v_2 + v_3 = 20$

$CO + (1/2)O_2 \longrightarrow CO_2$
$CH_4 + 2O_2 \longrightarrow CO_2 + 2H_2O$
He does not react with oxygen
v_1 ml of CO gives v_1 of CO_2
v_2 of CH_4 gives v_2 ml of CO_2

For combustion of oxygen, oxygen used is $\frac{v_1}{2}$

For combustion of methane, volume of oxygen used is $2v_2$

$$\frac{v_1}{2} + 2v_2 = 13$$

$$v_1 + v_2 = 14$$

From these equations we get $v_1 = 10$ ml, v_2 is 4 ml and v_3 is 6 ml.

HYDROGEN BONDS

For example in alcohols, a positively polarized hydrogen atom from one molecule is attracted to a negatively polarized oxygen of another alcohol molecule. This is intermolecular hydrogen bonding which increases the molecular weight and leads to higher boiling point. Generally an electronegative atom is in between two electropositive atoms or vice versa in a hydrogen-bonded system. $CHCl_3$ and HCN also have intermolecular hydrogen-bonding. In HCN the carbon is sp-hybridized and makes carbon electronegative and correspondingly hydrogen becomes more electropositive and a partial positive charge is created. Nitrogen gets a partial negative charge and NCH — NCH hydrogen-bonding is possible. Sometimes the hydrogen bond is formed within a molecule (intramolecular) as in the case of *o*-nitrophenol or *o*-nitrobenzoic acid. A six-membered ring is formed in the *o*-nitrophenol.

Such molecules are steam-volatile and have low boiling points and can be easily separated by steam distillation. 1,3-diketones are stable in *enol* form mainly because of the intramolecular hydrogen bonding.

If we compare diethyl ether and n-butyl alcohol (functional isomers), the alcohol has a boiling point of 117.7 degrees but ether has a boiling point of 34.6 degrees. Ethers do not form hydrogen bonds with themselves like alcohols. But their solubilities are similar to that of alcohols in water. This is because they form hydrogen bonds with water. In addition to carboxylic acids and alcohols, amides also form intermolecular hydrogen bonding and have higher boiling points. Boiling point of acetamide is 21.2 degrees Celsius. Primary and secondary amines have high boiling points due to intermolecular hydrogen bonding. When molecules are intermolecularly hydrogen-bonded we call them associated molecules. Peptides also have hydrogen bonds.

Solved Example 12

1,2-propanediol and 1,3-propanediol have higher boiling points than any of butyl alcohols of almost same molecular weight explain:

Answer

The two OH groups in these diols give rise to more hydrogen bonding than the monohydric alcohols.

If we compare the basicities of amines in gas phase, the basicity order is

$$(CH_3)_3N > (CH_3)_2NH > CH_3NH_2 > NH_3$$

But in aqueous solution, hydrogen-bonding takes place and we must see the number of hydrogens available in protonated form.

If we consider NH_4^+, $^+NRH_3$, $^+NR_2H_2$ and $^+NR_3H$ and $^+NR_4$, secondary amines $R_2N^+H_2$ have more hydrogens available than tertiary amines. They are more basic than tertiary amines.

We will be discussing reactions of alkyl halides by S_N1 and S_N2 processes. For S_N2 reactions polar aprotic solvent is used. This is because the anion should be available for nucleophilic substitution reaction. "The protonated form of the nucleophile is generated in a protic polar solvent which reduces its nucleophilicity and hence slows down the substitution reaction." In polar aprotic solvents like acetone and DMSO, the hydrogen-bonding is not possible since hydrogens suitable for hydrogen-bonding are not available.

INDUCTIVE, RESONANCE AND STERIC EFFECTS AND THEIR APPLICATIONS

Inductive Effects

Let us consider $CHCl_3$ and CHF_3. In the first compound, the three chlorines being electronegative attract electrons and the compound is polarized by the –I effect and has a dipole moment. The compound CHF_3 has the element with higher electronegativity and should be more polar by –I effect.

Groups like Cl, Br, F are –I groups and CH_3, C_2H_5, etc. are + I groups.

Acidity of carboxylic acids If we compare acetic acid, monochloroacetic acid, dichloroacetic acid and CCl_3COOH, the trichloroacetic acid is highly acidic due to inductive effect of three chlorines. Proton is easily removed.

In electrophilic aromatic substitution, halogens are ortho, para directing due to inductive effect.

If we consider carbocations $(CH_3)_3C^+$ is more stable than $(CH_3)_2CH^+$ which is more stable than $CH_3CH_2^+$. The increase of methyl groups leads to more neutralization of positive charge because of + I effect of CH_3 groups and hence tertiary carbocations are most stable. If we think of carbanions primary is most stable. This is because the negative charge is neutralized maximum when hydrogens replace alkyl groups. Hydrogen can be considered more electron-attracting compared to alkyl groups.

If we compare acidities of ethane, ethylene and acetylene, the sp-hybridized carbon is most electronegative and stabilizes the negative charge of anion leading to maximum acidity.

Inductive effect becomes weak as the distance to substituent increases.

If we consider $CH_3CH_2CHClCOOH$ and $CH_2ClCH_2CH_2COOH$, the first compound is more acidic since Cl is nearer to COOH group. In the second one chlorine is away and the –I effect reduces.

Let us consider the condition when HX adds to alkenes of the following formula $R—CH_2—CH{=}CH_2$.

If R is represented by H, Br or Cl, the rate of addition decreases when halogen replaces hydrogen. When H^+ adds according to Markovikov rule due to electronegativity of halogen, CH_2 of CH_2X has a positive charge, adjacent positive charge is created already by addition of HX. This has a destabilizing effect and the rate of addition reduces.

$$\overset{\delta-\ \ \delta+}{Br—CH_2}—\underset{\underset{H}{|}}{C}{=}CH_2 + H—X \rightarrow \overset{\delta-\ \ \delta+}{Br—CH_2}—\underset{\underset{H}{|}}{C^+}—CH_3$$

The successive carbons have positive charge, and this is destabilizing.

Aldehydes and ketones form hydrate which involves equilibrium between C=O form and hydrate form. As we replace hydrogens in acetone by fluorine, hexafluoroacetone has high equilibrium constant for hydrate formation due to the electronegative fluorines. Similarly chloral hydrate formation has high equilibrium constant compared to acetaldehyde hydrate due to the presence of electronegative chlorines.

Resonance Effects

If we compare acetic acid and peroxyacetic acid, resonance stabilization is not possible in the per acid due to the extra oxygen and hence it is much less acidic than acetic acid.

Phenol is more acidic than alcohols because the phenoxide ion is resonance-stabilized.

The resonance effect is mainly responsible for orientation of eletrophilic aromatic substitution in benzene rings with one substituent. Groups like CH_3 are *o-p*-directing whereas groups like CHO, NO_2 are *meta*-directing. More details will be covered in chapter on benzene and its reactions.

Steric Effects

The steric factor is playing mainly in Nucleophilic substitution reactions. Tertiary butyl chloride cannot undergo S_N2 reactions since the trajectory of the nucleophile approach to the tertiary carbon bearing the leaving group is hindered and the energetics involved in such a displacement are very unfavourable.

The triphenyl methyl radical (trityl radical) is very stable due to the bulky phenyl groups.

In Hoffmann elimination of quaternary salts to give alkenes, the less substituted alkene forms because of steric factor. This is discussed in detail in amines chapter.

When hydrogenation is carried out on double-bonded systems, the more substituted double bond hydrogenates slowly due to steric effect. Ease of adsorption on catalyst surface is affected by steric factor. Hydrogenation is easier for $CH_2{=}CHCH_3$ rather than $CH_3CH{=}CHCH_3$.

When p-isopropyl toluene is reacted with CH_3COCl in $AlCl_3$, the CH_3CO group is *ortho* to methyl group in the substituted toluene and not *ortho* to isopropyl due to steric effect.

Generally *keto* form is more stable than *enol* form, If *keto* form of acetone is compared with *enol* form, *keto* form is more stable, but 1,3-diketones are stabilized due to conjugation in *enol*. The following compound is stable as *enol* due to steric crowding (Figure 1.18).

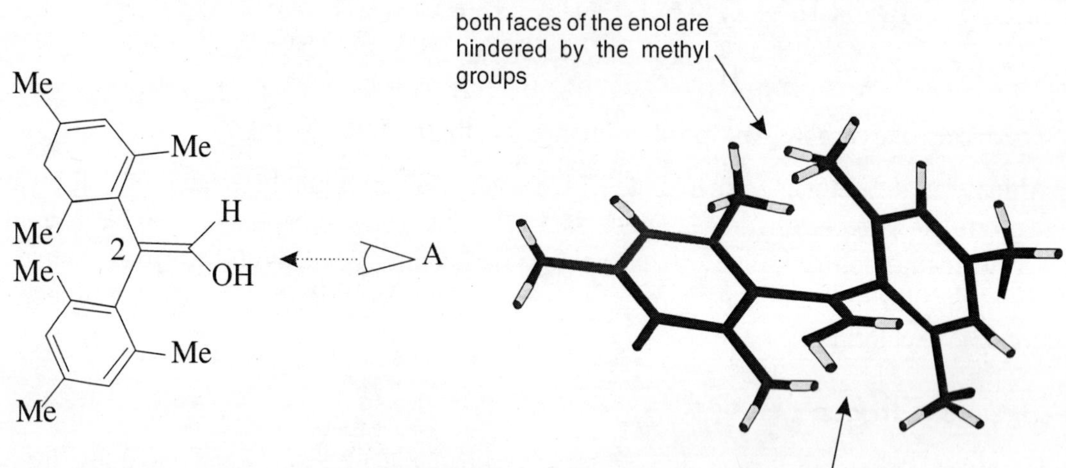

Figure 1.18 Enol difficult to protanate due to steric effect

When aldehydes and ketones react with HCN, cyanohydrin is formed which involves equilibrium between reactants and products. The equilibrium constants for aldehyde, cyanohydrin are higher since ketones have bulky alkyl groups.

POLARITY AND INDUCTIVE EFFECTS IN ALKYL HALIDES

The carbon halogen bond in an alkyl halide results from overlap of sp^3 hybrid of carbon with halogen orbital. C-X bond is polar due to the electronegativity of the halogen. The carbon bears a partial positive charge and halogen partial negative charge. The polarity can be measured in terms of dipole moments. If we consider CH_3F and CH_3Cl, the CH_3Cl has a higher dipole moment than CH_3F even though fluorine is more electronegative. This is because the C-Cl bond distance is greater than C-F bond distance. (C-F bond length in CH_3F is 1.39 angstroms but the C-Cl bond length in CH_3Cl is 1.78A). The alkyl halides are good electrophiles and react with nucleophiles like OH-, CN- in nucleophilic substitution reactions.

If we try to compare CH_3Cl and $CHCl_3$, $CHCl_3$ has greater dipole moment due to 3 Cl. If we compare $CHBr_3$ and CHI_3, the bromo compound will have greater dipole moment.

Reactive Intermediates

The carbocations (carbonium ions), carbanions, free radicals and carbenes are the reactive intermediates.

i. *Carbocations* The stability of carbocations can be explained by using the concept of hyperconjugation, which is explained below.

Hyperconjugation Let us consider the ethyl cation. The cation is sp^2-hybridized and the p-orbital of carbon of CH_2 is empty. The electrons of the neighbouring sigma bonds involving the three hydrogens are dragged to this empty p-orbital and there is sigma bond delocalization. The neighbouring CH_3 group has 3 beta hydrogens for delocalization. If one of the hydrogens of ethyl cation is replaced by methyl group we will have 6 beta hydrogens for delocalization ($(CH_3)_2CH$) and if both the hydrogens of CH_2 in ethyl cation are replaced by methyl groups, we have 9 delocalized structures ($(CH_3)_3C$) and the greater the number of delocalized sigma bonds, the more stable is the carbocation. The same reason applies for the stability of free radicals. Figure 11.19 indicates the hyperconjugation in ethyl cation.

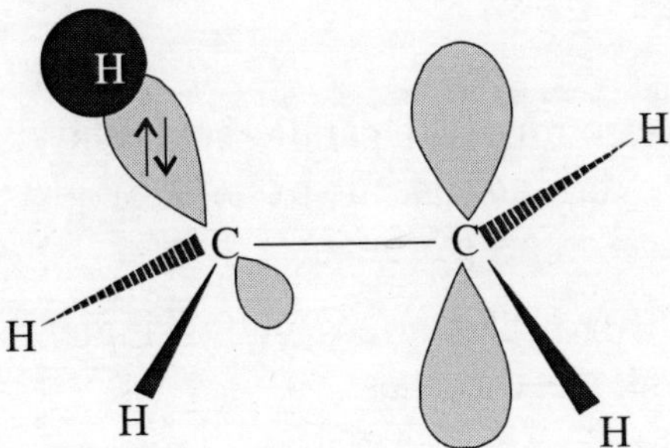

Figure 11.19 Hyperconjugation in ethyl cation.

Suppose we have allyl cation and benzyl cation we have resonance stabilization due to pi bond delocalization. These ions are more stable than the primary, secondary and tertiary carbocations. The relative order of stability is

Benzyl cation>allyl cation> Tertiary cation >Secondary cation>Primary cation.

In the case where the positive charge of carbon is next to a hetero atom, then the cation will be much stabilized by resonance.

ii. *Free radicals* Let us consider triphenyl methyl radical. This stable radical also called trityl radical is prepared from trityl chloride by treating with Ag^+.

The stability of triphenylmethyl radical is due to steric factors. The three phenyl rings are not coplanar but twisted out of plane by 30 degrees like a propellor. This radical is more delocalized than diphenylmethyl radical or benzyl radical. The central carbon is sterically shielded by twisted phenyl groups making it very hard for the molecule to react.

Triphenyl methyl derivatives with ortho substituents offer more stability by steric factor and the phenyl rings are twisted by 50 degrees or more.

In the reaction of triphenylmethyl chloride with RCH_2OH in pyridine, we get the ester and the mechanism operating is S_N1. The trityl chloride forms the carbocation which reacts with alcohol to form ether. Pyridine is used to remove proton from the oxonium ion. Pyridine is used to remove HCl formed. Tritylperchlorate which has the triphenyl methyl cation and perchlorate anion is a good Lewis acid catalyst.

iii. *Carbanion* The anion obtained from an alkyne by treatment with $NaNH_2$ is also a carbanion. $CH_3C\equiv\bar{C}$. This reactswith RI to give alkylated alkyne.

$$CH_3C\equiv\bar{C}: + RI \longrightarrow CH_3C\equiv CR + I^-$$

iv. *Carbenes* Carbenes have the formula $:CH_2$ or :CHPh or :CHR or $:CPh_2$. Carbenes are of two kinds.

i. Single carbene
ii. Triplet carbene

A triplet carbene has two electrons with parallel spins and singlet has spins paired. Triplet carbene has one electron in sp^2 hybrid and one in a p orbital. Singlet carbenes have pair of electrons in nonbonding sp^2 hybrid with an empty p orbital.

Bond angle for triplet is 130 to 150 degrees but for singlet it is 100 – 120 degrees. Triplet carbenes are more stable than singlet carbenes.

Also dihalocarbenes are useful intermediates. They are formed by alpha elimination of Cl^- from CCl_3^- in the presence of KOH. CCl_2 has electrons paired and exists as singlet. The intermediate CCl_3^- and $:CCl_2$ are important in the Reimer–Tiemann reaction of phenols.

Carbenes undergo insertion reactions. With alkenes they form cycloalkenes. If we use $CHCl_3$ and potassium, tert. butoxide the cyclo compound with two chlorines attached to the same carbon is formed.

Mainly singlet carbene reacts in these reactions since triplet carbene needs flipping of one electron spin which is slow. The addition of singlet carbenes is stereospecific. *Cis* alkenes give *cis* cyclopropanes. For triplet carbenes both *cis* and *trans* products are formed.

Diazomethane on irradiation can form carbene. Even though alcohols cannot be converted to ethers by diazomethane under ordinary conditions similar to phenols, they can be converted to ethers with diazomethane on irradiation with light. Carbene is produced which inserts in ROH to give $ROCH_3$. Carbenes can also insert in R_3CH to give $R_3C—CH_3$.

PROBLEMS

I. "EXPLAIN WHY" QUESTIONS AND SMALL STRUCTURE-BASED PROBLEMS

1. Enol ether (Figure Ex.1.1) undergoes hydration much faster than ethylene. Why?

Answer

Ethylene gives rise to $CH_3CH_2^+$, a primary carbonium ion, but the enol ether with H^+ forms a heteroatom-stabilized cation which is in resonance (2 in Figure Ex1.1) and hence reacts much faster.

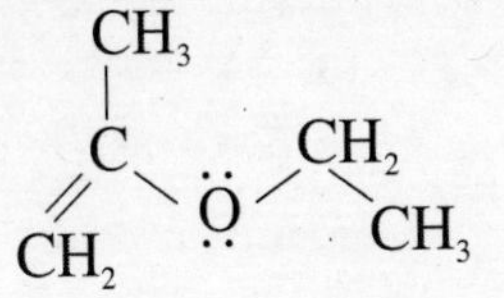

(1)

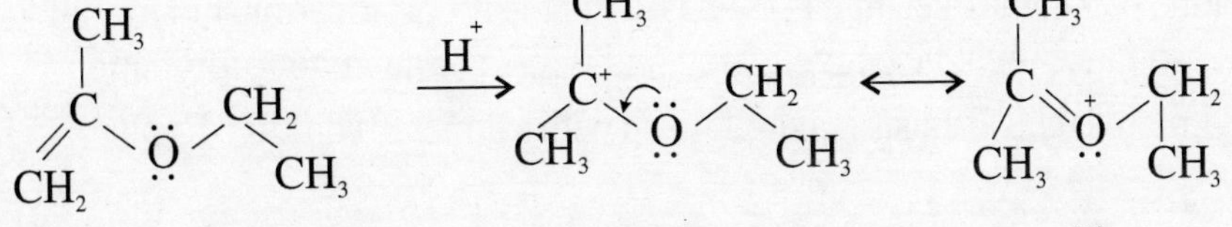

(1) (2) (3)

Figure Ex.1.1 Enol ether reaction in dil.H_2SO_4

2. Cyclohexene with NBS gives 3-bromocyclohexene. Why?

Answer

The allyl radical formed is very stable due to resonance.

3. If we consider ethyl alcohol, isopropyl alcohol and *t*-butyl alcohol, HBr addition is fastest with *t*-butyl alcohol. Why?

   Answer

   t-butyl alcohol forms the most stable carbocation and hence *t*-butyl bromide is formed at the fastest rate

4. When ethyl benzene is chlorinated in sunlight, it is the hydrogen of CH_2 in C_2H_5 group attached to the phenyl ring which is replaced by chlorine. Why?

   Answer

   The $C_6H_5CHCH_3$ radical is stabilized by resonance.

5. *Ortho*-nitrophenol is steam-distilled but *para*-nitrophenol cannot be purified by steam-distillation. Why?

   Answer

   Intramolecular hydrogen-bonding reduces boiling point and hence for *ortho* compounds steam-distillation is possible.

6. Trifluoroacetic acid has a lower pK_a than trichloroacetic acid. Why?

   Answer

   Fluorine is more electronegative and the –I effect makes trifluoroacetic acid more acidic.

7. Peroxyacetic acid has a higher pK_a than acetic acid. Why?

   Answer

   CH_3COO- has resonance structures but the ion $CH_3C(=O)\text{—O—}\bar{O}$ cannot be stabilized by resonance and hence the peroxy acid is weaker.

8. Aniline is a poorer base than CH_3NH_2. Why?

   Answer

   The lone pair in nitrogen in CH_3NH_2 is available for protonation, but in aniline the lone pair is in resonance with the benzene ring and is less available for protonation.

9. Triphenylmethyl radical exists in equilibrium with a dimer in solution. Explain.

   Answer

   This is due to the stability by resonance for this radical and the odd electron is delocalized in the three phenyl rings.

10. *Trans* 1, 2-dimethylcyclohexane or *trans*-1,3-dimethylcyclohexane are optically active but the *cis* forms of these cycloalkanes are optically inactive. Why?

    Answer

    Trans compounds are dissymmetric with 2-fold axis of symmetry, but *cis* compounds have plane of symmetry and are optically inactive.

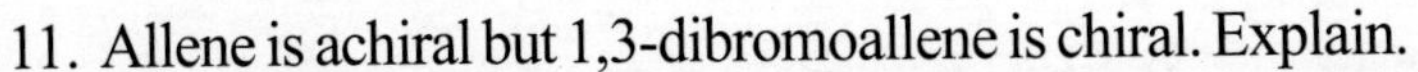

11. Allene is achiral but 1,3-dibromoallene is chiral. Explain.

Answer

Allene has a plane of symmetry and has S_4 axis and is optically inactive. 1,3-dibromoallene has two-fold axis and is optically active.

12. 1,3-cyclopentanediol has 3 stereoisomers but 1,4-cyclohexanediol has 2 stereoisomers. Explain.

Answer

1,3-cyclopentanediol has *cis* and *trans* forms and *trans* form is optically active with d and l forms. In the case of 1,4-cyclohexanediol there are only *cis* and *trans* forms with no optical activity.

13. For pentane the number of monochloroderivative isomers is 7 but the number of fractions distilled are 5. Explain.

Answer

The seven isomers include two sets of *d* and *l* forms which cannot be separated by distillation and hence fractions are 5.

II. Multiple Choice Questions (One out of four answers is correct)

1. The shape of one of the following molecules is linear.

   a) C_2H_4 b) C_2F_4 c) C_3O_2 d) H_2CO

2. One of the central atoms in one of the following molecules is sp^2-hybridized.

   a) CH_3OH b) $(CH_3)_3N$ c) PH_3 d) C_3H_6

3. Hybridization of central carbon in one of the following molecules is sp^3-hybridized

   a) COS b) $CH_2{=}CH{-}Cl$
   c) allene d) $CH_2{=}CH{-}CH_2{-}CH_2{-}CH_2{-}CH{=}CH_2$

4. One of the following molecules has hybridization for C_1 as sp^2 for C_2 as sp^2 and for C_3 as sp

   a) pentane b) 2-ethylpropene
   c) $H_2C{=}CH{-}C{\equiv}CH$ d) $CH_2{=}CH{-}CH_2{-}CH{=}CH_2$

5. One of the following species is sp_2-hybridized.

   a) CH_3^+ b) CH_3^- c) CH_2: triplet d) CH_3

6. The molecule $CH_3{-}CH_2{-}CH{=}CH{-}C{\equiv}CH$ has _____ sigma and _____ pi bonds.

   a) 4, 3 b) 3, 2 c) 1, 3 d) 11, 3

7. The molecule which is not planar is

   a) Naphthalene b) propylene c) pyrrole d) cyclooctatetraene

8. The most acidic among the following molecules is
   a) acetaldehyde b) 2,4-pentanedione c) phenol d) propionic acid
9. One of the following ions is most stable
   a) $CH_3CH_2^+$ b) $(CH_3)_2CH^+$ c) $(CH_3)_3C^+$ d) triphenylmethyl cation
10. The least stable cation among the following is
   a) $(C_2H_5)_3C^+$ b) $(CH_3)_2CH^+$ c) $CH_3\text{-}CH_2CH_2^+$ d) NH_4^+
11. The compound which is dissymmetric is
   a) 1,3-butadiene b) biphenyl
   c) allene d) 1,2-*trans*-dibromocyclopropane
12. The compound which is achiral is
   a) 1,3-dichloroallene b) Fiests acid
   c) 1-bromopropane d) 1,3-*trans*-dichlorocyclopropane
13. In pentane-2,3,4-triol one of the following is chiral.
   a) synsyn b) synanti c) antianti d) antisyn
14. One of the following is prochiral to HCN.
   a) CH_3COCH_3 b) $CNCH_2CH_2CN$ c) CH_3CH_3 d) CH_3CHO
15. The compound which has no stereogenic centre but is optically active is
   a) ethane b) tartaric acid c) 1,3-dichloroallene d) biphenyl
16. The compound which does not exhibit geometrical isomerism is
   a) 3-hexene b) maleic acid c) $CH_3CH=CHCl$ d) 1-pentene
17. The intramolecularly hydrogen-bonded molecule is
   a) *p*-nitrophenol b) *o*-nitrobenzoic acid c) *p*-nitrobenzoic acid d) C_2H_5OH
18. Plane of symmetry is present in
   a) 1,2-*trans*-dibromocyclopentane b) staggered form of mesotartaric acid
   c) 1,3-dichloroallene d) eclipsed form of meso 2,3-dibromobutane
19. Dipole moment is zero for
   a) 2-butene (*cis*) b) $(C_2H_5)(H)C=C(C_2H_5)(CH_2Cl)$
   c) $CH_3CH=CH_2$ d) $BrCH=CHBr$ (*trans*)
20. One of the following has 8 possible stereoisomers.
   a) ethanes b) glucose c) tartaric acid d) 2,3-dibromobutane

21. The enol form is least stable for
    a) 2-butanone b) $CH_3COCH_2COCH_3$ c) phenol d) CH_3CHO
22. Salicylaldehyde has low melting point because
    a) it has intermolecular hydrogen bonding b) it has intramolecular hydrogen bonding
    c) it exists as a dimer d) it exists as a trimer
23. The least acidic among the following is
    a) trichloroacetic acid b) fluoroacetic acid
    c) peroxyacetic acid d) $C_6H_5CH_2COOH$
24. pK_a is greatest for
    a) acetylene b) propionic acid c) peroxyacetic acid d) benzoic acid
25. pK_a is minimum for
    a) monochloroacetic acid b) aniline
    c) trifluoroacetic acid d) $ClCH_2CH_2COOH$
26. The most basic nitrogen compound among the following is
    a) aniline b) cyclohexylamine c) pyridine d) pyrrole
27. The acid with highest pK_a is
    a) butyric acid b) 2-chlorobutyric acid
    c) 3-chlorobutyric acid d) 4-chlorobutyric acid
28. Hydrogen-bonding is minimum for
    a) N,N-dimethyl acetamide b) benzamide
    c) acetamide d) methyl acetamide
29. The staggered form of mesotartaric acid is achiral because of
    a) reflection symmetry b) alternating axis c) inversion centre d) axis of symmetry
30. If I is CH_3CN, II is CH_3OCH_3, III is CH_3COOH and IV is CH_3COCH_3,
    then the order of increasing dipole moment is
    a) IV<III<II<I b) II<III<IV<I c) I<IV<III<II d) III<I<IV<II

KEY

1. c	2. d	3. d	4. c	5. c	6. d	7. d
8. d	9. d	10. d	11. d	12. c	13. d	14. d
15. c	16. d	17. b	18. d	19. d	20. b	21. a
22. b	23. c	24. a	25. b,c	26. b	27. d	28. a
29. c	30. b					

III. Screening Test Questions of Previous IIT-JEE Question Papers

1. Which has the maximum dipole moment?

   a) CH_3Cl b) CH_2Cl_2 c) $CHCl_3$ d) CCl_4 (2003)

2. Which of the following represents the correct decreasing order of acidity?

   a) Benzoic acid > acetylene > phenol > *m*-nitrophenol
   b) phenol > benzoic acid > *m*-nitrophenol > acetylene
   c) benzoic acid > *m*-nitrophenol > phenol > acetylene
   d) *m*-nitrophenol > benzoic acid > acetylene > phenol

3. Which of the following acids has the smallest Ka?

   a) $CH_3CHFCOOH$ b) FCH_2CH_2COOH
   c) $BrCH_2CH_2COOH$ d) $CH_3CHBrCOOH$ (2002)

4. Which of the following compounds exhibits stereoisomerism?

   a) 2-methylbutene-1 b) 3-methylbutyne-1
   c) 3-methylbutanoic acid d) 2-methylbutanoic acid

5. The nodal plane in the pi bond of ethylene is located in

   a) the molecular plane
   b) a plane parallel to the molecular plane
   c) a plane perpendicular to the molecular plane which bisects C-C sigma bond at right angles.
   d) a plane perpendicular to molecular plane containing C-C sigma bond

6. Identify the correct order of boiling points of the following compounds

   I. $CH_3CH_2CH_2CH_2OH$ II. $CH_3CH_2CH_2CHO$ III. $CH_3CH_2CH_2COOH$

   a) I>II>III b) 3>I>II c) I>III>II d) III>II>I

7. Which of the following has lowest dipole moment?

   a) 2-butene (*cis*) b) 2-butyne
   c) 1-butene d) $CH_2{=}CH{-}C{\equiv}CH$

8. The number of isomers for compound with molecular formula CBrClFI is

   a) 3 b) 4 c) 5 d) 2 (2000)

   The correct order of basicities of the following compound is

   a) $CH_3{-}C(NH_2){=}NH$ b) $CH_3CH_2NH_2$
   c) $(CH_3)_2NH$ d) CH_3CONH_2

   a) 2>1>3>4 b) 1>3>2>4 c) 3>1>2>4 d) 1>2>3>4

10. Among the following, the strongest base is
 a) aniline b) *p*-nitroaniline c) *m*-nitroaniline d) benzylamine (2000)
11. Which of the following exhibits geometrical isomerism?
 a) 1-phenyl 2-butene b) 3-phenyl 1-butene
 c) 2-phenyl 1-butene d) 1,1-diphenyl 1-propene
12. Which of the following has most acidic hydrogen?
 a) 3-hexanone b) 2,4-hexanedione c) 2,5-hexanedione d) 2,3-hexanedione
13. In the compound $CH_2{=}CH\text{-}CH_2\text{-}CH_2\text{-}C{\equiv}CH$. The C2-C3 bond is of the type
 a) sp-sp^2 b) sp^3-sp^3 c) sp-sp^3 d) sp^2-sp^3 (1999)
14. Tautomerism is exhibited by
 a) $C_6H_5CH{=}CHOH$ b) quinone
 c) $(CH_3)_3{-}C{-}\overset{O}{\overset{\|}{C}}{-}(CH_3)_3$ d) 1,2-cyclohexanedione (1998)
15. Among the following compounds the strongest acid is
 a) C_2H_2 b) C_6H_6 c) C_2H_6 d) CH_3OH (1998)
16. How many optical isomers are possible for butane2,3diol?
 a) 1 b) 2 c) 3 d) 4 (1997)
17. Among the following the most basic compound is
 a) aniline b) *p*-nitroaniline c) acetanilide d) benzylamine
18. The number of possible enantiomeric pairs that can be produced during monochlorination of 2-methylbutane is
 a) 2 b) 3 c) 4 d) 1
19. The correct statements about compounds A, B and C is

(a)	(b)	(c)
$COOCH_3$	$COOH$	$COOH$
H—C—OH	H—C—OH	H—C—OH
H—C—OH	H—C—OH	HO—C—H
$COOH$	$COOCH_3$	$COOCH_3$

 a) A and B are identical b) A and B are diastereoisomers
 c) A and C are enantiomers d) A and B are enantiomers

20. In the following compounds the order of acidity is

    I. phenol II. *p*-cresol III. *m*-nitrophenol IV. *p*-nitrophenol

    a) III>IV>I>II b) I>IV>III>II c) II>I>III>IV d) IV>III>I>II

21. Arrange in the order of increasing dipole moment.

    I. toluene II. *m*-dichlorobenzene III. *o*-dichlorobenzene IV. *p*-dichlorobenzene

    a) I<IV<II<III b) IV<I<II<III c) IV<I<III<II d) IV<II<I<III

22. Allylisocyanide has

    a) 9 sigma bonds and 4 pi bonds

    b) 8 sigma bonds, 3 pi bonds and 2 nonbonding electrons

    c) 8 sigma bonds and 5 pi bonds

    d) 8 sigma bonds, 3 pi bonds and 4 nonbonding electrons (1995)

23. The kind of delocalization involving sigma bonds is called ________ (1994)

24. Assertion-reason Question

    Assertion

    Acetate ion is more basic than methoxide ion.

    Reason

    Acetate ion is resonance-stabilized.

25. The isoelectronic structures among the following are

    I. CH_3^+, II. H_3O^+ III. NH_3 IV. CH_3^-

    a) I and II b) III and IV c) I and III d) II, III and IV (1993)

26. What is the decreasing order of strengths of bases

    I. OH^- II. NH_2^- III. $H—C\equiv C^-$ IV. $CH_3—\bar{C}H_2$

    a) IV>II>III>I b) III>IV>II>I c) I>II>III>IV d) II>III>I>IV

KEY

1. c	2. c	3. c	4. d	5. a	6. b	7. b
8. d	9. b	10. d	11. a	12. b	13. D	14. a,d
15. b	16. b	17. d	18. a	19. d	20. D	21. c
22. b	23. hyperconjugation	24. d	25. d	26. b		

IV. Multiple Choice Questions (More than one answer is correct)

1. Compounds which have 2 pi bonds are

   a) $CH_3CHCHCH_3$ b) $CH_3—C_2H_5$

   c) $CH_3CH=C=CH_2$ d) $CH_2=CH—CH_2—CH=CH_2$

2. Compounds which are optically active are

   a) alanine b) 1,2-*trans*-dichlorocyclopropane(*trans*)

   c) 1,2-*trans*-dichlorocyclopropane(*cis*) d) $[Co(en)_3]+$

3. The chiral molecules among the following are

   a) glycine b) 1,2-*trans*-dibromocyclopentane

   c) alanine d) *cis*-1,2-dichlorocyclopropane

4. Molecules which are chiral without chiral centre are

   a) $CH_3CHClCOOH$ b) 1,2-*trans*-dibromocyclopentane

   c) 1,3-dibromoallene d) alanine

5. Compounds having a meso form are

   a) benzoic acid b) $CH_3CH_2CBrCHBrCHQCH_3$

   c) $CH_3CHBrCHBrCH_3$ d) 2,3-butanediol

6. Compounds having no meso form are

   a) 2,3-butanediol b) $CH_3CHBrCHClCH_3$

   c) 1,2-*trans*-cyclopropane dibromide d) $CH_3CHCOOHCHCOOHCH_3$

7. Total stereoisomers are 3 for

   a) alanine b) $CH_3CHClCHClCH_3$ c) 2,3-butanediol d) glycine

8. No meso product is obtained in reactions

   a) tetramethylethylene + HCl b) hydrogenation of ethylene

   c) 2-butene(*trans*) + Br_2/CH_2Cl_2 d) 2-butene(*cis*) + dil.$KMnO_4$

9. The following reactions involve carbocations

   a) addition of HCl to propylene

   b) dehydration of t-butyl alcohol

   c) dehydrohalogenation of t-butyl chloride in alc. KOH

   d) halogenation of alkanes in UV light

10. The following reactions involve intermediates of free radicals

    a) halogenation of ethane

    b) addition of NBS to toluene

    c) the treatment of silver salt of carboxylic acids with bromine

    d) addition of HBr to propylene in the absence of peroxide

11. The following reactions involve carbenes or diradicals
    a) propylene + CH_2N_2
    b) $CH_3NH_2 + CHCl_3 + KOH$
    c) phenol + $CHCl_3$ + KOH
    d) esterification of benzoic acid
12. Hybridization of intermediate is sp^2 for
    a) propene + HCl reaction
    b) dehydrohalogenation of tertiary butylbromide
    c) addition of NBS to propylene
    d) $CH_3MgX + H_2O$
13. Intramolecular hydrogen-bonding is present in
    a) *o*-nitrobenzoic acid
    b) salicylic acid
    c) $C_2H_5COCH_2COCH_3$
    d) CH_3COCH_3
14. Two enol diastereoisomers are possible for
    a) $ClCH_2CH_2CHO$
    b) CH_3COCH_3
    c) $ClCH_2CH_2CH_2CHO$
    d) $CH_2{=}CHCl$
15. Trouton's rule is not obeyed by
    a) ethanol
    b) butyric acid
    c) acetone
    d) benzene
16. Dipole moment is nonzero for
    a) *p*-dibromobenzene
    b) *p*-chloronitrobenzene
    c) *o*-diiodobenzene
    d) *m*-dichlorobenzene
17. Which of the IUPAC names is correct for the compounds listed
    a) $(C_3H_7)CHBrCH{=}CHCH_3$ 4-bromo 6-heptene
    b) $CH_3CHClCH_2CH_3$ 2-chlorobutane
    c) $CH_2{=}C(CH_3){-}CH{=}CH_2$ 2-methyl 1,3-butadiene
    d) $CH_3CH_2CH_2CHOHCH_3$ 4-pentanol
18. The intermediate in the reactions has 67% p character
    a) propylene + HCl
    b) dehydration of ethanol
    c) $CH_3MgX + H_2O$
    d) acetylide ion + CH_3I in base
19. Empirical formula weight is 13 g for
    a) C_2H_2
    b) C_6H_6
    c) C_2H_4
    d) CH_3
20. Some of the following are isomers of butanone
    a) $CH_3CH_2CH_2CHO$
    b) $CH_2{=}CH{-}CH_2CH_2OH$
    c) $CH_2{=}CHOC_2H_5$
    d) $CH_3CH_2CH_2CH_2OH$

21. *cis-trans* isomerism is possible in
    a) 1,3-butadiene b) 3-hexene c) 4-octene d) propyne
22. Two of these pairs are not resonance structures
    a) $CH_2^- —CH=CH—CH_2+$ and $CH_2+— CH—CH—CH_2$
    b) CH_3COCH_3 and $CH_2=COHCH_3$
    c) CH_3CONH_2 and $CH_2=COHNH$
    d) two Kekule structures of benzene
23. Resonance is responsible for
    a) higher acidity of acetic acid relative to peracetic acid
    b) greater acidity of chloroacetic acid compared to acetic acid
    c) greater basicity of ethylamine relative to aniline
    d) higher pK_a of acetic acid relative to that of tribromoacetic acid
24. Three stereoisomers are possible for
    a) tartaric acid b) $CH_3CHClCHClCH_3$ c) lactic acid d) glycine
25. Some of the following statements about isomers of C_3H_6DCl are true
    a) 3 of 5 structural isomers are optically active.
    b) There are totally 8 isomers and 5 distilling fractions.
    c) All the structural isomers are chiral.
    d) The number of isomers and distilling fractions are same.

KEY

1. c,d	2. a,d	3. b,c	4. b,c	5. c,d	6. b,c	7. b,c
8. a,b,d	9. a,b,c	10. a,b,c	11. a,b,c	12. a,b,c	13. a,b,c	14. a,c
15. a,b	16. b,c,d	17. b,c	18. a,b	19. a,b	20. a,b,c	21. a,b,c
22. b,c	23. a,c	24. a,b	25. a,b			

V. Assertion–Reason Questions

The questions 1–20 consist of an *assertion* in column 1 and *reason* in column 2. Use the following key to choose an appropriate answer

a) If both assertion and reason are correct and reason is the correct explanation of assertion
b) If both assertion and reason are correct but reason is not the correct explanation of assertion
c) If assertion is correct but reason is incorrect
d) If assertion is incorrect but reason is correct

No.	Assertion	Reason
1.	1,3-diketones are stable in enol form.	They form intermolecular hydrogen bonds.
2.	Ethylether has a lower boiling point than n-butyl alcohol but the solubilities are the same.	Intermolecular hydrogen bonding increases the boiling point, but both the compounds interact to same extent with water and the solubility remains the same.
3.	Fluoromethane has smaller dipole moment than chloromethane.	The C—F bond distance is smaller than C—Cl bond distance.
4.	C_3O_2 has central carbon with 33% s character.	The central carbon is sp-hybridized.
5.	*Trans*-cyclooctene is chiral.	It is dissymmetric with rotational axis of symmetry.
6.	cis1,2-dibromoethylene has smaller dipole moment than *cis* 1,2-dibromo1,2-dichloroethylene.	Chlorine is more electronegative than bromine. The first compound has hydrogens instead of bromine.
7.	The *cis* form of 1,2-*cis* cyclopentane dicarboxylic acid is chiral but *trans* form is achiral.	*Cis* form has plane of symmetry but *trans* form is dissymmetric
8.	Acetone is more acidic than 2-methylpropene.	In the anion of acetone, negative charge is on oxygen but in anion of 2-methylpropene, negative charge is on carbon.
9.	$CH_3OCH_2^+$ is more stable than $(CH_3)_3C^+$.	The heteroatom cation is stabilized by resonance.
10.	C_4H_{10} has 5 monochloroderivatives with total isomers of monochlorinated butane equal to 6 but distilling fractions are 5.	There is one optically active compound with d and l forms having same boiling point.
11.	Triphenyl methyl chloride gave nonconducting solution in SO_2.	This is due to the formation of highly stable triphenyl methyl cation.
12.	$C_6H_{12}Cl_2$ has three meso disterioisomers in eclipsed form.	They are achiral due to plane of symmetry.
13.	C_2H_5OH reacts with diazomethane to give ether without a catalyst but phenol does not give ether.	Ethanol is less acidic than phenol.
14.	4-bromocyclohexene has enantiomers identical with lactic acid.	Both have one asymmetric carbon.

(*Contd.*)

No.	Assertion	Reason
15.	Tertiary butyl alcohol has pKa greater than that of methanol.	The increased number of alkyl groups offer steric hindrance to solvation by water and the stability of $(CH_3)_3CO^-$ is less.
16.	2,3-dimethyl butane has a higher boiling point than neopentane.	2,3-dimethylbutane is a more branched isomer.
17.	Cyclobutane and cyclopropane get hydrogenated to give cyclopentane and cyclobutane.	They have ring strain and get converted to open chain hydrocarbons.
18.	Among $CH_2{=}CHCH_2Cl$(II) and, $CH_2{=}CHCH_2CH_2Cl$(I) compound II loses chloride ion more readily.	The molecular weight of I is greater than that of II.
19.	Pyrrole is not basic but pyridine is basic.	Pyrrole has a 5-membered ring.
20.	CH_3CH_2COOH is prochiral to reagent $Br_2 + PBr_3$ but $(CH_3)_2CHCOOH$ is not.	The replacement of one hydrogen in CH_3CH_2COOH leads to chirality but replacement of hydrogen in isopropyl group does not create the chiral centre.

KEY

1. c	2. a	3. a	4. d	5. a	6. d	7. d
8. a	9. a	10. a	11. d	12. a	13. d	14. a
15. a	16. d	17. d	18. b	19. b	20. a	

VI. Comprehension Passages

Passage I

In organic chemistry, important reactive intermediates are Carbonium ions, carbanions and free radicals.

Answer the following questions

1. In the addition of HCl to ethylene, the reactive intermediate is

   a) ethyl cation b) ethyl radical c) ethyl anion d) carbine

2. In the addition of H^+ to an enol ether and ethylene, the reaction is much faster for enol ether because

   a) heteroatom cation is resonance stabilized

   b) heteroatom cation has hyperconjugation

   c) heteroatom is involved as a free radical

   d) a carbine is formed as intermediate in enol ether reaction

3. When ethylbenzene reacts with Cl_2 at 773 K the product is
   a) $C_6H_5CHClCH_3$
   b) $C_6H_5CH_2CH_2Cl$
   c) HCl is eliminated
   d) $C_6H_5CH_2CHCl_2$

KEY

1. a 2. a 3. a

Passage 2

Resonance decides about lower acidity in acids and lower basicity in amines compared to structures in which resonance does not operate. Resonance is different from tautomerism. In resonance forms the structures are imaginary. Tautomers exist with definite %.

Answer the following questions

1. The least acidic among the following acids is
   a) acetic acid
   b) peroxyacetic acid
   c) benzoic acid
   d) trichloroacetic acid
2. The compound most basic among the following is
   a) Aniline b) $C_6H_5CH_2NH_2$ c) pyrrole d) pyridine
3. The products of ozonolysis for *o*-xylene namely CHOCHO, $CH_3COCOCH_3$ CH_3COCHO indicate that
   a) Xylene has three different isomers
   b) *o*-xylene has two kekule structures
   c) The boiling point of *o*-xylene is different from that of *m*-xylene
   d) The melting points of isomers of xylene are different.

KEY

1. b 2. b 3. b

VII. Matrix Matching

Question 1

Column 1	Column 2
a) Carbene is intermediate	p) Halogenation of methane
b) Carbonium ion is intermediate	q) Addition of CH_2N_2 to propylene
c) Free radical is intermediate	r) Dehydration of alcohols
d) Due to more hyperconjugated structures	s) t-butyl radical is more stable than ethyl radical
e) Dehydrohalogenation of tertiary halides	t) Addition of NBS to propylene
	u) Follows E1 mechanism

Answer

$[a \rightarrow q], [b \rightarrow r], [c \rightarrow p, t], [d \rightarrow s], [e \rightarrow u]$

Question 2

Column 1	Column 2
a) acidity of trifluoroacetic acid is higher than that of trichloroacetic acid	p) staggered form of meso 2,3-dibromobutane
b) benzoic acid has pK_a less than that of peroxybenzoic acid	q) –I effect
c) geometrical isomerism is not possible	r) resonance effect
d) optical activity is removed due to centre of symmetry	s) tetraethyl ethylene
	t) propylene

Answer

$[a \rightarrow q], [b \rightarrow r], [c \rightarrow s, t], [d \rightarrow p]$

Question 3

Column 1	Column 2
a) Steam distillation	p) geometrical isomerism is not possible
b) acetylene	q) *o*-nitrophenol
c) 1,2-dichlorocyclopentadiene (*trans*)	r) acidity is due to 50% s character
d) *cis* 1,2dibromocyclopentane	s) 3 stereoisomers
	t) 2 diastereoisomers
	u) optically inactive due to plane of symmetry

Answer

$$[a \rightarrow q], [b \rightarrow p, r], [c \rightarrow s, t], [d \rightarrow u]$$

VIII. Structure-Based and Numerical Problems

1. A volatile liquid is obtained when a sample of natural rubber was pyrolysed. It contained C and H only. On ozonolysis it produced 2HCHO and CH_3COCHO. 1.36 g of sample produced 4.4 g of CO_2 and 1.44 g of water. The molecular weight was 66.5 g. Find the molecular formula of the hydrocarbon.

Answer

The volatile liquid is $CH_2{=}C(CH_3){-}CH{=}CH_2$

Ozonolysis of this in $(CH_3)_2S$ gives $2HCHO + CH_3COCHO$.

$\%C = 12 \times 4.4 \times 100/(44 \times 1.36) = 88.23$

$\%H = 2 \times 1.44 \times 100/(18 \times 1.36) = 11.77$

Atomic ratio of carbon = 88.23/12 = 7.35

For hydrogen it is 11.77

C:H = 7.35/7.35: 11.77/7.35 = 1:1.6 or 5:8

The empirical formula is C_5H_8 mwt/emp.wt = 66.5/68 = 1

molecular formula is also C_5H_8.

2. A hydrocarbon has 87% carbon. The molecular weight is with 96 grams. Catalytic reduction with Pd yields B by absorption of 2 equivalents of hydrogen formula C_7H_{16}. On ozonolysis of

A two fragments result. One fragment is acetic acid. The other one is an optically active carboxylic acid, C($C_5H_{10}O_2$). Identify A, B and C.

Answer

A has formula C_7H_{12}

$$CH_3C{\equiv}C\underset{\displaystyle CH_3}{\underset{|}{C}}HCH_2CH_3$$

(4-methylhex-2-yne)

Atomic ratio for carbon = 87/12

C:H = 7.25/7.25:13/7.25 = 1:1.79

C:H is 7:12.5 or 7:12

Emp. formula is C_7H_{12}

Since m.wt. is 96 the molecular formula is also C_7H_{12}

B is (3-methylhexane)

$$CH_3CH_2CH_2\underset{\displaystyle CH_3}{\underset{|}{C}}HCH_2CH_3$$

$$C\ CH_3CH_2\underset{\displaystyle CH_3}{\underset{|}{C}}HCOOH$$

3. An optically active alcohol with formula $C_{11}H_{16}O$ did not absorb hydrogen by Pd reduction. A with sulphuric acid dehydrates to give B which is optically inactive but exhibits geometrical isomerism. A Victor Mayer study gave the molecular weight as 146 grams. B on ozonolysis gave propanal and another compound with formula C_8H_8O (C). It gave positive answer for haloform reaction. Find A, B and C.

Answer

$$\text{A is } C_6H_5-\overset{\displaystyle CH_3}{\overset{|}{\underset{\displaystyle OH}{\underset{|}{C}}}}-CH_2CH_2CH_3$$

$$\text{B is } C_6H_5\underset{\displaystyle CH_3}{\underset{|}{C}}{=}CHCH_2CH_3$$

C is acetophenone.

4. 20 ml of hydrocarbon was exploded with 80 ml of oxygen in an eudiometer tube. On cooling, the volume reduced to 60 ml. On treatment with KOH, a contraction of 40 ml was observed. Find the molecular formula and the compound.

Answer

Volume of $CO_2 + O_2$(unused) = 60 ml

Volume of used oxygen = 80–20 = 60 ml

CO_2 absorbed = 40 ml

$20x = 40$ Hence $x = 2$

$20(x + y/4) = 60$

$x + y/4 = 3$

$y/4 = 3 - 2 = 1$

$y = 4$

The formula is C_2H_4 and the compound is ethylene.

5. An organic compound P having molecular formula $C_5H_{10}O$ is treated with dil.H_2SO_4 to give two compounds Q and R both of which answer positive for the iodoform test. The reaction of $C_5H_{10}O$ with dil.H_2SO_4 is 10^{15} times faster than the reaction of sulphuric acid with ethylene. Identify P, Q and R and give the reason for the faster rate of $C_5H_{10}O$ with sulphuric acid.

Answer

After we study aldehydes, ketones and ethers we will know that the possible compounds with formula $C_5H_{10}O$ are a ketone, an aldehyde an enol ether, an unsaturated alcohol and a cyclic ether. Among these structures only the enol ether given in.

Figure Ex.1.2 shows the compound P. $C_5H_{10}O$ can react with dil.H_2SO_4 to give a heteroatom cation which is stable due to resonance. The mechanism of addition of water is given below. The products 4 and 5 both give positive haloform reaction (This aspect will be discussed in carbonyl compounds chapter).

CH_3
C
CH_2 — Ö — CH_2 — CH_3

(1)

(2)

CH_3 | C(=CH_2)–Ö–CH_2–CH_3 → CH_3 | C^+(CH_3)–Ö–CH_2–CH_3

(1)

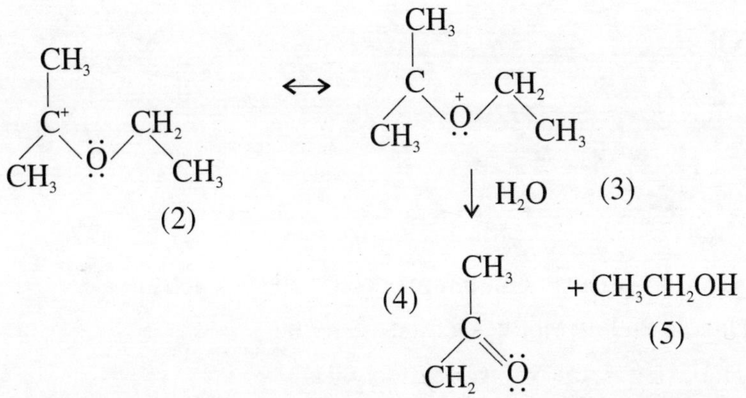

$CH_3-\overset{+}{C}(CH_3)-\ddot{O}-CH_2CH_3$ (2) $\leftrightarrow$ $CH_3-C(CH_3)=\overset{+}{O}-CH_2CH_3$ (3)

$\downarrow H_2O$

(4) $CH_3-C(CH_3)=\ddot{O}$ + CH_3CH_2OH (5)

Figure Ex.1.2 Reaction scheme of enol ether with dil.H_2SO_4

Enol ether with dil.H_2SO_4 reaction scheme is described as follows: (1) is the enol ether which reacts with hydrogen ion from acid. The ether reacts with H^+ to give the oxygen-stabilized carbocation (2). (2) is delocalized by resonance and gives (3). (3) with H_2O gives (4) acetone and (5) ethanol both of which answer positive for haloform reaction. The details of haloform reaction will be given in aldehydes and ketones chapter.

6. Mandelic acid weighing 28 mg was dissolved in 1 cc of ethanol and placed in a sample cell of path length = 10 cm, and the optical rotation was –4.35°. Find the specific rotation.

Answer

$$-\frac{4.35}{0.028 \times 1} = -155.4^\circ$$

7. Enantiomer A is 75% and enantiomer B is 25%. What is % enantiomeric excess?

Answer

$$\frac{75-25}{100} \times 100 = 50\%$$

8. Draw the eclipsed and staggered conformation for meso form of tartaric acid in Fisher Projection forms

Answer

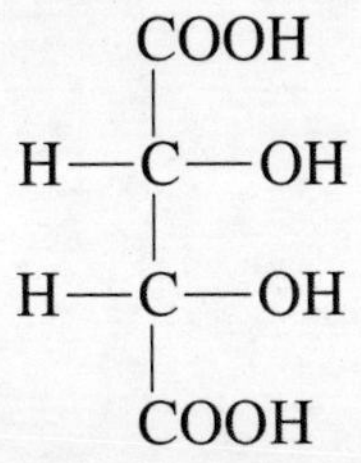

```
      COOH
       |
  H—C—OH
       |
  H—C—OH
       |
      COOH
```

(eclipsed form with plane of symmetry)

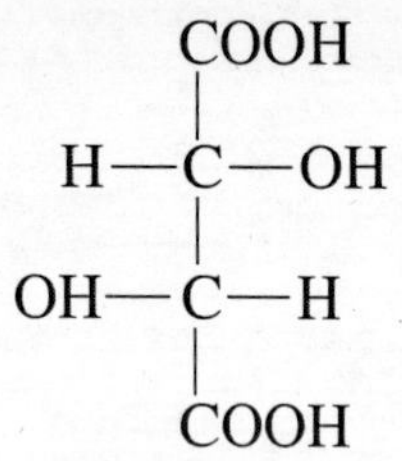

```
        COOH
         |
    H—C—OH
         |
  OH—C—H
         |
        COOH
```

(staggered form with inversion centre or center of symmetry)

9. Assign Cahn Ingold Prelog priorities for the following
   a) $CH{=}CH_2$, $CH(CH_3)_2$ — $C(CH_3)_3$ CH_2CH_3
   b) $C{\equiv}CH$, $C{=}CH_2$, $C(CH_3)_3$ phenyl
   c) CO_2CH_3, $COCH_3$, CH_2OCH_3, CH_2CH_3
   d) $C{\equiv}N$ — CH_2Br, CH_2CH_2Br, Br

Answer

highest → lowest priority

   a) $C(CH_3)_3$ $CH{=}CH_2$ $CH(CH_3)_2$ CH_2CH_3
   b) phenyl, acetylenic, $C(CH_3)_3$, $CH{=}CH_2$
   c) $COOCH_3$, $COCH_3$, CH_2OCH_3, CH_2CH_3
   d) Br, CH_2Br, CN, CH_2CH_2Br

10. Assign E, Z configuration to the following alkenes.

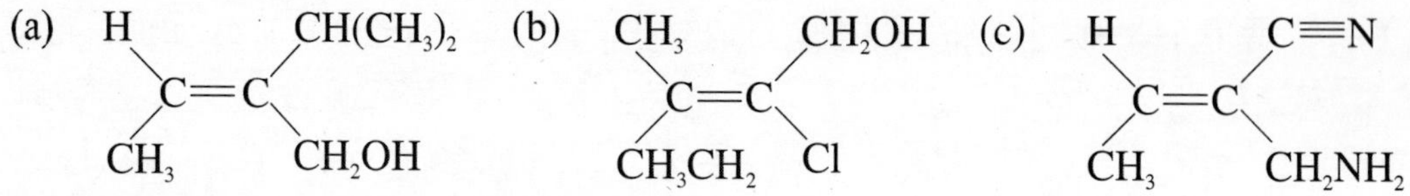

(a) H and CH_3 on one carbon, $CH(CH_3)_2$ and CH_2OH on the other:

```
(a)  H         CH(CH3)2
       \      /
        C=C
       /      \
    CH3        CH2OH
```

```
(b)  CH3       CH2OH
       \      /
        C=C
       /      \
  CH3CH2      Cl
```

```
(c)  H          C≡N
       \      /
        C=C
       /      \
    CH3        CH2NH2
```

Answers

(a) is Z form, (b) is Z form, (C) is E form).

11. Name the compound given below

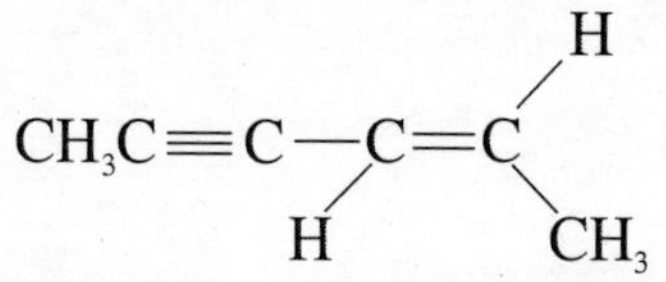

```
                      H
                     /
CH3C≡C—C=C
            /      \
           H        CH3
```

Answer

trans-2-hexen-4-yne.

12. Propanaldehyde can form two enols. Write the two enol forms.

Answer

The two enols arise due to stereochemical differences. It can exist in Z or E form as shown below.

```
      H   H
      |   |
  CH3C=C
          |
          OH
```

(Z) form

```
      H  OH
      |   |
  CH3C=C
          |
          H
```

(E) form

13. Two compounds have the same % of carbon and hydrogen namely 92.3 and 7.7. The first compound is acidic and is obtained by the hydrolysis of CaC_2. The second one is obtained by heating three moles of the first compound in a red-hot tube and it is aromatic. Find the two compounds after finding the empirical formula.

Answers

Empirical formula is CH. The compounds are acetylene and benzene.

14. An amino acid casein, a constituent of protein, has the following % for N, S, C, H and oxygen. The molecular weight is 150 g.C = 40.25, H = 7.43, N = 9.4, S = 21.5, O = 21.42. Find the molecular formula.

Answers

$C_5H_{11}SNO_2$.

15. 0.015 litres of a hydrocarbon was burnt in excess oxygen. On cooling the mixture, the volume was 0.0375 litres. On addition of caustic potash a further reduction in volume of 0.03 litres was observed. What is the hydrocarbon?

Answers

$CxHy + [x + (y/4)]O_2 \longrightarrow xCO_2 + (y/2)H_2O$

Contraction on treatment with KOH = 30 ml

15 ml of hydrocarbon will produce $15x$ CO_2

$15x = 30$

$x = 2$

Total contraction per ml = Vol. of hydrocarbon + Vol. of O_2 – (Vol. of CO_2 + Vol. of H_2O)

$=1 + (x + y/4) - x - 0 = 1 + (y/4)$

For 15 ml,

$15(1 + (y/4)) = 37.5$

$y = 6$

Formula is C_2H_6.

16. Find the % composition of

a) C_3H_7Cl b) CH_4ON_2 c) C_6H_8NCl

Answers

a. For C_3H_7Cl, the total weight is $12\times3 + 7\times1 + 35.5 = 78.5$ g

% carbon = $36\times100/78.5 = 45.9$

% hydrogen = $7\times100/78.5 = 8.9$

% chlorine = $1\times35.5\times100/78.5 = 45.2$

In a similar way % composition in (b) and (c) are found out as

(b) For CH_4ON_2, %C = 20 % H = 6.7% Oxygen = 26.6% Nitrogen = 46.7

(c) For C_6H_8NCl, %C = 55.6% H = 6.2% chlorine = 27.4% Nitrogen=10.8

17. A 7 mg sample of hydrocarbon with molar mass = 140.3 is found to yield 21.96 mg of CO_2 and 8.99 mg of water when burnt with oxygen. Find the molecular formula and also the volume of oxygen in litres used for combustion at 298 K and 1 atm. pressure.

Answers

$$\%C = \frac{12 \times 21.96 \times 100}{(44 \times 7)} = 85.55$$

$$\%H = \frac{2 \times 8.99 \times 100}{(18 \times 7)} = 14.26$$

$$\text{Atomic ratio for carbon} = \frac{85.55}{12} = 7.129$$

Atomic ratio for hydrogen = 14.26

C: H = 1:2

CH_2 is the empirical formula.

M.wt/emp.wt = 140.3/14 = 10

Hence molecular formula is $C_{10}H_{20}$.

$CxHy + [x + (y/4)]O_2 \rightarrow xCO_2 + (y/2)\, H_2O$

$C_{10}H_{20} + (10 + 20/4)\, O_2 \rightarrow 10CO_2 + 10H_2O$

One mole of hydrocarbon requires 15 moles of O_2 at STP.

0.007/140 will require 15×0.007/140 moles of O_2 at STP.

7.5×10^{-4} moles of oxygen at STP.

$7.5 \times 10^{-4} \times 22.4$ litres at STP.

$7.5 \times 10^{-4} \times 22.4 \times 298/273$ litres at 298 K = 0.018 litres

18. Isopentane is allowed to undergo free radical chlorination and the reacting mixture is separated by careful fractional crystallization.
 a) How many fractions of formula $C_5H_{11}Cl$ would you expect to collect?
 b) Write structures of isomers including enantiomers and specify which is R and which is S configuration.
 c) Give the structure of the fraction showing optical activity.

Answers

$CH_3CH_2—CH(CH_3)—CH_3$ is isopentane.

The monochloro derivatives are

1. $ClCH_2CH_2—CH(CH_3)CH_3$
2. $CH_3C^*H(Cl)—CH(CH_3)—CH_3$
3. $CH_3C(Cl)(CH_3)—CH_2CH_3$
4. $CH_3CH_2—C^*H(CH_3)—CH_2Cl$
5. $CH_3CH_2—C^*H(CH_2Cl)—CH_3$

4 and 5 are identical even though Cl attacks at C_4 or C_5 carbon.

Thus there are 4 fractions. Isomers are 6 (two sets of d and l forms due to optical activity of compounds 2 and 4).

The R and S forms of fraction 2 are given in Fischer Projection formula as

$$\begin{array}{c} H \\ | \\ (1)\ Cl—C—C_3H_7\,(2) \\ | \\ CH_3\ (3) \end{array}$$

The configuration results by clockwise rotation. Hence this is R configuration.

The S configuration isomer is

$$\begin{array}{c} H \\ | \\ (2)\ C_3H_7—C—Cl\ (1) \\ | \\ CH_3\ (3) \end{array}$$

The configuration results by anticlockwise rotation. Hence this is S configuration.

Isomer 4 has the following R and S configurations

$$\begin{array}{c} H \\ | \\ (2)\ C_2H_5—C—CH_3\ (3) \\ | \\ CH_2Cl\ (1) \end{array}$$

. This is R configuration.

$$\begin{array}{c} H \\ | \\ (3)\ CH_3—C—C_2H_5\,(2) \\ | \\ CH_2Cl\ (1) \end{array}$$

. This is S configuration.

There will be equal amounts of R and S leading to racemic mixture.

19. Describe the a) Dumas method and b) Kjeldahl method of determining % of nitrogen in an organic compound.

(The student must consult a standard organic chemistry text and describe these methods)

2

ALKANES

INTRODUCTION

Alkanes are the simplest form of organic molecules containing carbon and hydrogen only, with formula C_nH_{2n+2}. They are extremely unreactive and alkanes such as pentane and hexane are used as solvents, especially for purification of organic compounds. Alkanes burn and undergo combustion, and methane, propane and butane are used as domestic fuels and petrol is a mixture of alkanes containing largely isooctane. Pentane, hexane and isooctane have molecular formula as follows:

$$CH_3(CH_2)_3CH_3 \text{ (pentane)}$$

$$CH_3(CH_2)_4CH_3 \text{ (hexane)}$$

$$(CH_3)_3C—CH_2—C—(CH_3)_3 \text{ (isooctane)}$$

Homologous Series

A homologous series (*homos* in greek means "same as") is a family of compounds in which each member differs from the next by one methylene (CH_2) group. The members of homologous series are called homologs. Propane ($CH_3CH_2CH_3$) and butane ($CH_3CH_2CH_2CH_3$) are homologous.

PHYSICAL PROPERTIES AND STRUCTURE OF ALKANES

Boiling Points

The unbranched alkanes (pentane to heptadecane) are liquids while higher homologues are solids. The boiling points of unbranched alkanes increase as the number of carbons increase. Branched alkanes have lower boiling points than unbranched isomers since branched alkanes have smaller surface area.

For example, pentane has a boiling point of 36°C whereas 2-methylbutane has a boiling point of 28°C and 2,2-dimethylpropane boils at 9°C.

The boiling point of a compound of any homologous series increases as the molecular weight increases due to increase in van der Waals forces.

SOLVED EXAMPLE I

Match the boiling points with the appropriate alkanes

1. 106	a) Octane
2. 116	b) 2-methylheptane
3. 126	c) 2,2,3,3-tetramethylbutane
4. 151	d) Nonane

Answer

$1 \rightarrow$ c, $2 \rightarrow$ b, $3 \rightarrow$ a, $4 \rightarrow$ d

EXERCISE I

Match the boiling points with the appropriate alkanes

1. 69°C	a) Butane
2. 174°C	b) Decane
3. –0.5°C	c) Hexane

Answer

$1 \rightarrow$ c, $2 \rightarrow$ b, $3 \rightarrow$ a

Melting Points

Solid alkanes have low melting points. The melting points for any homologues series increase as their molecular weight increases. The increase in melting points is less regular than the increase in boiling point since packing influences the melting point of a compound. The melting points of alkanes with even number of carbons fall on a smooth curve. The two curves do not overlap.

Packing is less tight for alkanes with odd number of carbons and intermolecular forces decrease leading to decreased melting points.

SOLVED EXAMPLE 2

Boiling points increase regularly by increase of molecular weights, but melting points increase irregularly. Why?

Answer

The different shapes of molecules affect the packing and hence melting point increase is not regular.

Solubility in Water

Alkanes are insoluble in water.

Densities

The densities of alkanes increase with increasing molecular weight. Even a 30-carbon skeleton of an alkane (triacontane) has a density (0.8097 g/cc) which is less than that of water (0.9982 at 293 K). This means that a mixture of alkane and water will separate into 2 layers with less dense alkane floating on the top.

SOLVED EXAMPLE 3

All liquid and solid alkanes float in water Why?

Answer

They are less dense than water.

CHEMICAL PROPERTIES AND STRUCTURE

All hydrocarbons on combustion yield CO_2 and water. Unbranched alkanes have higher heat of combustion than the methyl branched isomers. Two successive hydrocarbons differing by a CH_2 group have their heat of combustion differing by 156 kcal/mole.

SOLVED EXAMPLE 4

Find the heat of combustion of 2-methylnonane given that heat of combustion of 2-methylheptane is 1306 kcal/mole.

Answer

The two hydrocarbons differ by two CH_2 groups.

Hence $\Delta H = 1306 + (2 \times 156) = 1618$ kcal/mole

The compound having lowest heat of combustion is the most stable one.

Among the hydrocarbons, 2,3,3-tetramethylbutane, methylheptane and octane, the most branched compound is the most stable one and has the lowest heat of combustion.

EXERCISE 2

Arrange the following compounds in increasing order of heat of combustion.

Pentane, isopentane, neopentane and hexane

Answer

Hexane > pentane > isopentane > neopentane

PREPARATION OF ALKANES

There are various methods of preparation of alkanes—carbide hydrolysis, Wurtz reaction, Kolbe electrolysis, Corey–House synthesis, carbene insertion and so on.

i. *Grignard reaction*

$RX + Mg \longrightarrow RMgX + H_2O \longrightarrow RH$

RMgX can be treated with any weak acid (alcohols, amines, carboxylic acids) to obtain hydrocarbon instead of using water.

ii. *Treatment with Li metal and hydrolysis*

$RX + Li \longrightarrow RLi$

$RLi + H_2O \longrightarrow RH$

iii. *Reduction of alkyl halides*

a) $RX + Zn + H^+ \longrightarrow RH + Zn^{2+} + X^-$ (Frankland method)

b) $4RX + LiAlH_4 \longrightarrow 4RH + LiX + AlX_3$

(or $RX + H^- \rightarrow RH + X^-$ (H – comes from $LiAlH_4$)

c) $RX + (n-C_4H_9)_3SnH \longrightarrow RH + (n-C_4H_9)_3SnX$ (dry ether)

iv. *Wurtz reaction*

$RX + 2Na + XR \longrightarrow R\text{-}R + 2NaX$

v. *Corey–House synthesis*

$RX + 2Li \longrightarrow RLi + RX$ in diethyl ether

$2RLi + CuI \longrightarrow R_2CuLi + LiI$

$R_2CuLi + R'X \longrightarrow R - R' + RCu + RLi$

Alkyl halide must be CH_3X or primary for good yield. The dialkyl cuprate can have methyl, primary, secondary or tertiary groups. Two alkyl groups being coupled need not be different.

SOLVED EXAMPLE 5

$CH_3I + Li + CuI \longrightarrow X + RI \longrightarrow$ Hexane

What is R group?

Answer

R is n-pentyl group

SOLVED EXAMPLE 6

What is the product of the reaction between $PhCHCl_2$ and $(CH_3)_2CuLi$ in tetrahydrofuran?

Answer

$PhCH(Me)_2$

SOLVED EXAMPLE 7

Vinyl iodide reacts with $(C_2H_5)_2CuLi$ to give A in diethyl ether.

Answer the following questions

a) What is the structure of A?
b) Will A exhibit geometric isomerism?
c) What are the different isomers of A?

Answer

a) A is $C_2H_5CH{=}CH_2$.
b) Since it is a terminal alkene, it cannot exhibit geometrical isomerism.
c) It has isomers which are 1-butene and 2-butene. 2-butene has *cis* and *trans* isomers.

vi. *Kolbe electrolysis* A concentrated solution of sodium or potassium salt of a carboxylic acid or mixture of carboxylic acids on electrolysis gives R—R or R—R′. Different side products are also obtained.

The mechanism of the reaction of $2C_2H_5COONa$ is

$$C_2H_5COO^- \longrightarrow C_2H_5CO_2^{\bullet} + e$$

$$C_2H_5CO_2^{\bullet} \longrightarrow C_2H_5^{\bullet} + CO_2$$

$$2C_2H_5^{\bullet} \longrightarrow C_4H_{10}$$

$$C_2H_5^{\bullet} + C_2H_5^{\bullet} \longrightarrow C_2H_6 + C_2H_4$$

$$C_2H_5^{\bullet} + C_2H_5CO_2^{\bullet} \longrightarrow C_2H_5COOC_2H_5$$

vii. Soda lime distillation RCOONa

$$RCOONa + NaOH \xrightarrow{CaO} R—H + Na_2CO_3$$

viii. Alkenes undergo hydrogenation by Sabatier–Senderson method with Ni catalyst at 473 K giving alkanes.

Alkynes can also be hydrogenated.

ix. Alcohols, aldehydes, ketones, fatty acids and their derivatives are reduced in the presence of red phosphorus and HI at 200°C.

$$ROH \longrightarrow RH + H_2O + I_2$$

$$RCHO \longrightarrow RCH_3 + H_2O + 2I_2$$

$$RCOR \longrightarrow RCH_2R + H_2O + 2I_2$$

$$RCOOH \longrightarrow RCH_3 + H_2O + 3I_2$$

$$RCOOR' \longrightarrow RCH_3 + R'H$$

$$(RCO)_2O \longrightarrow 2RCH_3$$

$$RCOCl \longrightarrow RCH_3$$

$$RCONH_2 \longrightarrow RCH_3$$

x. *Clemmenson reduction of carbonyl compounds* Zn–Hg and conc. HCl are added to the C=O compound. The product is an alkane and water.

$$CH_3COCH_3 + 2H_2 \xrightarrow{(Zn+Hg+HCl)} CH_3CH_2CH_3 + H_2O$$

Mechanism of Clemmenson reduction This reaction consists of the reduction of the carbonyl group of aldehydes and ketones to the methylene group with amalgamated zinc and concentrated hydrochloric acid. Ketones are reduced more often than aldehydes. Reduction of acetophenone under these conditions, by refluxing for several hours, forms ethyl benzene. Figure 2.1 shows an example.

$$C_6H_5COCH_3 \xrightarrow[\text{Con HCl}]{\text{Zn (Hg)}} C_6H_5CH_2CH_3$$

Figure 2.1 Clemmenson reduction

xi. *By hydrolysis of carbides* Aluminium carbide and beryllium carbide on hydrolysis give methane.

$$Al_4C_3 + 12H_2O \xrightarrow{\Delta} 3CH_4 + 4Al(OH)_3$$

$$Be_2C + 4H_2O \xrightarrow{\Delta} 2Be(OH)_2 + CH_4$$

xii. *Hydroboration of alkenes* Alkenes on hydroboration in acetic acid give alkanes.

$$2R-CH=CH_2 + B_2H_6 \xrightarrow{\text{acetic acid}} 2(RCH_2CH_2)_3B \rightarrow 6RCH_2CH_3$$

xiii. Wolff–Kishner reduction of carbonyl compounds

The C=O compound is treated with hydrazine, KOH and triethylene glycol at 175°C to get the hydrocarbon (hydrazine, KOH and triethylene glycol).

$$RCOCH_2CH_3 \longrightarrow RCH_2CH_2CH_3$$

Figure 2.2 shows the mechanism of Wolff–Kishner reduction.

$$R_2C{=}N{-}NH_2 + B{:} \longrightarrow R_2C{=}N{-}NH^- + BH^+$$

$$R_2C{=}N{-}NH^- \longrightarrow R_2C{-}\underset{\substack{|\\ H}}{N}{=}N^-$$

$$R_2\underset{\substack{|\\ H}}{C}N{=}N^- \longrightarrow R_2\bar{C}H + N_2$$

$$R_2\bar{C}H \longrightarrow + BH^+ \longrightarrow R_2CH_2 + B{:}$$

Figure 2.2 Mechanism of Wolff–Kishner reduction

xiv. *Catalytic reduction of thioacetal* Aldehydes react with thioacetals to give alkanes in the presence of Ni/H_2.

$$RCHO + R'SH \xrightarrow{Ni/H_2} R-CH(SR')_2 \rightarrow RCH_3$$

xv. *Carbene insertion* Diazomethane is first converted to $:CH_2$ and N_2. $:CH_2$, a carbene, inserts between C—C bonds in alkanes to create a new alkane.

$$C_5H_{12} + :CH_2 \rightarrow CH_3CH_2CH_2CH_2CH_2CH_3 + CH_3CH(CH_3)CH_2CH_2CH_3 + CH_3CH_2CH(CH_3)CH_2CH_3$$

The yields are 48%, 35% and 17% respectively.

PREPARATION OF ALICYCLIC HYDROCARBONS AND SOME OF THEIR REACTIONS

i. *Carbene insertion* Carbene insertion is mainly used to obtain cyclic hydrocarbons from alkenes. For example propylene reacts with carbene to give methylcyclopropane. We will learn later that Carbene insertion takes place in phenols C_6H_5OH to give ethers and in acids to give esters. We will also learn that Aliphatic alcohols undergo carbene insertion to produce ethers only in the presence of catalysts.

Carbenes are generally produced from diazomethane which has the following resonance structures.

$$H_2C{=}N^{+}{=}N^{-} \longleftrightarrow H_2C^{-}{-}N^{+}{\equiv}N \longleftrightarrow H_2C^{-}{-}N{=}N^{+}$$

The important step in insertion reaction is protonation of diazomethane which then loses nitrogen. Protonation is possible only when the reacting compound is acidic like carboxylic acids or phenols.

ii. *Dibromide reaction with Na* 1,4-dibromobutane reacts with Na to give cyclobutane and NaBr.

$$BrCH_2CH_2CH_2CH_2Br + 2Na \longrightarrow \text{Cyclobutane} + 2NaBr$$

iii. *Simmons–Smith reaction*

$$CH_3CH{=}CH_2 + CH_2I_2 \xrightarrow{Zn/Hg} \text{Methyl cyclopropane}$$

iv. Cycloalkenes on reduction with H_2/Ni give cycloalkanes.

v. 1,3-butadiene by a special reaction called Diels–Alder reaction with ethylene gives cyclohexene which on hydrogenation gives cyclohexane.

Cyclic hydrocarbons with 3 and 4 carbons have lot of strain and hence undergo the following reactions to give straight-chain compounds.

$$\text{Cyclopropane} + H_2 \xrightarrow{\text{Ni (catalyst)}} \text{Propane}$$

$$\text{Cyclobutane} + HBr \longrightarrow n\text{-butylbromide}$$

Similar reactions do not occur in cyclopentane or cyclohexane since strain is minimum in them.

Reactions of Saturated Hydrocarbons

Alkanes undergo the following reactions:

1. *Nitration* Nitration is done in nitric acid in gas phase at 400°C or in liquid phase. Propane gives a mixture of nitrated products.
2. *Sulphonation* alkanes having tertiary groups undergo sulphonation when treated with oleum ($H_2S_2O_7$). Optically active alkanes racemize and the reaction passes through a carbonium ion intermediate.

3. *Pyrolysis* Process conducted at high temperature (1200°C) and the reaction leads to a mixture of alkanes.
4. *Isomerization* Normal alkanes (butane) can be converted to isobutene by addition of $AlCl_3$ and HCl.
5. *Aromatization* n-hexane can be converted to benzene using Cr_2O_3–Al_2O_3 at 600°C.
6. *Halogenation* This is the most important reaction of alkanes and will be discussed in detail in the following sections.

HALOGENATION

i. Fluorine's reactivity with methane is very high compared to reaction with bromine or chlorine. The reaction involves free radical mechanism. Fluorine can react in the dark with alkanes.

Chain initiation

$$F_2 \rightarrow 2F\bullet$$

$$\Delta H = +38 \text{ k cal / mole} \quad Ea = +38 \text{ kcal / mole}$$

Propagation

$$F\bullet + CH_4 \longrightarrow HF + CH_3^\bullet \quad \Delta H = -32 \text{ kcal / mole} \; Ea = +1.2 \text{ kcal / mole}$$

$$CH_3^\bullet + F_2 \longrightarrow CH_3F + F\bullet \quad \Delta H = -70 \text{ kcal / mole} \; Ea = \text{small}$$

The overall heat of the propagation reaction is –102 kcal/mole and the low energy of activation account for the high reactivity of fluorine towards Methane.

ii. Iodination is reversible.

$$CH_4 + I_2 \longrightarrow CH_3I + HI$$

The HI formed must be removed by oxidation with HIO_3, HNO_3, HgO, etc. according to reaction

$$5HI + HIO_3 \longrightarrow 3I_2 + 3H_2O$$

iii. For chlorination the overall heat of propagation reaction is –24.5 kcal/mole and energy of activation is higher. Hence chlorine is less reactive.

iv. Due to very high energy of activation for the propagation step (18.6 kcal/mole) bromine is much less reactive.

The general mechanism of halogenation for higher alkanes is

Initiation

$$Cl_2 \text{(light or heat)} \longrightarrow 2Cl^-$$

Propagation

$$CH_3CH_2H + Cl\cdot \longrightarrow CH_3\dot{C}H_2 + HCl$$

$$CH_3\dot{C}H_2 + Cl{:}Cl \longrightarrow CH_3CH_2Cl + Cl\cdot$$

Termination

$$CH_3\dot{C}H_2 + Cl. \longrightarrow CH_3CH_2Cl$$

$$CH_3\dot{C}H_2 + CH_3\dot{C}H_2 \longrightarrow C_4H_{10}$$

$$Cl\cdot + Cl\cdot \longrightarrow Cl_2$$

Nature of Substitution Products in Halogenation

Halogenation of alkanes depends on the following factors.

1. Probability factor
2. Reactivity of hydrogens
3. Reactivity of halogens

1. *Probability factor* This depends on the number of inequivalent hydrogens in a compound. For example n-butane contains 6 equivalent 1° hydrogens and 4 equivalent 2° hydrogens. Hence the chance of removing 1°H relative to 2°H is 3:2.

2. *Reactivity of hydrogens* The order of reactivity of hydrogen is

$$3°\,H > 2°\,H > 1°\,H$$

This order arises due to the relative stabilities of the radicals which are obtained by homolytic fission and we discuss hyperconjugation below.

Tertiary radical > secondary radical > primary radical. In heterolytic fission of RX we have R^+ and X^- and R^+ is called the carbocation. The order of stability of carbocations is

Tertiary cation > secondary cation > primary cation. This is due to hyperconjugation, which is explained below.

Stabilities of free radicals and carbocations The relative stabilities of free radicals and carbocations can be explained by the phenomenon of hyperconjugation (no-bond resonance or sigma bond delocalization or Baker–Nathan effect).

Hyperconjugation Let us consider the ethyl cation in Figure 2.3. The cation is sp^2 hybridized and the p-orbital of carbon of CH_2 is empty. The electrons of the neighbouring sigma bonds involving the three hydrogens are dragged to this empty p-orbital and there is sigma bond delocalization. The neighbouring CH_3 group has 3 beta hydrogens for delocalization. If one of the hydrogens of ethyl cation are replaced by methyl we will have 6 beta hydrogens for delocalization ($(CH_3)_2CH$) and if both the hydrogens of CH_2 in ethyl cation are replaced by methyl groups we have 9 delocalized structures ($(CH_3)_3C$) and the greater the number of delocalized sigma bonds the more stable is the carbocation. The same reason applies for stability of free radicals. Figure 2.3 indicates hyperconjugation in ethyl cation.

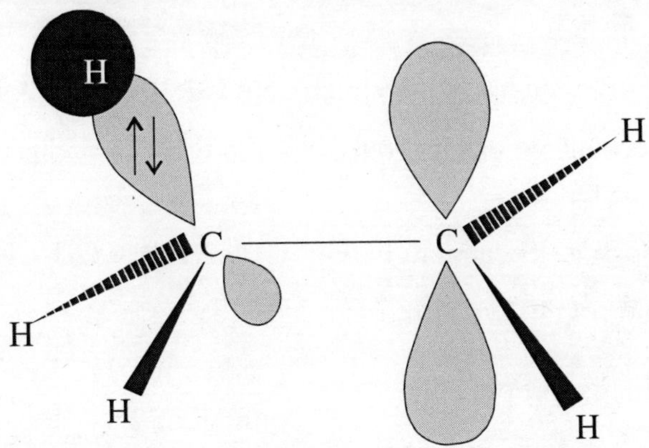

Figure 2.3 Hyperconjugation in ethyl cation

Suppose we have allyl cation and benzyl cation we have resonance stabilization due to pi bond delocalization. These ions are more stable than the primary, secondary and tertiary carbocations. The relative order of stability is

Benzyl cation > allyl cation > Tertiary cation > secondary cation > primary cation

Concept of resonance Electrons restricted to a particular region are called localized electrons. Localized electrons belong to a single atom or confined to a bond between two atoms. The following compounds have localized electrons.

$$CH_3\ddot{N}H_2$$

$$CH_3CH{=}CH_2$$

Delocalized electrons are shared by more than two atoms. Benzene has two important Kekule structures as shown in Figure 2.4.

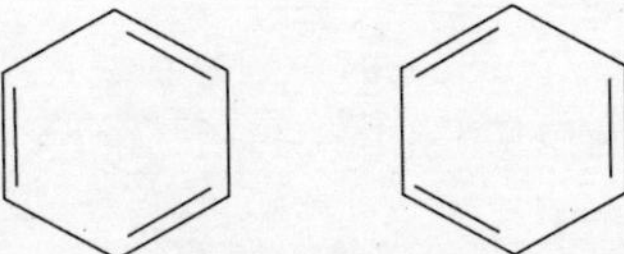

Figure 2.4 Kekule structures of benzene

A compound with delocalized electrons is said to have resonance. Resonance gives more stability compared to hyperconjugated structures.

Abstraction of tertiary hydrogens is more exothermic than that of secondary hydrogens, the abstraction of which in turn is more exothermic than that of primary hydrogens.

3. *Reactivity of halogens* Chlorine is more reactive than bromine and influenced by probability factors, but is less selective. Bromine is more selective but less reactive and less influenced by probability factors.

Let us consider chlorination.

Reactivity of primary: secondary: tertiary hydrogens is 1:3.8:5 for chlorination

Taking all this into account, we can find out the proportion of monohalogen products obtained in the chlorination of propane

$$CH_3CH_2CH_3 \text{ — (heat or light) } CH_3CH_2CH_2Cl + CH_3CHClCH_3$$

The relative proportion of the two isomers is

$$\text{n-propyl chloride/isopropyl chloride} = \frac{\text{Number of 1°H}}{\text{Number of 2°H}} \times \frac{\text{Reactivity of 1°H}}{\text{Reactivity of 2°H}}$$

$$= \frac{6}{2} \text{ X } \frac{1}{3.8} = \frac{6}{7.6}$$

$$\text{Hence\% of n-propyl chloride} = \frac{6}{(6+7.6)} \text{ x } 100 = 44.1$$

and% of isopropyl chloride is $100 - 44.1 = 55.9\,\%$

SOLVED EXAMPLE 8

Find the proportion of monosubstituted products from isobutane on chlorination:

Answer

Isobutane has the structure

```
        CH3
        |
  CH3—C—CH3
        |
        CH3
```

There is one tertiary hydrogen and 9 primary hydrogens. Hence,

```
          H
          |
% of ClCH2—C—CH3
          |
          CH3
```

$$= \frac{9\times1}{(9\times1)+(1\times5)} = \frac{9}{14} = 64\%$$

% of $(CH_3)_3C—Cl = 36$

Isobutyl chloride = 64% and tert. butyl chloride = 36%.

Thus we find that mixture of chlorides is produced when inequivalent hydrogens are present in an alkane. If halogenation reaction has to be used as a synthetic route single product must be obtained. For this the alkane must have all hydrogens equivalent (neopentane, cyclopropane)

Consider bromination. There is selectivity and tertiary compounds are preferentially formed. Bromination is said to be regioselective. If a reaction is 100% regioselective it is called regiospecific reaction. The reason for selectivity of bromine is as follows

Let us compare chlorination and bromination in propagation step.

Chlorination

$$Cl\cdot + CH_3CH_2CH_3 \longrightarrow CH_3CH_2CH_2^{\cdot} + HCl\ \Delta H = -8\,kJ/mol$$

$$Cl\cdot + CH_3CH_2CH_3 \longrightarrow HCl + CH_3CH\cdot CH_3^{\cdot}\ \ \Delta H = -17\,kJ/mol$$

$$Cl\cdot + CH_3CH(CH_3)CH_3 \longrightarrow CH_3C\cdot(CH_3)_2 + HCl\Delta H = -25\,kJ/mol$$

For bromination

$$Br\cdot + CH_3CH_2CH_3 \longrightarrow CH_3CH_2CH_2^{\cdot} + HBr\ \Delta H^{-} = 59\,kJ/mol$$

$$Br\cdot + CH_3CH_2CH_3 \longrightarrow HBr + CH_3CH\cdot CH_3^{\cdot}\ \ \Delta H^{-} = 50\,kJ/mol$$

$$Br\cdot + CH_3CH(CH_3)CH_3 \longrightarrow CH_3C\cdot(CH_3)_2 + HBr\ \Delta H^{-} = 42\,kJ/mol$$

Thus for bromination the steps are endothermic but for chlorination they are exothermic.

If steps are exothermic, the transition state energy is of similar order as reactants. So both the products, n-propyl chloride and isopropyl chloride, are formed. But in the case of endothermic bromination, transition state energy is of same order as that of product. Only the stabler product (tertiary bromide compared to secondary or primary) is formed. This is due to greater stability of tertiary radicals by hyperconjugation. According to Hammond–Leffler postulate, the structure of the transition state of an endothermic step resembles the products of the step more than it does the reactants. For exothermic steps, the transition state resembles that of the reactants.

Solved Example 9

What will be the exclusive product of the reaction of isobutane with bromine?

Answer

(*t*-butyl bromide)

The preferential formation of one product more compared to others is called regioselectivity. Thus bromination is highly regioselective.

The concept of selective radical bromination is used in allylic bromination of a compound containing a double bond. For example propylene when treated with bromine will give allyl bromide.

If concentration of bromine is large, addition at the double bond will be the competing reaction. To avoid this reaction, we use N-bromosuccinimide (NBS) which generates small amount of bromine required for allylic substitution. HBr produced in the substitution reaction reacts with NBS to maintain the low concentration of bromine.

SOLVED EXAMPLE 10

Give the structure of the principal halide in the bromination of methylcyclopentane

Answer

1-bromo-1-methylcyclopentane

SOLVED EXAMPLE 11

Which one of the following hydrocarbons gives a mixture of monosubstituted chlorides?

a) ethane b) cyclopropane

c) n-butane d) neopentane

Answer

c) n-butane

SOLVED EXAMPLE 12

Find the ratio of isomeric chlorinated products of n-butane. ($CH_3CH_2CH_2CH_3$)

Answer

$$\% \text{ of n-butyl chloride} = \frac{6\times1}{(6\times1)+(4\times3.8)} = \frac{6}{20.2} = 28\%$$

$$\% \text{ of isobutyl chloride} = 100 - 28 = 72\%$$

CONFORMATIONS OF ALKANES

Conformations are different spatial arrangements of a molecule that are generated by rotation about single bonds. The particular conformation a molecule adopts has a profound influence on its properties.

Ethane adopts two conformations—staggered and eclipsed—and we can draw the Newman projection formula. The staggered has *gauche* and *anti* forms. The torsion angle into 60 degrees we call it gauche and if the torsion angle is 180 degrees we call the form as anti. The two staggered conformations are equivalent in ethane. Generally staggered conformation is more stable with less energy (121 kJ/mole less in energy compared to eclipsed form).

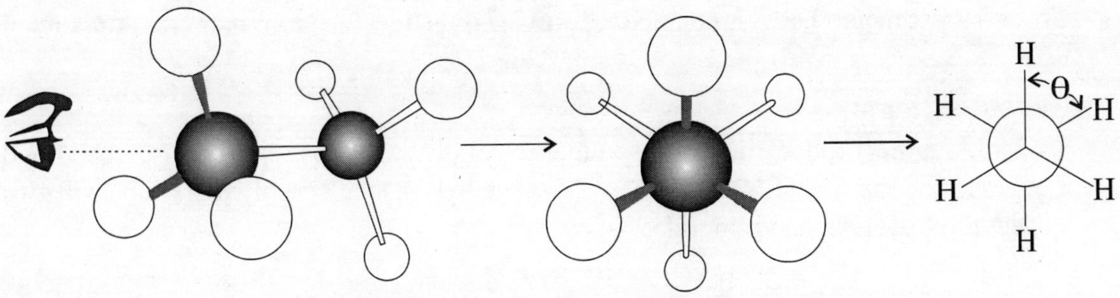

Figure 2.5 Ball-and-stick models of configurations of ethane

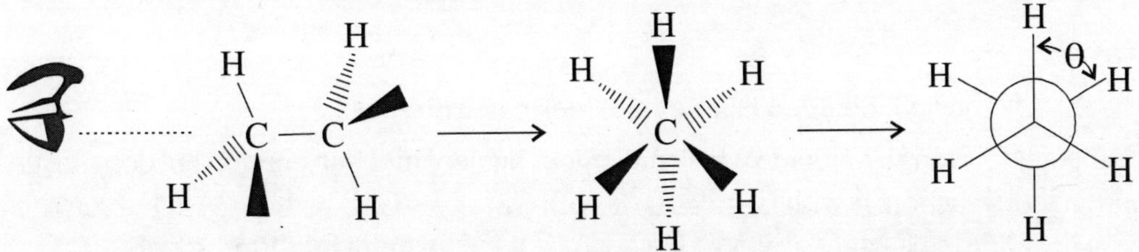

Figure 2.6 Line and wedge formulas of ethane configurations

PROBLEMS

I. "Explain Why" Questions and Small Structure-Based Problems

1. Propane has lower boiling point than ethanol. Explain.

   Answer

   The higher boiling point of ethanol is due to hydrogen bonding.

2. Nitro compounds have higher boiling points than alkanes of same molecular weight. Why?

   Answer

   Nitro compounds have dipole moments.

3. Arrange in decreasing order of boiling points for

   A. 2,2,4-trimethylpentane B. octane and C. 2-methylheptane

   Answer

   C > A > B

4. RCl is treated with Li in ether to form RLi. RLi reacts with water to form isopentane. In the Corey–House method, RCl is coupled to form 2,7 dimethyl octane. What is the structure of RCl?

Answer

The coupling product must be a symmetrical molecule whose carbon-to-carbon bond was formed between C(4) and C(5) of 2,7-dimethyl octane. The only RCl which will give this product is isopentyl chloride.

$$[ClCH_2CH_2CH(CH_3)CH_3]$$

5. Compound A reacts with $(CH_3)_2CuLi$ in diethyl ether to give methyl cyclohexane (B) and CH_3Cu and LiI. How many monochlorinated products are possible for B and give the structure of A

Answer

A is cyclohexyl iodide. B has 7 monochlorinated products

6. Compound A is an alkyl iodide which undergoes nucleophilic substitution but does not undergo elimination reaction. A with Li gives B. B with CuI gives C. C with D gives E. E on oxidation with hot alkaline $KMnO_4$ gives $C_6H_5CH_2COOH$, CO_2 and H_2O. Find A to E.

Answer

$C_6H_5CH_2I$ does not have beta carbon and so does not undergo elimination.

A is $C_6H_5CH_2I$.

B is $C_6H_5CH_2Li$

C is $(C_6H_5CH_2)_2CuLi$

D is $CH_2{=}CHI$ (Vinyl iodide)

E is $C_6H_5CH_2CH{=}CH_2$

7. Kolbe electrolysis of compound A gives n-hexane as the main product. How will you prepare n-hexane by Corey–House synthesis. Find structure of A.

Answer

A is CH_3CH_2COOK n-Hexane is prepared from $(CH_3CH_2CH_2CH_2)_2CuLi$ and C_2H_5I or $(CH_3CH_2CH_2CH_2CH_2)_2CuLi$ and CH_3I.

8. The % of monochlorinated products for A corresponding to primary hydrogen : secondary hydrogens : tertiary hydrogens is 42 : 35 : 23. The molecular weight of hydrocarbon is 72. How is the compound prepared by Corey–House synthesis?

Answer

Since distribution for primary : secondary : tertiary is 1 : 3.8 : 5, we have 9 primary hydrogens, 2 secondary hydrogens and 1 tertiary hydrogen. The structure of hydrocarbon is

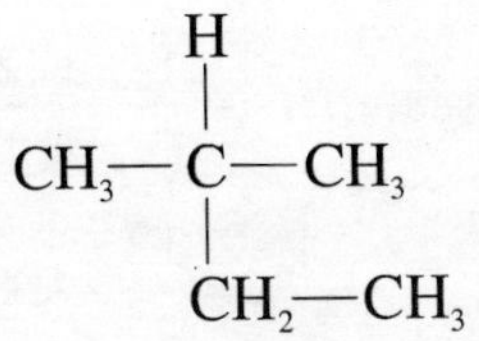

$$\begin{array}{c} H \\ | \\ CH_3-C-CH_3 \\ | \\ CH_2-CH_3 \end{array}$$

It is prepared from isopropyl iodide and $(C_2H_5)_2CuLi$ or C_2H_5I and $[CH_3CHCH_3]_2CuLi$. The second method is preferable since alkyl halide used is primary.

9. 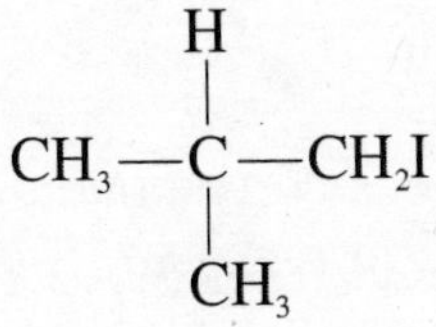 $\begin{array}{c} H \\ | \\ CH_3-C-CH_2I \\ | \\ CH_3 \end{array}$ is reacted with $[(CH_3)_2CH]_2$ CuLi to give A. Find the ratio of monosubstituted chlorides for primary, secondary and the tertiary hydrogen

Answer

There are 12 primary hydrogen, 2 secondary and 2 tertiary hydrogen. Hence, the product distribution for primary: secondary: tertiary is 39.7 : 34.5 : 25.8

10. A 5-membered hydrocarbon produces the 1° halide, 2° halide and 3° halide in the ratio 21 : 53 : 26 per cent. Identify the hydrocarbon. Find the isomer of this compound which gives only one monosubstitution product.

Answer

N-Pentane, Neopentane

11. Find the number of monochlorinated products for hydrocarbon A obtained by combining secondary butyl iodide with n-pentyl iodide in Corey–House synthesis.

Answer

The hydrocarbon is 3-methyloctane and it has the structure $CH_3CH_2CH(CH_3)CH_2CH_2CH_2CH_2CH_3$ and has 8 different hydrogens and will give 8 monosubstitution products.

12. Find the % of isomeric products of monochlorination of a) 2,3-dimethylbutane b) n-pentane

Answer

a) 2,3-dimethylbutane

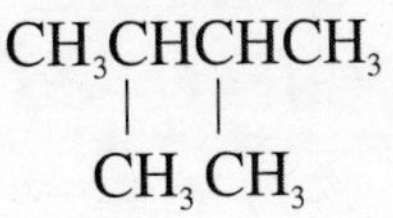

$$\begin{array}{c} CH_3CHCHCH_3 \\ |\quad | \\ CH_3\ CH_3 \end{array}$$

There are 12 primary hydrogens and 2 tertiary hydrogens.

$$\% \text{ of primary isomers} = \frac{12\times1}{(12\times1+2\times5)}\times100$$
$$= \frac{12}{22}\times100 = 55$$

$$\% \text{ of tertiary isomers} = 45$$

b) n-pentane

$$CH_3CH_2CH_2CH_2CH_3$$

There are 6 primary hydrogens There are 4 secondary hydrogens corresponding to $CH_3CHCl\ CH_2CH_2CH_3$ and two secondary hydrogens corresponding to $CH_3CH_2CHClCH_2CH_3$.

$$\frac{6\times1}{(6\times1+4\times3.6+2\times3.6)}\times100 = 21\%\ ClCH_2CH_2CH_2CH_2CH_3$$
$$\frac{4\times3.6}{27.6}\times100 = 53\% \text{ of } CH_3CHClCH_2CH_2CH_3$$
$$\frac{2\times3.6}{27.6}\times100 = 26\% \text{ of } CH_3CH_2CHClCH_2CH_3$$

13. Which process of chlorination or bromination gives better yield of halogenated product for 1-methylcyclohexane.

Answer

The hydrogen is tertiary and since bromination is selective to produce tertiary product in very high yield, bromination will give better yield.

14. An alkane with molecular weight 72 produces only one monochlorinated product. Write the Corey–House reagents to produce this alkane.

Answer

The alkane is neopentane which has all hydrogens equivalent.

```
       CH3
        |
CH3 –  C – CH3
        |
       CH3
```

This can be produced from $[(CH_3)_3C]_2CuLi$ and CH_3I.

The other possibility $(CH_3)_2CuLi$ and $(CH_3)_3Cl$ is ruled out since the halide is tertiary.

15. An alkene A on hydrogenation gives an alkane with molecular weight 100 which gives 6 monochlorinated products. Find the alkene. Can it exhibit geometric isomerism? What is the oxidation product of this alkene with alk. $KMnO_4$? What is the ozonolysis product in $(CH_3)_2S$?

Answer

The alkane with formula C_7H_{16} (100 is mol.wt.) giving 6 monosubstituted products is

$$\begin{array}{c} CH_3 \\ | \\ CH_3CCH_2CH_2CH_2CH_3 \\ | \\ H \end{array}$$

The corresponding alkene gives $(CH_3)COCH_3$ and $CH_3CH_2CH_2COOH$ on oxidation with alk. $KMnO_4$. On ozonolysis, it gives acetone and $CH_3CH_2CH_2CHO$. The alkene is $CH_3—\underset{\underset{CH_3}{|}}{C}=CHCH_2CH_2CH_3$ and cannot exhibit geometric isomerism.

II. Multiple Choice Questions (One out of four answers is correct)

1. 6 monochloro derivatives are possible for
   a) butane　　b) propane
   c) 2,3-dimethylbutane　　d) 2-methylhexane
2. The % of monochlorinated products $ClCH_2CH_2CH_3$ and $CH_3CHClCH_3$ are in the ratio
   a) 45 and 55　　b) 45 and 35　　c) 65 and 35　　d) 55 and 45
3. The compound 2-methylpentane can be produced from A and B in diethyl ether where A and B are
   a) CH_3I and $(CH_3)_2CuLi$
   b) $(CH_3CHCH_3)_2CuLi$ and $CH_3CH_2CH_2I$
   c) $(CH_3CH_2CH_2)_2CuLi$ and isopropyl iodide
   d) CH_3I and $(CH_3CH_2CH_2CH_2)_2CuLi$
4. Only one monochlorinated product is possible for
   a) cyclohexane　　b) 1-methylcyclohexane
   c) 1-ethylcyclohexane　　d) isobutene
5. An alkene A which gives an ozonolysis in reducing as well as oxidizing conditions on reduction gives B. B gives ——— monochlorinated products
   a) 3　　b) 2　　c) 4　　d) 5

6. Ethylene reacts with diazomethane to give A which gives two monochlorinated products primary halide and secondary halide in the percentage ratio

   a) 45 and 55 b) 45 and 35 c) 65 and 35 d) 55 and 45

7. Addition of HCl to cyclopropane gives

   a) $ClCH_2CH_2CH_2Cl$ b) cyclopropyl chloride
   c) 1,2 dichlorocyclopropane d) n-propyl chloride

8. The product of cyclobutane with hydrogen in NI on chlorination gives monosubstituted chlorides $CH_3CH_2CH_2CH_2Cl$ and $CH_3CHClCH_2CH_3$ in the percentage ratio

   a) 28 and 72 b) 72 and 28 c) 45 and 55 d) 55 and 45

9. On bromination the tertiary bromide $(CH_3)_2CBrCH_2CH_3$ is the product in high yield. The molecular weight of hydrocarbon is

   a) 58 b) 44 c) 84 d) 72

10. The insertion of carbene to ethylene produces a hydrocarbon which gives ——— monosubstituted halide

    a) 1 b) 2 c) 3 d) 4

11. Which one of the following alkanes by insertion of carbene gives a product which gives only one monosubstituted halide?

    a) neopentane b) propane c) butane d) pentane

12. Which one of the following alkanes reacts with chlorine to give more than one monochloride?

    a) ethane b) butane
    c) neopentane d) $(CH_3)_3CC(CH_3)_3$

13. Tertiary butyl bromide is obtained as the main product in the bromination of compound A. Find the two reactants used to produce A by Corey–House synthesis.

    a) $[(CH_3)_2CH]_2CuLi$ and CH_3I b) $(CH_3)_2CuLi$ + isopropyl iodide
    c) ethyl iodide + $(C_2H_5)_2CuLi$ d) n-propyl iodide and $(CH_3)_2CuLi$

14. Cyclobutane on hydrogenation in NI gives A. A can be prepared by Wurtz reaction of

    a) CH_3I and C_3H_7I b) CH_3I and CH_3I
    c) C_2H_5I and C_3H_7I d) C_4H_9I and C_2H_5I

15. Arrange the following cations in increasing order of stability.

    I) $(CH_3)_3C^+$ II) CH_3^+ III)$CH_3CH_2O^+$ IV) $C_6H_5CH_2^+$

    a) I$<$II$<$III$<$IV b) I$<$III$<$II$<$IV c) I$<$II$<$IV$<$III d) II$<$I$<$IV$<$III

16. The radical $(CH_3)_3C^-$ is stabilized by

    a) resonance b) tautomerism c) hyperconjugation d) hydrogen bonding

17. The hydrogenation of 2-hexene gives A. A can give ——— monosubstituted chlorides

a) 3 b) 2 c) 1 d) 4

18. $CH_3-\underset{\underset{CH_3}{|}}{C}=CHCH_3$ is reduced to hydrocarbon P.

The substrates used to prepare P by Corey–House synthesis will be

a) $[(CH_3)_2CH]_2CuLi$ and CH_3CH_2I

b) $[(CH_3)_2CH]_2CuLi$ and $CH_3CH_2CH_2I$

c) $[CH_3CH_2]_2CuLi$ and $CH_3CH_2CH_2CH_2I$

d) $[(CH_3)_2CH]_2CuLi$ and CH_3I

19. Wurtz reaction of CH_3I and C_2H_5I gives rise to one main product A along with side reactions. A is produced by

a) hydrogenation of cyclobutane b) hydrogenation of cyclopentane

c) hydrogenation of cyclopropane d) hydrogenation of ethylene oxide

20. $(C_2H_5)_2CuLi$ combines with C_4H_9I to give P. P has three monosubstituted chlorides (primary and two secondary chlorides) in the ratio

a) 42:16:42 b) 16:42:42 c) 20:30:50 d) 42:42:16

KEY

1. d	2. a	3. b	4. a	5. b	6. a	7. d
8. a	9. d	10. a	11. a	12. b	13. a	14. a
15. d	16. c	17. a	18. c	19. c	20. b	

III. Multiple Choice Questions (More than one answer is correct)

1. 6 monochloro derivatives are possible for

a) butane b) propane

c) 2-methylhexane d) 2,6-dimethylheptane

2. The % of monochlorinated products are in the ratio 45 : 55 and 36 : 64 for

a) propane b) isobutene c) pentane d) butane

3. The hydrocarbon with 5 carbons giving rise to 3 monosubstituted chlorides can be produced from A and B by Wurtz reaction where A and B are

a) CH_3I and C_4H_9I

b) C_2H_5I and $CH_3CH_2CH_2I$

c) $(CH_3CH_2CH_2)_2CuLi$ and isopropyl iodide

d) CH_3I and $(CH_3CH_2CH_2CH_2)_2CuLi$

4. Only one monochlorinated product is possible for
   a) cyclohexane b) 1-methylcyclohexane
   c) 1-ethylcyclohexane d) ethane
5. The compound which gives acetone on ozonolysis in reducing as well as oxidizing conditions when reduced gives A. Some statements about A are wrong. These are
   a) A gives 2 monochloro derivatives
   b) A gives 4 monochloro derivatives
   c) A has molecular weight of 86 grams
   d) The monochloro derivatives are in the ratio 10 : 90
6. Ethylene reacts with diazomethane to give A about which some statements given below are true
   a) A has two monochloro derivatives
   b) The two monochloro derivatives are in the ratio 45 : 55
   c) A can be obtained by hydrogenation of cyclopropane
   d) A has 3 inequivalent hydrogens
7. The addition reaction of HCl to cyclopropane yields straight-chain halide $CH_3CH_2CH_2Cl$. This type of reaction is not possible for
   a) cyclopentane b) cyclohexane
   d) cyclobutane d) ethylene oxide
8. The product of cyclobutane with hydrogen in Ni is A. A can be obtained by combining ________ and ________ in Wurtz reaction
   a) C_2H_5I and C_2H_5I b) CH_3I and $CH_3CH_2CH_2I$
   c) CH_3I and isopropyl iodide d) C_3H_7I and C_3H_7I
9. On bromination the tertiary bromide $(CH_3)_2BrCH_2CH_3$ is the product of hydrocarbon P. Some statements about P are true. They are $(CH_3)_2CBrCH_2CH_3$
   a) The molecular weight of hydrocarbon is 58 grams
   b) The hydrocarbon is isobutene
   c) The molecular weight of A is 84 grams
   d) The hydrocarbon is n-butane
10. The insertion of carbene in ethylene produces a hydrocarbon A. Some statements about A are true. These are
   a) A gives 2 monosubstituted chlorides
   b) A gives only one monosubstituted chloride
   c) A on hydrogenation on Ni gives propane
   d) With HCl, A gives 1-chloropropane

11. The cations stabilized by resonance are

a) $C_6H_5CH_2^+$ b) Allyl cation

c) *t*-butyl cation d) $CH_3CH_2O^+$

12. Some of the following alkanes react with chlorine to give more than one monochloride. They are

a) ethane b) butane

c) neopentane d) 1-methylcyclohexane

13. Tertiary butyl bromide is obtained as the main product in bromination of compound A. A can be synthesized by

a) Wurtz reaction of isopropyl iodide and CH_3I

b) Corey–House reaction of $[(CH_3)_2CH]_2CuLi$ and CH_3I

c) Corey–House reaction of $(CH_3)_2CuLi$ and $[(CH_3)_2CH]$

d) Wurtz reaction of n-propyl iodide and CH_3I

14. Cyclobutane on hydrogenation on Ni gives A. A can be prepared by Wurtz reaction of

a) CH_3I and C_3H_7I b) C_2H_5I and C_2H_5I c) C_2H_5I and C_3H_7I d) C_4H_9I and C_2H_5I

15. RMgX + Y gives alkane. Y can be

a) H_2O b) ROH c) RCOOH d) RCl

16. The radical $(CH_3)_3C\cdot$ is

a) sp^2 hybridized

b) planar

c) stabilized by hyperconjugation

d) stabilized by resonance

17. The hydrogenation of 2-hexene gives A. Some statements about A are wrong. These are

a) A can give 4 monosubstituted chlorides

b) A gives 3 monochlorides

c) A is prepared by Wurtz reaction of 2 moles of n-propyl iodide with Na

d) A is prepared by reacting C_3HI with C_2H_5I along with Na

18. $CH_3-\underset{\underset{CH_3}{|}}{C}=CHCH_3$ is reduced to hydrocarbon P. P is prepared from

a) $[(CH_3)_2CH]_2CuLi$ and CH_3CH_2I

b) $[(CH_3)_2CH]_2CuLi$ and $CH_3CH_2CH_2I$

c) $(CH_3)_2CHCH_2CH_2COONa$ with NaOH and CaO

d) $[(CH_3)_2CH]_2CuLi$ and CH_3I

19. Wurtz reaction of CH_3I and C_2H_5I gives rise to one main product A along with side reactions. A is produced by
    a) hydrogenation of cyclobutane
    b) hydrogenation of cyclopentane
    c) hydrogenation of cyclopropane
    d) Corey–House reaction of $(CH_3)_2CuLi$ with C_2H_5I in diethyl ether
20. $(C_2H_5)_2CuLi$ combines with C_4H_9I to give P. The statements true about P are
    a) P gives 3 monosubstituted chlorides
    b) P is obtained by hydrogenation of 1-hexene, 2-hexene or 3-hexene
    c) P gives 3 monosubstituted chlorides in the ratio 25:25:50
    d) P is obtained by hydrogenation of $CH_3—C(CH_3)═CHCH_3$

KEY

1. c, d	2. a, b	3. a, b	4. a,d	5. b, d	6. a, b, c	7. a, b, d
8. a, b	9. a, b	10. b, c, d	11. a, b, d	12. b, d	13. a, b	14. a, b
15. a, b, c	16. a, b, c	17. b, c	18. a, c	19. b, d	20. a, b	

IV. Assertion–Reason Questions

The questions 1–15 consist of an *assertion* in column 1 and *reason* in column 2. Use the following key to choose an appropriate answer

a) If both assertion and reason are correct and reason is the correct explanation of assertion
b) If both assertion and reason are correct but reason is not the correct explanation of assertion
c) If assertion is correct but reason is incorrect
d) If assertion is incorrect but reason is correct

No.	Assertion	Reason
1.	Addition of HCl to cyclopentane does not give $CH_3CH_2CH_2CH_2CH_2Cl$ similar to cyclopropane giving $CH_3CH_2CH_2Cl$.	The number of carbons in cyclopentane is 5.
2.	Hydrogenation of cyclopropane gives propane.	Cyclopropane has ring strain and this is relieved by formation of open chain compound.
3.	The cation $CH_3CH_2O^+$ is more stable than $(CH_3)_3C^+$	t-butyl cation is stabilized by resonance.

(*Contd.*)

No.	Assertion	Reason
4.	Wurtz reaction can be used to produce methane.	For Wurtz reaction to occur there must l minmum two carbons in the product.
5.	Corey House reaction of $(CH_3)_2CuLi$ with t-butyl iodide is a good method of synthesizing neopentane.	Alkyl iodides used should not be seconda or tertiary halides in this synthesis.
6.	Bromination of isobutane produces t-butyl iodide in very high yield.	This is due to selectivity of bromination.
7.	The boiling point of 2,2,3,3-tetramethylbutane is lower than that of nonane.	Branched alkanes are more symmetrical aı have lower boiling point.
8.	2,2,3,3-tetramethylbutane has lower heat of combustion than nonane.	It has the highest branching and is mo stable.
9.	Boiling points increase regularly by increase of mol.wts.but melting points increase irregularly.	The different shapes of molecules affect tl packing and hence m.pt increase is n regular.
10.	$C_6H_5CH_2^+$ is more stable than isopropyl cation	$C_6H_5CH_2^+$ is stabilized by resonance.
11.	2-methyl hexane and n-heptane produce the same number of monochlorides.	They contain different numbers inequivalent hydrogens.
12.	n-pentane and 2-methylbutane do not produce the same number of monochlorides.	Their molecular weights are same.
13.	Ethane, cyclobutane, cyclohexane are suitable to produce monohalogen derivatives by halogenation to the extent of 100%.	All these compounds contain hydrogeı which are all equivalent.
14.	A mixture of alkane and water separates into two layers with alkane floating on top.	Alkanes have lower density than water.
15.	Boiling point of C_2H_5OH is lower than that of propane.	Alcohols have hydrogen-bonding leading increase of boiling point.

KEY

1. (b)	2. (a)	3. (c)	4. (d)	5. (d)	6. (a)	7. (a)
8. (a)	9. (a)	10. (a)	11. (d)	12. (b)	13. (a)	14. (a)
15. (d)						

V. Comprehension Passages

Passage I

Alkanes undergo one of the important reactions, namely halogenation.

Fluorine's reactivity is very high compared to reaction by bromine or chlorine. The reaction involves free radical mechanism Fluorine can react in the dark with alkanes.

Chain Initiation

$F_2 \longrightarrow 2F\bullet$

$\Delta H = +\ 38\,\text{kcal/mole}$

$Ea = +\ 38\,\text{kcal/mole}$

Propagation

$F\bullet + CH_4 \longrightarrow HF + CH_3^{\bullet}$

$\Delta H = +32\,\text{kcal/mole}$

$Ea = +1.2\,\text{kcal/mole}$

$CH_3^{\bullet} + F_2 \longrightarrow CH_3F + F.$

$\Delta H = +70\,\text{kcal/mole}$

$Ea = \text{Small}$

The overall heat of the propagation reaction is –102 kcal/mole and the low energy of activation accounts for the high reactivity of fluorine towards methane.

Iodination is reversible.

$$CH_4 + I_2 \rightarrow CH_3I + HI$$

The hi formed must be removed by oxidation with HIO_3, HNO_3, HgO, etc. For reaction with HIO_3 we have the equation

$$5HI + HIO_3 \longrightarrow 31_2 + 3H_2O$$

The main important halogenation reactions are chlorination and bromination.

For chlorination the overall heat of propagation reaction is –24.5 kcal/mole and Ea is higher. Hence chlorine is less reactive. Due to very high energy of activation for propagation step (18.6 kcal/mole), bromine is much less reactive.

Nature of substitution products

Halogenation of alkanes depends on 1) probability factor 2) reactivity of hydrogens 3) reactivity of halogens

Probability factor This depends on the number of inequivalent hydrogens in a compound. For example n-butane contains 6 equivalent 1° hydrogens and 4 equivalent 2° hydrogens. Hence the chance of removing 1°H relative to 2°H is 3 : 2.

Reactivity of hydrogens The order of reactivity of hydrogen is $3^0H > 2^0H > 1^0H$

This order arises due to the relative stabilities of the radicals which are obtained by hemolytic fission and we discuss hyperconjugation below.

Tertiary radical > secondary radical > primary radical. Compared to chlorination bromination is selective and gives very high yield corresponding to tertiary hydrogen. For synthetic purposes the compound should have all hydrogens equivalent to produce only ne monohalogen derivative

Answer the following questions

1. Arrange in decreasing order the number of monochlorinated products for

   I. ethane II. 2,3-dimethylbutane III. n-hexane IV. 2-methylheptane

   a) IV>II>III>I b) IV>III>II>I c) I>II>III>IV d) II>I>III>IV

2. Cyclobutane on hydrogenation in NI gives A. The ratio of primary chloride to secondary chloride for A on halogenation is

   a) 28 : 72 b) 72 : 28 c) 45 : 55 d) 55 : 45

3. One of the following chlorides cannot be prepared in the laboratory in maximum yield by chlorination of hydrocarbon.

   a) n-butyl chloride b) neopentyl chloride
   c) cyclopropyl chloride d) ethyl chloride

KEY

1. b **2.** a **3.** a

Passage 2

There are various methods of preparation of alkanes. Carbide hydrolysis, Wurtz reaction, Kolbe electrolysis, Corey–House synthesis, carbene insertion and so on.

Be_2C on hydrolysis gives methane. In Wurtz reaction of RX and R′X are treated with Na. Dibromides of type $Br(CH_2)_n$—Br with Na give cycloalkanes

Answer the following questions

1. $BrCH_2CH_2CH_2Br$ with Na gives A. A on reduction with hydrogen gives

   a) cyclobutane b) propane c) butane d) methane

2. Carbene insertion to butane gives A. A gives——— monochloro derivatives
   a) 1 b) 2 c) 3 d) 4
3. 2-methylhexane is prepared by reacting——— with——— in Corey–House synthesis
   a) $[(CH_3)_2CH]CuLi$ with $CH_3CH_2CH_2CH_2I$
   b) $[(CH_3)_2CH]CuLi$ with CH_3CH_2I
   c) $[(CH_3)_2CH]CuLi$ with $CH_3CH_2CH_2CH_2I$
   d) $[(CH_3)_2CH]I$ with $(CH_3CH_2CH_2CH_2)CuLi$

KEY

1. b 2. c 3. a

VI. Matrix Matching

Question 1

Column 1	Column 2
a) cyclopropane	p) $BrCH_2CH_2CH_2Br + 2Na$
b) Methane	q) Carbene insertion to ethylene
c) C_2H_5MgX +acid	r) Al_4C_3 hydrolysis
d) Cyclobutane $+H_2$	s) Be_2C hydrolysis
	t) C_2H_6
	u) butane

Answer

$[a \rightarrow p,q], [b \rightarrow r,s], [c \rightarrow t], [d \rightarrow u]$

Question 2

Column 1	Column 2
a) 2-monosubstitition products on halogenation	p) ethane
b) only one monosubstitution product	q) propane
c) Wurtz reaction of $2CH_3I + 2Na$	r) butane
d) Ratio of monochlorinated products for isobutane	s) cyclopentane
	t) 64:36

Answer

$[a \rightarrow p,r], [b \rightarrow p,s], [c \rightarrow p], [d \rightarrow t]$

Question3

Column 1	Column 2
a) t-butyl radical	p) cyclopropane +HCl
b) n-propyl chloride	q) propane on chlorination gives 45% of this product
c) Cyclopentane	r) 9 hyperconjugated structures
d) $C_6H_5CH_2^+$	s) $BrCH_2CH_2CH_2CH_2CH_2Br$
	t) stable due to resonance

Answer

$[a \rightarrow r][b \rightarrow p,q][c \rightarrow s][d \rightarrow t]$

VII. Structure-Based and Numerical Problems

1. Compound A is obtained as the product of Wurtz reaction between 2 moles of B with Na. B is the only monosubstitution product of ethane. B isomerizes to C by treatment with $AlCl_3$ with trace of water at 573 K. C undergoes oxidation with $KMnO_4$ to give D. D easily undergoes dehydration in sulphuric acid to give E. Ozonolysis of E in reducing conditions gives acetone and formaldehyde. Identify A to E.

Answer

1) $2C_2H_5Cl(B) + 2Na \longrightarrow CH_3CH_2CH_2CH_3(A)$

$$C_2H_6 + Cl_2 \longrightarrow C_2H_5Cl$$

$$\underset{(B)}{CH_3CH_2CH_2CH_3} \xrightarrow[573\ K]{AlCl_3} \underset{(C)}{CH_3-\overset{\overset{H}{|}}{\underset{\underset{CH_3}{|}}{CH}}-CH_3}$$

$$\text{Isobutene} \xrightarrow{KMnO_4} (CH_3)_3C-OH\ \ (D)$$

$$\text{t-butyl alcohol} \longrightarrow (-H_2O) \longrightarrow (CH_3)_2C{=}CH_2(E)$$

$$E + O_3(\text{red. Conditions}) \longrightarrow HCHO + \text{acetone}$$

2. Compound A is produced in haloform reaction of methyl ketones as yellow precipitate. A reacts with sodium arsenite in NaOH to give B, C and D. B reacts with Zn diethyl ether in the presence of Cu to give CC. CC reacts with DD to give E. Find the possible

monosubstitution products of E. DD on ozonolysis in reducing conditions produces HCHO and ethyl methyl ketone. DD on reduction with Ni catalyst gives G. G gives 4 monochlorinated products. G gives a selectively one monobrominated product H. Identify A to H.

Answer

In haloform reaction CHI_3 is the yellow precipitate.

$$CHI_3\,(A) + Na_3AsO_3 + NaOH \rightarrow CH_2I_2(B) + NaI(C) + Na_3AsO_4(D)$$
$$ICH_2I + Zn(\text{diethyl ether, Cu}) \rightarrow ICH_2ZnI(CC)$$

$$(B)\ ICH_2ZnI + CH_2{=}\underset{\displaystyle CH_3}{\overset{\displaystyle C_2H_5}{\overset{|}{\underset{|}{C}}}}\quad (DD) \rightarrow \text{1-ethyl-methylcyclopropane (E)}$$

E gives 5 monosubstitution products on chlorination.

$$(DD) + O_3 \rightarrow HCHO + \text{ethyl methyl ketone}$$

$$D + H_2 \longrightarrow CH_3{-}\underset{\displaystyle H}{\overset{\displaystyle C_2H_5}{\overset{|}{\underset{|}{C}}}}{-}CH_3(G)$$

G on selective bromination gives

$$CH_3{-}\underset{\displaystyle Br}{\overset{\displaystyle C_2H_5}{\overset{|}{\underset{|}{C}}}}{-}CH_3\ (H)$$

3. Compound A with IUPAC name 1-methylbicyclo (5,1,0)octane is formed by reaction of B with C in ether. BB is produced in haloform reaction of methyl ketones as yellow precipitate. BB reacts with sodium arsenite in NaOH to give B, CC and DD. C on ozonolysis in reducing conditions gives $CH_3CO(CH_2)_5CHO$. C on reduction gives E. E gives 4 monochlorinated products. Find A, B, C, D, E, BB, CC and DD.

Answer

A is 1-methylbicyclo (5, 1, 0) octane

B is ICH_2I

C is 1-methylcycloheptene

BB is CHI_3

CC is Na_3AsO_4

DD is NaI

E is 1-methylcycloheptane

4. Compound A exhibits geometric isomerism and on reduction gives B. B is obtained by reduction of C with H_2 in nickel. C is prepared by action of Na on a dibromide D. B with chlorine in sunlight gives two monosubstituted chlorides in the percentage ratio 28:72. Find A to D. What are the isomers of A and B. Find the number of monosubstituted chlorides for isomer of B and find the ratio of the monosubstituted chlorides. B can be prepared from $(C_2H_5)_2CuLi$ and C_2H_5I.

Answer

A is 2-butene

B is butane

C is cyclobutane

D is $BrCH_2CH_2CH_2CH_2Br$.

1-butene is isomer of A and isobutane is isomer of B. This gives 2 monochlorides in the % ratio 36:64

5. The dibromide A with molecular weight 230 on treatment with Zn gives B. It contains two more carbons and 4 hydrogens more than that of C. C is obtained by carbene insertion to ethylene. Find the structural isomers of B including alkenes. Does it have diastereoisomers? Which alkene isomers exhibit geometric isomerism. Which structural isomer exhibits optical activity? Why does not *cis*-1,2-dimethylcyclopropane show optical activity?

Answer

The dibromide A is

$$BrCH_2-\underset{\underset{CH_3}{|}}{\overset{\overset{CH_3}{|}}{C}}-CH_2Br$$

The cycloalkane B is 1,1-dimethylcyclopropane. Its isomers are *cis* and *trans*-1,2-dimethylcyclopropane. The alkene isomers are 1-pentene and

2-pentene. 2-pentene exhibits geometric isomerism. *trans*-1,2-dimethylcyclopropane is optically active due to dissymmetry. C is cyclopropane. *cis–trans* isomers are also called diastereoisomers. *cis*-1,2-dimethylcyclopropane does not show optical activity due to plane of symmetry.

6. Compound A is obtained from CH_3I and C_2H_5I on treatment with Na along with other side products. B on reduction with H_2 gives A. B decolorizes $KMnO_4$. B with CH_2I_2 and Zn/Hg

gives C. D contains one carbon less than B and reacts with 1,3 butadiene on heating to give E. E on reduction with H_2/Ni gives F. Find the number of monosubstituted chlorides for F.

Answer

A is $CH_3I + C_2H_5I + 2Na \rightarrow C_3H_8 + 2NaI$

B is $CH_3CH{=}CH_2 + H_2 \rightarrow$ Propane

C is $CH_3CH{=}CH_2 + CH_2I_2$ (Zn/Hg) $\rightarrow$ methylcyclopropane

D is ethylene

1, 3 butadiene + ethylene $\xrightarrow{\Delta}$ cyclohexene (Diels–Alder reaction) (E)

Cyclohexene (E) + H_2 $\rightarrow$ cyclohexane (F). It gives only one monosubstituted chloride.

7. Write all the isomers of compounds I and II with formulae C_3H_6 and C_4H_8. Name them as A, B, C, D, E, F, G. The compounds B, E, F and G decolorize $KMnO_4$. How is A prepared? What is the reaction of A with HCl? How does B react with NBS? How are C and D prepared? What is the reaction of $ClCH_2CH_2CH_2Cl$ with Mg? What is the reaction of $ClCH_2CH_2CH_2CH_2Cl$ with Mg?

Answer

A is cyclopropane

B is propylene

C is cyclobutane

D is methylcyclopropane

E is 1-butene

F is 2-butene (*cis*)

G is 2-butene (*trans*)

A is prepared from $BrCH_2CH_2CH_2Br$ with Zn or Mg or Na

A with HCl gives n-propyl chloride.

Cyclobutane is prepared from $Br(CH_2)_4Br$ with Zn. If Mg is added to the dichloride we get $ClMgCH_2CH_2CH_2CH_2MgCl$

8. A and B are two isomers with formula C_6H_{14}. A undergoes selective bromination to give C in very high yield. B gives three monochlorinated products. A and B are obtained by hydrogenation of D and E. Write other isomers of D and E. Which of these isomers exhibit geometric isomerism. Find the ratio of monosubstituted chlorides of A and B in %.

Answer

$$CH_3{-}\overset{\displaystyle CH_3}{\underset{\displaystyle H}{\overset{|}{\underset{|}{C}}}}{-}CH_2CH_2CH_3 \quad (A) \qquad CH_3CH_2CH_2CH_2CH_2CH_3 \quad (B)$$

$$CH_3-\overset{\displaystyle CH_3}{\underset{\displaystyle Br}{C}}-CH_2CH_2CH_3 \quad (C) \qquad (CH_3)_2C{=}CHCH_2CH_3 \quad (D)$$

$CH_3CH{=}CHCH_2CH_2CH_3$ (E)

Other isomers of (D) and (E) are

$CH_3CH{=}C(CH_3)CH_2CH_3$

$CH_3CH_2CH{=}CHCH_2CH_3$

$CH_2{=}CHCH_2CH_2CH_2CH_3$

The isomers $CH_3CH{=}C(CH_3)CH_2CH_3$ and $CH_3CH_2CH{=}CHCH_2CH_3$ exhibit geometric isomerism.

For A, the monosubstituted chlorides (primary) are F

$$CH_3-\overset{\displaystyle CH_2Cl}{\underset{\displaystyle H}{C}}-CH_2CH_2CH_3 \quad (F) \qquad CH_3-\overset{\displaystyle CH_3}{\underset{\displaystyle H}{C}}-CH_2CH_2CH_2Cl \quad (G)$$

Secondary chlorides are

$$CH_3-\overset{\displaystyle CH_3}{\underset{\displaystyle H}{C}}-CHClCH_2CH_3 \quad (H) \qquad CH_3-\overset{\displaystyle CH_3}{\underset{\displaystyle H}{C}}-CH_2CHClCH_3 \quad (I)$$

And tertiary chloride is

$$CH_3-\overset{\displaystyle CH_3}{\underset{\displaystyle Cl}{C}}-CH_2CH_2CH_3 \quad (J)$$

The ratios of % of F, G, H, I and J will be 21 : 11 : 25 : 25 : 18

3

ALKENES

PHYSICAL PROPERTIES

Alkenes are insoluble in water and soluble in benzene, ether and chloroform. They are less dense than water. Boiling point increases with increasing carbons. Branching reduces the boiling point. Substituted alkenes possess dipole moments. Among *cis* and *trans* compounds, *cis* has a greater dipole moment. Symmetrical *trans* alkenes have zero dipole moment.

Boiling points The boiling points increase with increasing carbon atoms and there is a 20–30 degree rise for each added carbon. Branching reduces boiling points.

HEATS OF COMBUSTION AND STRUCTURE

More substituted alkenes are more stable and this is reflected in the numerically smaller value of heat of combustion. For example, in the different butenes, $CH_3CH_2CH{=}CH_2$, $CH_3CH{=}CHCH_3$(*trans*) and $(CH_3)_2C{=}CH_2$, the heats of combustion are –2710, –2712 and –2703 kJ/mol. This is due to the fact that two factors, Namely steric repulsion and hyperconjugation, operate together. Steric repulsion does not figure in 1-butene but among *cis* and *trans*, 2-butenes, steric repulsion is more for *cis* compound and steric repulsion destabilizes a molecule. Hyperconjugation stabilizes a molecule. Hyperconjugation is higher in 2-butene compared to 1-butene.

The final order of stability among 1- and 2-butenes is *trans*-2-butene > *cis*-2-butene >1-butene.

Heats of hydrogenation also reflects this point. If we see heats of hydrogenation for different alkenes, the more substituted ones are more stable.

$$CH_2{=}CH_2\ (\Delta H_{hyd} = -137.2\,kJ)$$

$$CH_3CH{=}CH_2\ (\Delta H_{hyd} = -125.9\,kJ)$$

$$CH_3CH{=}CH_2\ (\Delta H_{hyd} = -126.8\,kJ)$$

$$CH_3CH{=}CHCH_3\ (\Delta H_{hyd} = -119.7\,kJ)$$

$$(CH_3)_2C{=}CH_2\ (\Delta H_{hyd} = -118.8\,kJ)$$

In general, stabilities will be in the order

$R_2C{=}CR_2 > R_2C{=}CHR > R_2C{=}CH_2 > RCH{=}CHR'(trans) > RCH{=}CHR'(cis) > RCH{=}CH_2 > CH_2$

METHODS OF PREPARATION

1. Dehydrohalogenation of Alkyl Halides

Heating alkyl halides with alcoholic KOH produces alkenes

$$RCH_2CH_2X + \text{ethanolic KOH} \longrightarrow RCH{=}CH_2 + KX + H_2O$$

The dehydrohalogenation can take place by two mechanisms, E1 and E2. E1 elimination is unimolecular. Secondary and tertiary halides undergo E1 mechanism and the reaction condition is to use a very weak base.

Rate = k [alkylhalide]

The steps of the E1 reaction are

$$-\underset{H}{\overset{X}{C}}-\underset{H}{C}\longrightarrow X^- + -\overset{+}{C}-\underset{|}{\overset{|}{C}}-\ \text{(Slow)}$$

(Carbocation)

$$\text{Carbocation} + :B^- \longrightarrow C{=}C + BH\ \text{(fast)}$$

E2 elimination is a single step involving two molecules, and is bimolecular. It occurs when we use a strong base, and primary halides easily undergo E2 mechanism. Secondary and tertiary halides give rise to steric crowding. The compound must have a β hydrogen. CH_3Cl, benzyl bromide and neopentyl bromide are compounds that lack β hydrogen and hence cannot undergo dehydrohalogenation by E2 mechanism.

Solved Example 1

Which of the following halides undergo elimination to give alkene?

a) $C_6H_5CH_2Cl$ b) $C_6H_5CH_2CH_2Cl$ c) CH_3Br d) $(CH_3)_3CCH_2Cl$

Answer

b and d

Solved Example 2

Which of the following can never undergo elimination by E1 or E2 ?

a) $C_6H_5CH_2Cl$ b) $(CH_3)_3CCH_2Cl$

Answer

a

b can undergo elimination by E1 in ionizing solvent since rearrangement of carbocation is possible. Under E2 conditions both these cannot undergo elimination

Reactions happening under E1 mechanism will yield rearranged products due to shift of atoms or groups leading to more stable carbocations. Neopentyl bromide in a solvent ethanol or formic acid (ionizing solvent) undergoes E1 reaction producing

$$(CH_3)_2C{=}CHCH_3$$

The mechanism is

$$-\overset{CH_3}{\underset{CH_3}{\overset{|}{\underset{|}{\;}}}}-CH_2Br \longrightarrow CH_3-\overset{CH_3}{\underset{CH_3}{\overset{|}{\underset{|}{C}}}}-CH_2^{+} \longrightarrow CH_3-\overset{CH_3}{\overset{|}{C^{+}}}-CH_2CH_3$$

Primary carbocation

Tertiary carbocation

$\downarrow -H^{+}$

$$CH_3-\underset{CH_3}{\underset{|}{C}}{=}CHCH_3$$

Wherever possible, rearrangements happen due to higher stability of secondary or tertiary carbocations. Methyl shift takes place in this example.

Methyl shift In the dehydrohalogenation of neopentyl bromide, the primary carbocation rearranges to tertiary carbocation by methyl shift.

Hydride shift Consider dehydration of alcohols. n-butyl alcohol gives the n-butyl cation and by hydride shift changes to $CH_3CH_2CH^+CH_3$ (secondary cation) and produces 2-butene as the major product by loss of proton. The scheme is

$$CH_3CH_2CH_2CH_2OH^+ \longrightarrow \underset{\text{(a)}}{CH_3CH_2CH_2\overset{+}{C}H_2}$$

(a) is rewritten as

$$CH_3CH_2-\underset{H}{\overset{H}{C}}-\overset{+}{C}H_2 \longrightarrow CH_3CH_2-\underset{+}{\overset{H}{C}}-\overset{+}{C}H_3 \longrightarrow CH_3CH{=}CHCH_3$$

Hydride shift

Zaytzeff rule In dehydrohalogenations, Zaytzeff rule is followed. This rule states that β elimination proceeds in the direction that leads to the more substituted alkene. In the previous example, 2-butene is more substituted compared to 1-butene.

Exceptions to Zaytzeff rule The exceptions to Zaytzeff rule are as follows:

Case 1 Carrying dehydrohalogenation in the presence of bulky bases like potassium *t*-butoxide

Case 2 If alkyl fluorides are used, % of lower substituted alkenes is more.

Case 3 Delocalization in the presence of conjugated double bond also leads to more % of lower substituted alkene.

Explanation for case 1 If we carry out dehydrohalogenation in potassium *tert.* butoxide, tertiary halides dehydrohalogenate to give less substituted product (Hoffmann rule).

$$(CH_3)_3C\bar{O} + CH_3CH_2C(CH_3)_2Br \xrightarrow{(75°C\ (t\text{-butyl alcohol}))} \text{2-Methyl 2-butene (27.5\%)} +$$
$$\text{2-Methyl 1-butene (72.5\%)}$$

This is due to the steric bulk of the base and to the fact that in *tert.* butyl alcohol the halide is associated with solvent molecules and becomes more bulky.

The following table shows the distribution of Zaytzeff and Hoffmann products on elimination for the halide $CH_3CH_2CH_2CH_2CHXCH_2CH_3$.

Substrate $CH_3CH_2CH_2CH_2CHXCH_2CH_3$	Base and solvent	Percent composition of alkene		
		1-hexene	2-hexene	
			cis	*trans*
X = I	MeO- and MeOH	19	18	63
X = Cl	MeO- and, MeOH	33	17	50
X = F	MeO- and MeOH	69	9	21
X = $OSO_2C_6H_5$	MeO-, and MeOH	33	23	44
X = I	*t*-BuO- and, *t*-BuOH	78	7	15
X = Cl	*t*-BuO-, and *t*-BuOH	91	4	5
X = F	*t*-BuO-, and *t*-BuOH	97	1	1
X = $OSO_2C_6H_5$	*t*-BuO- and, *t*-BuOH	83	14	4

There are two general trends, which are recognized.

1. Poorer leaving groups favour elimination by Hoffmann rule. This is indicated by the increasing % of terminal olefine as we change the halogen from iodide to fluoride.
2. Stronger bases favour formation of less substituted alkenes due to steric factor.

We will see in the chapter on Amines (chapter 11) that elimination of amine oxides always gives less substituted alkenes in more %.

***Explanation for case* 2** Let us consider $CH_3C(F)(H)—CH_2—CH_2CH_2CH_3$. On elimination it gives more 1-hexene relative to 2-hexene in CH_3OK. In the transition state, the C—H bond-breaking prefers the primary carbanion relative to secondary carbanion due to its higher stability and hence the lower substituted alkene is in more %.

The following table indicates that increased amounts of 1-hexene arise as we go from iodide to fluoride.

Halogen	% 2-hexene	% 1-hexene
$C_6H_{13}I$	81	19
$C_6H_{13}Br$	72	28
$C_6H_{11}Cl$	67	33
$C_6H_{13}F$	30	70

***Explanation for case* 3** Delocalization in the presence of conjugated double bond also leads to more % of lower substituted alkenes.

Consider the dehydrobromination of

$$CH_2{=}\underset{\displaystyle H}{\underset{|}{C}}{-}CH_2{-}\underset{\displaystyle Cl}{\underset{|}{CH}}{-}CH(CH_3)_2 \quad (A)$$

The two possibilities of loss of proton from the bromide by beta elimination give rise to the following products

$$CH_2{=}\underset{\displaystyle H}{\underset{|}{C}}{-}CH_2^{-}CH{=}C(CH_3)_2 \qquad (B)$$

Or

$$CH_2{=}\underset{\displaystyle H}{\underset{|}{C}}{-}CH{=}^{-}CHCH(CH_3)_2 \qquad (C)$$

C is less substituted but conjugated with the other double bond and is more stable. Hence it is present in more amount.

2. Dehydration of Alcohols

Alcohols dehydrate in the presence of sulphuric acid to give alkenes. The order of dehydration is *tert.* alcohol > secondary alcohol > primary alcohol.

Mechanism of dehydration

It involves reaction between alcohol and acid

$$-\underset{\displaystyle H}{\underset{|}{C}}-\overset{|}{\underset{\displaystyle OH}{\underset{|}{C}}}- +H{:}B \longleftrightarrow -\underset{\displaystyle H}{\underset{|}{C}}-\overset{|}{\underset{\displaystyle OH_2^+}{\underset{|}{C}}}- +{:}B-$$

(Protonated alcohol)

$$\text{protonated alcohol} \rightarrow \text{carbocation}$$

$$\text{carbocation} + {:}B- \rightarrow \text{alkene} + HB$$

Dehydration leading to change in ring size Cyclic compounds undergo dehydration and relative stabilities of carbocations lead to change in ring size as shown in Figure 3.1.

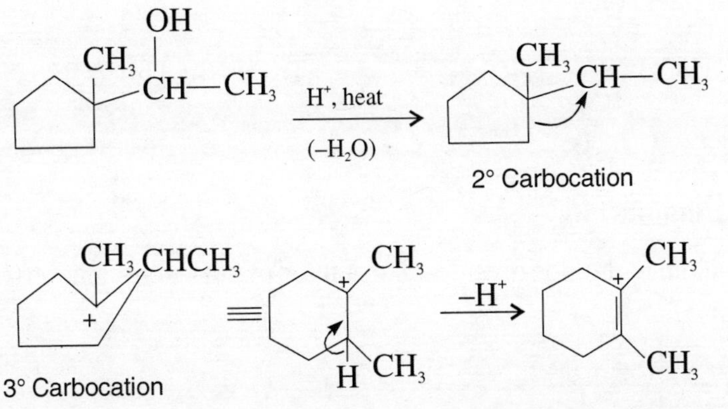

OH | CH3—C(cyclopentyl)—CH—CH3 →(H+, heat; (–H2O)) 2° Carbocation

3° Carbocation ≡ → (–H+)

Figure 3.1 Dehydration of (1-methyl)cyclopentylethanol

3. Dehalogenation of Vicinal Halides

$$RCHBrCH_2Br + Zn(alcohol)heat \longrightarrow RCH{=}CH_2 + ZnBr_2$$

4. Partial Reduction of Alkynes

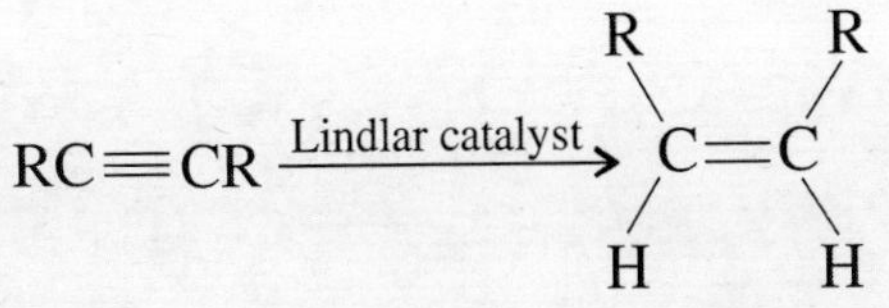

$$RC{\equiv}CR \xrightarrow{\text{Lindlar catalyst}} \text{(R)(H)C=C(R)(H)}$$

cis-alkene

$$RC{\equiv}CR \xrightarrow[\text{(ethanol)}]{Na/NH_3} \text{(R)(H)C=C(H)(R)}$$

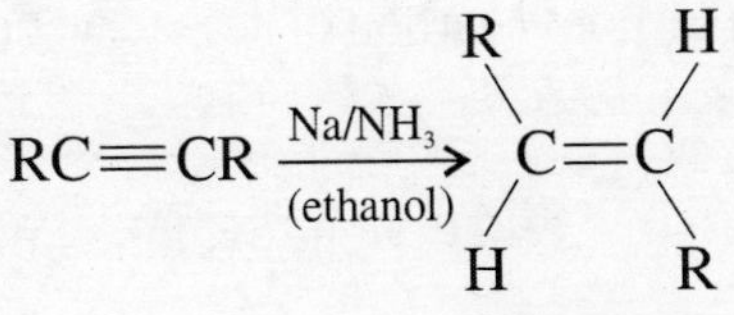

Trans-alkene

The mechanism is discussed in detail in alkynes chapter.

5. Kolbe Electrolysis of Dicarboxylic Acid Salts by Electrolysis

Succinate anion decarboxylates to give ethylene and 2 moles of CO_2.

6. Cope Elimination

A tertiary amine oxide on heating to 150°C gives alkene and a hydroxylamine.

$$RCH_2CH_2\overset{O^-}{\underset{CH_3}{N^+}}—CH_3 \xrightarrow{\text{heat}(150^\circ C)} RCH{=}CH_2 + (CH_3)_2NOH$$

7. Hoffmann Elimination

Quaternary ammonium hydroxide gives less substituted olefin by elimination (Hoffmann rule).

$$\left[CH_3CH_2\underset{^+N(CH_3)_3}{CH}—CH_3\right]OH^- \xrightarrow{\text{(heat 150°C)}} CH_3CH{=}CHCH_3(5\%) + CH_3CH_2CH{=}CH_2(95\%) + (CH_3)_3N + H_2O$$

The mechanism will be discussed in amines chapter.

8. Wittig Reaction

Ketones or aldehydes react with triphenyl phosphonium ylide in solvents like DMSO or tetrahydrofuran to give alkenes.

$$RCOR' + (C_6H_5)_3P^+—\overset{A}{\underset{B}{C^-}} \longrightarrow \overset{R}{\underset{R'}{C}}{=}\overset{A}{\underset{B}{C}} + (C_6H_5)_3PO$$

The reaction is an example of regiospecific reaction. One product is formed in 100% yield. The double bond is formed between carbonyl carbon of ketone or aldehyde and negatively charged carbon of ylide.

$$\text{Cyclohexanone} + (C_6H_5)_3P^+CH_2^- \rightarrow \text{methylene cyclohexane} + (C_6H_5)_3PO$$

9. Pyrolysis of Xanthates

$$Ph_2CHCH_2OH \xrightarrow{Na} Ph_2CHCH_2O^-Na^+ \xrightarrow{CS_2} Ph_2CHCH_2—O—\overset{S}{\overset{\|}{C}}—S^-Na^+$$

$$\downarrow CH_3I$$

$$Ph_2C{=}CH_2 + SCO + HSCH_3 \xleftarrow{\Delta(200°C)} Ph_2CHCH_2—O—CS_2CH_3$$

10. Pyrolysis of Esters

This is called Tschugaev reaction

$$CH_3CH_2OCOMe \xrightarrow{773K} R_2C{=}CH_2 + MeCOOH$$

REACTIONS OF ALKENES

Catalytic hydrogenation is stereoselective and syn addition takes place.

Addition of Hydrogen Halides

The addition follows Markovnikov rule for unsymmetrical alkenes. In such a case the hydrogen adds to carbon having greater number of substituents. Thus 1-butene gives 2-bromobutane (80%). The addition is polar and goes through carbonium ion formation. The reactivity of HX is

$$HF < HCl < HBr < HI$$

$$R_2C{=}R_2C + HX \longrightarrow R_2CH{-}CR'_2X$$

Addition of HX can be of two kinds.

1. *Carbocation reaction* If R′ is more substituted than R as given below

$$CH_3CH_2\underset{\displaystyle H}{\underset{|}{C}}{=}\underset{\displaystyle CH_3}{\underset{|}{C}}{-}CH_3$$

then the product is

$$CH_3CH_2{-}\overset{\displaystyle H}{\overset{|}{\underset{\displaystyle H}{\underset{|}{C}}}}{-}\overset{\displaystyle CH_3}{\overset{|}{\underset{\displaystyle X}{\underset{|}{C}}}}{-}CH_3$$

In some cases rearrangements of carbocations may take place and a mixture of halides may result. For example 3-methyl-1-butene with HCl gives, 2-chloro-3-methylbutane and 2-chloro-2-methylbutane

2. *Peroxide effect in HBr addition* In the presence of peroxides, the addition of HBr proceeds by free radical mechanism and the product is anti to Markovnikov addition. 1-butene + HBr in peroxide gives 1-bromobutene as the only product. This is called peroxide effect. The alkene first forms the radical.

$$HBr \longrightarrow H. + Br$$

$$CH_3CH_2CH{=}CH_2 + Br \longrightarrow CH_3CH_2\dot{C}H{-}CH_2Br \text{ and } CH_3CH_2\underset{\displaystyle Br}{\underset{|}{C}}H{-}\dot{C}H_2$$

(A) (B)

(A) is secondary and more stable

$$CH_3CH_2{-}\dot{C}H{-}CH_2Br + H \longrightarrow CH_3CH_2{-}CH_2CH_2Br$$

Since $CH_3CH_2\dot{C}H-CH_2-Br$ is more stable, it is preferentially formed and the addition of one more mole of HBr gives the product 1-bromobutane.

Stereochemistry of addition of halogens Halogenation of alkenes proceeds by formation of cyclic halonium ion. Evidence for this can be inferred from the following reactions of *cis*-2-butene and *trans*-2-butene to give stereochemically different products. *cis*-2-butene gives a mixture of two enantiomers (racemic) but *trans*-2-butene gives the meso compound. This indicates that the same carbocation $CH_3CHBrCHCH_3$ cannot be formed for the two isomers.

The Br_2 splits into Br^+ and Br^- and the nucleophile attacks the side opposite the bridging group cyclic halonium ion to give anti-addition product. When *cis*-2-butene reacts with Br_2 it gives rise to a racemic product. If *trans*-2-butene reacts with Br_2 it gives rise to meso product.

Reaction 1

H_3C, H / C=C / H, CH_3 —(Br_2, CCl_4)→ Br, CH_3, H / C–C / Br, H, CH_3

trans-2-Butene — (2R,3S)-2,3-Dibromobutane (a meso compound)

Reaction 2

H_3C, H / C=C / H_3C, H —(Br_2, CCl_4)→ Br, CH_3, H / C–C / H, CH_3, Br + Br, CH_3, H / C–C / H, CH_3, Br

cis-2-Butene — (2R,3R) — (2S,3S)

Figure 3.2 Stereochemistry of addition of Br_2 to the butenes

It is well known that a mixture of two enantiomers gives rise to a racemic product. This is due to cyclic bromonium ion formation by addition of Br^+ and the second Br^- adds in the opposite direction.

The mechanisms of addition of Br_2 to *cis*- and *trans*-2-butenes are given in Figure 3.2.

Reaction of Alkenes with SeO_2 or Dilute Alkaline $KMnO_4$

Unlike the bromine addition which is anti, the OH groups attach in syn fashion to give dihydroxy compound with dil. $KMnO_4$ or SeO_2.

cis-2-butene gives the meso compound and *trans*-2-butene gives a mixture of two enantiomers (racemic).

Hydration of Alkenes

Concentrated sulphuric acid adds on to alkenes to give alkyl hydrogen sulphates. This alkyl hydrogen sulphate can hydrolyse to give alcohol and sulphuric acid. Hydration of alkenes is restricted to monosubstituted and disubstituted alkenes of type $RCH{=}CHR$.

The hydration of alkenes to give alcohol proceeds by carbocation mechanism. Let us consider the addition of hydronium ion to 2-methylpropene.

$$(CH_3)_2C{=}CH_2 + H_3\overset{+}{O} \longleftrightarrow \text{(slow) Tert. butyl cation + Water}$$

$$\text{Tert. butyl cation + HOH} \longrightarrow \text{Tert. butyloxonium ion}$$

$$\text{t-butyloxonium ion — + Water} \longrightarrow \text{Tert. butyl alcohol + Hydronium ion}$$

Hydroboration Oxidation of Alkenes

Hydroboration is a reaction in which a boron hydride R_2BH adds on to $C{=}C$ bond. A new carbon–hydrogen bond and a carbon–boron bond result as given below:

This procedure gives only one product according to anti-Markovnikov addition.

$$3CH_3CH{=}CH \xrightarrow{THF:BH_3} (CH_3CH_2CH_2)_3B \xrightarrow{H_2O_2,\ OH^-} 3CH_3CH_2CH_2OH + Na_3BO_3$$

Alkenes Vicinal Halohydrins

Aqueous solutions of bromine, chlorine and iodine react with alkenes to form halohydrins. A halonium intermediate is formed.

$$X_2 + H_2O \longrightarrow XOH$$

$$CH_2CH{=}CH_2 + OH^- X^+ \longrightarrow CH_3\underset{\displaystyle OH}{\underset{|}{C}H}{-}CH_2X$$

Markovnikov rules is followed, and off is the negative parted X is the positive part.

Epoxidation of Alkenes

Alkenes react with a peroxy acid to give epoxides.

$$\text{1-dodecene} + \text{peracetic acid} \longrightarrow \text{1,2-epoxydodecane}$$

The addition is syn.

Addition to Dihalocarbenes

Alkenes react with dihalocarbenes generated from CHX_3 in the presence of $(CH_3)_3COK$ in *t*-butanol by α elimination to give dibromoalkenes by syn addition.

$$Br_3C - H + {}^-OC(CH_3)_3 \longrightarrow Br_3C^- + HOC(CH_3)_3$$

$$(Br)(Br)(Br)C^- \longrightarrow (-Br^-) \longrightarrow :CBr_2 + Br^-$$

Dibromo-carbene

This is α elimination since Br^- leaves from the same carbon. In dehydrohalogenation of alkyl halides, CH_3CH_2X, the β hydrogen of CH_3 leaves and is known as β elimination.

$$(CH_3)(H)C{=}C(CH_3)(H) + :CBr_2 \longrightarrow (CH_3)(H)C{-}C(CH_3)(H) \text{ bridged by } C(Br)(Br)$$

Ozonolysis of Alkenes

It is very important for structural study of compounds containing pi bonds, alkenes, alkynes, and conjugated, and cumulated and nonconjugated dienes. Compounds having exocyclic double bonds, cycloalkenes, cyclic dienes and aromatic compounds all undergo ozonolysis.

With ozone, Zn and water, the alkene undergoes oxidation to give carbonyl compounds.

$$\underset{\text{(Propylene)}}{CH_3CH{=}CH_2} + O_3 \xrightarrow[\text{or dimethyl sulphide}]{(Zn,\ H_2O)} CH_3CHO + CH_2O$$

Thus use of Zn and H_2O and dimethyl sulphide corresponds to reducing condition product.

In the presence of H_2O_2 (oxidizing conditions), propylene gives CH_3COOH and HCOOH.

For alkenes ozonolysis is conducted by adding ozone, Zn and water. Zn prevents the oxidation of aldehydes by reacting with oxidants (excess ozone and H_2O_2). Alternatively a modern method of using methanol and reduction by dimethylsulphide is used. This is referred to as workup under reducing conditions. Workup under oxidizing conditions means the H_2O_2 present oxidizes the aldehyde to the corresponding acid.

$$CH_3CH{=}CH_2 + O_3\ \text{(Zn, acetic acid)} \longrightarrow CH_3CHO + HCHO$$

$$(CH_3)_2C{=}CH_2 + O_3\text{(Zn, acetic acid)} \longrightarrow \text{Acetone} + HCHO$$

(workup under reducing conditions)

$$(CH_3)_2C{=}CH_2 + O_3\text{(water)} \longrightarrow \text{Acetone} + HCOOH$$

(workup under oxidising conditions in presence of H_2O_2)

Ketones do not undergo oxidation to acids.

Hence alkenes of the type $R_2C{=}CR_2$ give same products on ozonolysis in reducing and oxidizing conditions.

Alkenes with O_3 and $NaBH_4$ give alcohols

$$R{-}CH{=}CHR' + O_3 + NaBH_4 \longrightarrow RCH_2OH + R'CH_2OH$$

Cycloalkenes on ozonolysis give rise to open-chain dicarbonyl compounds. Cyclopentene gives $CHOCH_2CH_2CH_2CHO$ in reducing conditions and $HOOCCH_2CH_2CH_2COOH$ in oxidizing conditions.

Products of ozonolysis in case of dienes depends on the type of diene.

There are three types of dienes—nonconjugated, conjugated and cumulated.

$CH_2{=}C{=}CH_2$ is cumulated and gives 2HCHO + CO_2 in reducing conditions. Conjugated diene like 1,3-butadiene gives 2HCHO and $CHOCH_2CHO$ in reducing conditions. Nonconjugated diene like $CH_2{=}CHCH_2CH_2CH{=}CH_2$ gives 2HCHO and $CHOCH_2CH_2CHO$ in reducing conditions.

Hydrogenation

Catalytic hydrogenation using Pt, Pd, Ni or Rh gives the alkane. The reaction is exothermic. Different isomers may yield the same alkene as given below:

2-methyl-1-butene, 2-methyl-2-butene and 3-methyl-1-butene, all yield n-butane.

Catalytic hydrogenation is stereoselective and syn addition takes place.

PROBLEMS

I. "Explain Why" Questions and Small Structure-Based Problems

1. When $(CH_3)_3CCHOHCH_3$ undergoes dehydration, the main product is tetramethylethylene. Explain.

Answer

The secondary carbocation undergoes methyl shift to the tertiary carbocation which loses a proton to give tetramethyl ethylene. This is illustrated below.

$$CH_3-C(CH_3)(CH_3)-C^{+}H-CH_3 \longrightarrow CH_3-C^{+}(CH_3)-CH(CH_3)-CH_3 \xrightarrow{-H^+} (CH_3)_2C=C(CH_3)_2$$

Tetramethylethylene

$$CH_3-C(CH_3)(CH_3)-CH(OH)-CH_3 \xrightarrow{-OH^-} CH_3-C(CH_3)(CH_3)-C^{+}H-CH_3$$

$$\downarrow$$

$$CH_3-C(CH_3)=C(OH)-CH_3 \xleftarrow{-H^+} CH_3-C^{+}(CH_3)-CH(CH_3)-CH_3$$

2. Acid-catalysed dehydration of neopentyl alcohol gives 2-methyl-2-butene as major product. Explain.

Answer

$$CH_3-\overset{\overset{\displaystyle CH_2OH}{|}}{\underset{\underset{\displaystyle CH_3}{|}}{C}}-CH_3 \rightarrow CH_3-\overset{\overset{\displaystyle CH_2^+}{|}}{\underset{\underset{\displaystyle CH_3}{|}}{C}}-CH_3 \rightarrow CH_3-\overset{+}{\underset{\underset{\displaystyle CH_3}{|}}{C}}-CH_2CH_3(-H+) \rightarrow (CH_3)_2C=CHCH_3$$

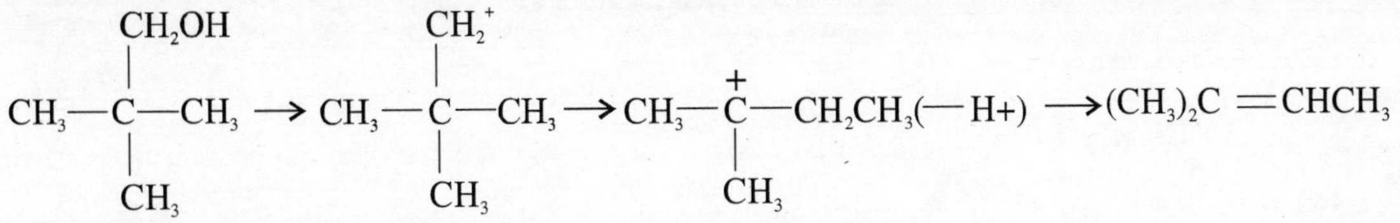

This primary carbocation rearranges to tertiary carbocation and then a proton is lost to give 2-methyl-2-butene.

3. Among the following two alkenes A and B, one undergoes hydration 7000 times faster than the other, at 298 K. Which alkene hydrates faster and why?

A is *Trans*-cyclopropyl $-CH=CHCH_3$ B is cyclopropyl$-\overset{\overset{\displaystyle CH_3}{|}}{C}=CH_2$

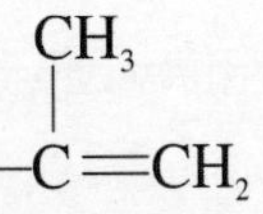

Answer

B can form tertiary carbocation. Hence it reacts much faster in the hydration reaction.

4. What is the product of hydroboration oxidation of 2-methylpropene?

Answer

2-methyl-1-propanol. The addition is of anti-Markovnikov type.

5. 2-methylpropene reacts with bromine in water. What is the product?

Answer

1-bromo-2-methylpropanol

6. Which alkene among the following on hydrogenation gives isobutane? Can this alkene exhibit geometrical isomerism?

1. 1-butene 2. butene 3. 2-methylpropene

Answer

2-methylpropene. No.

7. *trans*-2-methylcyclohexanol, by acid-catalysed dehydration, gives 1-methylcyclohexene as the major product, but *trans*-1-bromo-2-methylcyclohexane on dehydrohalogenation gives 3-methylcyclohexene as the major product. Explain.

Answer

In the first case acid catalysis takes place by E1 mechanism and the more stable tertiary carbonium ion leads to 1-methylcyclohexene. In the dehydrohalogenation, E2 mechanism operates to give 3-methylcyclohexene.

8. When treated with conc. H_2SO_4, 3,3-dimethyl-2-butanol dehydrates much slower than 2,3-dimethyl 2-butanol to give same product, 2,3-dimethyl-2-butene. Explain.

$$\begin{array}{ccccccc} & & H & & CH_3 & & \\ & & | & & | & & \\ CH_3 & - & C & - & C & - & CH_3 \qquad (A) \\ & & | & & | & & \\ & & OH & & CH_3 & & \end{array}$$

Answer

$$\begin{array}{ccccccc} & & CH_3 & & CH_3 & & \\ & & | & & | & & \\ CH_3 & - & C & - & C & - & CH_3 \qquad (B) \\ & & | & & | & & \\ & & OH & & H & & \end{array}$$

A gives a secondary carbocation which has to rearrange to tertiary cation. But B gives a tertiary cation directly and hence it reacts faster.

9. 3-bromo-2,2,4,4-tetramethylpentane does not give product by E2 elimination. Explain Also find the product if ionizing solvent is present.

$$\begin{array}{ccccccccc} & & CH_3 & & H & & CH_3 & & \\ & & | & & | & & | & & \\ CH_3 & - & C & - & C & - & C & - & CH_3 \\ & & | & & | & & | & & \\ & & CH_3 & & Br & & CH_3 & & \end{array}$$

3-bromo-2,2,4,4-tetramethylpentane

Answer

No E2 elimination is possible, since β hydrogen is absent. But in ionizing solvent E1 process is available and we get the carbocation given below.

$$\begin{array}{ccccccccc} & & CH_3 & & H & & CH_3 & & \\ & & | & & | & & | & & \\ CH_3 & - & C & - & \overset{}{\underset{+}{C}} & - & C & - & CH_3 \\ & & | & & & & | & & \\ & & CH_3 & & & & CH_3 & & \end{array}$$

This is secondary cation. Methyl shift takes place to give

$$\begin{array}{ccccccccc} & & CH_3 & & H & & CH_3 & & \\ & & | & & | & & | & & \\ CH_3 & - & \underset{+}{C} & - & C & - & C & - & CH_3 \\ & & & & | & & | & & \\ & & & & CH_3 & & CH_3 & & \end{array} \xrightarrow{-H^+} \begin{array}{ccccccccc} & & & & & & CH_3 & & \\ & & & & & & | & & \\ CH_3 & - & C & = & C & - & C & - & CH_3 \\ & & | & & | & & | & & \\ & & CH_3 & & CH_3 & & CH_3 & & \end{array}$$

10. Suggest why the conversion of $CH_2=CHCH(CH_3)_2$ into $CH_3CHOHCH(CH_3)_2$ by conc. H_2SO_4 is not a good synthetic route.

Answer

The reaction goes through carbonium ion mechanism. The carbocation formed initially $CH_3CH^+CH(CH_3)_2$, will rearrange to tertiary cation, $CH_3CH_2C^+(CH_3)_2$, which will give rise to the more stable tertiary alcohol, $CH_3CH_2C(OH)-(CH_3)_2$

II. Multiple Choice Questions (One out of four answers is correct)

1. Arrange in the order of increasing % of terminal alkene by elimination for the following bases towards 4-methyl-2-pentyl iodide.
   I. Potassium *tert.* butoxide II. potassium tricyclohexylmethoxide III. Potassium propanoxide
   a) I<II<III b) III<II<I c) II<I<III d) III<I<II
2. Among the following compounds, the strongest acid is
   a) acetylene b) ethylene c) benzene d) methanol
3. The intermediate in the addition of HCl to propene in presence of peroxide is
   a) $CH_3CH.CH_2Cl$ b) $CH_3\overset{+}{C}HCH_3$ c) $CH_3CH_2CH_2$ d) $CH_3CH_2\overset{+}{C}H_2^-$
4. The compound having smallest numerical value for heat of hydrogenation per mole is
   a) 1-butene b) trans-2-butene c) *cis*-2-butene d) 1,3-butadiene
5. Alkene obtained from $Me-C(OH)(H)-C(CH_3)_3$ as major product is
   a) $CH_2=C(Me)-C(Me)_3$ b) $Me_2C=C(Me)_2$
   c) $Me_2CH-CH=CH_2$ d) $Me_2C=CHMe$
6. One of the following alkenes is relatively more reactive with HX
   a) ethylene b) propylene c) 2-butene d) $(CH_3)_2C=CH_2$
7. When $H_2C=CHCF_3$ reacts with HCl the most stable intermediate is
   a) $CH_3\overset{+}{C}H.CF_3$ b) $\overset{+}{C}H_2.CH_2CF_3$ c) $\overset{+}{C}H_2CH_2CF_3$ d) $CH_3\overset{+}{C}H.CF_3$
8. The dehydrohalogenation of 2-bromobutane with alcoholic KOH gives
   a) 2-butene(*cis*) only b) 1-butene only
   c) 2-butene(*trans*) as the major product d) 1-butene as the major product

9. The ozonolysis of an olefin gives only $CH_3COC_2H_5$. The olefin is

a) $CH_3CH{=}CH_2$ b) tetramethylethylene

c) $CH_3C(C_2H_5){=}C(C_2H_5)CH_3$ d) $CH_3CH = CHCH_3$

10. The ozonolysis of $CH_3CH{=}CHCH_3$ in $(CH_3)_2S$ followed by addition of $NaBH_4$ gives

a) $2CH_3CHOCHCH_3$ b) $2CH_3COCH_3$ c) $2HCHO$ d) $2CH_3CH_2OH$

11. When 2-butene is treated with $CHBr_3$ in potassium *t*-butoxide followed by addition of $LiAlH_4$ in ether, the product obtained is

a) 1,2-dimethyl dibromocyclopropane b) *trans*-1,2-dimethylcyclopropane

c) cyclopropane d) *cis*-1,2-dimethylcyclopropane

12. HCHO is not the ozonolysis product of one of the following unsaturated compounds in Zn and water.

a) $CH_3CH{=}CHCH_3$ b) $CH_2{=}C(CH_3)_2$

c) $CH_2{=}CH_2$ d) $CH_2{=}CH{-}CH_2{-}CH{=}CH_2$

13. Tertiary butyl bromide heated with acetate ion followed by ozonolysis in the presence of H_2O_2 gives

a) isobutylene b) propylene c) CH_3CHO d) acetone and CO_2 and H_2O

14. Ozonolysis of $CH_2{=}CH{-}CH_2CH_2CH_2CH_2CH{=}CH_2$ in $(CH_3)_2S$ gives one of the products which is same as the ozonolysis product of

a) $CH_3CH{=}CHCH_2CH{=}CHCH_3$ b) $CH_3CH{=}C{=}CHCH_3$

c) Cyclohexa-1,2-diene d) Cyclopenta-1,2-diene

15. When styrene is treated with NBS in the presence the of aqueous DMSO in water, the product is

a) $C_6H_5CHBrCH_3$ b) $C_6H_5CH_2CH_2Br$ c) $C_6H_5CHOHCH_2Br$ d) $C_6H_5CHBrCH_2Br$

16. Cyclohexanone and acetone are the ozonolysis products in the presence of O_3, Zn and water from the following alkene

a) isopropylidenecyclohexane b) cyclohexene

c) cyclohexadiene d) no suitable answer

17. An unsaturated hydrocarbon absorbs 2 equivalents of hydrogen by catalytic hydrogenation to give alkane which gives only one monosubstituted halide by free radical chlorination. The unsaturated hydrocarbon on ozonolysis gives 2 moles of butanediol. The hydrocarbon is

a) cyclooctene b) cyclooctane c) 1,5-cyclooctadiene d) cycloheptene

18. An alkene A on ozonolysis in Zn/H_2O gives formaldehyde and $(CH_3)_2CHCHO$. By Simmons–Smith reaction, A gives a cyclic hydrocarbon B. A and B are
    a) isopropylcyclopropane, 3-methyl-1-butene
    b) 2-butene, isopropylcyclopropane
    c) 1-butene, cyclopropane
    d) 3-methyl-1-butene, isopropylcyclopropane
19. 1,1-dimethylcyclohexane results by catalytic hydrogenation of
    a) 3,3-dimethylcyclopentene   b) 4,4-dimethylcyclopentene
    c) 3,4-dimethylcyclohexene   d) 3,3-dimethylcyclohexene
20. (1-methyl)cyclopentyl chloroethane with alc.KOH gives A. A on ozonolysis in $(CH_3)_2S$ gives
    a) 1,2-dimethylcyclohexene   b) 1,2-dimethylcyclopentane
    c) 1,2-dimethylcycloheptane   d) 1,2-dimethylcyclobutane

KEY

1. d	2. d	3. b	4. d	5. b	6. d	7. d
8. c	9. c	10. d	11. d	12. a	13. d	14. c
15. c	16. a	17. c	18. d	19. d	20. a	

III. Screening Test Questions of Previous IIT-JEE Question Papers

1. The nodal plane in the σ bond of ethylene is located in
   a) the molecular plane
   b) a plane parallel to the molecular plane
   c) a plane perpendicular to the molecular plane which bisects the carbon–carbon sigma bond at right angle
   d) A plane perpendicular to the molecular plane which contains the carbon–carbon sigma bond (2002)
2. Which of the following hydrocarbons has the lowest dipole moment
   a) *cis*-2-butene   b) 2-butyne
   c) 1-butyne   d) $CH_2{=}CH{-}C{=}CH$ (2002)
3. The reaction of propene with HOCl proceeds via the addition of
   a) H+ in first step   b) Cl^+ in first step
   c) OH^- in first step   d) Cl^- and OH^- in a single step (2001)

4. Hydrogenation of $CH_3—C{\equiv}C—C(Me)—CH{=}C(Me)$ (with H shown on the C of C=C and on the CH) in the presence of poisoned palladium catalyst gives
   a) an optically active compound b) an optically inactive compound
   c) a racemic mixture d) a diastereomeric mixture (2001)
5. The number of isomers for a compound with formula $C_2BrClFI$ is
   a) 3 b) 4 c) 5 d) 6 (2001)
6. In the presence of peroxides, HCl and HI do not give anti-Markovnikov product with alkenes because
   a) Both are highly ionic.
   b) One is oxidizing and other is reducing.
   c) One of the steps is endothermic in both cases.
   d) All steps are exothermic in both cases. (2000)
7. Which of the following compounds will exhibit geometrical isomerism?
   a) 1-phenyl-2-butene b) 3-phenyl-1-butene
   c) 2-phenyl-1-butene d) 1,1-diphenyl-1-propene (2000)
8. Which one of the following alkenes will react fastest under catalytic hydrogenation conditions?
   a) *cis*-dialkyl ethylene b) $\begin{matrix} R & & H \\ | & & | \\ C & = & C \\ | & & | \\ H & & R \end{matrix}$
   c) trialkylethylene d) tetralkylethylene 2000
9. The reaction of $CH_3—CH{=}CH—C_6H_4OH$ with HBr gives
   a) $CH_3—CHBrCH_2—C_6H_4O$ b) $CH_3—CH_2CHBr—C_6H_4OH$
   c) $CH_3—CHBrCH_2—C_6H_4Br$ d) $CH_3—CH_2CHBr—C_6H_4Br$ (1998)
10. Which of the following compounds will show geometrical isomerism
    a) 2-butene b) propene
    c) 1-phenylpropene d) 2-methyl-2-butene (1998)
11. During debromination of mesodibromobutane, the major compound formed is
    a) n-butane b) 1-butene c) *cis*-2-butene d) *trans*-2-butene (May 1997)
12. The number and type of bonds between carbons in CaC_2 are
    a) one σ and one π bond b) one σ and two π bonds
    c) one σ and one and half π bonds d) one σ bond (1995)

13. The compound

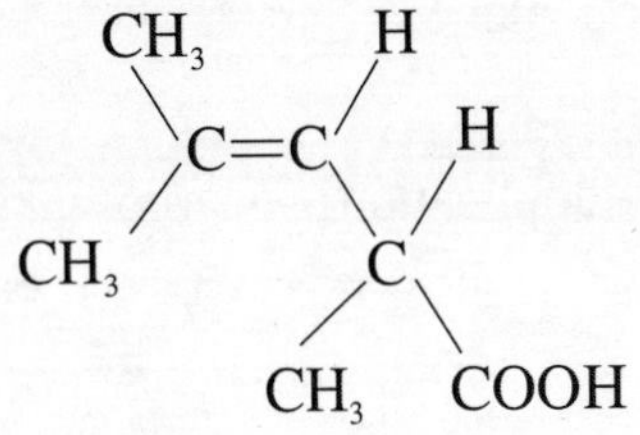

$$\begin{array}{c} CH_3 \quad\quad H \\ \diagdown C{=}C \diagup \quad H \\ \diagup \quad\quad \diagdown \diagup \\ CH_3 \quad\quad C \\ \quad\quad \diagup \diagdown \\ \quad\quad CH_3 \quad COOH \end{array}$$

Exhibits

a) geometric isomerism
b) geometric and optical isomerism
c) optical isomerism
d) tautomerism (1995)

14. The predominant product formed when 3-methyl-2-pentene reacts with HOCl is

a) $CH_3CH_2{-}\overset{Cl}{\underset{CH_3}{C}}{-}\overset{OH}{CH}{-}CH_3$
b) $(CH_3)_3C{-}C{-}CHOHCH_3$
c) $CH_3CH_2\overset{Cl}{C}{-}\overset{Cl}{CH}{-}CH_3$
d) $CH_3CH_2\overset{CH_3}{\underset{CH_3}{C}}OH\ CHClCH_3$ (1995)

15. Which one of the following has the smallest heat of hydrogenation per mole?

a) 1-butene
b) *trans*-2-butene
c) *cis*-2-butene
d) 1,3-butadiene

KEY

1. a	2. b	3. b	4. b	5. d	6. b	7. d
8. c	9. b	10. a,c	11. d	12. b	13. c	14. d
15. d						

IV. Multiple Choice Questions (More than one answer is correct)

1. 3-chloro-3-methylhexane can be prepared from some of the following alkenes on addition of HCl

a) 3-methyl-2-hexene
b) 3-methyl-3-hexene
c) 1-methyl-2-hexene
d) 2-ethyl-2-propylethylene

2. 2-pentene with HBr in ether gives
   a) 2-bromopentane
   b) 3-bromopentane
   c) 1-bromocyclobutane
   d) 1-bromopentane
3. The following products arise as major compounds due to the application of Markovnikov rule. They are
   a) allylbromide by adding NBS to propylene
   b) isopropyl chloride from propylene by adding HCl
   c) *t*-butyl chloride from 2-methylpropene and HCl
   d) hydroboration oxidation of propylene leading to n-propanol
4. The following products arise as major compounds due to operation according to anti-Markovnikov fashion. They are
   a) addition of HBr in the presence of peroxide to 1-pentene to give n-pentylbromide
   b) isopropyl chloride from propylene by adding HCl
   c) addition of HCl to 2-methylpropene leading to t-butyl chloride
   d) hydroboration oxidation of propylene to n-proponal
5. The meso product is obtained in
   a) addition of Br_2 to *trans*-2-butene
   b) addition of $KMnO_4$ to *cis*-3-hexene
   c) oxidation by SeO_2 of *cis*-2-butene
   d) addition of $KMnO_4$ to *trans*-3-hexene
6. Free radical substitution reaction takes place in the following reactions
   a) halogenation of propane
   b) bromination of propylene with NBS
   c) HBr addition in peroxide to 1-pentene
   d) HCl addition to propylene
7. Carbocation or heteroatom cation intermediate is formed in
   a) dehydrohalogenation in very strong base of n-butyl chloride
   b) dehydration of *t*-butyl alcohol
   c) addition of HCl to 2-methylpropene
   d) addition of NaOH to methoxymethyl chloride
8. syn addition takes place in case of
   a) addition of Br_2 in CCl_4 to *trans*-2-butene
   b) addition of $KMnO_4$ to *trans*-2-butene
   c) addition of SeO_2 to *trans*-3-hexene
   d) hydrogenation of 2-butene
9. Racemic mixture is formed as a product in
   a) addition of Br_2 to *cis*-2-butene
   b) addition of SeO_2 to *trans* 2-butene
   c) addition of $KMnO_4$ to *trans*-3-hexene
   d) addition of Br_2 to *trans* 2-butene

10. Ozonolysis of the following alkenes in dimethyl sulphide yields HCHO as one of the products
    a) $C_2H_5CH{=}CH_2$
    b) 2-butene
    c) methylene cyclohexane
    d) $(CH_3)CH{=}CH_2$
11. 3,3-dimethyl-2-butanol on dehydration gives
    a) 3,3-dimethyl-1-butene
    b) 2,3-dimethyl-2-butene
    c) 2-methyl-2-pentene
    d) 1-methyl-3-pentene
12. The alkene with terminal CH_2 group is not prepared in major % by which of the following reactions?
    a) Dehydrohalogenation of $CH_3CH_2CHClCH_3$ by alc.KOH
    b) Dehydrohalogenation of $CH_3CH_2CHClCH_2CH_3$ by alc.KOH
    c) $(C_2H_5)_4N + OH^-$ on heating
    d) Dehydrohalogenation of 2-fluorohexane in CH_3OK
13. Which of the following compounds cannot produce an alkene due to unavailability of beta hydrogen?
    a) benzyl bromide
    b) CH_3Br
    c) neopentyl chloride in a nonionizing solvent
    d) $C_6H_5CH_2CH_2Cl$
14. Which of the following compounds or ions can produce an alkene by treatment with alc.KOH or heating?
    a) C_2H_5Br
    b) neopentyl chloride in formic acid
    c) $(CH_3)_4N^+OH^-$
    d) $C_6H_5CH_2Br$
15. *t*-butyl alcohol can be prepared by
    a) reaction of 2-methylpropene with conc. H_2SO_4
    b) oxymercuration–demercuration of 2-methylpropene
    c) hydroboration oxidation of 2-methylpropene
    d) treatment of 2-butene with conc. H_2SO_4
16. When $(C_2H_5)_2CHCHOHCH_3$ is dehydrated,
    a) the major product is $(C_2H_5)_2C{=}CHCH_3$
    b) the major product does not exhibit geometric isomerism
    c) the major product exhibits geometric isomerism
    d) the minor product does not exhibit geometric isomerism
17. The dehydrohalogenation of $CH_3CHC_2H_5CHClCH_3$ by alc.KOH produces
    a) a terminal alkene and two more alkenes with *cis* and *trans* geometry
    b) alkenes and all of them have dipole moments
    c) alkenes and none of them has a dipole moments
    d) alkenes with same stability

18. The ozonolysis of the which of the following alkenes gives same product in oxidizing or reducing conditions
   a) tetramethylethylene b) tetraethylethylene
   c) dicyclopentylethylene d) propylene
19. The *trans* isomer of which of the following compounds shows optical activity?
   a) 1,2-dichlorocyclopropane b) 2-butene
   c) 1,2-cyclopropane dicarboxylic acid d) 1,2-dibromocyclopentane
20. CO_2 is one of the products of oxidation of
   a) $CH_3CH{=}C{=}CH_2$ in $(CH_3)_2S$ by ozonolysis
   b) propylene with alkaline $KMnO_4$
   c) 1,4-pentadiene in ozonolysis
   d) 1,3-butadiene on ozonolysis
21. HCl addition to which alkene of these gives 1-chloro-1-methylcyclohexane?
   a) 1-methylcyclohexene b) 3-methylcyclohexene
   c) methylene cyclohexane d) 1-methylcyclopentene
22. The products obtained by addition of HCl to 1-butene are
   a) $CH_3CH_2CH_2CH_2Cl$ b) $CH_3CHClCH_2CH_3$
   c) $(CH_3)_3CCl$ d) $CH_2{=}CHCH_2Cl$
23. The ozonolysis products of $CH_3CH_2CH(CH_3)CH{=}C(CH_3)CH_2CH_2CH_3$ in reducing conditions will be
   a) 2-methylbutanal b) 2-pentanone c) 2-butanone d) Propanone
24. 1,3-pentadiene on ozonolysis in reducing conditions gives
   a) HCHO b) CHOCHO c) CH_3CHO d) acetone
25. The ozonolysis products under oxidizing and reducing conditions are different for
   a) tetramethyl ethylene b) propene c) 2-butene d) 3-hexene

KEY

1. a,b,d	2. b,c	3. b,c	4. a,d	5. a,b,c	6. a,b,c	7. b,c,d
8. b,c,d	9. a,b,c	10. a,c,d	11. a,b	12. a,b	13. a,b,c	14. a,b
15. a,b	16. a,b,d	17. a,b	18. a,b,c	19. a,b,d	20. a,b	21. a,c
22. a,b	23. a,b	24. a,b,c	25. b,c,d			

V. Assertion–Reason Questions

The questions 1–15 consist of an *assertion* in column 1 and *reason* in column 2. Use the following key to choose an appropriate answer

a) If both assertion and reason are correct and reason is the correct explanation of assertion

b) If both assertion and reason are correct but reason is not the correct explanation of assertion

c) If assertion is correct but reason is incorrect

d) If assertion is incorrect but reason is correct

No.	Assertion	Reason
1.	Addition of bromine to 1-butene gives two optical isomers. (IIT 1998)	The dibromo compound $CH_2BrCHBrCH_2CH_3$ has one asymmetric centre.
2.	1-butene with HBr in peroxides gives 1-bromobutane. (IIT 2000)	It involves the formation of a primary radical.
3.	The compound $(CH_3)_3CCH_2Cl$ does not undergo elimination by E2 mechanism.	The compound is a primary compound.
4.	1-butene has one isomer but 2-butene has two isomers.	2-butene exhibits geometric isomerism.
5.	*Trans* 1,2-dibromocyclopentane obtained from 1,2-cyclopentadiene and bromine is optically active.	It has centre of inversion.
6.	1-methylcyclohexanol on dehydration gives 1-methylhexane but 1-chloro1-methyl cyclohexane on dehydrohalogenation produces mainly methylene cyclohexane.	The first compound has OH and second one has Cl as substiutent.
7.	(1-methyl)cyclopentylchloroethane on dehydrohalogenation produces mainly a cyclic compound with exocyclic double bond. Whose ozonolysis products will be HCHO and aldehyde-containing cyclopentyl group.	The carbocation from the halide rearranges to give 1,2-dimethyl cyclohexene which will give only one ozonolysis product.
8.	The ozonolysis products are different in oxidizing-reducing conditions for 2-butene but they are same in the case of tetramethylethylene.	The product formed in reducing conditions, a ketone, is resistant to further oxidation.
9.	When *trans* 2-butene reacts with $KMnO_4$ in dilute alkaline condition, the racemic diol is formed.	The addition of OH group is syn to alkene when $KMnO_4$ is used.

(*Contd.*)

No.	Assertion	Reason
10.	When *cis* 2-butene reacts with Br_2 in CCl_4, meso product is formed.	The addition of Br_2 (in terms of Br^+ and Br^-) is anti and product expected is racemic.
11.	3-methyl 1-butene with HCl gives major product as 2-chloro 3-methylbutane.	The secondary carbocation formed rearranges to tertiary cation to give finally higher yield of 2-chloro 2-methylbutane.
12.	The rate of $RCH_2CH{=}CH_2$ by HX is lower for chlorocompound relative to compound where Cl or Br.	The addition of Cl or Br produces a +ve charge on carbon next to the C of double bond which already has $\delta+$. This destabilizes, and the effect is more with more electronegative atom.
13.	2,3-dimethyl 2-butene on ozonolysis in reducing or oxidizing conditions gives acetone only.	The double bond is between carbon 2 and carbon 3.
14.	2-methylpropene with HCl gives t-butyl chloride but with HBr in peroxide it gives $(CH_3)_2CHCH_2Br$	In the presence of peroxide, anti-Markovnikov addition takes place.
15.	Heat of combustion of tetramethyl ethylene is numerically smaller than that of $CH_3CH{=}CH_2$	Substituted alkenes are more stable.

KEY

1. a	2. c	3. b	4. a	5. c	6. b	7. d
8. a	9. a	10. d	11. d	12. a	13. b	14. a
15. a						

VI. Comprehension Passages

Passage I

Alkenes are prepared by various methods. Some important ones are

i. dehydration of alcohols using H_2SO_4 or H_3PO_4
ii. dehydrohalogenation with alc. KOH or potassium *t*-butoxide.

Answer the following questions

1. 3,3-dimethyl 2-butanol on dehydration gives
   a) 2,3-dimethyl-2-butene and 2,3-dimethyl-1-butene
   b) 2-methyl-2-pentene and 1-methyl-2-pentene
   c) 3-pentene
   d) 2-pentene

2. 2-methyl-2-butene and 2-methyl-1-butene are the products of dehydration of
   a) 2-butanol b) 2-methyl-1-butanol c) 2-pentanol d) 2-hexanol
3. When *t*-pentyl alcohol is dehydrated, the major product by ozonolysis in reducing conditions gives
   a) CH_3CHO + acetone b) 2 moles of acetone
   c) CH_3CH_2CHO and HCHO d) 2 moles of HCHO

KEY

1. a 2. b 3. a

Passage 2

Alkenes undergo addition reactions with halogens X_2 and HX. They also undergo oxidation with dil. $KMnO_4$ to give diols. Stereochemistry plays an important role in these reactions

Answer the following questions

1. meso 2,3-dibromobutane is obtained by adding Br_2 to
   a) 1-butene b) 2-pentene c) *cis*-2-butene d) *trans*-2-butene
2. Racemic mixture of 2,3-dihydroxybutane is obtained when dil.$KMnO_4$ is added to
   a) *trans*-2-butene b) *cis*-2-butene c) 1-butene d) 2-pentene
3. $CH_3CH_2CH_2Br$ is obtained when HBr in peroxide is added to
   a) 2-methylpropene b) Propylene c) 2-butene d) 2-pentene

KEY

1. d 2. a 3. b

Passage 3

Ozonolysis reaction of unsaturated compounds takes place in reducing conditions (Zn/acetic acid or $(CH_3)_2S$) or in oxidizing conditions in the presence of H_2O_2 to give different products which help in structural analysis to assign double bond positions properly.

Answer the following questions

1. (1-methyl)cyclopentyl chloroethane on dehydrohalogenation gives A. A on ozonolysis gives
   a) HCHO + 2-pentanone b) CH_3CHO, 2-butanone
   c) CH_3CHO + propanone d) $OHC(CH_2)4CHO$

2. The ozonolysis products $CH_3CH_2CH(CH_3)CHO$ and 2-butanone are obtained in reducing conditions for
   a) 2,5-dimethyl-4-heptene b) 3-heptene
   c) 3-methyl-2-heptene d) 2-methyl-2-hexene
3. Compound A gives two monochlorides on halogenation in sunlight. It is obtained from B by reduction. B is prepared from $Br(CH_2)_5Br$ and Zn. C is an unsaturated compound with two hydrogens less than B. C on ozonolysis gives only one product, which is
   a) $CH_3CH_2CH_2CHO$ b) HCHO
   c) CO_2 d) $CHOCH_2CH_2CHO$

KEY

1. d 2. a 3. d

VII. Matrix Matching

Question I

Column 1	Column 2
a) $CH_2{=}C{=}CH_2 + O_3$ (reducing conditions)	p) $HCHO + CO_2$
b) $CH_3CH{=}C{=}CH_2 + O_3$ (reducing conditions)	q) CH_3CHO, CO_2, HCHO
c) *cis* 2-butene + Br_2	r) meso-2,3-dibromobutane
d) *trans* 2-butene + Br_2	s) racemic mixture of enantiomers of 2,3-dibromobutane
e) $CH_3CH{=}CH_2$ + HBr (peroxide)	t) $CH_3CH_2CH_2Br$
	u) free radical mechanism

Answer

$[a \rightarrow p]\,[b \rightarrow q]\,[c \rightarrow s]\,[d \rightarrow r]\,[e \rightarrow t, u]$

Question 2

Column 1	Column 2
a) NBS + propylene (500°C)	p) $BrCH_2$—CH=CH_2
b) propylene + hot alk.$KMNO_4$	q) free radical reaction
c) 2-chloropentane + CH_3OK	r) CH_3COOH +CO_2
d) 2-fluoropentane + CH_3OK	s) 2-pentene
	t) 1-pentene
	u) carbanion mechanism

Answer

$[a \rightarrow p, q][b \rightarrow r][c \rightarrow s][d \rightarrow t, u]$

Question 3

Column 1	Column 2
a) *cis* 3-hexene	p) dipole moment is zero
b) *trans* 3-hexene	q) reacts with SeO_2 to give mesodiol
c) $CH_3CH{=}CH_2$	r) reacts with hot alk.$KMnO_4$ to evolve CO_2 along with acetic acid as other product
d) tetraethylethylene	s) Ozonolysis in $(CH_3)_2S$ gives CH_3CHO + HCHC
	t) Ozonolysis products are same in reducing or oxidizing conditions

Answer

$[a \rightarrow q][b \rightarrow p][c \rightarrow r, s][d \rightarrow t]$

VIII. Structure-Based and Numerical Problems

1. Obtain styrene by two routes starting from triphenylphosphine and RX and aldehyde.

Answer

$(C_6H_5)_3P + CH_3Br \rightarrow S_N2$ type reaction yields $(C_6H_5)_3P^+CH_3Br$

Deprotonation of this salt by base gives ylide.

$$(C_6H_5)_3P^+CH_2^-H + Y- \rightarrow (C_6H_5)_3P^+CH_2^- + HY$$

$$\text{Benzaldehyde} + (C_6H_5)_3P^+CH_2^- \rightarrow \text{Styrene}$$

Alternatively formaldehyde can react with $(C_6H_5)_3P^+CHC_6H_5^-$ to give styrene.

2. Three isomeric alkenes A, B and C with formula C_5H_{10} are hydrogenated to yield the same saturated hydrocarbon D which gives 4 monosubstituted chlorides. A and B give the same tertiary alcohol by oxymercuration–demercuration. B and C give different primary alcohols by hydroboration oxidation. Find A, B and C. Find the ratio of monosubstituted chlorides of D. Find the ozonolysis products under reducing conditions of A, B and C.

Answer

A and B give tertiary alcohols and hence must have $=C(CH_3)_2$ grouping. B and C give primary alcohols and must have $\underset{\displaystyle CH_3}{\overset{|}{CH_2}}=\overset{|}{C}-$

Reaction of A and B is according to Markovnikov rule and B and C react by anti-Markovnikov rule. The hydrocarbon D giving 4 monochlorides is $CH_3-\underset{\underset{\displaystyle CH_3}{|}}{\overset{\overset{\displaystyle H}{|}}{C}}-CH_2CH_3$

A can be $CH_3CH=\underset{\underset{\displaystyle CH_3}{|}}{C}-CH_3$

B can be $CH_2=\underset{\underset{\displaystyle CH_3}{|}}{C}-CH_2CH_3$

C can be $CH_2=CHCH(CH_3)_2$

The 4 monochlorides are in the ratio 28.3:33.9:23.2:14.1 % corresponding to $ClCH_2CH(CH_3)CH_2CH_3$, $CH_3CH(CH_3)CHClCH_3$, $CH_3CClCH_3CH_2CH_3$ and $(CH_3)_2CHCH_2CH_2Cl$.

Ozonolysis products of A in reducing conditions are CH_3CHO and acetone. B gives ozonolysis products, namely HCHO, $CH_3COCH_2CH_3$.

C gives HCHO $+(CH_3)_2CHCHO$.

3. A compound of formula C_7H_{12} (A) is optically active and belongs to S configuration. This compound on complete reduction gives C_7H_{16} (B) which is optically active. B gives 7 mono substitution products and one selective monobrominated product (C). Partial reduction of A gives D (with Lindlar Pd). Is D optically active? A gives on ozonolysis CO_2 and $COOHCH_2CHCH_3CH_2CH_3$. E is another compound with formula C_6H_{10} (optically active) and by complete reduction gives F which produces 4 monosubstitution products on chlorination. By bromination F gives one major product which cannot undergo S_N2 reactions. Is F

optically active? The ozonolysis of E gives CHOCOOH and $COOHCHCH_3CH_2CH_3$. If E on partial reduction (Lindlar Pd) gives G will it be optically active? Identify A to G.

Answer

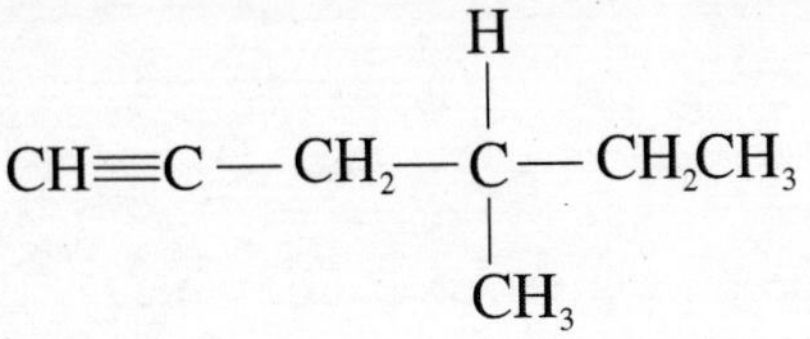

$$CH{\equiv}C-CH_2-\underset{CH_3}{\overset{H}{\overset{|}{\underset{|}{C}}}}-CH_2CH_3 \qquad (A)$$

B is

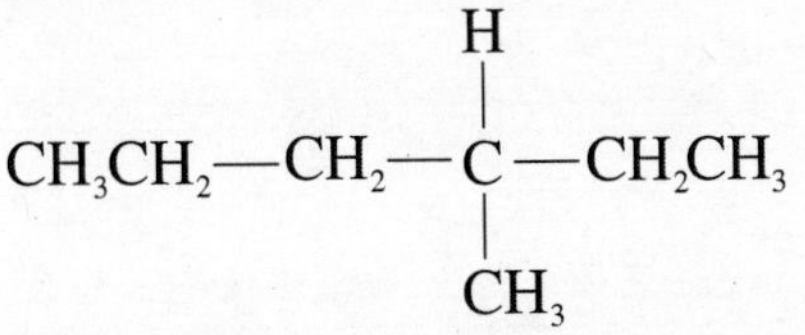

$$CH_3CH_2-CH_2-\underset{CH_3}{\overset{H}{\overset{|}{\underset{|}{C}}}}-CH_2CH_3 \qquad (B)$$

There are 7 inequivalent hydrogens

A on ozonolysis gives $CO_2 + CH_3CH_2CH(CH_3)CH_2COOH$

The monobrominated product of B is

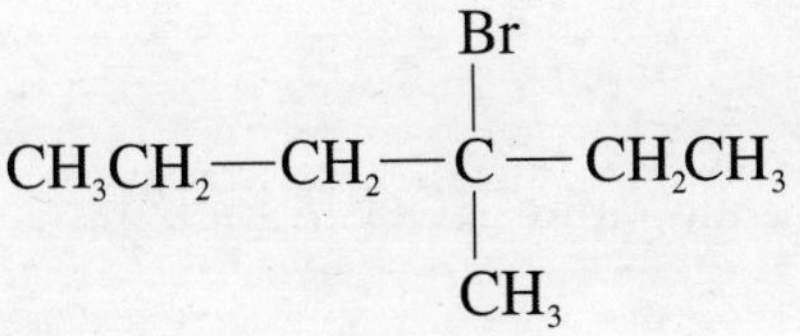

$$CH_3CH_2-CH_2-\underset{CH_3}{\overset{Br}{\overset{|}{\underset{|}{C}}}}-CH_2CH_3 \qquad (C)$$

A with Lindlar Pd gives

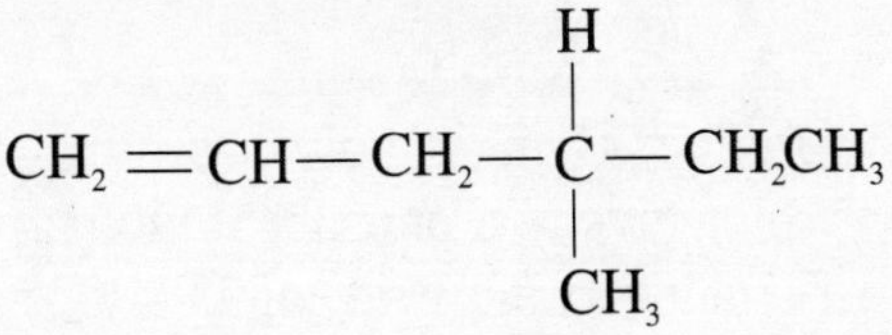

$$CH_2=CH-CH_2-\underset{CH_3}{\overset{H}{\overset{|}{\underset{|}{C}}}}-CH_2CH_3 \qquad (D)$$

E is

$$CH{\equiv}C-\underset{CH_3}{\overset{H}{\overset{|}{\underset{|}{C}}}}-CH_2CH_3$$

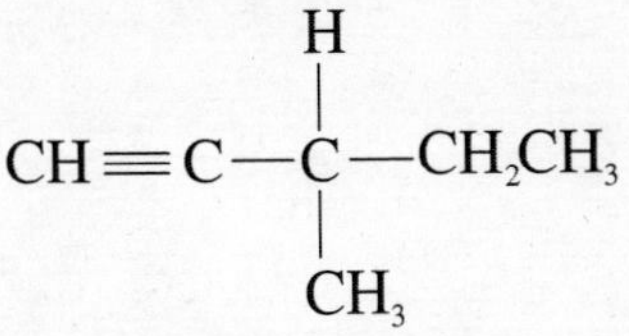

This on reduction completely gives

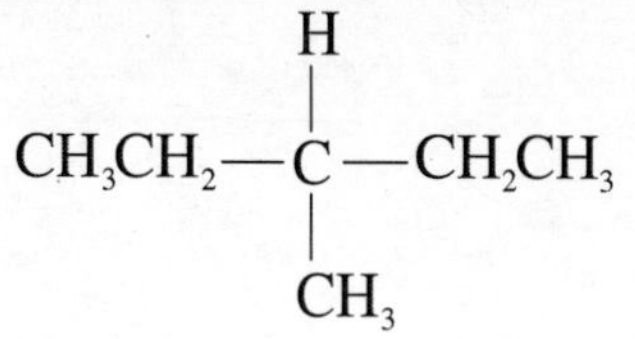

$$CH_3CH_2-\underset{\underset{CH_3}{|}}{\overset{\overset{H}{|}}{C}}-CH_2CH_3 \quad (F)$$

F is optically inactive. One monobrominated product is tertiary halide which is

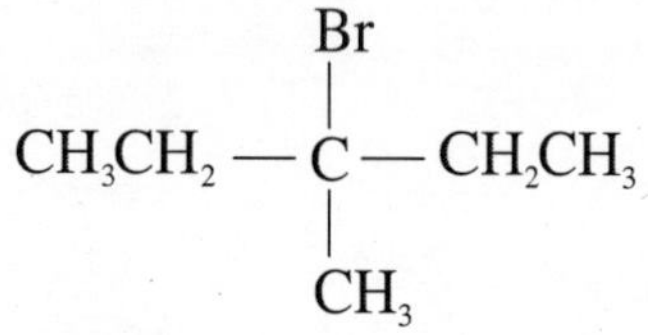

$$CH_3CH_2-\underset{\underset{CH_3}{|}}{\overset{\overset{Br}{|}}{C}}-CH_2CH_3$$

Partial reduction of E with Lindlar Pd gives G

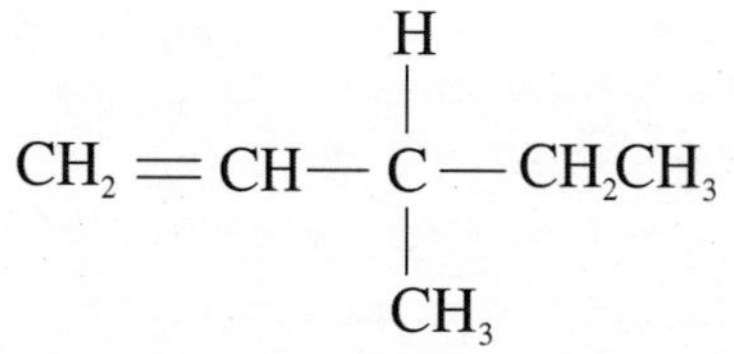

$$CH_2=CH-\underset{\underset{CH_3}{|}}{\overset{\overset{H}{|}}{C}}-CH_2CH_3 \qquad (G)$$

G is optically active.

4. (1-methyl)cyclopentyl chloroethane reacts with alcoholic KOH in ionizing solvent to give A. A on ozonolysis gives B. Find A and B.

Answer

(A) 1,2-dimethyl cyclohexene

(B) $CH_3CO(CH_2)_4COCH_3$

5. Compound A on dehydration gives B as major product. B on ozonolysis produces acetone and acetaldehyde. B on hydrogenation gives C which gives 4 monosubstitution products on chlorination. A is also obtained by solvolysis of D at a very slow rate of 0.00002 relative to ethyl halide under S_N2 conditions. D cannot undergo E2 elimination. Find A to D.

Answer

$$\text{A is } CH_3-\underset{\underset{CH_3}{|}}{\overset{\overset{CH_3}{|}}{C}}-CH_2OH$$

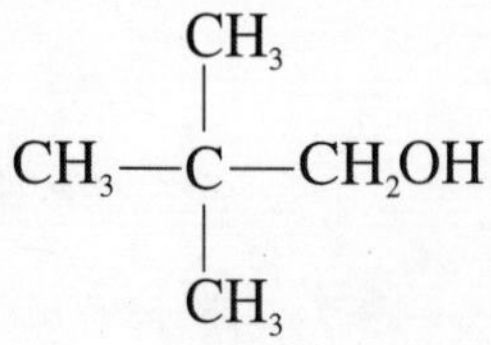

B is $CH_3—\underset{\underset{CH_3}{|}}{C}=CHCH_3$

C is $CH_3—\underset{\underset{CH_3}{|}}{\overset{\overset{H}{|}}{C}}—CH_2CH_3$

D is $CH_3—\underset{\underset{CH_3}{|}}{\overset{\overset{CH_3}{|}}{C}}—CH_2Cl$

Since there is no B hydrogen, this halide D does not undergo E2 elimination.

6. An alkene A produced formaldehyde and 4,4-dimethyl-2-pentanone. Find the alkene. If this alkene undergoes HBr addition in the absence of peroxides, what is the product? This product is obtained on monobromination from an alkane. Identify the alkane and report how many monosubstitution products are possible on chlorination for this alkane. Which chlorinated product cannot undergo dehydrohalogenation by E2 mechanism?

Answer

A is $CH_2=\overset{\overset{CH_3}{|}}{C}—CH_2—C(CH_3)_3$ that produces 4,4-dimethyl-2-pentanone.

Ozonolysis in reducing conditions of A gives HCHO and $(CH_3)_3CCH_2COCH_3$ (4,4-dimethyl-2-pentanone)

This alkene on HBr addition in the absence of peroxides follows Markovnikov rule to give

$$CH_3=\underset{\underset{Br}{|}}{\overset{\overset{CH_3}{|}}{C}}—CH_2—C(CH_3)_3$$

This halide is obtained from the alkane $CH_3—\underset{\underset{Br}{|}}{\overset{\overset{CH_3}{|}}{C}}—CH_2—C(CH_3)_3$

This alkane has 4 inequivalent hydrogens to give 4 monochlorinated products.

$$CH_3-\underset{H}{\overset{CH_3}{\underset{|}{\overset{|}{C}}}}-CH_2-\underset{CH_3}{\overset{CH_2Cl}{\underset{|}{\overset{|}{C}}}}-CH_3$$ cannot undergo dehydrohalogenation by E2 mechanism

since beta hydrogen is missing.

7. Dehydration of 2,2,3,4,4-pentamethyl-3-pentanol gave two alkenes A and B. Ozonolysis of the lower-boiling alkene gives formaldehyde and 2,2,4,4-tetramethyl-3-pentanone. The other alkene on ozonolysis gives formaldehyde and 3,3,4,4-tetramethyl-2-pentanone. Hydrogenation of A gives C. It easily undergoes monobromination to give D. B on hydrogenation gives E which gives 5 different monosubstitution products on chlorination. Identify A to E

Answer

A is $$CH_3-\underset{CH_3}{\overset{CH_3}{\underset{|}{\overset{|}{C}}}}-\underset{CH_2}{\underset{\|}{C}}-\underset{CH_3}{\overset{CH_3}{\underset{|}{\overset{|}{C}}}}-CH_3$$

B is $$CH_3-\underset{CH_3}{\overset{CH_3}{\underset{|}{\overset{|}{C}}}}-\underset{CH_3}{\overset{CH_3}{\underset{|}{\overset{|}{C}}}}-\underset{CH_2}{\underset{\|}{C}}-CH_3$$

C is $$(CH_3)_3-C-\underset{CH_3}{\overset{H}{\underset{|}{\overset{|}{C}}}}-C(CH_3)_3$$

D is $$(CH_3)_3-C-\underset{CH_3}{\overset{Br}{\underset{|}{\overset{|}{C}}}}-C(CH_3)_3$$

$$\text{E is } (CH_3)_3-\underset{\underset{CH_2}{|}}{\overset{\overset{CH_3}{|}}{C}}-\underset{\underset{CH_3}{|}}{\overset{\overset{H}{|}}{C}}-C-CH_3$$

8. Compound A has the formula C_7H_{14} and does not decolorize $KMnO_4$. It reacts with bromine in light to give the monobrominated product B. C is another isomer of A and decolorizes $KMnO_4$ and on ozonolysis gives HCHO and OCH—$CH_2C(CH_3)_3$. With chlorine, A gives many monosubstitution products. B in the presence of sodium ethoxide in ethanol gives D. D on ozonolysis gives $CH_3CO(CH_2)_4CHO$. Identify A to C.

Answer

A is a cyclic compound since it does not decolorize $KMnO_4$.

It is [cyclohexane ring with CH_3] (1-methylcyclohexane).

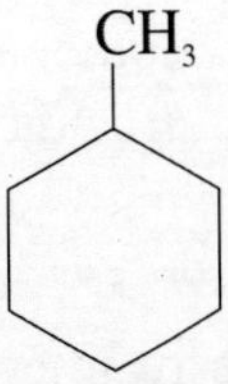

B is [cyclohexane ring with CH_3 and Br on the same carbon] (1-methyl-1-bromocyclohexane).

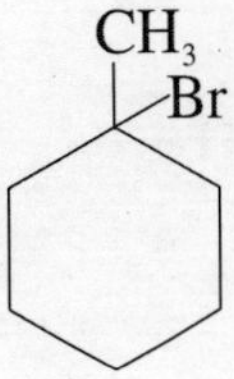

C is an olefinic compound and has structure, $CH_2{=}\underset{\underset{H}{|}}{C}H-CH_2-C(CH_3)_3$. Since the products are HCHO and OCH — $CH_2C(CH_3)_3$, 1-methyl-1-bromocyclohexane (B) on dehydrohalogenation gives 1-methylcyclohexene (D) which on ozonolysis gives $CH_3CO(CH_2)_4CHO$.

D is [cyclohexene ring with CH_3 on the double-bonded carbon] (1-methylcyclohexene).

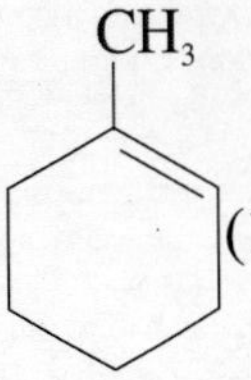

9. Substance H, with formula $C_{10}H_{19}Br$, gives a mixture of two isomers, I and J of formula $C_{10}H_{18}$, on treatment with base. If the base used is KOH in ethanol, the main product is I. However if the base used is potassium *tert.* butoxide in *tert.* butyl alcohol, the main product is J. Both I and J are converted to K, of formula $C_{10}H_{22}$, on treatment with an excess of hydrogen in the presence of a catalyst. Compound I on treatment with hot aqueous permanganate solution followed by acidification, yields a mixture of two compounds, L to M, whose structures are shown below. The ratio of L : M produced in this reaction is 2:1. Given this information, answer each of the following questions.

   a) Suggest structures for the compounds H to K.
   b) What products are expected upon ozonolysis of J followed by work-up with zinc and water?
   c) Compound M is formed as the sole product of the reaction of a different alkene, N, of formula C_4H_6, with hot aqueous permanganate solution. Suggest a structure for N.
   d) Compound L is the sole organic product obtained by treating yet another alkene O, of formula C_4H_8 with hot aqueous permanganate solution followed by an acidic work-up. Suggest a structure for O. Notice that L contains three carbon atoms while O contains four carbon atoms.

   What happened to the fourth carbon atom during the permanganate treatment?

L and M are $H_3C—\overset{\overset{\large :O:}{||}}{C}—CH_3$ and $HOOC—CH_2CH_2—COOH$

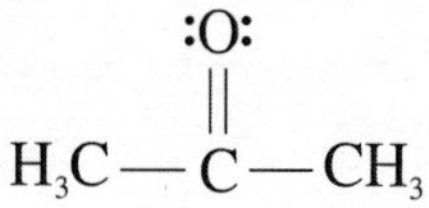

Answer

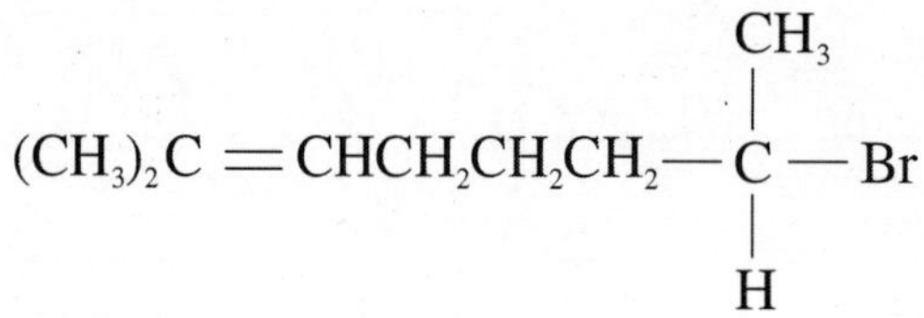

$$(CH_3)_2C=CHCH_2CH_2CH_2—\underset{\underset{\large H}{|}}{\overset{\overset{\large CH_3}{|}}{C}}—Br \qquad (H)$$

H with KOH in ethanol gives I as the main product. I is

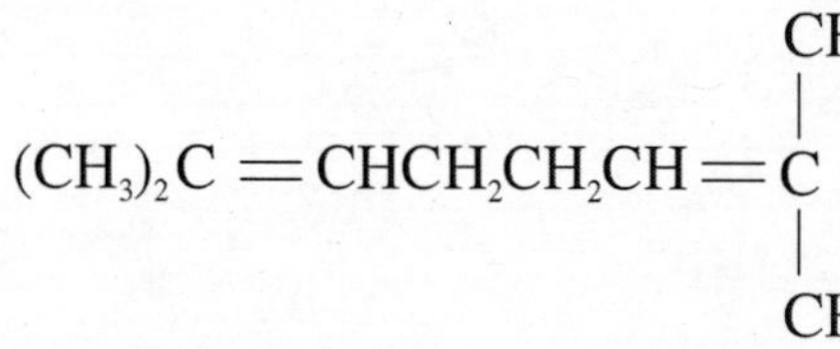

$$(CH_3)_2C=CHCH_2CH_2CH=\underset{\underset{\large CH_3}{|}}{\overset{\overset{\large CH_3}{|}}{C}} \qquad (I)$$

H with $(CH_3)_3COK$ in *t*-butanol undergoes E2 elimination to give

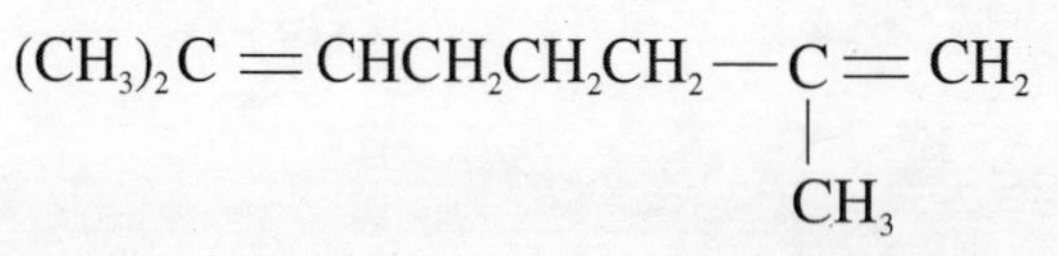

$$(CH_3)_2C=CHCH_2CH_2CH_2—\underset{\underset{\large CH_3}{|}}{C}=CH_2 \qquad (J)$$

$$\text{I or J} + 2H_2 \rightarrow \text{alkane}$$

Alkane is

$$(CH_3)_2CHCH_2CH_2CH_2-\underset{\underset{CH_3}{|}}{\overset{\overset{CH_3}{|}}{CH}} \qquad (K)$$

$$\text{I} + \text{hot aq.}KMnO_4 \rightarrow 2(CH_3)_2CO(L) + COOHCH_2CH_2COOH(M)$$

$$L : M = 2 : 1$$

$$\text{J} + \text{Ozone} + \text{Zn}(H_2O) \rightarrow (CH_3)_2CO + OCHCH_2CH_2COCH_3 + HCHO$$

$$\text{Cyclobutene(N)} + \text{hot AQ.}KMnO_4 \rightarrow COOHCH_2CH_2COOH(L)$$

$$(CH_3)_2C=CH_2(O) + \text{hot aq.}KMnO_4 \rightarrow (CH_3)_2CO(L) + CO_2 + H_2O$$

The extra carbon is in the form of CO_2.

10. An unknown compound K with formula C_6H_{12} decolorizes bromine water and is isoxidized by hot acidified $KMnO_4$ to a resolvable carboxylic acid C_4H_9COOH. The alkene K on hydrogenation gives L. L is prepared from $(CH_3CHCH_2CH_3)_2CuLi$ and CH_3CH_2I in ether. How many monosubstitution products are possible by chlorination of L. Find K and L and the carboxylic acid. Is L optically active?

Answer

K is $CH_2=CH-\underset{\underset{CH_3}{|}}{CH}-CH_2CH_3$

L is $CH_3CH_2-\underset{\underset{CH_3}{|}}{CH}-CH_2CH_3$

L is optically inactive and gives 4 monochloro products. The acid is

$HOOC-\underset{\underset{CH_3}{|}}{CH}-CH_2CH_3$

11. Compound X has molecular weight of 56 g. It decolorizes $KMnO_4$. When HBr is added it gives Y. Y with AgOH gives an alcohol which can give a very stable carbocation. On hydrogenation it gives a saturated compound Z which gives one monobrominated compound in high yield. Can X exhibit geometric isomerism?

Answer

X is $CH_3—\underset{\underset{CH_3}{|}}{C}=CH_2$

Y is *t*-butyl bromide isobutane.

Z is isobutene which selectively gives *t*-butyl bromide in good yield.

X cannot exhibit geometric isomerism.

12. Give the structures of isomeric alkyl bromides of formula $C_5H_{11}Br$. And answer the following questions.
    i. Which isomer undergoes E1 elimination at fastest rate?
    ii. Which of the isomers cannot produce E2 product?
    iii. Which isomers yield a single alkene by E2 elimination?
    iv. Which isomer yields a complex mixture of alkenes by E2 elimination.

    The isomers are

    $CH_3CH_2CH_2CH_2CH_2Br$ (A)

    $CH_3—\overset{\overset{CH_3}{|}}{\underset{\underset{CH_3}{|}}{C}}—CH_2Br$ (B)

    $(CH_3)_2CHCH_2CH_2Br$ (C)

    $(CH_3)_2CBrCH_2CH_3$ (D)

    $CH_3CHBrCH_2CH_2CH_3$ (E)

    $CH_3CH_2CHBrCH_2CH_3$ (F)

    i. D undergoes E1 elimination at fastest rate since it is tertiary compound.
    ii. B cannot produce E2 product since no beta hydrogen is present.
    iii. A and C yield single alkene by elimination since there is only one beta hydrogen and these are $CH_3CH_2CH_2CH=CH_2$,$(CH_3)_2CHCH=CH_2$

       F also yields $CH_3CH_2CH = CHCH_3$ as the only alkene either by loss of beta hydrogen on the left or right of Br.
    iv. E yields $CH_3CH=CH_2CH_2CH_3$ and $CH_2=CHCH_2CH_2CH_3$, a mixture of alkenes by E2 elimination since they have beta hydrogens on both sides of bromine.

13. Write the structure of all isomeric alcohols with formula $C_5H_{12}O$ and answer the following questions.

i. Which one of these undergoes acid-catalysed dehydration most readily?

ii. Draw the structure of the most stable carbocation.

iii. Which alkenes may be derived from carbocation in (ii)?

iv. Which alcohols will yield carbocation by hydride shift?

v. Which alcohols will yield carbocation by methyl shift?

vi. What are the different alkenes formed totally?

Answer

Alcohols having the formula $C_5H_{12}O$, are the following

A is $CH_3CH_2CH_2CH_2CH_2OH$

B is $CH_3-\underset{\underset{CH_3}{|}}{\overset{\overset{OH}{|}}{C}}-CH_2CH_3$

C is $CH_3\underset{\underset{OH}{|}}{C}HCH_2CH_2CH_3$

D is $(CH_3)_2CHCH_2CH_2OH$

E is $(CH_3)_2CH\underset{\underset{OH}{|}}{C}HCH_3$

F is $CH_3CH_2\underset{\underset{OH}{|}}{C}HCH_2CH_3$

i. B undergoes dehydration most readily since it gives rise to tertiary carbocation.

ii. The tertiary carbocation is $CH_3-\underset{\underset{CH_3}{|}}{\overset{+}{C}}-CH_2CH_3$ and this is the most stable carbocation.

iii. The alkenes derived from this cation are

$$(CH_3)_2C{=}CHCH_3 \text{ and } CH_2{=}\underset{\displaystyle CH_3}{\underset{|}{C}}HCH_2CH_3$$

iv. $(CH_3)_2CH\,CH_2CH_2OH$ is one alcohol that yields carbocation by hydride shift. The cation formed is

$$(CH_3)_2{-}\underset{\displaystyle H}{\underset{|}{C}}{-}CH_2{-}\overset{+}{C}H_2$$

By hydride shift the cation formed is $(CH_3)_2{-}\underset{\displaystyle H}{\underset{|}{C}}{-}\overset{\displaystyle H}{\overset{|}{\underset{+}{C}}}{-}CH_3$. This ion shifts once more leading to cation in (ii). The alkene formed is $(CH_3)_2C{=}CHCH_3$

$$CH_3CH_2\underset{\displaystyle OH}{\underset{|}{C}}H{-}CH_2CH_3$$

v. $(CH_3)_2CH\,\underset{\displaystyle OH}{\underset{|}{C}}HCH_3$ gives the cation $CH_3{-}\overset{\displaystyle CH_3}{\overset{|}{\underset{\displaystyle H}{\underset{|}{C}}}}{-}\overset{+}{\underset{\displaystyle H}{\underset{|}{C}}}{-}CH_3$ Hydride shift leads to

tertiary cation $CH_3{-}\overset{\displaystyle CH_3}{\overset{|}{\underset{+}{C}}}{-}CH_2CH_3$

$$CH_3CH_2\underset{\displaystyle OH}{\underset{|}{C}}H{-}CH_2CH_3$$

gives $CH_3{-}CH_2{-}\overset{\displaystyle H}{\overset{|}{C}}{-}CH_2CH_3$

Hydride shift gives $CH_3CH_2{-}\underset{+}{C}H_2{-}\overset{\displaystyle H}{\overset{|}{C}}{-}CH_3$

vi. The various alkenes formed are

I. $CH_3CH_2CH_2CH{=}CH_2$ from $CH_3CH_2CH_2CH_2CH_2OH$

II. $\begin{array}{l} CH_2{=}\underset{\displaystyle CH_3}{\underset{|}{C}}CH_2CH_3 \end{array}$ from tertiary alcohol

III. $(CH_3)_2C{=}CHCH_3$

IV. The alcohol $CH_3{-}\overset{\displaystyle H}{\overset{|}{\underset{\displaystyle OH}{\underset{|}{C}}}}{-}CH_2CH_2CH_3$ gives $CH_3CH{=}CHCH_2CH_3$ in *cis* and *trans* forms and also $CH_2{=}CHCH_2CH_2CH_3$.

V. $(CH_3)_2CHCH_2CH_2OH$ gives $(CH_3)_2CHCH{=}CH_2$

VI. $(CH_3)_2CH\underset{\displaystyle OH}{\underset{|}{C}}HCH_3$ gives $(CH_3)_2C{=}CHCH_3$.

VII. $CH_3CH_2\underset{\displaystyle OH}{\underset{|}{C}}HCH_2CH_3$ gives $CH_3CH_2CH{=}CHCH_3$.

14. Compound A has the formula $C_{11}H_{16}$ and contains a benzene ring. It gives 4 monosubstituted chlorides. Give the structure of the monochlorides. One of the monochlorides undergoes dehydrohalogenation by E2 mechanism to give $C_6H_5CH{=}CHCH(CH_3)_2$ and $C_5H_6CH_2CH{=}C(CH_3)_2$. Which alkene is in higher yield? Find A. Find the ozonolysis products in $(CH_3)_2S$ of the two alkenes?

Answer

A is $C_6H_5CH_2CH_2CH(CH_3)_2$,

The 4 monochlorides are

$$C_6H_5CH_2CH_2{-}\overset{\displaystyle H}{\overset{|}{\underset{\displaystyle Cl}{\underset{|}{C}}}}{-}CH(CH_3)_2$$

$C_6H_5CHClCH_2CH(CH_3)_2$,

$$C_6H_5CH_2CH_2\overset{\overset{CH_3}{|}}{\underset{\underset{CH_3}{|}}{C}}Cl \text{ and}$$

$$C_6H_5CH_2CH_2\overset{\overset{H}{|}}{\underset{\underset{CH_2Cl}{|}}{C}}-CH_3$$

$C_6H_5CHClCH_2CH(CH_3)_2$ on dehydrohalogenation gives $C_6H_5CH{=}CHCH(CH_3)_2$ and $C_6H_5CH_2CH{=}C(CH_3)_2$. The first one is conjugated and hence produced in more amount even though it is less substituted. The Hoffmann rule is followed.

Ozonolysis products are C_6H_5CHO and $CHOCH(CH_3)_2$.

For the second one the products of ozonolysis are $C_6H_5CH_2CHO$ and acetone.

4

ALKYNES AND DIENES

ALKYNES

NOMENCLATURE

$$H_3C—CH_2—C{\equiv}CH$$

1-butyne

If another functional group is present it has the priority

$$\underset{5}{HC}{\equiv}\underset{4\,3}{CCH_2}—\underset{2}{CH(OH)}—\underset{1}{CH_3}$$

5-pentyne-2-ol

PHYSICAL PROPERTIES

The C—H bond of acetylene is shorter than that of ethylene and ethane. Due to electron density being closer to the nucleus. C—H bond of acetylene is stronger than that of ethylene due to the fact that % of s orbital is more for acetylene compared to ethylene or ethane. Boiling points are similar to that of alkenes and alkanes. Densities are much lower than that of water and alkynes are insoluble in water.

ACIDITY OF ALKYNES

- pKa values of acetylene, ethylene and ethane are 26, 45 and 52. Thus acetylene has highest acidity due to the fact that the unshared pair of electrons in the anion occupies sp hybrid where s character is high (50%).

- Terminal alkynes are similar to acetylene in acidity 3,3-dimethyl-1-butyne has pK_a 25.5.
- Acetylene can be converted to anion by $NaNH_2$ which is a strong base, but not by OH^-, which is a weaker base.
- Terminal alkynes give precipitate with ammoniacal silver nitrate or ammoniacal cuprous chloride.

METHODS OF PREPARATION OF ACETYLENE AND OTHER ALKYNES

i. *Hydrolysis of calcium carbide* Acetylene can be prepared from sodium acetylide, but it is not used synthetically since sodium acetylide itself is prepared from acetylene.

ii. Heating haloform with Ag powder gives acetylene.

$$2CHI_3 + 6Ag \longrightarrow CH{\equiv}CH + 6AgI$$

iii. By the electrolysis of a concentrated solution of potassium fumarate or potassium maleate.

$$NaO_2CCH{=}CHCO_2Na \longrightarrow 2H_2O \longrightarrow CH{\equiv}CH + 2NaOH + H_2$$

iv. *Alkylation of acetylene* Acetylene is converted to sodium acetylide by $NaNH_2$ and then alkylated.

$$HC{\equiv}CNa + CH_3CH_2CH_2CH_2Br — HC{\equiv}CCH_2CH_2CH_2CH_3$$

1-pentyne

This reaction follows S_N2 mechanism. The reaction is therefore mainly feasible in primary and methyl halides. Secondary and tertiary halides react by elimination.

For example

$$HC{\equiv}C^- + \text{t-butylbromide} \longrightarrow \text{acetylene} + \text{2-methylpropene} + Br^-$$

Details of S_N1 and S_N2 reactions will be presented in chapter on alkyl halides.

v. *Alkynes from elimination reactions*

a) double dehydrohalogenation of a gemdihalide

$$RCH_2CX_2R' + NaNH_2 \longrightarrow R{-}C{\equiv}C{-}R' + 2NH_3 = + 2NaX$$

vi. Double dehydrohalogenation of vicinal halide

$$RCHXCHXR' + 2NaNH_2 \longrightarrow R{-}C{\equiv}C{-}R' + 2NH_3 + 2NaX$$

This procedure is very useful to prepare terminal alkynes.

vii. Alkenyl halides undergo dehydrohalogenation in strong basic conditions) ($2NaNH_2$ + water) to give alkynes.

$$(CH_3)_2CHCHClC{=}CH_2 \text{ gives } (CH_3)_2CHCH_2C{\equiv}CH$$

REACTIONS OF ALKYNES

1. *Partial hydrogenation of alkynes* For semi-hydrogenation to alkenes catalysts used are

a) Pd supported on barium sulphate
b) Nickel boride catalyst
c) Lindlar palladium on calcium carbonate to which quinoline and lead acetate are added to deactivate the catalyst so that hydrogenation stops at alkene stage.
d) Na or Li/NH_3 in ethanol

Hydrogenation to alkenes is stereoselective. *cis* or *Z* alkene is obtained. 2-butyne with Lindlar Pd gives *cis*-2-butene. But use of Na or Li/NH_3 gives *trans*-2-butene. The mechanism is given in Figure 4.1.

$$Li + R—C\equiv C—R \longrightarrow RC=CR \text{ (Radical anion)} \xrightarrow{H—\ddot{N}H—Et} RC=CHR \text{ (Vinylic radical)}$$

$$\text{Vinylic radical} \xrightarrow{Li} \textit{trans}\text{-Vinylic anion} \xrightarrow{H—\ddot{N}H—Et} \textit{trans}\text{-Alkene}$$

Radical anion

Vinylic radical

Vinylic radical

trans-Vinylic anion

trans-Alkene

Figure 4.1 Mechanism of alkyne reduction in Li/NH_3

Alkenyl radical formation is rate-determining and it can be in Z or E form and E form is more stable due to less steric crowding. Hence *trans* product is formed.

2. *Addition of HBr* Addition is according to Markovnikov rule. 1-hexyne gives with HBr 2-bromo-1-hexene. Excess hydrogen halides give gemdihalides.

$$RC\equiv CR' + 2HF \longrightarrow R—CH_2—CF_2—R'$$

(Both halogens attached to same carbon lead to gemdihalide.)

3. *Hydration of alkynes in presence of Hg^{2+} and sulphuric acid* Addition of aq. sulphuric acid in the presence of Hg^{2+} ion ($HgSO_4$ or HgO) hydrates the alkyne. Markovnikov rule is followed to give enols which convert to the carbonyl compound. Only acetylene gives acetaldehyde. All other alkynes give ketones.

If internal alkynes which have different alkyl groups on both sides of triple bond are used ($CH_3CH_2C\equiv CCH_3$), two ketones are produced.

4. *Hydroboration oxidation of alkynes (addition of water) (Anti Markovnikov)* Alkynes undergo hydroboration oxidation in BH_3, THF and in the presence of OH^-, H_2O_2 and water. For symmetric alkynes the product is a ketone. Enol forms which converts to ketone. For 3 moles of 2-butyne the product is three moles of $CH_3CH_2COCH_3$. If the alkyne is a terminal alkyne then product is an aldehyde. In hydroboration oxidation the nucleophile is H^- and this creates the carbonyl group on the terminal carbon. For example propyne reacts as follows by hydroboration oxidation.

i. ($CH_3C\equiv CH + BH_3$ (THF)

ii. $OH^-, H_2O_2, H_2O \longrightarrow CH_3CH{=}CHOH$ (enol) converts to aldehyde CH_3CH_2CHO For producing enol one equivalent of BH_3 has to be used to stop reaction at alkene stage. In internal alkynes steric factor prevents further addition of BH_3 and stops at alkene stage. For terminal alkynes a special reagent diisoamyl (secondary isoamyl) borane is added.

5. *Halogen addition to alkynes* Alkynes react with 2 moles of X_2 to give tetrahaloalkanes.

$$CH_3C\equiv CH + 2Cl_2 \longrightarrow \underset{\text{1,1,2,2-tetrachloropropane}}{CH_3C(Cl_2)CHCl_2}$$

6. *Vinylation*

$$CH\equiv CH + CH_3O^- \longrightarrow {}^-CH{=}CHCH_3 \longrightarrow CH_3OH \longrightarrow CH_2{=}CHOCH_3 + CH_3O^-$$

The enol ethers give rise to heteroatom ions which are resonance-stabilized and with dil. H_2SO_4 react to give ketones and alcohols.

7. *Ozonolysis of alkynes* $RC\equiv CR'$ add on to ozone in water to give carboxylic acids. If the alkyne is a terminal alkyne like propyne, the products are CH_3COOH and carbonic acid, but carbonic acid decomposes to CO_2 and H_2O. Acetylene is exceptional to give CHOCHO and HCOOH

If alkynes contain double bonds, the double bonds react faster in ozone. Hence triple bond can be retained and the other unsaturated part will get oxidized to the acid.

$$R'C\equiv C(CH_2)_3CH{=}CHR'' \longrightarrow R'C\equiv C(CH_2)_3COOH + R''COOH$$

8. *Polymerization* When passed through a heated tube acetylene polymerizes to give benzene. Propyne polymerizes to 1,3,5-trimethylbenzene. 2-butyne polymerizes to hexamethylbenzene.

DIENES

There are three types of dienes namely cumulated $CH_2{=}C{=}CH_2$, conjugated $CH_2{=}CH{-}CH{=}CH_2$ and nonconjugated dienes $CH_2{=}CH(CH_2)_3CH{=}CH_2$. Conjugated dienes are stabilized by resonance. They are more stable than nonconjugated dienes and cumulated dienes are least stable.

HYBRIDIZATION IN DIENES

In case of cumulated dienes the central carbon is flanked by two double bonds and hence this carbon is sp-hybridized.

HEATS OF HYDROGENATION

The nonconjugated dienes have heat of hydrogenation twice the value of the corresponding monosubstituted alkene. 1-butene has ΔH (hydrogenation) equal to –136 kJ/mol and 1,4-pentadiene having two double bonds has ΔH (hydrogenation) equal to –253 kJ/mol. But a conjugated diene like 1,3-butadiene has ΔH (hydrogenation) equal to –236 kJ/mol. Extra stability is about 3.6 kcal/mol due to resonance. Bond lengths are shorter for dienes compared to alkenes.

METHODS OF PREPARATION

1. Dehydration of diols

$$HOCH_2CH_2CH_2CH_2CH_2OH \longrightarrow H_2SO_4H_2C{=}CH{-}CH{=}CH_2$$

2. Dehydrohalogenation of dihalides

$$CH_3CHXCH_2CH_2X \longrightarrow (-2HX)\ H_2C{=}CH{-}CH{=}CH_2$$

3. Dehydrohalogenation of allylic halides

$$H_2C{=}CHCHXCH_3 \longrightarrow (\text{-HX}) \longrightarrow H_2C{=}CH{-}CH{=}CH_2$$

4. Cyclic dienes are prepared from dibromides

   Trans 1,2 bromocyclohexane by E_2 elimination two times successively with $NaNH_2$ or *t*-BuOK gives cyclohexa-1,3-diene.

   Cycloalkene dibromides by successive elimination of two moles of HX gives rise to dienes as shown in Figure 4.2.

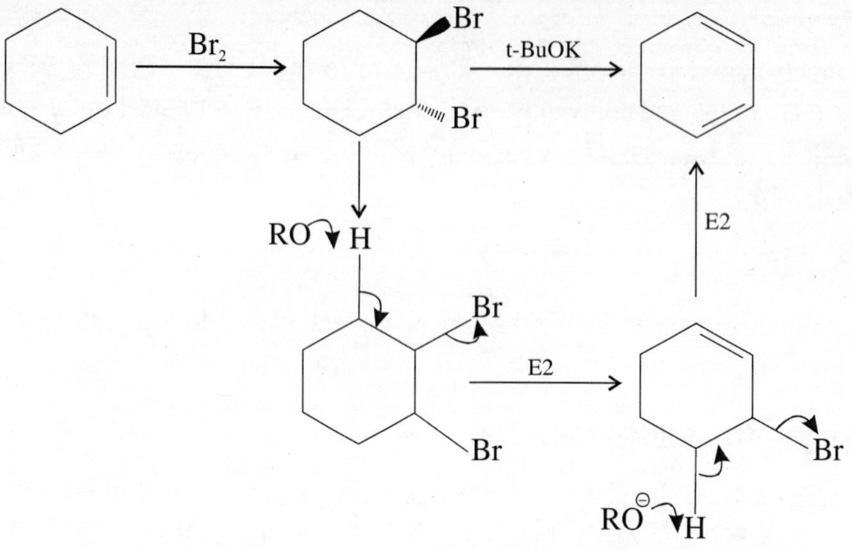

Figure 4.2 Synthesis of a diene by double E2 elimination

REACTIONS OF DIENES

Diels–Alder Reaction

Figure 4.3 shows the Diels–Alder Reaction

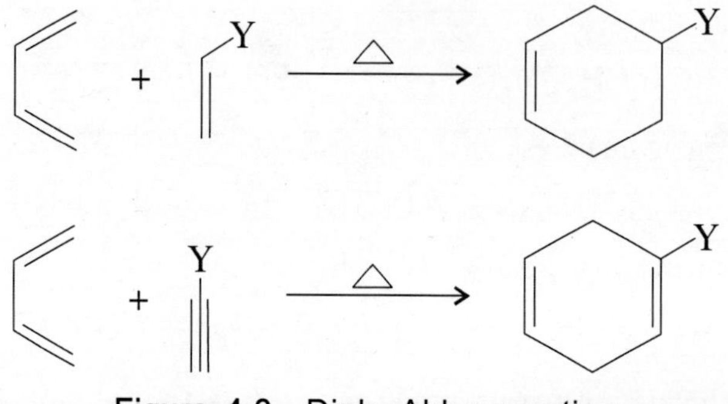

Figure 4.3 Diels–Alder reaction

The [4+2]-cycloaddition of a conjugated diene and a dienophile (an alkene or alkyne) is an electrocyclic reaction that involves the 4π-electrons of the diene and 2π-electrons of the dienophile. The driving force of the reaction is the formation of new σ-bonds, which are energetically more stable than the π-bonds.

In the case of an alkynyl dienophile, the initial adduct can still react as a dienophile if not too sterically hindered. In addition, either the diene or the dienophile can be substituted with cumulated double bonds, such as substituted allenes.

With its broad scope and simplicity of operation, the Diels–Alder reaction is the most powerful synthetic method for unsaturated six-membered rings.

CONFORMATIONS OF CONJUGATED DIENES

- In order to have a conjugated system of atomic orbitals, in which all of the 2p (AOs) are parallel (e.g., all $2p_z$), all 4 of the carbons of the diene system must be coplanar, as we will see further in a moment.
- This means that there are only two ground-state conformations of such a conjugated diene. One is called s-*trans* and the other s-*cis* (Figure 4.4).
- The *cis* or *trans* designation is based upon the relationship around the central, formally single C—C bond.

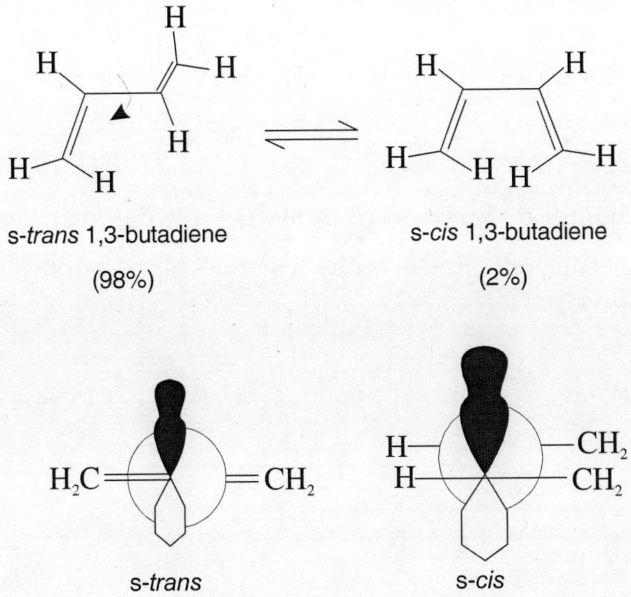

s-*trans* 1,3-butadiene (98%) ⇌ s-*cis* 1,3-butadiene (2%)

s-*trans* s-*cis*

Figure 4.4 s-*cis* and s-*trans* forms of 1,3-butadiene

- The s-*trans* conformation is more thermodynamically stable because the s-*cis* conformation has a steric repulsion between the two *inside hydrogens*.
- It should also be noted that in some cyclic systems, this inside repulsion is not present, and the molecule is rigidly compelled to have the *s-cis* conformation (no other stable conformation exists). In other cyclic cases, the *s-trans* conformation may be the only one available; i.e., there is 0% s-*cis* (Figure 4.5).

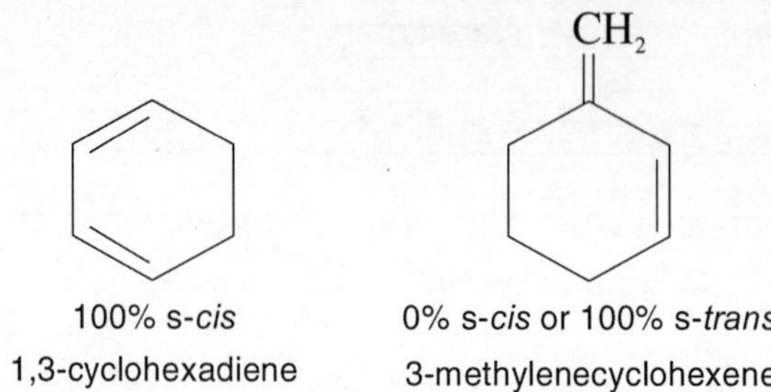

Figure 4.5 Examples of 100% s-*cis* form and 100% s-*trans* form

The compound on the right-hand side which is 0% *cis* will not undergo Diels–Alder reaction. But the s-*cis* form undergoes Diels–Alder reaction as presented in Figure 4.6.

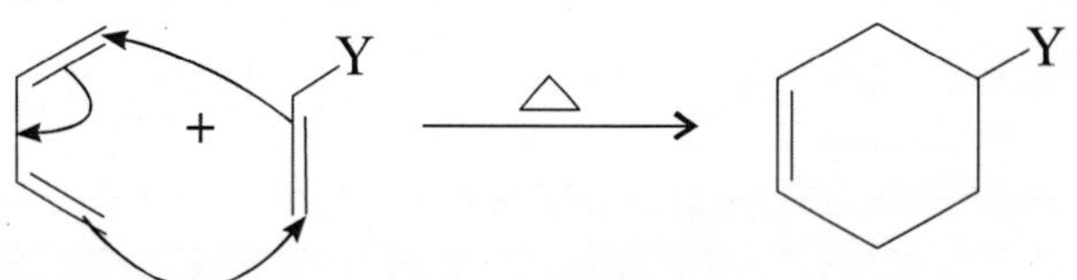

Figure 4.6 Mechanism of the Diels–Alder Reaction

Overlap of the molecular orbitals (MOs) is a requirement apart from the diene being in s-*cis* conformation (Figure 4.7).

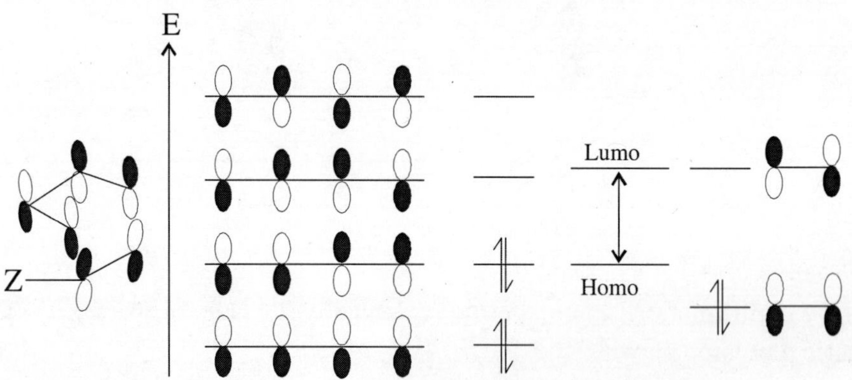

Figure 4.7 Overlap of MOs in Diels–Alder reaction

Overlap between the highest occupied MO of the diene (HOMO) and the lowest unoccupied MO of the dienophile (LUMO) is thermally allowed in the Diels–Alder reaction, provided the orbitals are of similar energy. The reaction is facilitated by electron-withdrawing groups on the dienophile, since this will lower the energy of the LUMO. Good dienophiles often bear one or two of the following substituents: CHO, COR, COOR, CN, C = C, Ph, or halogen. The diene component should be as electron-rich as possible.

The reaction is stereoselective. Figure 4.8 gives examples of Diels–Alder reaction involving different dienophiles.

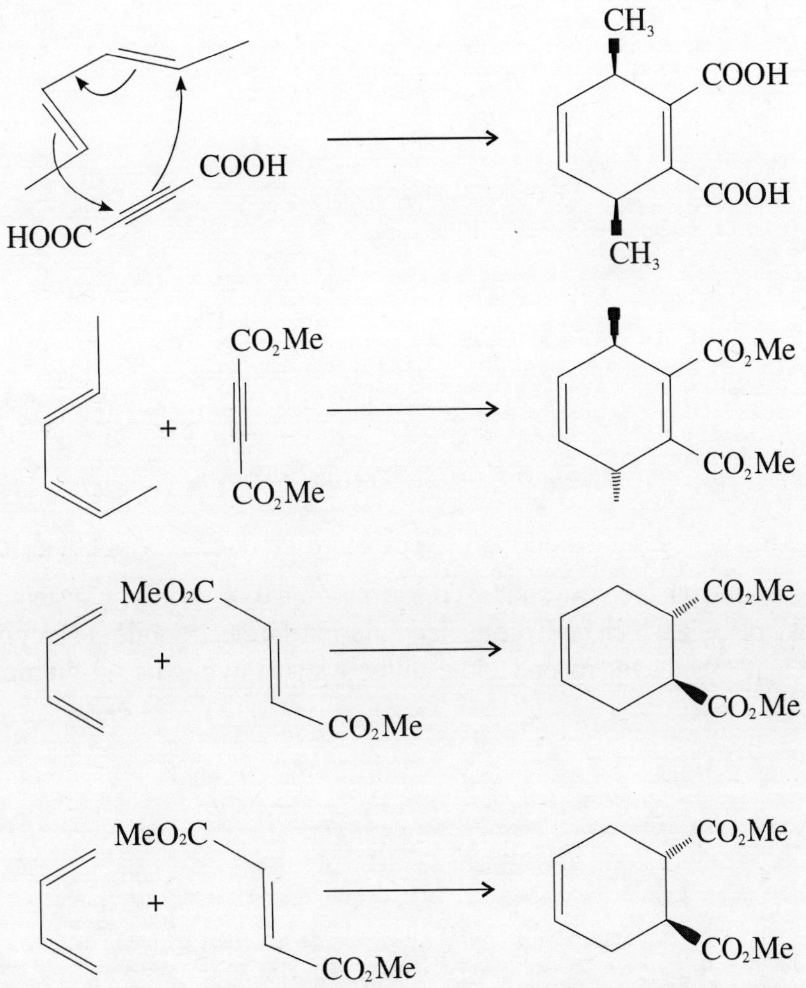

Figure 4.8 Examples of Diels–Alder addition reactions

Electrophilic Addition of Br_2 to Butadiene

The reaction of bromine with dienes can produce regioisomers; the outcome can be altered depending on the conditions used. If the reaction is done at lower temperatures, the bromine just adds across one of the double bonds to give a 1,2-dibromide. This is the kinetic product.

The 1,4-dibromide is formed only when the reactants are heated and is the thermodynamic product. The mechanism is the electrophilic attack on the diene to give a bromonium ion, which opens to give the dibromide. 1,2-dibromide can then react further because it can undergo nucleophilic substitution (Figure 4.9).

1,2-dibromide

Low temperature—Kinetic Product

1,4-dibromide

High temperature—Thermodynamic Product

Figure 4.9 Low temperature and high temperature products of 1,3-butadiene with Br_2

Bromide is a good nucleophile and leaving group, and with an allylic system like this, S_N1 reaction can take place in which both the nucleophile and the electrophile are bromine. Bromide can attack where it left or at the far end of the allylic system, giving the 1,4-dibromide which is more stable.

I. "Explain Why" Questions and Small Structure-Based Problems

1. Although both butyl bromide and 4-bromo-1-butene are primary halides, the latter undergoes dehydrobromination more rapidly. Explain.

Answer

The structure of 4-bromo-1-butene is $CH_2{=}CHCH_2CH_2Br$. This will give rise to resonance-stabilized cation. On dehydrobromination this gives $CH_2{=}CH{-}CH{=}CH_2$ which is conjugated. 1-bromobutane gives 1-butene which is not conjugated. And ordinary carbocation is generated.

2. 3-methylene cyclohexene does not undergo Diels–Alder reaction. Why?

Answer

It is in s-*trans* conformation. And Diels–Alder reaction takes place when the compound is in s-*cis* conformation.

3. The trienes

$$\underset{H}{\overset{H_3C}{>}}C{=}C{=}C{=}C\underset{H}{\overset{CH_3}{<}} \quad \text{and} \quad \underset{H}{\overset{H_3C}{>}}C{=}C{=}C{=}C\underset{CH_3}{\overset{H}{<}}$$

have different dipole moments. Why?

Answer

They are *cis-trans* isomers.

4. 1,3-cyclopentadiene and 1,3-cyclohexadiene are very reactive in Diels–Alder reaction. Why?

Answer

These dienes are locked in s-*cis* conformation and hence are very reactive.

5. 2,4-hexadiene never reacts by Diels–Alder reaction. Why?

Answer

It is in s-*trans* form since the s-*cis* form is unstable due to steric interaction of terminal CH_3 groups.

6. $^{14}CH_3CH{=}CH_2$ is brominated with NBS. The product consists of labelled carbon on carbon of CH_3 and carbon of CH_2Br. Why?

Answer

The reaction product consists of allyl bromide with labelled carbon on $^{14}CH_2{=}CHCH_2Br$ and $H_2C{=}CH^{14}CH_2Br$ since the free radical has resonance structures.

7. 2-butyne isomerizes to 1-butyne by $NaNH_2$ in NH_3. The intermediate was isolated and ozonolysis was performed and the products were CH_3CHO, CO_2 and HCHO. Explain.

Answer

The intermediate is 1,2-butadiene, $CH_3CH{=}C{=}CH_2$, which gives these ozonolysis products.

8. Propyne gives precipitate with ammoniacal $AgNO_3$ but 2-butyne does not give precipitate. Explain.

Answer

Propyne is a terminal alkyne with acidic hydrogen and hence gives precipitate.

9. When acetylene is hydrated with Hg and H_2SO_4, we get acetaldehyde, but propyne gives acetone. Why?

Answer

In acetylene, addition of OH on either carbon gives aldehyde only, but in propyne, Markovnikov addition leads to enol of a ketone which tautomerizes to the ketone.

10. Cyclooctatetraene readily reacts with potassium to form dianion. The salt has the structure $2K^+$ and $C_8H_8^{2-}$. Explain.

Answer

The dianion has 10 p electrons and it is planar satisfying Huckel rule criteria and is aromatic. Hence this reaction occurs.

II. MULTIPLE CHOICE QUESTIONS (ONE OUT OF FOUR ANSWERS IS CORRECT)

1. One of the following alkynes does not give precipitate with ammoniacal cuprous chloride

   a) acetylene b) 2-butyne c) propyne d) 1-butyne

2. 1,1-dichlorobutane is converted to 1-butyne using

   a) alc. KOH b) 2 moles of $NaNH_2$
   c) sulphuric acid d) one mole pot. *t*-butoxide

3. Diels–Alder reaction of s-*cis* butadiene with ________ gives cyclohexene.

   a) ethylene b) propylene c) ethane d) propyne

4. The least reactive diene in Diels–Alder reaction is

   a) 1,3-cyclopentadiene b) 1,3-cyclohexadiene
   c) furan d) 3-methylene cyclohexene

5. One of the following alkynes exhibits *cis-trans* isomerism.

   a) propyne b) 2-butyne c) 1-butyne d) 3-penten-1-yne

6. Alkyne A on complete reduction gives a saturated compound which on bromination gives mainly $(C_2H_5)_2CBrCH_3$. One of the following statements related to the alkyne is true.

   a) The alkyne is optically active.
   b) The alkyne exhibits geometric isomerism.
   c) The alkyne on hydration in Hg^{2+} and H_2SO_4 gives an aldehyde.
   d) The alkyne gives different reduction products if we use Lindlar Pd or Na/NH_3.

7. One of the following statements related to 2,3,4-hexatriene is true.

   a) It shows optical activity.
   b) Ozonolysis yields HCHO as one product.
   c) It has two isomers differing in dipole moments.
   d) On complete hydrogenation, it gives a hydrocarbon which gives only one monochloride.

8. The compound which undergoes reduction with Lindlar Pd to give 1,3-butadiene is
   a) 1-butyne
   b) 2-butyne
   c) $CH_2{=}CHC{\equiv}CH$
   d) $CH_3CH{=}CHC{\equiv}CH$
9. One of the various isomers containing cyclohexenyl ring with formula C_7H_{10} has conjugated double bonds and on ozonolysis gives HCHO as one of the products in $(CH_3)_2S$. This isomer is
   a) 1-methyl-1,4-cyclohexadiene
   b) 1-methyl-1,3-cyclohexadiene
   c) 3-methylenecyclohexene
   d) 4-methylenecyclohexene
10. Diels–Alder reaction can be used to distinguish
    a) 1-hexyne and 2-hexyne
    b) 3-octyne and 4-octyne
    c) 1,3-cyclopentadiene and 1,3-cyclohexadiene
    d) 1,3-cyclopentadiene and 3-methylenecyclohexene
11. Compound A has formula C_8H_{12} and gives on ozonolysis in reducing conditions HCHO, $CHOCH_2CH(CH_3)$—COOH and CH_3COOH. One statement about this molecule is true
    a) It has two triple bonds.
    b) Reduction with Lindlar Pd will give a different product relative to reduction by Na/NH_3.
    c) It is optically active.
    d) The complete reduction of this compound gives a hydrocarbon which is optically active.
12. One of the ozonolysis products of this alkyne is same as that for 2-butyne under oxidizing conditions.
    a) acetylene
    b) 3-hexyne
    c) 4-octyne
    d) propyne
13. Which of the following compounds has the incorrect name
    a) (Z)2-methyl 1-pentene
    b) 3,4-diethyl-3-hexene
    c) 2,3-dimethyl-1,3-butadiene
    d) 3-methylenecyclohexene
14. The diene which is optically active is
    a) $CH_2{=}C{=}CH_2$
    b) $CH_3CH{=}C{=}CH{-}C_2H_5$
    c) $Cl_2C{=}C{=}CCl_2$
    d) $CH_2{=}CH{-}CH{=}CH_2$
15. One of the following is not the proper method for preparing alkynes.
    a) $CaC_2 + H_2O$
    b) $CHI_3 + Ag \longrightarrow$ Acetylene + AgI
    c) $CH{\equiv}CH + NaOH \longrightarrow CH{\equiv}CNa + CH_3I \longrightarrow$ Propyne
    d) $CH{\equiv}CCH_3 + NaNH_2 \longrightarrow NaC{\equiv}CCH_3 + CCH_3 \longrightarrow CH_3C{\equiv}CCH_3$

16. An unsaturated compound of formula C_5H_6 on ozonolysis gives CH_3COOH, COOHCOOH and CO_2. It gives a precipitate with ammoniacal $AgNO_3$. The compound has one of the following set of characteristics.
    a) It has two triple bonds and is a terminal alkyne and does not exhibit geometrical isomerism.
    b) It has one triple bond and on partial reduction with Lindlar Pd gives *cis* alkene.
    c) It is optically active.
    d) The completely saturated compound gives 5 monosubstituted chlorides ($CH_3C{\equiv}C{-}C{\equiv}CH$).
17. When propyne reacts with CH_3MgBr, it gives C and D and addition of CH_3I to D gives E. C, D and E are
    a) CH_4, $CH_3C{\equiv}CMgBr$, and $CH_3C{\equiv}CCH_3$
    b) C_2H_6, $CH_3C{\equiv}CMgBr$, and $CH_3C{\equiv}CCH_3$
    c) CH_4, $CH_3C{\equiv}CMgBr$, and $CH_3C{\equiv}CC_2H_5$
    d) CH_4, $C_2H_5C{\equiv}CMgBr$, and $CH_3C{\equiv}CCH_3$
18. Identify a reagent in the following list, which distinguishes 1-butyne and 2-butyne.
    a) bromine in CCl_4 b) H_2 Lindlar catalyst
    c) dilute H_2SO_4, $HgSO_4$ d) ammoniacal Cu_2Cl_2 solution (2002)
19. Propyne and propene can be distinguished by
    a) conc. H_2SO_4 b) Br_2 in CCl_4 c) dil. $KMnO_4$ d) $AgNO_3$ in ammonia (2000)
20. In the compound $CH_2{=}CH{-}CH_2{-}CH_2{-}C{\equiv}CH$ the $C_2{-}C_3$ bond is of the type
    a) sp-sp^2 b) sp^3-sp^3 c) sp-sp^3 d) sp^2-sp^3 (1999)
21. The product obtained by addition of Hg^{2+} and H_2SO_4 (dilute) to 1 butyne is
    a) $CH_3CH_2COCH_3$ b) $CH_3CH_2CH_2CHO$
    c) CH_3CH_2CHO + HCHO d) CH_3CH_2COOH +HCOOH (1999)

KEY

1. b	2. b	3. a	4. d	5. d	6. a	7. c
8. c	9. c	10. d	11. c	12. d	13. a	14. b
15. c	16. a	17. a	18. d	19. d	20. d	21. a

III. MULTIPLE CHOICE QUESTIONS (MORE THAN ONE ANSWER IS CORRECT)

1. Some of the following alkynes do not give precipitate with ammoniacal cuprous chloride. They are
    a) acetylene b) 2-butyne c) propyne d) ($CH_3C{\equiv}C{-}C{\equiv}CH$)

2. Stereoisomerism is possible in

a) $CH_3C \equiv C - CH(OH) - CH = CH_2$

b) 2,4-hexadiene

c) propylene

d) $CH_2 = CH\ (CH_2)_3\ CH_3$ attached to C=C (structure: $CH_2{=}CH$ and $(CH_2)_3CH_3$ on a $C{=}C$ with H, H)

$$\begin{array}{cc} CH_2{=}CH & (CH_2)_3\,CH_3 \\ | & | \\ C & {=}\ C \\ | & | \\ H & H \end{array}$$

3. Diels–Alder reaction is possible for the following dienes

a) 1,3-cyclopentadiene

b) 1,3-cyclohexadiene

c) 3-methylene cyclohexene

d) 2,4-hexadiene

4. Compounds having diastereoisomers are

a)
$$\begin{array}{cc} HC{\equiv}C & CH_3 \\ | & | \\ C & {=}\ C \\ | & | \\ H & H \end{array}$$

b) 2,3,4-hexatriene

c) allene

d) $CH_3CH = C = CH_2$

5. CO_2 is one of the products of ozonolysis in

a) $CH_3CH = C = CH_2$

b) $CH_3C \equiv CH$

c) $CH_3C \equiv C - C \equiv CH$

d) 2-butyne

6. Alkyne A with formula C_6H_8 on complete reduction gives a saturated compound which on bromination gives mainly $(C_2H_5)_2CBrCH_3$. Some of the following statements related to the alkyne is true.

a) The alkyne is optically active.

b) The alkyne does not exhibit geometric isomerism.

c) The alkyne on ozonolysis in reducing conditions gives CO_2, $HOOC - \underset{CH_3}{\overset{H}{C}} - CH_2CHO$ and HCHO.

d) The alkyne gives different reduction products if we use Lindlar Pd or Na/NH_3.

7. Some of the statements related to 2,3,4-hexatriene are true.
   a) It shows optical activity.
   b) The carbons 2 and 3 have sp hybridization.
   c) It has two geometric isomers.
   d) On complete hydrogenation, it gives a hydrocarbon which gives 3 monochlorides.
8. 1,3-butadiene can be obtained
   a) by reacting NBS with $H_2C{=}CHCH_2CH_3$ and dehydrohalogenating the product with alc.KOH
   b) when 2 molecules of acetylene are treated with $CuCl/NH_4Cl$ and the resulting vinyl acetylene is reduced with Lindlar Pd
   c) when two ethylene molecules are heated
   d) when propyne is mixed with CH_3I
9. Some of the following reactions lead to stereoisomeric products
   a) 1-hexyne + HBr $\longrightarrow CH_2Br{=}CH(CH_2)_3CH_3$
   b) 3-hexyne + HCl $\longrightarrow CH_3CH_2C(Cl){=}CH\,CH_2CH_3$
   c) 1-butyne + $Br_2(CH_2Cl_2) \longrightarrow CH_3CH_2C(Br){=}CHBr$
   d) 5-decyne + $Li(NH_3) \longrightarrow$ trans alkene
10. Diels Alder reaction is not possible for
    a) 3-methylenecyclohexene
    b) furan
    c) 1,3-butadiene (*trans*)
    d) 1,3-cyclohexadiene
11. Compound A has formula C_8H_{12} and gives, on ozonolysis in reducing conditions, HCHO, $CHOCH_2CH(CH_3){-}COOH$ and CH_3COOH. One statement about this molecule is true.
    a) It has two triple bonds.
    b) Reduction with Lindlar Pd will give the same product relative to reduction by Na/NH_3.
    c) It is optically active.
    d) The complete reduction of this compound gives a hydrocarbon which is optically active.
12. Propyl methyl ketone is obtained as a product or one of the products in
    a) hydration of 1-pentyne in Hg^{2+} and H_2SO_4
    b) ozonolysis of $CH_3CH_2CH_2C(CH_3){=}CHCH{=}CH_2$ in reducing conditions
    c) hydration of 2-pentyne in Hg^{2+} and H_2SO_4
    d) $(CH_3)_2CH{-}C(CH_3){=}CH{-}CH{=}CH_2$ by ozonolysis in reducing conditions

13. HCHO is one of the products of ozonolysis in $(CH_3)_2S$ for

a) allene
b) $CH_2{=}CH{-}CH_2{-}CH{=}CH_2$
c) $CH_2{=}CH{-}CH{=}CH_2$
d) $CH_3CH{=}C{=}CHCH_3$

14. Ozonolysis in Zn/acid does not produce a dicarboxylic acid in some of the following alkynes

a) $CH_3C{\equiv}CCH_2CH_2C{\equiv}CC_2H_5$
b) 2-butyne
c) $CH_2{=}CH{=}CH_2$
d) $CH_2{=}CH{-}C{\equiv}CH$

15. An alkyne $C_{10}H_{10}$ gives by oxidative cleavage a tricarboxylic acid with structure

$$\begin{array}{l} COOHCH_2CH{-}\ CH_2COOH \\ \qquad\qquad\ \ | \\ \qquad\quad\ CH_2COOH \end{array}$$

Some statements about this alkyne are true

a) Alkyne has 3 triple bonds.
b) Alkyne loses 3 molecules of CO_2 on oxidation.
c) Ammoniacal $AgNO_3$ gives a precipitate with the alkyne.
d) Alkyne is optically active.

16. Formaldehyde is one of the products in ozonolysis under reducing conditions of some of the dienes having cyclohexyl ring with formula C_7H_{10}. These dienes are

a) 3-methylenecyclohexene
b) 3-methyl1,4-cyclohexadiene
c) 4-methylenecyclohexene
d) 1-methyl-1,3-cyclohexadiene

17. $CHOCH_2CHO$ is one of the ozonolysis products in reducing conditions for some dienes having cyclohexyl ring with formula C_7H_{10}. They are

a) 1-methyl-1,4-cyclohexadiene
b) 2-methyl-1,3-cyclohexadiene
c) 3-methyl-1,4-cyclohexadiene
d) 3-methylene cyclohexene

18. Some of the dienes with formula C_6H_{10} given below do not exhibit geometric isomerism

a) 1,5-hexadiene
b) 2-methyl-1,4-pentadiene
c) 4-methyl-1,3-pentadiene
d) 2,4-hexadiene

19. The optically active compounds among the following are

a) $CH_3CH{=}C{=}CHC_2H_5$

b) $$\begin{array}{c} \qquad\qquad\quad H \\ \qquad\qquad\quad | \\ CH_3CH_2{-}C{-}C{\equiv}CH \\ \qquad\qquad\quad | \\ \qquad\qquad\quad CH_3 \end{array}$$

c) $CH_3C{\equiv}C\,CH_3$

d) $CH_2{=}CH{-}CH_2{-}CH_3$

KEY

1. a,c,d	2. a,b,d	3. a,b	4. a,b	5. a,b,c	6. a,b,c	7. b,c,d
8. a,b	9. b,c,d	10. a,c	11. a,b	12. a,c	13. a,b,c	14. b,c,d
15. a,b,c	16. a,c	17. a,c	18. a,b,c	19. a, b		

IV. Assertion–Reason Questions

The questions 1–15 consist of an *assertion* in column 1 and *reason* in column 2. Use the following key to choose an appropriate answer

a) If both assertion and reason are correct and reason is the correct explanation of assertion
b) If both assertion and reason are correct but reason is not the correct explanation of assertion
c) If assertion is correct but reason is incorrect
d) If assertion is incorrect but reason is correct

No.	Assertion	Reason
1.	Propyne with Hg^{2+} and H_2SO_4 gives acetone.	This is due to operation of Markonikoff rule to get enol corresponding to acetone.
2.	$CH_3C{\equiv}C{-}CH_2{-}CH{=}CH_2$ on ozonolysis in reducing conditions gives CH_3CHO, $COOHCH_2CHO$ and CO_2.	A triple bond gives acid, triple and double bonds on the two ends give $COOHCH_2CHO$ and terminal CH_2 gives HCHO.
3.	Propyne gives precipitate with ammoniacal $AgNO_3$.	It has a triple bond.
4.	1-butyne can be distinguished from 2-butyne by adding ammoniacal CuCl.	1-butyne has acidic hydrogen and gives precipitate with ammoniacal CuCl.
5.	*trans* 1,2-dibromocyclopentane obtained from 1,2-cyclopentadiene and bromine is optically active.	It has a two-fold rotation axis only and is dissymmetric.
6.	3-methylene cyclohexene does not react byDiels–Alder reaction.	It has a cyclohexene ring involved in conjugation.
7.	3-phenylpropyne on hydroboration with BH_3, H_2O_2, OH^- gives $C_6H_5CH_2CH_2CHO$.	The reaction follows Markonikoff rule.
8.	Acetylide ion with tert-butyl bromide gives t-butyl acetylene.	Since this is an S_N2 reaction, t-butyl compounds lead to elimination products.
9.	The allene 2,3-pentadiene is chiral.	The molecule has rotational axis of symmetry only.
10.	1,3-cyclopentadiene reacts fast in Diels–Alder reaction.	The diene is in s-*cis* conformation.

(*Contd.*)

No.	Assertion	Reason
11.	2,3-dimethyl 1,3-butadiene reacts with maleic anhydride in Diels–Alder reaction but 2,3-di-tert.butyl 1,3-butadiene does not react with maleic anhydride.	The steric factor of t-butyl prevents the s-*cis* conformation to be formed.
12.	2-butyne by hydroboration oxidation gives one ketone but 2-pentyne gives two ketones.	2-pentyne is an unsymmetrical alkyne with different groups on both sides of triple-bonded carbon.
13.	When 2-butyne isomerizes to 1-butyne in KNH_2, NH_3 and water the intermediate formed when separated evolves CO_2 on ozonolysis.	Allene $CH_3CH{=}C{=}CH_2$ is formed which on ozonolysis gives off CO_2.
14.	Heat of hydrogenation of 1,4-pentadiene is –61 kcal/mol but that of 1,3-butadiene is –57 kcal/mol.	Conjugation reduces the heat of hydrogenation.
15.	When HCl is added to 1,3-butadiene at 78 degrees Celsius, 75% of 3-chloro 1-butene is formed.	This is due to higher stability of allyl cation relative to a primary carbocation.

KEY

1. a	2. d	3. b	4. a	5. a	6. b	7. c
8. d	9. a	10. a	11. a	1 2. a	13. a	14. a
15. a						

V. Comprehension Passages

Passage I

Alkynes are prepared by some of these procedures. Carbide hydrolysis gives alkynes if it is Ca or Na. Alkaline metal carbide of formula M_2C_3 gives by hydrolysis propyne. CHI_3 reacts with Ag to give acetylene. Alkylation of alkynes is done by treating acetylene with $NaNH_2$ followed by addition of primary alkyl iodide.

Answer the following questions

1. C≡CNa reacts with *t*-butyl iodide to give
   a) $CH{\equiv}C(C(CH_3)_3$ b) $CH{\equiv}CCH_3$
   c) acetylene, 2-methylpropene and I^- d) no reaction

2. Sodium acetylide reacts with $CH_3CH_2CH_2I$ to give A. A on ozonolysis gives
   a) CO_2, $CH_3CH_2CH_2COOH$ b) CH_3COOH, C_2H_5COOH
   c) CH_3COCH_3, CH_3COOH d) 2 moles of C_2H_5COOH
3. Mg_2C_3 on hydrolysis gives A. A on hydration gives
   a) CH_3COOH b) acetone
   c) $CH_3COCH_2CH_3$ d) $CH_3COCH_2CH_2CH_3$

KEY

1. c **2. a** **3. b**

Passage 2

Alkynes undergo hydroboration oxidation to give carbonyl compounds. Alkynes on hydration with Hg^{2+} and H_2SO_4 give carbonyl compounds.

Answer the following questions

1. 1-butyne on hydration with Hg^{2+} and H_2SO_4 gives
   a) $CH_3CH_2COCH_2CH_3$ b) $CH_3COCH_2CH_3$
   c) $CH_3CH_2COCH_2CH_2CH_3$ d) $CH_3COOH + CO_2$
2. Hydration in Hg^{2+} and H_2SO_4 gives aldehyde for
   a) acetylene b) propyne c) 1-butyne d) 2-butyne

KEY

1. b **2. a**

Passage 3

Dienes are of three types. Cumulated dienes of type $CH_2{=}C{=}CH_2$, conjugated dienes of type $CH_2{=}CH{=}CH{=}CH_2$ and isolated dienes of type $CH_2{=}CHCH_2CH{=}CH_2$. Conjugated dienes in s-*cis* form react only with dienophiles by Diels–Alder reaction.

Answer the following questions

1. One of the following dienes has central carbon sp hybridized
   a) $CH_2{=}CH{-}CH_2CH{=}CH_2$ b) $CH_3CH{=}C{=}CHCH_3$
   c) $CH_2{=}C{-}CH{=}CH_2$ d) $CH_2{=}CH{-}C(CH_3)_2{-}CH{=}CH_2$
2. The diene which does not undergo Diels Alder reaction is
   a) *cis* 1,3-butadiene b) 1,3-cyclopentadiene
   c) furan d) 3-methylenecyclohexene

3. $H_2C{=}CHCH_2CH_3$ with NBS followed by reaction with alc.KOH gives
   a) $CH_2{=}CH{-}CH{-}CH{=}CH_2$ b) $CH_2{=}CH{-}CH{-}CH_2{-}CH{=}CH_2$
   c) $CH_2{=}C{=}CH_2$ d) $CH_2{=}CH{-}CH_2{-}CH_2{-}CH{=}CH_2$

KEY

1. b 2. d 3. a

VI. Matrix Matching

Question 1

COLUMN 1	COLUMN 2
a) HCHO is one of the products of ozonolysis of	p) allene
b) sp-hybridization	q) $CH_2{=}CH{-}CH{=}CH_2$
c) 50% s character	r) $CH_2{=}CH{-}C{\equiv}CH$
d) Diels–Alder reaction	s) s-*cis* 1,3-butadiene + maleic anhydride
e) hydroboration oxidation of 2-butyne	t) $CH_3CH_2COCH_3$

Answer

$[a \rightarrow p,q,r]\ [b \rightarrow p,r]\ [c \rightarrow p,r]\ [d \rightarrow s]\ [e \rightarrow t]$

Question 2

COLUMN 1	COLUMN 2
a) Precipitate with ammoniacal $AgNO_3$	p) acetylene
b) CaC_2 on hydrolysis	q) propyne
c) Mg_2C_3 on hydrolysis	r) 1-butyne
d) Hydration product is acetaldehyde	s) *cis* 1,3-butadiene
e) Diels–Alder reaction	

Answer

$[a \rightarrow p,q,r]\ [b \rightarrow p]\ [c \rightarrow q]\ [d \rightarrow p]\ [e \rightarrow s]$

Question 3

Column 1	Column 2
a) CHI_3 + Ag	p) CH_3CH_2CHO
b) propyne by hydroboration oxidation	q) CH_3COCH_3
c) propyne with Hg^{2+} and sulphuric acid	r) acetylene
d) 2-butyne on ozonolysis	s) CH_3COOH

Answer

$[a \rightarrow r]\ [b \rightarrow p]\ [c \rightarrow q]\ [d \rightarrow s]$

VII. Structure-Based and Numerical Problems

1. An unsaturated compound with formula C_6H_{10} (A) decolorizes $KMnO_4$, does not form precipitate with ammoniacal $AgNO_3$ and takes up 2 moles of hydrogen to give a hydrocarbon with formula C_6H_{14}(B). On bromination it gives a single bromide in high yield which gives rise to a tertiary radical. The compound contains 5 inequivalent hydrogens. B can be obtained from $[(CH_3)_2CH]_2CuLi$ and $CH_3CH_2CH_2I$ in ether. Find A and B. What are the ozonolysis products of A?

   Answer

   A is $(CH_3)_2CHC{\equiv}CCH_3$

   B is 2-methylpentane ozonolysis products of A, $(CH_3)_2CHCOOH$ and CH_3COOH.

2. An alkyne A with formula C_8H_{12} is optically active and on complete reduction gives 4-methylheptane. On partial reduction with Lindlar Pd it gives a compound B which is a diene. Find the ozonolysis products of the diene in reducing conditions. Is the diene optically active. Is it a stereoisomer? Find the ozonolysis products of A in Zn/HCl.

   Answer

   The alkyne is

   A is $CH_3CH{=}CH{-}C(H)(CH_3){-}C{\equiv}CCH_3$; B is $CH_3C(H){=}C(H){-}C(H)(CH_3)C(H){=}C(H){-}CH_3$

   It is (Z,Z) 4-methyl-2,5-heptadiene and is a geometric isomer with no optical activity.

3. A, B and C are three hydrocarbons. C gives 4 monosubstituted chlorides and has molecular weight 86 g. C is the hydrogenation product of B. B is the partial hydrogenation product of A. A and B are optically active. Find A, B and C. What are the ozonolysis products of A and B in Zn/HCl? What is the hydration product of A in Hg^{2+}/H_2SO_4 or by hydroboration oxidation?

Answer

C has molecular weight 86 g and has 4 monochlorinated products

$$\text{C is } CH_3CH_2\underset{\substack{|\\ CH_3}}{\overset{\substack{H\\ |}}{C}}CH_2CH_3;$$

$$\text{B is } CH_2{=}CH{-}\underset{\substack{|\\ CH_3}}{\overset{\substack{H\\ |}}{C}}{-}CH_2CH_3;$$

$$\text{A is } CH_3{\equiv}C{-}\underset{\substack{|\\ CH_3}}{\overset{\substack{H\\ |}}{C}}{-}CH_2CH_3;$$

This is optically active.

The corresponding alkene B is also optically active.

$$CH_2{=}CH{-}\underset{\substack{|\\ CH_3}}{\overset{\substack{H\\ |}}{C}}{-}CH_2CH_3$$

A on ozonolysis gives CO_2 and $COOH\underset{\substack{|\\ CH_3}}{CH}CH_2CH_3$.

B on ozonolysis gives HCHO and $CHO\underset{\substack{|\\ CH_3}}{CH}CH_2CH_3$.

A on hydration in Hg^{2+} and H_2SO_4 gives $CH_3COCH(CH_3)CH_2CH_3$.

By hydroboration oxidation it gives $CHOCH_2CH(CH_3)CH_2CH_3$.

4. Three compounds A, B and C have formula C_5H_8. All decolorize bromine water and show positive test with dil. $KMnO_4$. A gives precipitate with ammoniacal silver nitrate. B and C do not give precipitate. Both A and B yield pentane on complete hydrogenation. But C absorbs one mole of hydrogen to give C_5H_{10}. B on oxidative cleavage with hot basic $KMnO_4$ gives after acidification CH_3COOH and CH_3CH_2COOH. C with ozone gives $OCHCH_2CH_2CH_2CHO$. Find A, B and C.

Answer

$CH{\equiv}CC_3H_7$ (A) is the terminal alkyne that gives precipitate with ammoniacal silver nitrate.

$CH_3C{\equiv}CC_2H_5(B) + O_2(KMnO_4) \longrightarrow CH_3COOH + CH_3CH_2CH_2COOH$

And B reduces completely on hydrogenation to give n-pentane.

C takes one mole of hydrogen only. Hence it is a cycloalkene and cycloalkenes ozonize to give dicarbonyl compounds.

Cyclopentene gives $OCHCH_2CH_2CH_2CHO$ on ozonolysis.

B has geometric isomers and C exists in one form only.

5. Compound A with formula C_8H_{12} in the presence of Hg^{2+} and dil. H_2SO_4 gives B. A when treated with $NaNH_2$, NH_3 and CH_3I gives C. C when treated with Lindlar Pd catalyst gives D. D on ozonolysis gives cyclohexyl aldehyde and CH_3CHO. B on Clemmenson reduction gives E. E produces 6 monosubstitution products on chlorination. It produces selectively the monobrominated product. B undergoes haloform reaction which leads to cyclohexane carboxylic acid (F). Compound A gives precipitate with ammoniacal silver nitrate. On ozonolysis A gives F and CHOCOOH. C on complete reduction gives an alkane (G) which has one more CH_2 group than E. Find A to G. How many monosubstituted chlorinated products are possible with G.

Answer

A is cyclohexyl—$C{\equiv}CH$ (A)

On hydration it gives cyclohexylmethylketone (B)

$A + NaNH_2(NH_3) + CH_3I \rightarrow$ cyclohexyl—$CH{\equiv}CCH_3$ (C)

C on reduction by Lindlar Pd gives cyclohexyl —$CH{=}CHCH_3$(*cis*) (D)

B undergoes haloform reaction to give cyclohexane carboxylic acid (F). On Clemmenson reduction, the ketone gives cyclohexylethane (E). This has one tertiary hydrogen to give monobrominated product and has 6 different hydrogens to give 6 monochlorinated products.

C on complete reduction gives cyclohexyl propane (G). This gives 7 monochlorinated products.

6. A has the formula C_6H_{10}. It gives precipitate with solution containing $Ag(NH_3)_2OH$. On catalytic hydrogenation it yields B (C_8H_{14}). B is optically inactive. Find A and B.

Answer

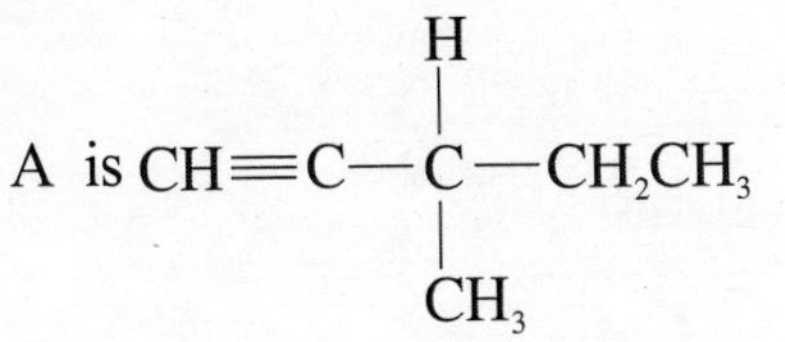

A is $CH{\equiv}C{-}\overset{\overset{\large H}{|}}{\underset{\underset{\large CH_3}{|}}{C}}{-}CH_2CH_3$

On complete hydrogenation of A we get

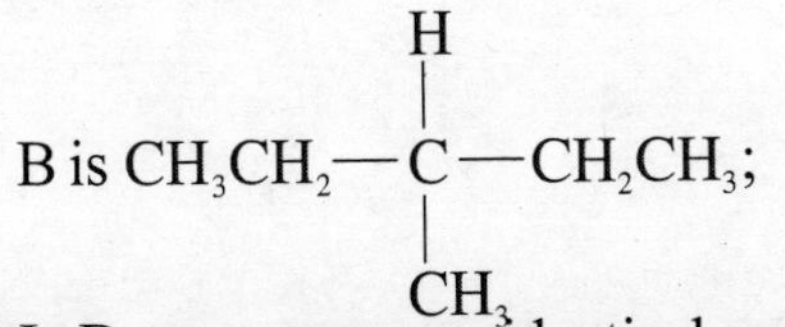

B is $CH_3CH_2{-}\overset{\overset{\large H}{|}}{\underset{\underset{\large CH_3}{|}}{C}}{-}CH_2CH_3$;

In B, two groups are identical and there is no optical activity.

7. Calcium carbide hydrolyses to give A. A reacts with $NaNH_2$, NH_3 to give B. C is produced from n-butyl alcohol by addition of HBr in acid. C reacts with B to give D. A and D give precipitate with solution containing $Ag(NH_3)_2OH$. D on partial hydrogenation with Lindlar Pd gives E. E gives on ozonolysis F and G in reducing conditions.

Answer

A is $CaC_2 + H_2O \longrightarrow CH{=}CH$

B is $CH{\equiv}CH + NaNH_2(NH_3) \longrightarrow CH{\equiv}CNa$

D is $CH{\equiv}CNa + CH_3CH_2CH_2CH_2Br\,(C) \longrightarrow HC{\equiv}CCH_2CH_2CH_2CH_3$

E is $D + \text{Lindlar Pd} \longrightarrow CH_2{=}CH(CH_2)_3CH_3$

$CH_2{=}CH(CH_2)_3CH_3 + O_3(Zn/HCl) \longrightarrow HCHO(F) + CHO(CH_2)_3CH_3(G)$

8. A bottle in a laboratory is labelled as "C6-alkyne mixture A and B". A chemist separates this mixture by careful distillation. When she reacts A with Li/NH_3, C is obtained. When A reacts with $HgSO_4$ and aq. H_2SO_4, two ketones D ($C_6H_{12}O$) and E ($C_6H_{12}O$) are obtained. When B reacts with Li/NH_3 it gives F. F with m-chloroperbenzoic acid gives G ($C_6H_{12}O$) which when exposed to aq. acid gives meso-diol H ($C_6H_{14}O_2$). A and B by complete hydrogenation give C_6H_{14} which gives 3 monosubstitution products on chlorination. Find A to H.

Answer

C is A+Li/NH_3 ⟶ trans alkene

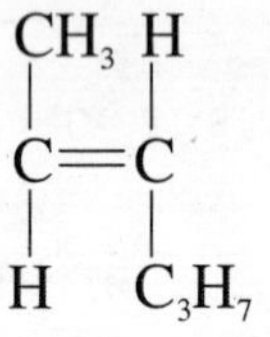

$$\begin{matrix} CH_3 & H \\ | & | \\ C & = C \\ | & | \\ H & C_3H_7 \end{matrix}$$

B is $CH_3CH_2C{\equiv}CCH_2CH_3$

B + Li/(NH_3) ⟶

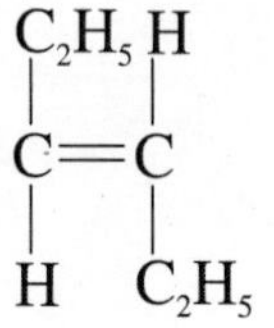

$$\begin{matrix} C_2H_5 & H \\ | & | \\ C & = C \\ | & | \\ H & C_2H_5 \end{matrix} \qquad (F)$$

F + mClperbenzoic acid (epoxidation) produces epoxide

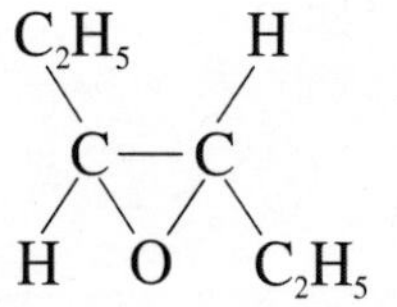

$$\begin{matrix} C_2H_5 & & H \\ & C - C & \\ H & O & C_2H_5 \end{matrix} \qquad (G)$$

$$\begin{matrix} & C_2H_5 & C_2H_5 & \\ H- & C - & C & -H \\ & | & | & \\ & HO & OH & \end{matrix} \qquad (H)$$

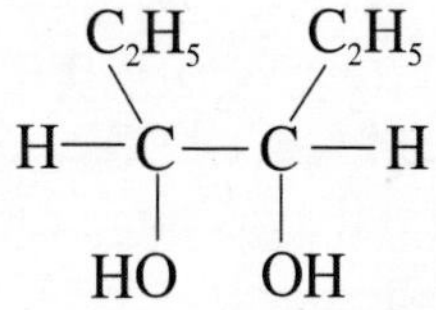

(Mesodive)

A or B on complete reduction give $CH_3CH_2CH_2CH_2CH_2CH_3$ which has 3 inequivalent hydrogens and gives 3 monochlorinated products on chlorination.

9. Compound A easily undergoes solvolysis to give B. B dehydrates easily to give C. Acetylide ion reacts with A to give acetylene, halide ion and C. D, an isomer of A, reacts with acetylide ion in ammonia to give E which on ozonolysis gives CO_2, H_2O and butanoic acid. D is obtained as one of the monosubstitution products from F. Find A to F.

Answer

A is $(CH_3)_3CCl$.

$(CH_3)_3CCl + H_2O \longrightarrow (CH_3)_3C{-}OH$ (B)

$(CH_3)_3C{-}OH$ on dehydration gives

$$CH_3-\underset{\displaystyle CH_3}{\underset{|}{C}}=CH_2 \quad (C)$$

$$(CH_3)_3-C-Cl + HC\equiv\bar{C} \longrightarrow CH\equiv CH + CH_3-\underset{\displaystyle CH_3}{\underset{|}{C}}-CH_2 + Cl^-$$

$$\underset{(F)}{CH_3-CH_2-CH_2-CH_3} \xrightarrow{Cl_2} \underset{\displaystyle Cl}{\underset{|}{CH_2}}-CH_2-CH_2-CH_3 \quad (D)$$

$$HC\equiv C^- + \underset{\displaystyle Cl}{\underset{|}{CH_2}}-CH_2-CH_2-CH_3 \longrightarrow HC\equiv C-CH_2CH_2CH_2CH_3 \quad (E)$$

$$\downarrow O_3$$

$$CO_2 + H_2O + CH_3CH_2CH_2COOH$$

5

AROMATIC HYDROCARBONS, BENZENE REACTIONS AND ELECTROPHILIC AROMATIC SUBSTITUTION

In the early days of organic chemistry, the word *aromatic* was used to describe fragrant substances such as benzaldehyde (from cherries, peaches and almonds), toluene (from tolubalsam), benzene (from coal distillate), benzoic acid and benzyl alcohol (from gum benzoin). But it was soon recognized that these substances differed in their chemical behaviour from most of the other organic compounds. Nowadays we use the word *aromatic* to refer to benzene and its structural relatives. In addition to benzene, toluene, benzaldehyde, the steroidal hormone estrone, the analgesic morphine and drugs such as valium (tranquilizer) are aromatic.

SOURCES OF AROMATIC HYDROCARBONS

The study of aromatic compounds started with the discovery of a hydrocarbon by Michael Faraday in 1825. This was called bicarburet of hydrogen (now called benzene).

In 1834, the German chemist Mitscherlich synthesized benzene by heating benzoic acid with calcium oxide. From vapour density measurements he showed that this compound had the molecular formula C_6H_6. Simple aromatic compounds come from two sources: coal and petroleum. Coal is an enormously complex mixture made up primarily of large arrays of benzenelike rings linked together. When heated to 1273 K in the absence of air, thermal breakdown occurs and a mixture of volatile products called coal tar boils off. This on fractional distillation gives benzene, toluene, xylene and naphthalene, indene and a host of other aromatic compounds. Petroleum unlike coal contains few aromatic compounds and has mainly alkanes. During petroleum refining however aromatic molecules are formed, when alkanes are passed over a catalyst at about 773 K under high pressure. For example, heptane is converted into toluene by dehydrogenation and cyclization.

STRUCTURE OF BENZENE

Even though benzene is unsaturated it does not undergo reactions characteristic of alkenes. And when treated with bromine it reacts very slowly to only one monobromo derivative, the bromobenzene. Fe is useful as a catalyst. The monobromo compound further reacts to give three disubstituted isomers. Actually it is possible to write 4 disubstituted products (two 1,2-dibromobenzenes) but one 1,3-dibromobenzene and 1,4-dibromobenzene. Kekule showed that the two 1,2-dibromobenzenes are the same and they oscillate between each other. He said that they couldn't be separated. Other structures were also proposed for benzene. The Dewar structures are given in Figure 5.1.

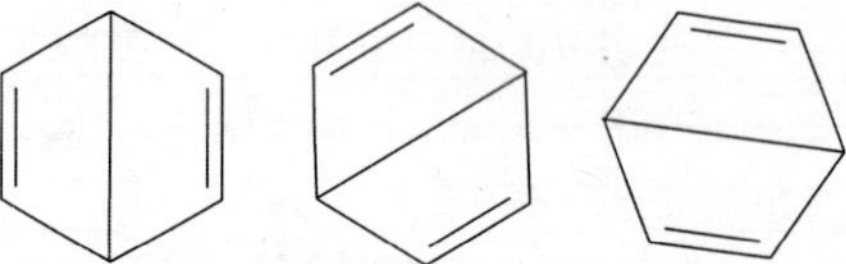

Figure 5.1 Dewar structure

Dewar benzene or **bicyclo[2.2.0]hexa-2,5-diene** is the bicyclic isomer of benzene with the molecular formula C_6H_6. The compound is named after James Dewar who, it is said, (incorrectly) proposed this structure for benzene in 1867.

The compound itself was first synthesized in 1962 as a *tert*-butyl derivative and then as the unsubstituted compound in 1963 by photolysis of *cis* 1,2-dihydro derivative of phthalic anhydride followed by oxidation with lead tetraacetate. Unlike benzene, Dewar benzene is not flat; the two cyclobutene rings make an angle. The compound has nevertheless considerable strain energy and reverts back to benzene with a chemical half-life of two days. This thermal conversion is relatively slow because it is symmetry-forbidden based on orbital symmetry arguments.

The three Dewar structures along with the two Kekulé structures are the main resonance structures in the valence bond theory description of benzene.

It was also thought that benzene might be represented as cyclohexatriene (Figure 5.2).

Figure 5.2 Cyclohexatriene structure

But such a structure should be highly reactive, and so did not account for the unreactive nature of benzene. Other structures that were proposed at the time were Armstrong's centroid structure and Ladenburg's prismane, which are given in Figure 5.3.

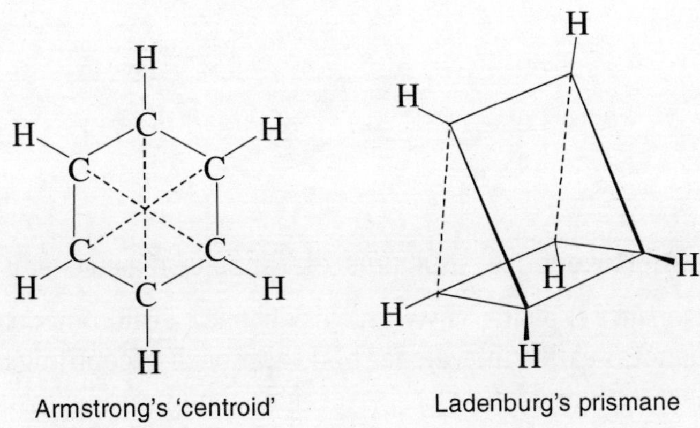

Figure 5.3 Armstrong and Ladenberg structures

Two important questions arose

1. Why is benzene not reactive like cycloalkenes?

   Cycloalkene reacts with $KMnO_4$ to give open-chain dicarboxylic acid. It reacts with hydrogen ions in acid to give cyclohexanol. It reacts with HCl to give chlorocyclohexane. Benzene does not react with any of these reagents.

2. Why does it give substitution product and not addition product?

Stability of Benzene

The heat of hydrogenation data for benzene indicate its stability. When we compare the heat of hydrogenation of benzene with those of cyclohexene and 1,3-cyclohexadiene, we have the following results.

Reactant	Product	ΔH (hydrogenation) kJ/mol
Cyclohexene	Cyclohexane	–118
1,3-cyclohexadiene	Cyclohexane	–230
Benzene	Cyclohexane	–206

If benzene were 1, 3, 5 cyclohexatriene, the ΔH (hydrogenation) must be around –354 kJ/ mole. But the actual value is –206 kJ/mol and it is stabilized by about 150 kJ. This energy is called the stabilization energy or resonance energy or delocalization energy of benzene. We have already seen resonance structures for allyl ion, 1,3-butadiene and other compounds. Here benzene has 2 resonance structures which give rise to this resonance energy. The double bonds in benzene are conjugated. And both structures are equivalent. Thus the double bond exists in all the bonds [1, 2], [2, 3], [3, 4], [4, 5], [5, 6] and [6, 1]. Another evidence for the delocalized structure is the bond length. All C—C bonds have same bond length equal to 1.39° intermediate between single bond length (1.54 Å) and double bond length (1.34 Å). The two resonance structures of benzene are shown in Figure 5.4.

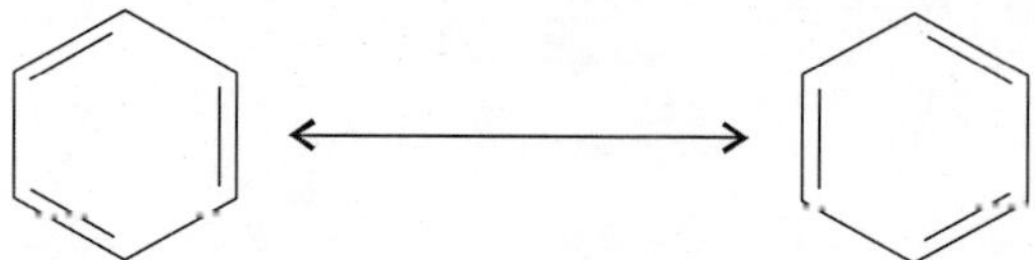

Figure 5.4 Resonance structures of benzene (Kekule structures)

The concept of resonance explains why we have only one 1,2-dibromoderivative for benzene. The concept of resonance is called the valence bond approach. According to molecular orbital theory, the six p orbitals are equivalent. The six p orbitals combine to form six molecular orbitals and the six p electrons are delocalized in the benzene ring. The orbitals are divided into bonding and antibonding orbitals. The three bonding orbitals have the six p electrons and the energy of the six p electrons (delocalized) is lower than the total energy of the three ethylene molecules. This energy is called delocalization energy (150 kJ/mole) or 36 kcal/mole. Thus we get the following facts about the structure of benzene:

1. It is a cyclic conjugated molecule.
2. It is unusually stable with a heat of hydrogenation of 150 kJ/mole less negative than we might expect for cyclic triene.
3. It has two Kekule resonance structures and all C—C bond lengths are same, equal to 1.39Å .
4. It is planar and has hexagonal structure with all bond angles equal to 120°.
5. It undergoes substitution reactions and no electrophilic addition reactions, which will destroy conjugation.

Hückel (4n+2) Rule

According to Eric Hückel a molecule is said to be aromatic if it is planar, with monocyclic ring with conjugation and a total of 4n + 2 pi electrons where *n* takes up any integer like 0,1,2, etc. Molecules with 2,6,10,14 ... pi electrons are aromatic. Molecules with 4,8,12,16 ... pi electrons are antiaromatic since delocalization of their sigma electrons leads to increase in energy. The rule applies to heterocyclic ring systems also.

Thus benzene (6p electrons), pyridine (6 p electrons without lone pair of nitrogen) and pyrrole (6 p electrons) including the lone pair of nitrogen are aromatic. The rule works for anions or cations with total p electrons (4n + 2). Thus cyclopentadiene anion, cylopropenyl cation, cycloheptatrienyl cation (tropylium ion) are all aromatic. The following compounds, cyclobutadiene (4p and electrons) cyclooctatetraene (8 p electrons) are not aromatic. Cyclodecapentene is not aromatic even though it has 10 pi electrons as shown in Figure 5.5.

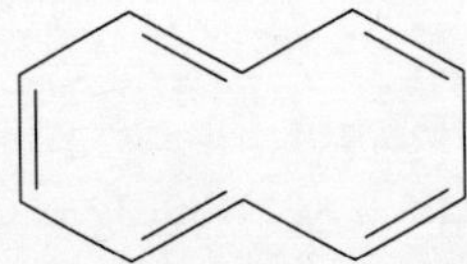

Figure 5.5 Non-aromatic cyclodecapentene

It is not flat and not planar. Hence it is not aromatic.

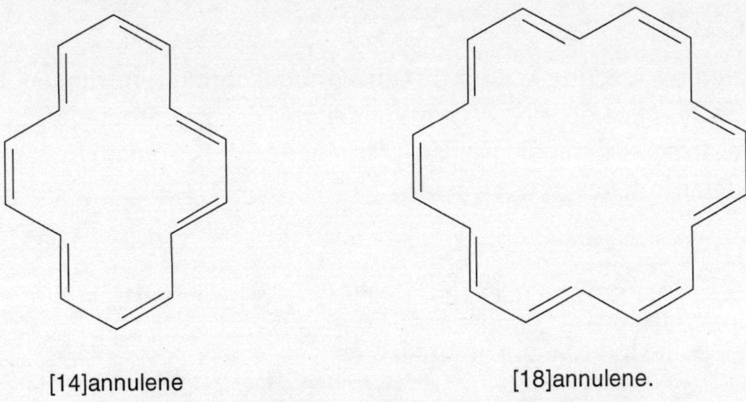

Figure 5.6 [14]annulene and [18]annulene

Only [14] and [18] annulenes are aromatic (Figure 5.6). Others are not aromatic. The [10] annulenes are not aromatic since they are not planar. Annulenes are those monocyclic compounds having alternate single and double bonds.

SOME CONSEQUENCES OF AROMATICITY

1. Cyclopentadiene has $pK_a = 16$ but 1,4-pentadiene has $pK_a = 45$. When cyclopentadiene loses a proton, it is stabilized as anion obeying Hückel 4n+2 rule. No such stability is possible for anion of 1,4-pentadiene.
2. Pyridine is basic but pyrrole is not basic. In pyridine the lone pair is not involved in the sextet formation according to Hückel rule, so lone pair is available for protonation. But in pyrrole the lone pair of electrons are also involved in sextet formation.
3. Generally *keto* forms are more stable than *enols*. But in case of phenol, the *enol* form is stable because of the six electrons in benzene ring.
4. Pyridine is a better base than aniline since the lone pair of electrons of pyridine are not delocalized like those of aniline in the benzene ring.

ELECTROPHILIC AROMATIC SUBSTITUTION REACTIONS

In electrophilic aromatic substitution reaction the electrophile E^+ reacts with the aromatic ring and substitutes for one of the hydrogens by choosing proper reagents. We can halogenate or nitrate or sulphonate or alkylate or acylate the ring. In all these reactions arenium ion is formed in the first step.

In the second step, a proton is removed and the electrophile attaches.

Nitration Reaction

Benzene is nitrated by reacting with a mixture of concentrated nitric and sulphuric acids. The electrophile in this reaction is NO_2^+ (nitronium ion). The diagrammatic scheme (Figure 5.7) illustrates the mechanism of nitration reaction. Nitric acid accepts a proton from sulphuric acid and acts as a base; sulphuric acid acts as an acid.

The first step is as follows:

$$2H_2SO_4 + HNO_3 \longleftrightarrow NO_2^+ + 2HSO_4^- + H_3O^+$$

The various steps are indicated in Figure 5.8

$$\text{C}_6\text{H}_6 + HNO_3 \xrightarrow[\text{heat}]{H_2SO_4} \text{C}_6\text{H}_5NO_2 + H_2O$$

Figure 5.7 Nitration of Benzene

Step 1:

An acid/base reaction. Protonation of the hydroxy group of the nitric acid. This provides a better leaving group.

Step 2:

Loss of the leaving group, a water molecule provides the nitronium ion, the reactive electrophile.

Step 3:

The electrophilic nitronium ion reacts with the nucleophilic C=C of the arene. This is the rate-determining step as it destroys the aromaticity of the arene.

Step 4:

Water functions as a base to remove the proton from the sp^3 carbon bearing the nitro group and reforms the C=C and the aromatic system.

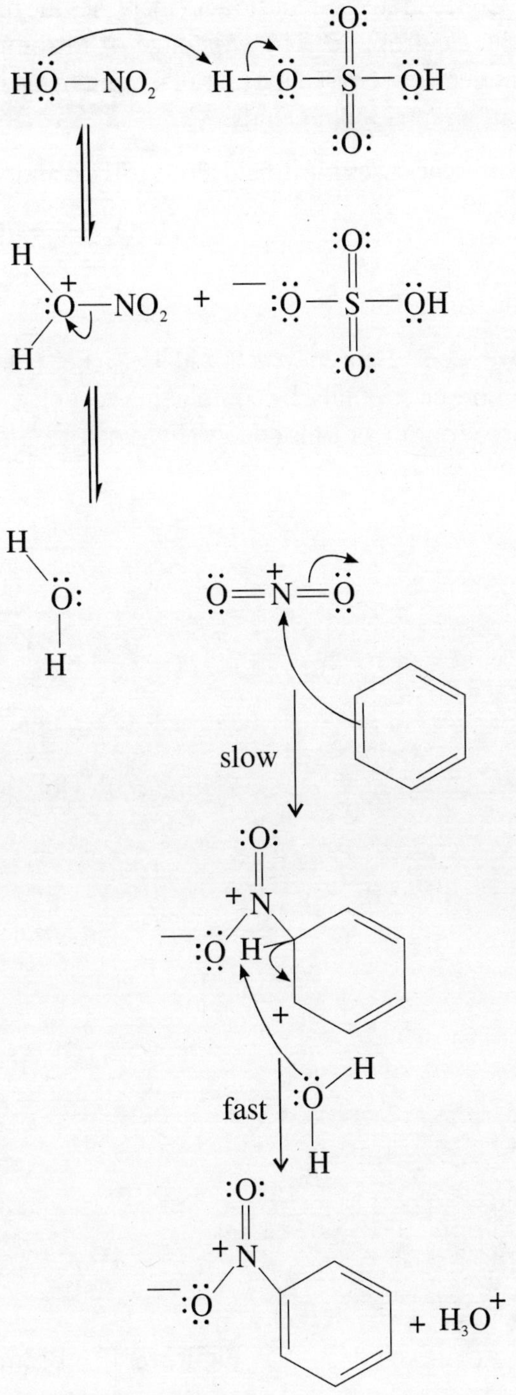

$HO-NO_2$ H—O—S—OH

$H_2O^+-NO_2$ + $^-O-SO_2-OH$

H—O—H O=N$^+$=O

slow

fast

$+ H_3O^+$

Figure 5.8 Mechanism for nitration of benzene

Concentrated sulphuric acid increases the rate of the reaction by increasing the concentration of the electrophile. The existence of nitronium ion has been demonstrated by cryoscopic measurements. Crystalline salts of nitronium ion like nitronium tetrafluoroborate in organic solvents nitrate aromatic compounds.

In some cases nitric acid alone is used and the first step is

$$2HNO_3 \longleftrightarrow NO_2^+ + NO_3^- + H_2O$$

Halogenation

Bromination Benzene reacts with Br_2 in Fe* (catalyst) to yield bromobenzene. The catalyst makes Br_2 more electrophilic by complexing aq. $FeBr_4^-Br^+$ which reacts as if it is Br^+. The Br^+ attacks the benzene ring to form the doubly allylic non-aromatic carbocation intermediate as given in Figure 5.9.

$$\overset{\delta-}{Br}-\overset{\delta+}{Br} \xrightarrow{Fe\ Br_3} Br_3\overset{\delta-}{Fe}\text{---}Br\text{---}\overset{\delta+}{Br}$$

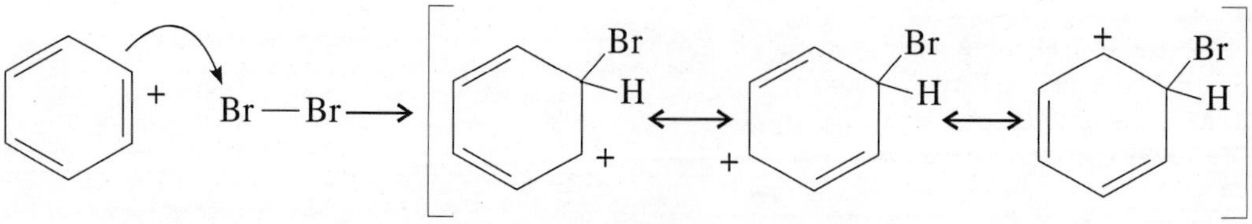

Figure 5.9 Halogenation of benzene

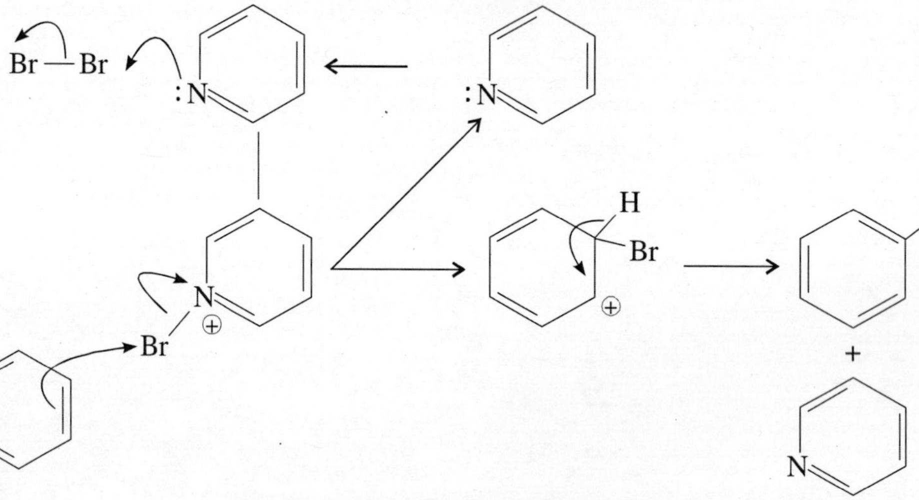

Figure 5.10 Bromination of benzene

* Br_2 and Fe react in the mixture (*in situ*) to from $FeBr_3$ which complexes with benzene. $FeBr_3$ or $FeCl_3$ should not be used since they react with moisture in air and inactive the catalyst.

In the next step Br^- does not add but a base present in solution abstracts H^+ to yield the brominated product. If addition of bromine occurs, 150 kJ /mol stabilization energy would be lost for the aromatic ring and the reaction will be endothermic. But substitution reaction leads to retention of stability of aromatic ring and is exothermic.

Another way to produce a more electrophilic source of bromine is to use pyridine, which forms pyridine–Br positively charged complex as shown in Figure 5.10. Pyridine is regenerated.

Fluorination Fluorine is highly reactive and so poor yields of monofluorinated products are obtained. Aromatic fluorides can be produced by thallation of aromatic ring according to

$$ArH + TI\ (OCOCF_3)_3 \longrightarrow Ar—TI(OCOCF_3)_2 \longrightarrow (KF,BF) \longrightarrow ArF$$

Another method is to nitrate benzene to nitrobenzene and reduce by Sn / HCl to aniline. Then aniline is treated with $NaNO_2 + H_2SO_4$ to diazotize. The diazonium salt can be treated with HBF_4 and on heating gives fluoro compound. The scheme is

$$\text{Benzene} + HNO_3 + H_2SO_4 \longrightarrow \text{Nitrobenzene} + Sn / HCl \longrightarrow \text{Aniline} + NaNO_2 + H_2SO_4$$

$$\downarrow$$

$$C_6H_5F + N_2 + H_2SO_4 \xleftarrow{HBF_4} C_6H_5\overset{+}{N_2}HSO_4$$

Chlorination Chlorination is done by adding Fe to chlorine to form $FeCl_3$ which forms complex with chlorine. Chlorination is similar to bromination in mechanism.

Iodination Iodination is done by using promoters like H_2O_2 and copper salts like $CuCl_2$. The promoters oxidize iodine to more electrophilic species according to the equation

$$I_2 + 2Cu^{2+} \rightarrow 2I + 2Cu^+$$

I^+ attacks the benzene ring and the cation formed is similar to that shown in Figure 5.9 (replace Br by I).

The diazonium salt $ArN_2^+\ HSO_4^-$ can be made to react with KI to give iodobenzene. (Diazonium salt will be discussed in detail in amines chapter.)

The diazonium ion $C_6H_5N_2^+$ reacts with HBF_4 to give C_6H_5F, BF_3 and N_2.

Iodination can be carried out with ICl in the presence of $ZnCl_2$ (Lewis acid) or acetyl hypoiodite and trifluoroacetyl hypoiodite.

Iodination can be carried out using I_2 in the presence of nitric acid.

Sulphonation

Aromatic rings can be sulphonated by reaction with fuming sulphuric acid (mixture of sulphuric acid + SO_3) the reactive electrophile is HSO_3^+ or neutral SO_3 depending on the reaction conditions.

Sulphonation is reversible. It is favoured in strong acid. But desulphonation is favoured in hot dilute aqueous acid. If we pass steam through the sulphuric acid and benzene, the high concentration of water shifts the equilibrium towards the left and desulphonation takes place. The mechanism is given in Figure 5.11 and the reaction is given in Figure 5.12.

Step 1:
The p electrons of the aromatic C═C act as a nucleophile, attacking the electrophilic S, pushing charge out onto an electronegative O atom. This destroys the aromaticity giving the cyclohexadienyl cation intermediate.

slow

Step 2:
Loss of the proton from the sp3 C bearing the sulphonyl group reforms the C═C and the aromatic system.

fast

Step 3:
Protonation of the conjugate base of the sulphonic acid by sulphuric acid produces the sulphonic acid.

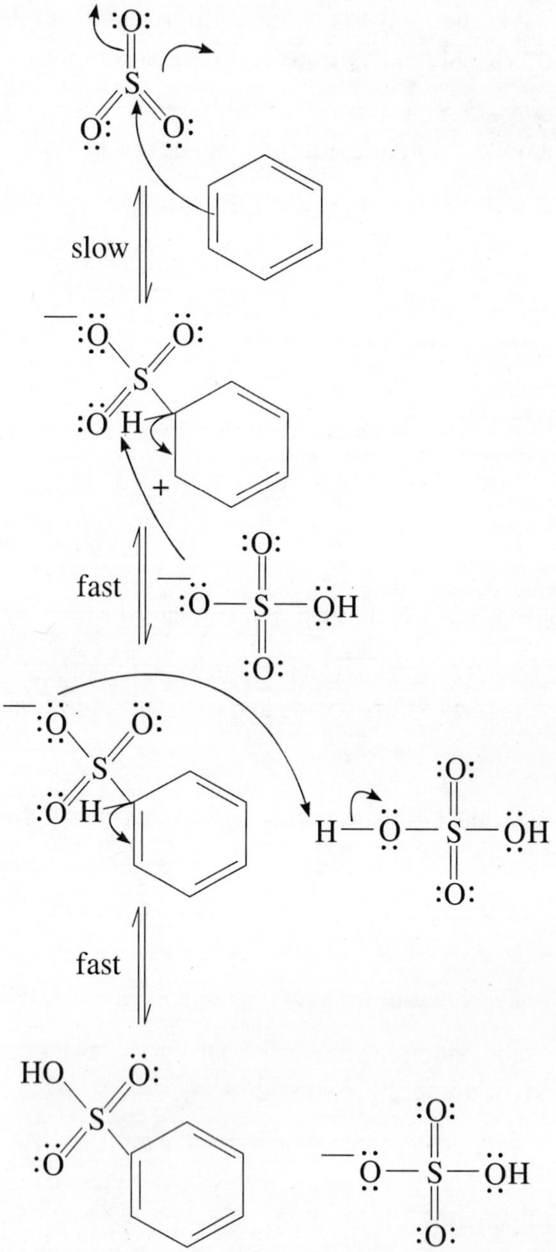

fast

Sulphuric acid acts as a base as well as an acid.

Figure 5.11 Mechanism for sulphonation of benzene

$$C_6H_6 + H_2SO_4 \underset{\text{heat}}{\rightleftharpoons} C_6H_5SO_3H + H_2O$$

Figure 5.12 Sulphonation of benzene

Points to remember

- Overall transformation: Ar—H to Ar—SO_3H a sulphonic acid.
- Reagent : for benzene, H_2SO_4 / heat or SO_3 / H_2SO_4 / heat (= fuming sulphuric acid).
- Electrophilic species : SO_3 which can be formed by the loss of water from the sulphuric acid.
- Unlike the other electrophilic aromatic substitution reactions, sulphonation is reversible.
- Removal of water from the system favours the formation of the sulphonation product.
- Heating sulphonic acid with aqueous sulphuric acid can result in the reverse reaction, desulphonation.
- Sulphonation with fuming sulphuric acid strongly favours formation of the product, the sulphonic acid.

Sulphonic acids are very strong and can be isolated from the reaction mixture by adding excess NaCl so that we get crystalline sodium benzene sulphonate.

Alkylation

Friedel–Crafts Alkylation In this reaction, the electrophile is a carbocation. For this reaction only alkyl halides (fluorides, chlorides, bromides or iodides) can be used. Aryl and vinyl halides cannot be used. Aryl cations and vinylic carbocations have very high energy. The reaction does not happen if the benzene ring is substituted by strong electron-withdrawing groups. When C_6H_5R is made to react with alkyl halide in the presence of Lewis acid, $AlCl_3$, and if R is —NR_3^+, —NO_2, —CN, —SO_3H, —CHO, —$COCH_3$, —COOH, —$COOCH_3$, —NH_2, —NHR and —NR_2, no reaction happens. The mechanism of the Lewis acid-catalysed reaction is given in Figures 5.13 and 5.14.

The general reaction scheme is shown below. For alkylation

$$C_6H_6 \xrightarrow[\text{AlCl}_3\text{ catalyst reflux, anhydrous conditions}]{\text{RCl}} C_6H_5R$$

Figure 5.13 Friedel–Crafts alkylation

Friedel–Crafts alkylation involves the alkylation of an aromatic ring with an alkyl halide using a strong Lewis acid catalyst. With anhydrous ferric chloride as a catalyst, the alkyl group attaches at the former side of the chloride ion. The general mechanism is shown below.

$$R{-}Cl + FeCl_3 \longrightarrow R^+ + FeCl_4^-$$

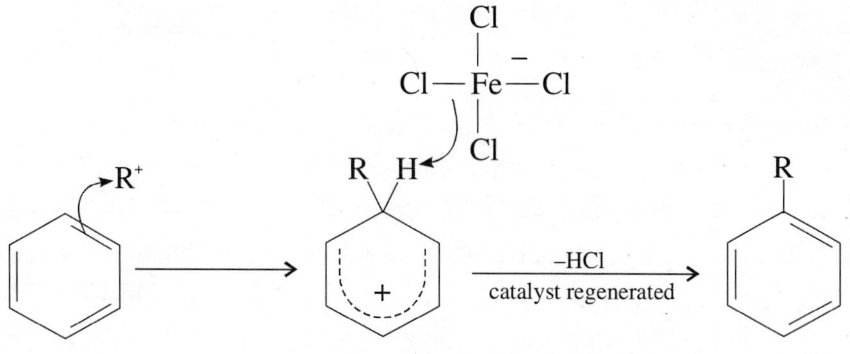

Figure 5.14 Mechanism of Friedel–Crafts alkylation

This reaction has one big disadvantage, namely that the product is more nucleophilic than the reactant due to the electron-donating alkyl chain. Therefore, another hydrogen is substituted with an alkyl chain, which leads to over-alkylation of the molecule. Also, if the chlorine is not on a tertiary carbon, carbocation rearrangement reaction will occur. This is due to the relative stability of the tertiary carbocation over the secondary and primary carbocations. The limitations of this reaction are listed below.

Limitations of Friedel–Crafts alkylation

1. One limitation is that skeletal rearrangements occur particularly when primary halides are used. Benzene + n-butyl chloride ($AlCl_3$ 0°C) gives *sec*-butylbenzene (65%) and n-butylbenzene (35%) (Rearrangement of carbocations).
2. Polyalkylations may occur due to increased activation by introduction of alkyl group.

Steric hindrance can be exploited to limit the number of alkylations, as in the *t*-butylation of 1,4-dimethoxybenzene (Figure 5.15).

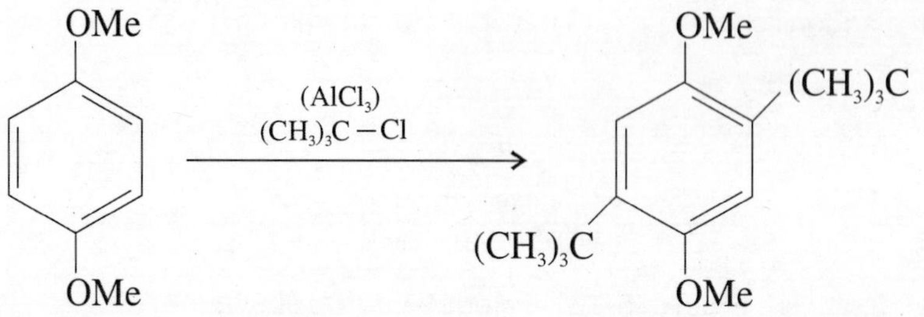

Figure 5.15 Steric hindrance and Friedel–Crafts alkylation

Alkylations are not limited to alkyl halides. Friedel–Crafts reactions are possible with any carbocationic intermediate such as those derived from alkenes and a protic acid,

Friedel–Crafts alkylation can be carried out using

1. alkene and HF or acid like H_3PO_4.
2. alcohol and BF_3 at 60°C.

Friedel–Crafts acylation Friedel–Crafts acylation (Figure 5.16) is the acylation of aromatic rings with an acyl chloride using a strong Lewis acid catalyst. Friedel–Crafts acylation is also possible with acid anhydrides. Reaction conditions are similar to the Friedel–Crafts alkylation mentioned above. This reaction has several advantages over the alkylation reaction. Due to the electron-withdrawing effect of the carbonyl group, the ketone product is always less reactive than the original molecule, so multiple acylations do not occur. Also, there are no carbocation rearrangements, as the carbonium ion is stabilized by the resonance structure in which the positive charge is on the oxygen.

Figure 5.16 Friedel–Crafts acylation

The viability of the Friedel–Crafts acylation depends on the stability of the acyl chloride reagent. Formyl chloride, for example, is too unstable to be isolated. Thus, synthesis of benzaldehyde via the Friedel–Crafts pathway requires that formyl chloride be synthesized in situ. This is accomplished via the Gattermann–Koch reaction, accomplished by reacting benzene with carbon monoxide and hydrogen chloride under high pressure, catalysed by a mixture of aluminium chloride and cuprous chloride.

In a simple mechanistic view (Figure 5.17), the first step consists of dissociation of a chlorine atom from acyl chloride to form an acyl cation. This is followed by nucleophilic attack of the arene towards the acyl group. Finally, a chlorine atom reacts to form HCl, and the $AlCl_3$ catalyst is regenerated.

Figure 5.17 Reaction mechanism for acylation (*Continues*)

O^{+} $[AlCl_4]$ C R

O C R + HCl + $AlCl_3$

Figure 5.17 Reaction mechanism for acylation

The acyl cation or acylium ion is stabilized by resonance as shown below.

$$R-\overset{+}{C}=\ddot{O}: \longleftrightarrow R-C\equiv\overset{+}{O}:$$
(A) (B)

In (B) all the atoms have octet of electron and hence this structure is stabilized.

BIRCH REDUCTION

Benzene can be reduced to 1,4-cyclohexadiene by treating it with an alkali metal (Na, Li or K) in liquid NH_3 and an alcohol. The mechanism for reduction is given in Figure 5.18.

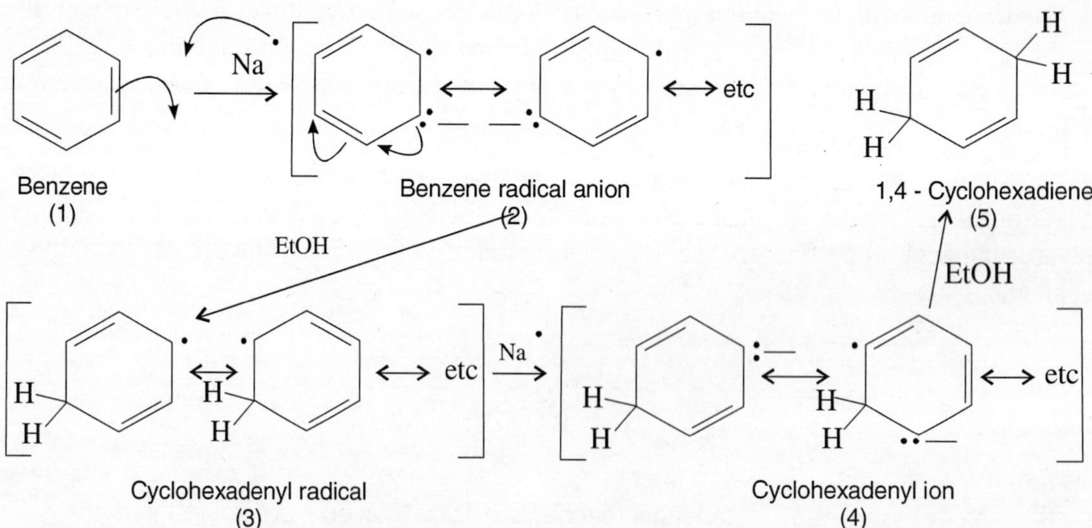

Figure 5.18 Birch reduction

EFFECT OF *ORTHO, META* AND *PARA* DIRECTING GROUPS IN MONOSUBSTITUTED BENZENES

Only one product can form when an electrophilic substitution occurs in benzene but what happens if we carry out a reaction on the aromatic ring which is already substituted? Substituents already present on the ring have two effects.

1. Substituents affect the reactivity of the aromatic ring. Some substituents activate the ring making it more reactive than benzene and some deactivate the benzene ring making it less reactive than benzene. For example in aromatic nitration, OH group makes the compound 1000 times more reactive than benzene while a NO_2 group deactivates the ring by 2×10^7 relative to benzene.
2. Substituent affects the orientation of the reaction. The three disubstituted products *ortho*, *meta* and *para* are usually not formed in equal amounts. Instead the nature of the substituent already present on the benzene ring determines the second substitution.

Substituents are classified as meta directing deactivators, ortho para directing deactivators and ortho para directing activators. No meta directing activators are known. All meta directing groups are deactivating while most of the ortho para directing groups are activating (except halogens). Table 5.1 gives these groups in decreasing order of activating and deactivating groups.

Table 5.1 Activating and deactivating groups

Ortho para directors	Meta directors
NH_2, NHR, NR_2, OH, -O	Strongly deactivating $—NO_2$, $—NR_3^+$, $—CF_3$, $—CCl_3$
Moderately activating $—NHCOCH_3$, NHCOR, $—OCH_3$, —OR	
Weakly activating $—CH_3$, $—C_2H_5$, —R, $—C_6H_5$	
Weakly deactivating —F, —Cl, —Br, —I	

THEORY OF ORIENTATION-INDUCTIVE AND RESONANCE EFFECTS

The inductive effect of a substituent 'Y' arises from the electrostatic interaction of the polarized bond to Y with the positive charge which is developed in the ring when it is attacked by an electrophile. If Y is a more electronegative atom or group than carbon, the ring will be the positive end of the dipole as shown in Figure 5.19.

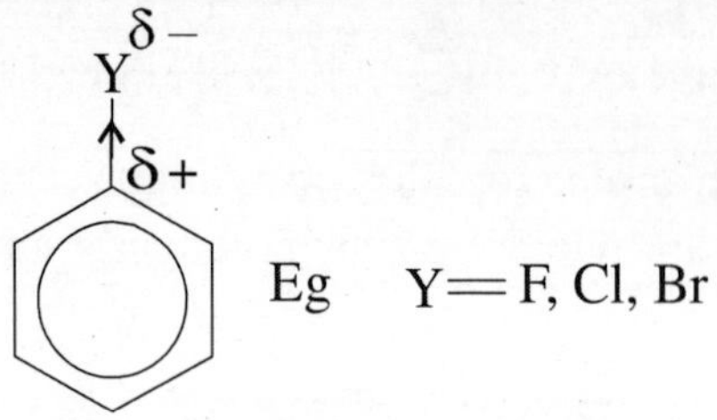

Eg Y = F, Cl, Br

Figure 5.19 Electrostatic interaction of electrophile in aromatic substitution

Attack by an electrophile will be retarded since this will lead to an additional full positive charge on the ring. The halogens are more electronegative than carbon and exert –I effect. The groups like —NR_3^+, NO_2, SO_2OH, CX_3, COOH, COOR have –I effect because the atom directly attached to the ring bears a full or partial positive charge (Figure 5.20).

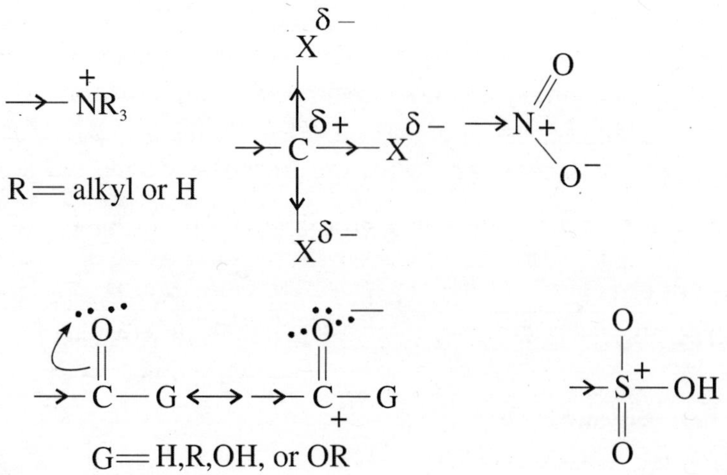

R = alkyl or H

G = H,R,OH, or OR

Figure 5.20 –I effects of –NR_3^+, NO_2, SO_2OH, CX_3, COOH, COOR

The resonance effect of Y refers to the possibility that the presence of Y may increase or decrease the resonance stabilization of the intermediate arenium ion. The Y substituent may cause one of the three contributors to resonance hybrid for arenium ion to better or worse than when Y is a hydrogen atom. If Y bears lone pair of electrons (one or more) this may lead to extra stability of arenium ion by giving one more resonance contributor with positive charge on Y.

The electron-donating resonance effect applies with decreasing strength in the order shown below:

$$—NH_2, > —NR_2, > —OH, > —OR > —X$$

X is a halogen, which is the least electron-donating group.

This order also refers to the activating ability of the groups. The halogens seem to act peculiarly. This is because these atoms are larger than carbon and hence orbitals containing non-bonding electron pairs are away from nucleus and do not overlap well with 2p orbital of carbon.

META DIRECTING GROUPS

All meta directing groups have either a partial or full positive charge on atom directly attached to the ring. Let us consider trifluoromethyl group. Due to the strongly electronegative fluorines, CF_3 group is strongly electron-withdrawing. It is strongly deactivating and a powerful *meta* director. The trifluoromethyl group affects reactivity by causing transition state leading to arenium ion highly unstable. This is due to the withdrawal of electrons from the developing carbocation leading to enhanced positive charge on the ring (Figure 5.21).

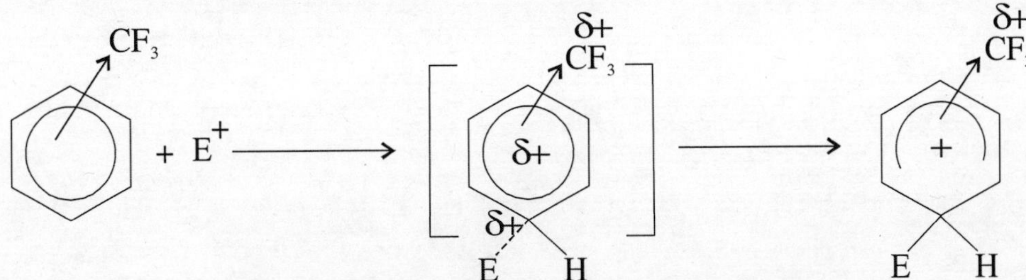

Figure 5.21 Electrophile reacting with $C_6H_5CF_3$

The arenium ion formed by ortho or para attack has a contributing structure very unstable since the positive charge is located on the ring carbon bearing the electronegative substituent. Such structures are not available for meta attack. Thus arenium ion by meta attack is most stable. Thus nitration with nitration mixture (sulphuric and nitric acids) gives 100% *meta*-nitro(trifluoromethyl)benzene.

We have to remember that *meta* substitution is the least unfavourable among the three unfavourable pathways. This means that substitution is faster at *meta* relative to *ortho* or *para* positions but it is slower relative to substitution in benzene.

Other groups acting like trifluoromethyl group are nitro, carboxyl, etc.

ORTHO-PARA DIRECTING GROUPS

Except for alkyl and phenyl substituents other o-p directing groups belong to systems having lone pair of electrons. This feature namely, an unshared pair of electrons adjacent to the ring determines orientation and influences reactivity. The main factor responsible is electron-releasing resonance effect.

Let us consider aniline. When we brominate aniline, the two ortho position and one para position get brominated to yield 1,3,5-tribromoaniline. Thus NH_2 is highly activating. There is a small electron-withdrawing inductive effect. But resonance effect predominates. The inductive effect is not much because electronegativity difference between carbon and nitrogen is not much since carbon is sp^2-hybridized. We can understand the resonance effect by drawing structures for ortho, para and meta attacks.

***Ortho* attack**

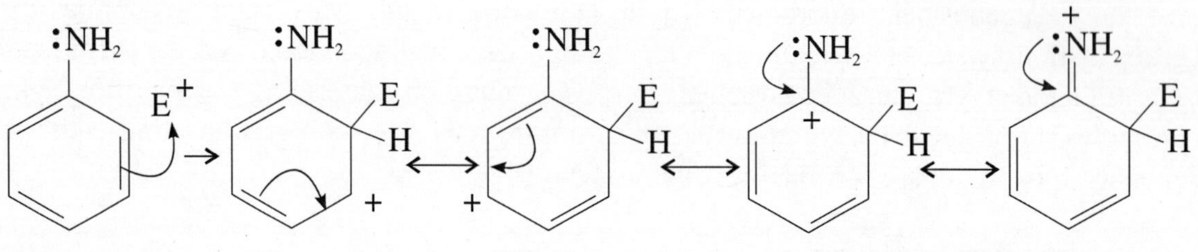

Relatively stable contributor

***Meta* attack**

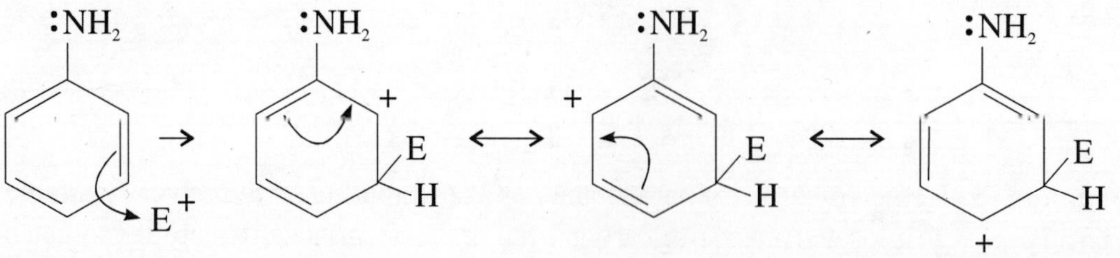

***Para* attack**

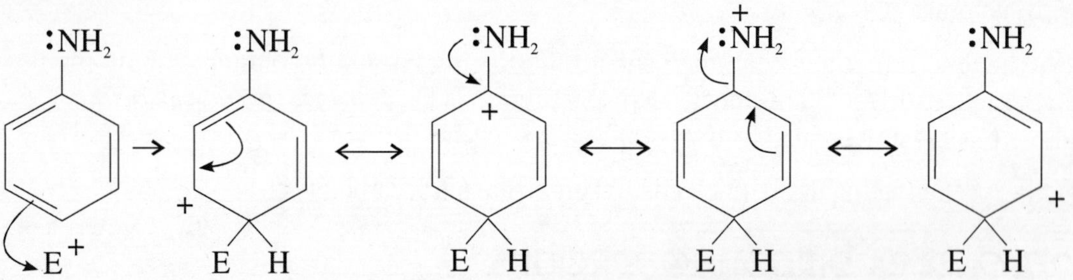

Relatively stable contributor

Figure 5.22 Electrophilic substitution in aniline

From Figure 5.22 we see that there are four resonance structures for *ortho* or *para* case, but only three for *meta*-substitution. Secondly the structures for *ortho* and *para* form an extra bond from nitrogen to carbon of ring. The extra bond also brings in an octet of electrons for all atoms in each structure. Thus these structures give high stability.

REACTIVITY OF HALOGENS

Let us consider chlorobenzene. By inductive effect, chlorine withdraws electron from ring and deactivates it. But when substitution occurs, the resonance contributors at *ortho* and *para* are stabilized by donation of electron pair by resonance. Oxygen is more elecronegative than halogens like chorine and bromine but OH is activating and halogens are deactivating. This is because resonance structures involving oxygen are due to overlap of 2p orbital of carbon with oxygen 2p orbital. But for chlorine, 3p orbital has to overlap with 2p of carbon which is less effective. Fluorobenzene is highly reactive in spite of its high electronegativity. This is because it overlaps well with carbon 2p. The two orbitals of carbon and fluorine are of similar size.

Side Chain Oxidation

Benzene is not easily oxidized by $KMnO_4$, but the aromatic compounds containing alkyl group attached to benzene ring easily undergo oxidation and the product is benzoic acid. The oxidation involves the formation of benzylic radicals and hence the aromatic compounds having benzylic hydrogen (toluene, n-butylbenzene, $C_6H_5CH_2Cl$, $C_6H_5CH_2OH$) all oxidize to benzoic acid by $KMnO_4$. If *t*-butylbenzene is reacted, there is no oxidation product since it has no benzylic hydrogen.

Oxidation of Benzene Ring

The benzene ring in an alkyl benzene can be converted to a carboxylic acid by ozonolysis followed by treatment with H_2O_2.

$$R—C_6H_5 \xrightarrow[(2)\ (H_2O_2)]{(1)\ O_3, CH_3COOH} R—\overset{\overset{\displaystyle O}{\|}}{C}—OH$$

The following are some examples of ortho, para and meta directing influences of substituents attached to aromatic ring leading to synthetic applications.

1. Conversion of toluene to *p*-nitrobenzoic acid
2. Conversion of toluene to *meta*-nitrobenzoic acid
3. Conversion of toluene to 1-bromo-2-trichloromethylbenzene
4. Conversion of aniline to *p*-nitroaniline
5. Nitration of *m*-bromochlorobenzene
6. *m*-chloroethylbenzene from benzene
7. Conversion of benzene to *m*-bromobenzoic acid
8. Conversion of benzene to 2-bromo-4-nitrotoluene

9. 2-chloro-4-methylphenol from benzene
10. Conversion of benzene to *p*-chlorostyrene

These conversions take place as follows:

1. Toluene $\xrightarrow{(HNO_3 + H_2SO_4)}$ *p*-nitroalkene $\xrightarrow{(KMnO_4, OH^-, \text{heat}, H_3O^+)}$ *p*-nitrobenzoic acid.

2. Toluene $\xrightarrow{(KMnO_4, OH^-, H_3O^+}$ benzoic acid $\xrightarrow{(HNO_3 + H_2SO_4)}$ *m*-nitrobenzoic acid.

3. Toluene $\xrightarrow{Br_2 + Fe}$ *o*-bromotoluene $\xrightarrow{Cl_2(hv)}$ 1-bromo-2- trichloromethylbenzene.

4. Aniline $\xrightarrow{CH_3COCl \text{ (reflux)}}$ $C_6H_5NHCOCH_3$ $\xrightarrow{(HNO_3 + H_2SO_4)}$ *o*-nitroacetanilide + *p*-nitroacetanilide

   p-nitroacetanilide $\xrightarrow{H_2O, H_2SO_4, \text{heat}}$ *p*-nitroaniline.

5. When *m*-bromochlorobenzene is nitrated, we get 62% 4-nitro-3-bromochlorobenzene + 37% 3-bromo-6-nitrochlorobenzene and 1% of 3-bromo-2-nitrochlorobenzene.

6. Benzene $\xrightarrow{(CH_3COCl + AlCl_3)}$ $C_6H_5COCH_3$ $\xrightarrow{Cl_2 + Fe}$ *m*-chloroacetophenone $\xrightarrow{Zn(Hg)\ HCl}$ *m.*-chloroethylbenzene.

7. Benzene $\xrightarrow{(CH_3Cl + AlCl_3)}$ toluene $\xrightarrow{(KMnO_4)}$ benzoic acid $\xrightarrow{Br_2 + Fe}$ *m*-bromobenzoic acid.

8. Benzene $\xrightarrow{(CH_3Cl + AlCl_3)}$ toluene $\xrightarrow{SO_3 + H_2SO_4}$ *p*-toluenesulphonic acid $\xrightarrow{NaOH(H_3O^+)}$ *p*-methylphenol

   p-methylphenol $\xrightarrow{Cl_2 + Fe}$ 2-chloro-4-methylphenol

9. Benzene $\xrightarrow{(CH_3Cl + AlCl_3)}$ toluene $\rightarrow SO_3 + H_2SO_4 \rightarrow$ *p*-toluenesulphonic acid $\xrightarrow{NaOH(H_3O^+)}$ *p*-methylphenol

   p-methylphenol $\xrightarrow{Cl_2 + Fe}$ 2-chloro-4-methylphenol

10. Benzene $\xrightarrow{(CH_3CH_2Cl + AlCl_3)}$ ethylbenzene $\xrightarrow{Cl_2 + Fe}$ *p*-chloroethylbenzene $\xrightarrow{}$ (*ortho* isomer) *p*-chloroethylbenzene

    p-chloroethylbenzene $\xrightarrow{NBS}$ *p*-chloro$C_6H_4CHBrCH_3$ $\xrightarrow{\text{Alc. KOH}}$ *p*-chloro C_6H_4—CH=CH_2

AROMATIC NUCLEOPHILIC SUBSTITUTION

Aryl halides do not undergo the conventional S_N1 and S_N2 reactions (Figure 5.23). They do not undergo S_N1 reactions because aryl cations are relatively unstable. The C—X bond of aryl and vinyl halides are shorter in length and stronger than benzylic and allylic halides. More energy is required for bond breaking. The short and strong C—X bonds arise because the carbon atom is sp^2-hybridized and electrons of carbon orbital are closer to nucleus than those in sp^3 hybridization. They do not undergo S_N2 reactions due to steric factor since the incoming nucleophile has to approach in the plane of carbon–carbon double bond to carry out a backside displacement.

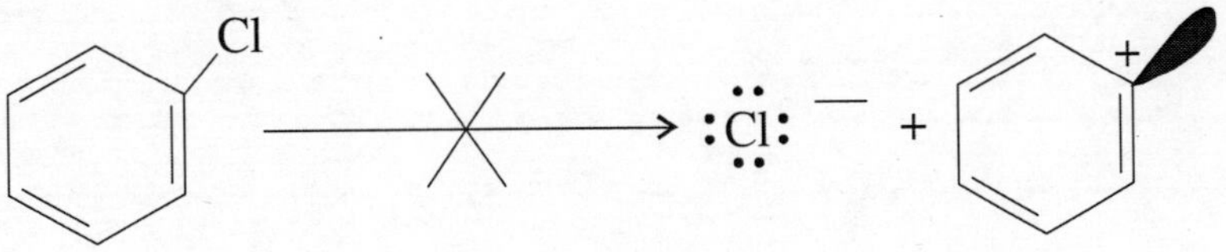

Figure 5.23 Chlorobenzene and S_N2 reaction

Aromatic nucleophilic substitution readily happens when the aryl carbon is bonded to halogen susceptible to nucleophilic attack. Nucleophilic substitution can occur if strong electron-withdrawing groups are present at *ortho* or *para* to halogen atom.

When *ortho*-nitrochlorobenzene reacts in the presence of aq. $NaHCO_3$ at 130°C in the presence of water, *o*-nitrophenol is produced. The presence of nitro groups in the aryl halide in the presence of $NaHCO_3$ needs slightly less temperature (109°C) to give phenol.

If three nitro groups are present and the temperature is 298 K, *m*-nitrochlorobenzene does not undergo this reaction.

Unsubstituted aryl halides in the presence of NaOH at 340°C at 2500 psi (pressure) followed by acidification gives phenol.

All these reactions follow addition, and elimination mechanism involving benzyne intermediate.

The mechanism is given in Figure 5.24.

In this mechanism Br^- is lost by 1,2-elimination to generate the highly reactive *benzyne* intermediate which traps nucleophilic species. This mechanism requires a strong base (for example, $NaNH_2$, BuLi) and is commonly used with haloaromatic substrates.

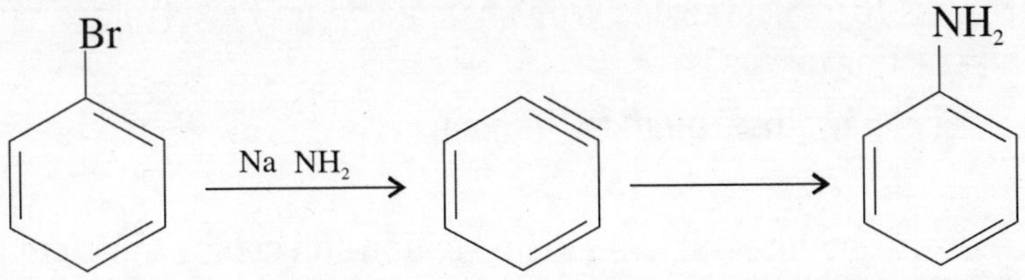

Figure 5.24 Benzyne intermediate in the conversion of bromobenzene to aniline

Evidence for the elimination mechanism comes from the observation that isotopically labelled substrates undergo scrambling of the label in the product, indicating the intermediacy of a symmetrical species. Furthermore 2,6-disubstituted halobenzenes will not undergo base-mediated reaction due to the requirement for an *ortho*-hydrogen.

diazotization; $-CO_2$, $-N_2$

benzyne

Furan

Diels–Alder adduct

Figure 5.25 Diels–Alder cyclization—trapping of benzyne.

Benzynes may be trapped with dienes, undergoing ***Diels–Alder Cycloaddition*** (Figure 5.25).

The benzyne triple bond has one p bond by p-p overlap and one pi bond formed by sp^2-sp^2 overlap. The latter pi bond is in the plane of benzene ring and is weak.

I. "Explain Why" Questions and Small Structure-Based Problems

1. Cyclooctatetraene readily reacts with potassium to form dianion. The salt has the structure $2K + C_8H_8^{2-}$. Explain.

   Answer

   The dianion has 10 p electrons and it is planar satisfying Hückel rule criteria and is aromatic. Hence this reaction occurs.

2. Azulene has a very large dipole moment. Explain.

   Answer

   Azulene consists of the cycloheptatrienyl cation and cyclopentadienyl anion which are both aromatic. Hence the dipole moment is large.

3. Cylopentadiene has pKa = 16 but 1,4-pentadiene has pKa=45. Why?

   Answer

   When cyclopentadiene loses a proton, it is stabilized as anion obeying Hückel 4n+2 rule. No such stability is possible for anion of 1,4-pentadiene.

4. Pyridine is basic but pyrrole is not basic. Explain.

   Answer

   In pyridine the lone pair is not involved in the sextet formation according to Hückel rule, so lone pair is available for protonation. But in pyrrole the lone pair of electrons are also involved in sextet formation.

5. Pyridine is a better base than aniline. Explain.

   Answer

   The lone pair electrons of pyridine are not delocalized like those of aniline in the benzene ring.

6. 3-chlorocyclopropene gives a precipitate with $AgBF_4$ and gives a stable positive ion. Explain.

   Answer

   In solution it exists as the cyclopropenium cation which is aromatic according to Hückel rule. AgCl is the precipitate.

7. Cycloprapanone is highly reactive but cyclopropenone is very stable.

   Answer

   The unsaturated ketone has positive charge on carbonyl carbon and we have cyclopropenium ion which is very stable.

8. Cycloheptatrienone is stable but cyclopentadienone easily reacts by Diels–Alder reaction. Explain.

   Answer

   Cycloheptatrienone has cycloheptatrienyl cation which is stable by Hückel rule, but if carbon of C=O has positive charge in cyclopentadienone, the cyclopentadienyl cation is unstable.

9. 5-chloro-1,3-cyclopentadiene (Figure Ex 5.1) undergoes S_N1 solvolysis very slowly even though it is doubly allylic. Explain.

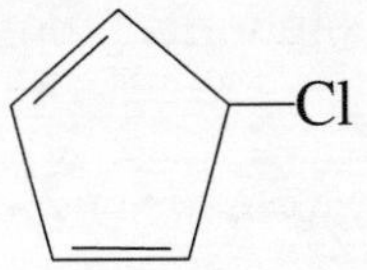

–Cl

Figure Ex.5.1 5-chloro-1,3-cyclopentadiene

Answer

The cation formed is cyclopentadienyl which is unstable according to Hückel rule. Hence the solvolysis is not fast like allylic ions.

10. Friedel–Crafts acylation followed by Clemmenson reduction is a better procedure than Friedel–Crafts alkylation to produce alkylated benzenes. Explain.

Answer

Friedel–Crafts alkylation generates many carbocations and they rearrange, and hence a single product is not possible.

II. Multiple Choice Questions (One out of four answers is correct)

1. The heat of hydrogenation for benzene is numerically ________ lower relative to that of 1,3-cyclohexadiene.
   a) 24 kJ b) 24 kcal c) 48 kJ d) 48 kcal
2. One of the following is aromatic according to Hückel rule
   a) cyclopentadiene b) cyclopentadienyl anion
   c) cyclopentadienyl radical d) cyclopentadienyl cation
3. One of the following system is aromatic according to Hückel rule
   a) cyclopentadiene b) cyclooctatetraene
   c) cyclobutadiene d) pyrrole
4. One of the following is antiaromatic
   a) benzene b) furan
   c) cyclopentadienyl anion d) [4]annulene
5. One of the following system is aromatic according to Hückel's rule.
   a) pentalene b) heptalene
   c) cyclopentadiene d) cycloheptatrienyl cation
6. In the nitration of benzene with mixture of concentrated nitric and sulphuric acids, one of the following statements is not true.
   a) Sulphuric acid acts as a base.
   b) Nitric acid acts as the base.
   c) The reaction is electrophilic substitution reaction.
   d) NO_2^+ is the electrophile.
7. The existence of nitronium ion in nitration of benzene can be demonstrated by
   a) mass spectrometry b) cryoscopy
   c) steam distillation d) no suitable answer

8. Friedel–Crafts alkylation happens without rearrangement for one of the following halides
   a) n-propyl halide b) $(CH_3)_3CCH_2Cl$
   c) $CH_3CH_2CH_2CH_2Cl$ d) C_2H_5Cl
9. Friedel Crafts alkylation is not possible for one of the following halides
   a) $CH_3CH_2CH_2CH_2Cl$ b) *t*-butyl chloride c) allyl chloride d) chlorobenzene
10. The synthetically useful alkyl halide for Friedel Crafts alkylation is
    a) n-butyl chloride b) n-propyl chloride c) vinyl chloride d) *t*-butyl chloride
11. The Friedel Crafts reaction cannot be carried out with
    a) n-butyl chloride + $AlCl_3$ b) ethylene + HF
    c) n-propyl alcohol + sulphuric acid d) chlorobenzene + $AlCl_3$
12. One of the following substrates can be used in Friedel–Crafts alkylation
    a) aniline b) nitrobenzene c) cyanobenzene d) toluene
13. The Friedel Crafts alkylation of neopentyl chloride with benzene in $AlCl_3$ gives
    a) neopentyl benzene b) *t*-butyl benzene
    c) n-pentyl benzene d) $C_6H_5-\overset{\displaystyle CH_3}{\underset{\displaystyle C_2H_5}{C}}-CH_2$
14. One of the following alkyl halides does not undergo rearrangement in Friedel–Crafts alkylation of benzene.
    a) *n*-butyl chloride b) vinyl bromide c) *n*-propyl chloride d) ethyl chloride
15. When *m*-nitroacetophenone is reduced with H_2, Pd/C in ethanol product is
    a) *m*-ethylnitrobenzene b) *m*-acetyl aminobenzene
    c) *m*-ethylaniline d) no reaction
16. In halogenation of benzene by Fe and Br_2, $FeBr_3$ formed acts as a
    a) catalyst b) electrophile c) Lewis Base d) nucleophile
17. The compound which does not undergo Friedel–Crafts acylation is
    a) toluene b) benzene c) pyridine d) nitrobenzene
18. Benzene does not undergo addition reaction because
    a) there is no localized double bond b) pi electrons of benzene are delocalized
    c) resonance stabilization energy is high d) contribution polar resonance hybrids is poor
19. Chlorobenzene is converted to phenol by NaOH under pressure and the mechanism involves
    a) carbocation b) transition state c) benzyne intermediate d) free radicals

20. Iodobenzene is prepared by reaction of
    a) benzene + I_2 + Fe
    b) benzene diazonium chloride with KI
    c) benzene + KI
    d) iodine at high temperature
21. One of the following cannot act as a base according to Hückel rule
    a) aniline
    b) pyridine
    c) pyrrole
    d) methylamine
22. *m*-chloroethylbenzene is prepared from benzene as follows:
    a) Chlorinate benzene and do Friedel–Crafts alkylation using ethyl chloride in $AlCl_3$.
    b) Alkylate with ethyl chloride in $AlCl_3$ followed by chlorination.
    c) Friedel–Craft's acylation is done and chlorinated and finally reduced by Clemmenson reduction.
23. Aniline is converted to *o*-nitroaniline with high percentage by
    a) nitrating with HNO_3
    b) converting to acetanilide followed by nitration, and hydrolysis
    c) converting to acetanilide, sulphonating and nitrating and then heating with water, sulphuric acid and treating with alkali
    d) no suitable method
24. When *m*-bromochlorobenzene is nitrated, the product is
    a) 5-nitro-3-chlorobromobenzene (100%)
    b) 4-nitro-3-chlorobromobenzene (100%)
    c) 4-nitro-3-chlorobromobenzene, 2-nitro-3-chlorobromobenzene and 5-nitro-3-chlorobromobenzene
    d) no suitable answer
25. The catalyst used in Birch reduction of benzene is
    a) Pd/H_2
    b) Ni/H_2
    c) Na/NH_3 in ethanol
    d) $NaBH_4$
26. One of the following statements is not true
    a) Aryl halides do not undergo S_N1, S_N2 reactions.
    b) The COOH group is *meta* directing.
    c) In Friedel–Crafts acylation the important species is acylium ion.
    d) Benzyl chloride does not undergo nucleophilic substitution.
27. The ozonolysis of Birch reduction product of A gives CH_3COCHO, CHOCHO and $CH_3COCOCH_3$. A is
    a) *p*-xylene
    b) *o*-xylene
    c) *m*-xylene
    d) toluene

28. The catalyst used in nitration of benzene is
    a) nitric acid b) nitronium ion c) sulphuric acid d) no suitable answer
29. One of the following is not used as catalyst in Friedel–Crafts reaction
    a) $AlCl_3$ b) BF_3 c) HF d) Fe
30. Nitrobenzene is used as—Friedel–Crafts alkylations.
    a) catalyst b) electrophile c) solvent d) no suitable answer

KEY

1. a	2. b	3. d	4. d	5. d	6. a	7. b
8. d	9. d	10. d	11. d	12. d	13. d	14. d
15. c	16. a	17. d	18. c	19. c	20. b	21. c
22. c	23. c	24. c	25. c	26. d	27. b	28. c
29. d	30. c					

III. Screening Test Questions of Previous IIT-JEE Question Papers

1. Identify the correct order of reactivity of the following towards electrophilic aromatic substitution
   I. Benzene II. toluene III. chlorobenzene IV. nitrobenzene
   a) I>II>III>IV b) II>I>III>IV c) IV>III>II>I d) II>III>I>IV (2002)
2. Toluene when treated with Fe / Br_2 gives p-bromobenzene as major product because the CH_3 group
   a) is *para* directing b) is *meta* directing
   c) activates the ring by hyperconjugation d) deactivates the ring (1999)
3. An aromatic molecule will
   a) have $4n\pi$ electrons b) $4n+2\,\pi$ electrons
   c) be planar d) be cyclic (1999)
4. Benzyl chloride is prepared from toluene by chlorination with
   a) SO_2Cl_2 b) $SOCl_2$ c) Cl_2 d) NaOCl (1998)
5. In the reaction of *p*-chlorotoluene with KNH_2 in liq. NH_3 the major product is
   a) *o*-toluidine b) *m*-toluidine and *p*-toluidine
   c) *p*-toluidine d) *p*-chloroaniline (July 1997)
6. Nitrobenzene can be prepared from benzene by using a mixture of HNO_3 and H_2SO_4. In the nitrating mixture HNO_3, acts as a
   a) base b) reducing agent c) acid d) catalyst (May)

7. Among the following statements of nitration of aromatic compounds the false one is
   a) The rate of nitration of benzene is same as that of hexadeuterobenzene.
   b) The rate of nitration of toluene is greater than that of benzene.
   c) The rate of nitration of benzene is greater than that of hexadeuterobenzene.
   d) Nitration is an electrophilic substitution reaction.
8. Arrange the following in order of increasing dipole moment:
   I. Toluene II. *m*-dichlorobenzene III. *o*-dichlorobenzene IV. *p*-dichlorobenzene
   a) I<IV<II<III b) IV<I<II<III c) IV<I<III<II d) IV<II<I<III (1996)
9. What is the decreasing order of reactivity among the following towards electrophilic aromatic substitution Chlorobenzene (I) benzene (II) anilium chloride (III) toluene (IV)
   a) II>I>III>IV b) III>I>II>IV c) IV>II>I>III d) I>II>III>IV (1995)
10. Bond dissociation energy needed to form benzyl radical is ________ than that required for forming methyl radical from methane.

KEY

1. c	2. a,c	3. b,c,d	4. a,c	5. b	6. a	7. c
8. b	9. c	10. less				

IV. Multiple Choice Questions (More than one answer is correct)

1. Some of the following are the consequences of Hückel rule
   a) Pyrrole is not a base but pyridine is basic.
   b) Cyclopentadiene has pKa = 16 but 1,4-pentadiene has pKa = 45.
   c) Generally *keto* forms are more stable than *enols*. But in case of phenol, the *enol* form is stable because of the six electrons in the benzene ring.
   d) Aniline is more basic than methyl amine.
2. Alkylation of benzene is possible by
   a) RCl with $AlCl_3$ added to C_6H_6 when R is any group other than aryl or vinyl group.
   b) Benzene + C_2H_5OH in BF_3
   c) Benzene + ethylene + HF
   d) Benzene + nitrobenzene
3. Aldehyde is the product when
   a) acetylene is treated with Hg^{2+}/H_2SO_4
   b) propyne is treated with H_2O_2 and diborane
   c) 1-pentyne is involved in hydroboration oxidation
   d) propyne + Hg^{2+}/H_2SO_4

4. The % of p character is 50 in
   a) acetylene b) central carbon in allene
   c) central carbon in propylene d) central carbon in propane
5. Alkylation of acetylene takes place when we use
   a) $NaNH_2$ followed by RI where R is primary to acetylene
   b) $NaNH_2 + C_6H_5CH_2Cl$ + acetylene
   c) $NaOH + C_6H_5CH_2I$ + acetylene
   d) $NaNH_2$ + tertiary halide + acetylene
6. Ethylbenzene is prepared from benzene by
   a) adding ethylene + HF b) C_2H_5Cl and $AlCl_3$
   c) ethanol + benzene + BF_3 d) ethylnitrite + benzene
7. Benzene can be converted to a single alkylated derivative in the case of Friedel–Crafts reaction with
   a) C_2H_5Cl b) *t*-butyl chloride c) *n*-propyl chloride d) *n*-butyl iodide
8. Benzene cannot be converted to *m*-chloroethylbenzene by
   a) adding $C_2H_5I + AlCl_3$ followed by Cl_2 / Fe addition
   b) converting benzene to nitrobenzene, then reduction to aniline, diazotization and adding Cl_2 / Fe. Then do Friedel–Crafts alkylation in $AlCl_3$.
   c) benzene + $CH_3COCl + AlCl_3$ followed by Cl_2 / Fe followed by Clemmenson reduction.
   d) benzene + $C_2H_5COCl + AlCl_3$, then Cl_2 / Fe followed by Clemmenson reduction
9. Nitrobenzene is reduced to aniline by
   a) Sn / HCl b) Fe / HCl c) $NaBH_4$ d) Ni and H_2
10. Ozonolysis of the following compounds in reducing conditions gives CH_3COCHO as one of the products
    a) *o*-xylene b) 1,4-dimethylbenzene
    c) benzene d) Birch reduction product of benzene
11. Friedel–Crafts alkylation does lead to rearranged products in case of
    a) ethyl chloride addition b) *t*-butyl chloride addition
    c) n-propyl chloride addition d) n-butyl chloride addition
12. Some of the following observations are true
    a) Deuterium isotope effect is not observed for electrophilic aromatic substitution.
    b) *m*-dinitrobenzene is reduced to *m*-aminonitrobenzene using NH_3 and H_2S.
    c) Friedel–Crafts acylation is possible using HCOCl.
    d) Lewis acids like HF, BF_3 are used in Friedel–Crafts reaction.

13. 2 moles of HCHO are formed in ozonolysis in reducing conditions of
    a) ethylene b) 1,3-butadiene
    c) $CH_2{=}CH{-}CH_2{-}CH{=}CH_2$ d) 2-butene
14. Partial reduction by Lindlar Pd or Na/NH_3 gives same product for
    a) 2-butyne b) 2-pentyne c) propyne d) acetylene
15. The hydration of some of the following alkynes produces more than one ketone as product.
    a) acetylene b) propyne c) 2-pentyne d) 2-hexyne
16. The products of ozonolysis of some of these substrates in reducing conditions undergo Cannizarro reaction.
    a) ethylene b) styrene c) 2-butene d) 3-hexene
17. The compounds which undergo oxidation by $KMnO_4$ to benzoic acid are
    a) benzaldehyde b) benzyl chloride c) *n*-butyl benzene d) *t*-butyl benzene
18. The product of some of the following reactions is benzoic acid.
    a) benzaldehyde + peracetic acid
    b) $C_6H_5COCH_3 + NaOH + I_2$ followed by acidification
    c) benzaldehyde + Tollen's reagent
    d) $C_6H_5COC_6H_5 + NaOH + I_2$
19. Among the following, the compounds that are optically active are
    a) 1,3-dichloroallene b) allene
    c) 1-chloro-3-bromoallene d) butadiene
20. Styrene is produced from some of the following reactions
    a) phenyl acetylene with Lindlar Pd b) Ethyl benzene with ZnO on heating
    c) Benzene + ethylene + $AlCl_3$ d) Benzene + n-propyl chloride + $AlCl_3$

KEY

1. a,b,c	2. a,b,c	3. a,b,c	4. a,b	5. a,b	6. a,b,c	7. a,b
8. a,b,d	9. a,b	10. a,b	11. a,b	12. a,b,d	13. a,b,c	14. c,d
15. c,d	16. a,b	17. a,b,c	18. a,b,c	19. a,c	20. a,b	

V. ASSERTION–REASON QUESTIONS

The questions 1–16 consist of an *assertion* in column 1 and *reason* in column 2. Use the following key to choose an appropriate answer.

a) If both assertion and reason are correct and reason is the correct explanation of assertion
b) If both assertion and reason are correct but reason is not the correct explanation of assertion
c) If assertion is correct but reason is incorrect
d) If assertion is incorrect but reason is correct

No.	**Assertion**	**Reason**
1.	Cyclodecapentane is not aromatic even though it has 10 pi electrons.	It is not planar.
2.	Many hydrocarbons have pKa>45; cyclopentadiene has pKa=16.	It is acidic because the removal of proton le: to cyclopentadienyl anion which obeys Huc rule.
3.	Cyclopropanone is very reactive but methylcyclopropenone is stable.	In methyl cyclopropenone CO group be polar one of the resonance structures has a + charge on carbon, the cation has 2π electr and obeys Hückel rule.
4.	Normally allylic cations undergo S_N1 solvolysis easily but 5-chloro 1,3-cyclopentadiene which is doubly allylic undergoes S_N1 solvolysis very slowly.	The cation formed is cyclopentadienyl cat and is not stable according to Hückel rule.
5.	The first step in electrophilic aromatic substitution is same as that of addition of alkenes but aromatic substitution is slower than alkene addition.	The intermediate benzenium ion is less sta than benzene. Its formation has high entha of activation. Loss of aromaticity energetically more unfavourable than the l of pi bond.
6.	There is no kinetic isotope effect observed in case of electrophilic aromatic substitution.	This shows that C—H bond breaking is not rate-determining step.
7.	The nitration of perdeuterated benzene proceeds at different rate relative to that of benzene.	The C—H or C—D bond breaking is not rate-determining step.
8.	If iodination is carried out in ICl for benzene, the electrophile is Cl^+.	In ICl, Cl is more electronegative and system polarizes as I^+Cl^-.

(*Contd.*)

No.	Assertion	Reason
9.	Friedal–Crafts acylation followed by Clemmenson reduction is a better procedure than Friedel–Crafts alkylation to prepare alkylated benzenes.	Friedel–Crafts alkylation gives rise to rearranged products resulting in the formation of stabler carbocations whereas acylation does not produce rearranged products.
10.	In Friedel–Crafts alkylation, multiple alkylation happens but acylation stops after first stage.	Alkyl groups are activating but acyl groups deactivate the benzene ring.
11.	$R—C^+{=}O^-$ is less stable than the $R—C{\equiv}O^+$ ion.	In acylium ion every atom has an octet of electrons. The ion is stabilized by interaction of vacant carbon orbital with lone pair orbital of oxygen.
12.	*p*-bromobenzoic acid cannot be prepared by the following steps. Toluene is oxidized and then bromination is done by Br_2 and $FeBr_3$.	COOH is *meta* directing and bromine will attach to the *meta* position.
13.	Toluene is oxidized to benzoic acid by $KMnO_4$ but t-butyl benzene cannot undergo this oxidation.	*t*-butyl has higher molecular weight.
14.	*p*-bromotoluene reacts with NaOH and water giving *m*-methylphenol and *p*-methylphenol whereas *m*-bromotoluene gives *o*-methyl, *m*-methyl and *p*-methylphenols.	The *m*-bromocompound has two benzyne intermediates.
15.	Arylhalides are unreactive towards S_N1 and S_N2 reactions.	Steric factor prevents S_N2 reactions. High stability of C—X bond in arylhalides prevents S_N1 reactions.
16.	The product of t-butylchloride, $AlCl_3$ and benzene is not oxidized by $KMnO_4$.	t-butylbenzene creates t-butyl cations.

KEY

1. a	2. a	3. a	4. a	5. a	6. a	7. d
8. d	9. a	10. b	11. a	12. a	13. b	14. a
15. a	16. b					

VI. Comprehension Passages

Passage 1

Hückel rule predicts aromaticity of compounds.

According To Eric Hückel a molecule is said to be aromatic if it is planar, with monocyclic ring with conjugation and a total of $4n + 2$ pi electrons where n takes up any integer like 0, 1, 2 etc. Molecules with 2, 6, 10, 14 … pi electrons are aromatic. Molecules with 4n pi electrons (4, 8, 12, 16) are antiaromatic since delocalization of their sigma electrons leads to increase in energy. The rule applies to heterocyclic ring systems also. Some molecules due to nonplanarity are antiaromatic. There are consequences of aromaticity namely basicity and acidity of compounds, reactivity by S_N1, stability of cyclic ketones etc.,

Answer the following questions

1. One of the following compounds is not aromatic.
   a) naphthalene b) azulene c) pentalene d) cyclopentadiene anion
2. One of the following statements is wrong
   a) Aniline is less basic than CH_3NH_2.
   b) Pyrrole is not basic but pyridine is basic.
   c) Cycloheptatrienone is very stable.
   d) Azulene has zero dipole moment.
3. The antiaromatic compound among the following is
   a) Cyclooctatetraene b) cyclobutadiene
   c) cycloheptatrienyl cation d) pyrrole

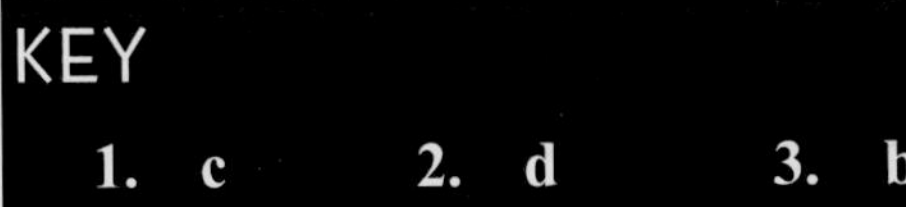

KEY

1. c **2.** d **3.** b

Passage 2

Benzene undergoes various reactions like nitration, halogenation, sulphonation and Friedel–Crafts reactions. Aromatic nucleophilic substitution happens by benzyne mechanism.

Answer the following questions

1. Friedel–Crafts alkylation is synthetically useful for one of the following alkyl halides
   a) *n*-butyl chloride b) *n*-propylchloride
   c) neopentylchloride d) ethylchloride

2. In sulphonation of benzene one of the following statements is not true.
   a) It is a reversible reaction.
   b) Sulphonation–desulphonation is useful from synthesis point of view.
   c) When benzene is sulphonated, the water is removed by forming azeotrope with benzene.
   d) Kinetic isotope effect is found in sulphonation of benzene.
3. One of the following statements about aromatic nucleophilic substitution is untrue.
   a) If an aryl halide contains two *ortho* substituents then nucleophilic substitution does not take place.
   b) *o*-dichlorobenzene with KNH_2 mainly forms *m*-chloroaniline
   c) *p*-dichlorobenzene with KNH_2 mainly gives *p*-chloroaniline
   d) *m*-dichlorobenzene with KNH_2 does not give any product

KEY

1. d **2. d** **3. d**

Passage 3

Substituents are classified as *meta* directing deactivators, *ortho, para* directing deactivators and *ortho-para* directing activators. No *meta* directing activators are known. All *meta* directing groups are deactivating. Most of the *ortho, para* directing groups are activating (except halogens). The following table gives in decreasing order of activating and deactivating groups.

Ortho, para directors	**Meta directors**
NH_2, NHR, NR_2, OH, —O	Strongly deactivating $—NO_2$, $—NR_3^+$, $—CF_3$, $—CCl_3$
Moderately activating $—NHCOCH_3$, NHCOR, $—OCH_3$, —OR	
Weakly activating $—CH_3$, $—C_2H_5$, —R, $—C_6H_5$	
Weakly deactivating —F, —Cl, —Br, —I	

Among the reactions of aromatic compounds, oxidation with $KMnO_4$ is important. All compounds having benzylic hydrogen on oxidation with $KMnO_4$ give benzoic acid.

Answer the following questions

1. One of the following compounds does not give benzoic acid on oxidation with $KMnO_4$
   a) benzyl alcohol
   b) benzaldehyde
   c) toluene
   d) $(C_2H_5)_3C—C_6H_5$
2. *m*-chloroethylbenzene is prepared by one of the following methods.
   a) Benzene is reacted with CH_3COCl in $AlCl_3$. Then it is chlorinated with Fe and Cl_2 and then Zn/HCl are added.
   b) Benzene is alkylated with C_2H_5Cl and then chlorinated.
   c) Benzene is first chlorinated and then alkylated by Friedel–Crafts reaction.
   d) Benzene is reacted with CH_3COCl in $AlCl_3$. Zn/HCl are added and then chlorination is done.
3. One of the following statements about electrophilic aromatic substitution is true.
   a) Kinetic isotope effect is there for electrophilic aromatic substitution.
   b) Nitrobenzene is a suitable solvent for Friedel–Crafts acylation and benzene is not.
   c) Nitration of benzene has NO_2^+ intermediate whose structure is angular.
   d) At 373 K, when PhMe is sulphonated, 79% of *o*-toluene sulphonic acid is formed.

KEY

1. d 2. a 3. b

VII. Matrix Matching

Question I

	Column 1		Column 2
a)	Huckel rule predicts aromaticity for	p)	14 annulene
b)	Birch reduction	q)	pyridine
c)	Synthetically useful alkyl halide for Friedel–Crafts reaction	r)	azulene
d)	Halogenation of benzene	s)	Na/NH_3 is the reducing agent
e)	Benzene to ethyl benzene in HF	t)	t-butyl chloride
		u)	$Fe+X_2$
		v)	ethylene

Answer

$[a \rightarrow p,q,r]\,[b \rightarrow s]\,[c \rightarrow t]\,[d \rightarrow u]\,[e \rightarrow v]$

Question 2

Column 1	Column 2
a) No Friedel–Crafts reaction for	p) aniline
b) Benzene to cumene by Friedel–Crafts reaction	q) nitrobenzene
c) R—C_6H_5 ozonolysis in CH_3COOH, H_2O_2	r) isopropylchloride
d) Pentalene	s) RCOOH
	t) not aromatic

Answer

$[a \to p,q]\ [b \to r]\ [c \to s]\ [d \to t]$

Question 3

Column 1	Column 2
a) 3-methylene 1,4-cyclohexadiene	p) antiaromatic
b) Benzene + Birch reduction + Ozonolysis in Zn/acid	q) not aromatic since methylene is outside the ring and has sp^3 carbon
c) Benzene + Ethylene oxide + $AlCl_3$	r) $2CHOCH_2CHO$
d) Toluene on sulphonation gives 78% of *p*-toluene sulphonic acid at 373K	s) $PhCH_2CH_2OH$
	t) Thermodynamic control prevails and the compound is less sterically hindered.
e) Cyclobutadiene	

Answer

$[a \to q][b \to r][c \to s][d \to t][e \to p]$

VIII. Structure-Based and Numerical Problems

1. A and B are isomers. A is aromatic and B is not aromatic. A reacts with Br_2 in CCl_4 to give C. C on debromination with KNH_2 gives D. D on ozonolysis gives benzoic acid and CO_2 and water. B reacts with bromine to give dibromo compound. B reacts with potassium to give a salt, the dianion of which is aromatic. Find A, B, C and D.

Answer

A is $C_6H_5CH{=}CH_2$;

B is Cyclooctatetraene;

C is $C_6H_5CHBrCH_2Br$;

D is $C_6H_5C{\equiv}CH$ 'D' on Ozonolysis gives C_6H_5COOH, CO_2 and H_2O

$C_8H_8 + 2K \rightarrow 2K + C_8H_8^{2-}$ —Cyclooctaterenedianion has 10 π electrons, which is aromatic.

2. When benzene reacts with $CHCl_3$ in the presence of $AlCl_3$, the product is triphenylmethane. Write the mechanism for this reaction.

Answer

$$CH_2Cl_2 + AlCl_3 \rightarrow \overset{+}{C}HCl_2\ AlCl_4^-$$

$$C_6H_6 + CHCl_2AlCl_4 \longrightarrow [C_6H_6(CHCl_2)]^+ \longrightarrow C_6H_5CHCl_2 + HCl + AlCl_3$$

$$C_6H_5CH_2Cl_2 + 2\,C_6H_6 \xrightarrow{AlCl_3} (C_6H_5)_3C\text{–}H$$

Figure Ex.5.2 Mechanism of benzene reacting with $CHCl_3$ in $AlCl_3$

$C_6H_4CHCl_2 + 2C_6H_6 \rightarrow (AlCl_3) \rightarrow$ triphenylmethane.

3. Compound A on treatment with CaO gives B on heating. C is a solvent and D is a Lewis acid. C reacts with B to give $C_6H_4CHCl_2$ and E and D. $C_6H_4CHCl_2$ reacts with 2 moles of B in D to give F. F forms only one monochlorinated product G. Find A to G.

Answer

A is Benzoic acid

B is Benzene

C is $CHCl_3$

D is $AlCl_3$

E is HCl

$(C_6H_5)_3C\text{–}H$ (F) $\quad (C_6H_5)_3C\text{–}Cl$ (G)

Figure Ex 5.3 Triphenyl methane and triphenylmethyl chloride.

4. Compound A on treatment with CaO gives B. B with CH_3Cl and $AlCl_3$ gives BB. BB on oxidation with $KMnO_4$ gives A. B with succinic anhydride in the presence of $AlCl_3$ gives C. C in the presence of Zn amalgam and HCl on refluxing gives D. D with $SOCl_2$ at 353 K gives E. E with $AlCl_3$ in CS_2 gives F. Identify A to F.

Answer

A is Benzoic acid;

B is Benzene;

BB is Toluene;

C is 3-Benzoyl propanoic acid;

D is 4-phenylbutanoic acid;

E is 4-phenylbutanoyl chloride.

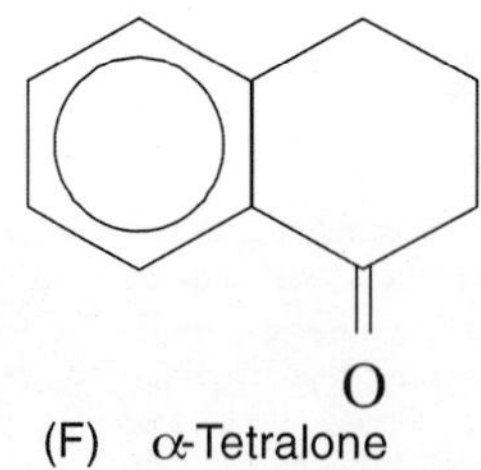

Figure Ex.5.4 Structure of cyclohexanone with aromatic ring (F)

5. CaC_2 on hydrolysis gives A. Three moles of A heated in a red-hot tube give B. B reacts with Na in liq. NH_3 in ethanol to give C. C on ozonolysis gives 2 moles of D. C reacts with NBS to give E. E with $(CH_3)_2CuLi$ gives F with formula C_7H_{10}. Identify A to F.

Answer

A is Acetylene;

B is Benzene;

C is 1,4-cyclohexadiene;

D is $CHOCH_2CHO$;

E is 3-bromo-1,4-cyclohexadiene. The Corey synthesis with $(CH_3)_2CuLi$ gives 3-methyl-1,4-cyclohexadiene, which is F.

6. A is obtained by the dehydration of ethyl alcohol. Compound A reacts with Cl_2 in CCl_4 to give B. B on dehalogenation with KNH_2 gives C. On heating 3 moles of C in a red-hot tube gives D. D with CH_3Cl and $AlCl_3$ gives E. E on Birch reduction gives F. F on ozonolysis followed by reduction with Zn and water gives CH_3COCH_2CHO and $OHCCH_2CHO$. Find A to F.

Answer

A is Ethylene;

B is $ClCH_2CH_2Cl$;

C is $CH\equiv CH$;

D is C_6H_6;

E is $C_6H_5CH_3$;

F is 1,4-Cyclohexadiene;

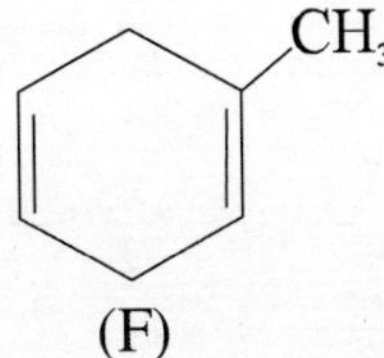

Figure Ex.5.5 The structure of Birch reduction product of toluene

7. Cyclohexanol on dehydration gives A. A on reduction with Ni gives B. B is also obtained by the hydrogenation with nickel under high pressure of C. D is a hydrocarbon which gives two monochlorides E and F in the ratio 54 : 46 percent. F reacts with C in $AlCl_3$, to give G. G with NBS gives H. H with NaOH gives I. I on dehydration gives J. J on ozonolysis in Zn and H_2O gives H and formaldehyde. H on reduction with Zn (Hg) gives K. C reacts with CH_3CH_2Cl in $AlCl_3$ gives K.

Answer

A is cyclohexene;

B is cyclohexane;

C is benzene;

D is propane;

E is n-propyl chloride;

F is isopropyl chloride;

G is $C_6H_5CH(CH_3)_2$;

H is $C_6H_5C—Br(CH_3)_2$;

I is $C_6H_5C—OH(CH_3)_2$;

J is $C_6H_5—\underset{\displaystyle CH_3}{\underset{|}{C}}=CH_2$;

K is $C_6H_5CH_2CH_3$.

8. A reacts with CH_3CH_2Cl in $AlCl_3$ to give B. B with NBS gives C. C with alcoholic KOH gives D. D is obtained by reduction of E with Lindlar Pd. D with A in HF gives F. E on ozonolysis gives G and CO_2 and water. G is also obtained by oxidation of B with $KMnO_4$. Find A to G.

Answer

A is Benzene

B is $C_6H_5CH_2CH_3$

C is $C_6H_5CHBrCH_3$

D is $C_6H_5CH{=}CH_2$

E is $C_6H_5C{\equiv}CH$

F is $(C_6H_5)_2CH_2CH_3$

G is C_6H_5COOH

6

ALKYL HALIDES AND ARYL HALIDES

ALKYL HALIDES

NOMENCLATURE

Alkyl halides are described as primary (1°), secondary (2°) or tertiary (3°) depending on how many alkyl substituents are attached to the C—X unit. The following examples illustrate the point.

- Fluoromethane or methyl fluoride
- Chloroethane or ethyl chloride (Primary)
- 2-iodopropane or isopropyl iodide (Secondary)
- 2-bromo-2-methylpropane or *tert*. butyl bromide (Tertiary)

PHYSICAL PROPERTIES

The polar bond creates a molecular dipole that raises the melting points and boiling points compared to alkanes.

Structure

- The alkyl halide functional group consists of an sp^3-hybridized C atom bonded to a halogen, X, via a σ bond.
- The carbon–halogen bonds are typically quite polar due to the electronegativity and polarizability of the halogen.

REACTIVITY

- The halogens (Cl, Br and I) are good leaving groups.
- The polarity makes the C atom electrophilic and prone to attack by nucleophiles via S_N1 or S_N2 reactions.
- Bases can remove β-hydrogens and cause 1,2-elimination to form alkenes via E1 or E2 reactions.
- Insertion of a metal like Mg creates an organometallic compound.

PREPARATION OF ALKYL HALIDES

Reactions of Alcohols with HX

One of the methods of preparation of alkyl halides is by nucleophilic substitution reaction between alcohols and HX. The reaction involves use of catalysts, if alcohols are not tertiary or secondary.

If alcohol is tertiary then HCl or HX reacts very easily since tertiary carbocation is very stable. Secondary alcohols can also react with HX to form alkyl halides. If alcohol is primary then reaction with dry NaBr cannot produce the halide since OH^- is a poor leaving group. But NaBr in acid produces ROH_2^+ which is a good leaving group and Br^- replaces it.

If ROH reacts with HCl, a catalyst is necessary. Since Cl^- is a weaker nucleophile than bromide or iodide (due to less polarizability) Cl^- does not react with primary or secondary alcohols. A Lewis acid $ZnCl_2$ is added to the reaction mixture; it forms a complex with alcohol and creates a better leaving group than H_2O.

$$R-\ddot{O}(H){:} + ZnCl_2 \longrightarrow R-\overset{+}{\ddot{O}}(H)-\overset{-}{Zn}Cl_2$$

$$Cl^- + R-\overset{+}{O}(H)-\overset{-}{Zn}Cl_2 \longrightarrow RCl + [Zn(OH)Cl_2]^-$$

$$[Zn(OH)Cl_2]^- + H^+ \longrightarrow ZnCl_2 + H_2O$$

The $ZnCl_2$ solution is called Lucas reagent and used for distinguishing primary, secondary and tertiary alcohols.

Halogenation of Alkanes

This has been discussed extensively in Chapter 2.

Reactions of Alcohols with Halogenating Agents

The following are the various reactions of alcohols with halogenating agents.

Reactions of alcohols with other halogenating agents ($SOCl_2$, PX_3)

$$SOCl_2 + -\overset{|}{\underset{|}{C}}-OH \xrightarrow[\text{Et}_3\text{N or pyridine}]{} -\overset{|}{\underset{|}{C}}-Cl + SO_2 + H-Cl$$

$$PCl_3 + -\overset{|}{\underset{|}{C}}-OH \longrightarrow 3\left(-\overset{|}{\underset{|}{C}}-Cl\right) + H_3PO_3$$

$$PBr_3 + -\overset{|}{\underset{|}{C}}-OH \xrightarrow[\text{Et}_3\text{N or pyridine}]{} 3\left(-\overset{|}{\underset{|}{C}}-Br\right) + H_3PO_3$$

To sum up,

- Alcohols can also be converted to alkyl chlorides using thionyl chloride ($SOCl_2$) or phosphorous trichloride (PCl_3).
- Alkyl bromides can be prepared in a similar reaction using PBr_3.
- Used mostly for 1° and 2° ROH.
- In the case of $SOCl_2$, a base is used to "mop-up" the acidic by-product.
- Common bases are triethylamine (Et_3N), or pyridine (C_6H_5N).
- In each case the —OH reacts first as a nucleophile, attacking the electrophilic centre of the halogenating agent.
- A displaced halide ion then completes the substitution displacing the leaving group.
- Note that it is not —OH that leaves, but a much better leaving group.

The advantage of these reagents is in that the reaction is not under the strongly acidic conditions like using HCl or HBr.

NUCLEOPHILIC SUBSTITUTION (S_N1 OR S_N2)

The next section describes the substitution reactions of alkyl halides.

Substitution reactions of alkyl halides are as follows

- Alkyl chlorides, bromides and iodides are good substrates for substitution reactions.
- A variety of nucleophiles can be used to generate a range of new functional groups.
- Figure 6.1 reflects some of the more important reactions we may encounter.

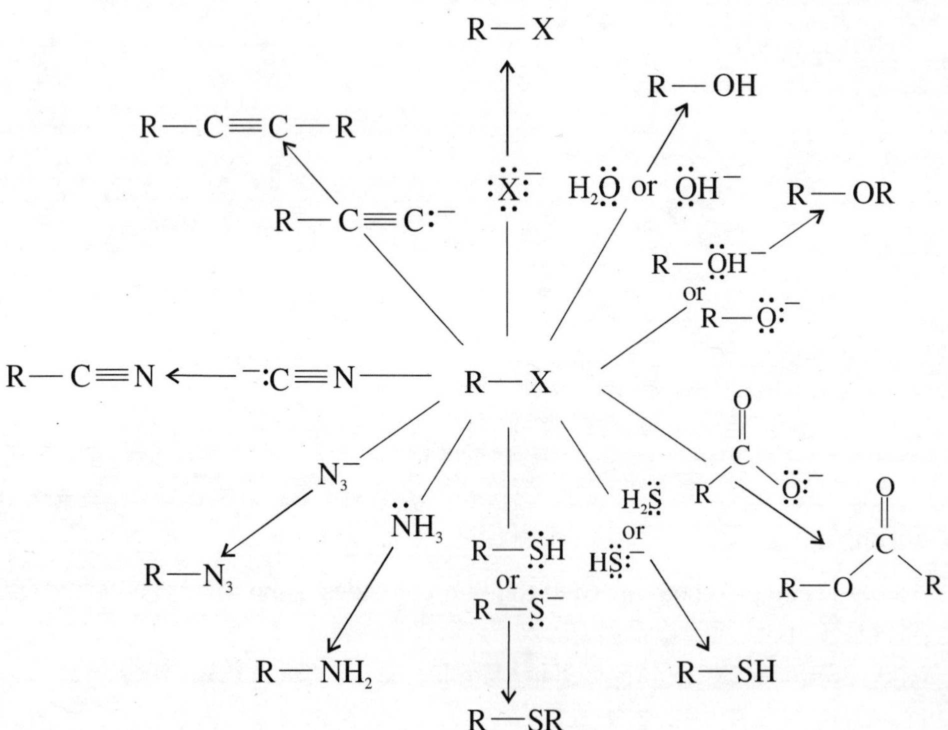

Figure 6.1 Various reactions of alkyl halides

Alkyl halides generally react by S_N1 or S_N2 pathways. The details are described below.

S_N2 *reactions* Nucleophilic substitution reactions occur when a nucleophile displaces the halogen from an alkyl halide. An example of this is the reaction of ethyl bromide with hydroxide ion, OH^-.

$$CH_3CH_2Br + OH^- \rightarrow CH_3CH_2OH + Br^-$$

The OH^- is the nucleophile, and the Br^- ion is the leaving group. The rate law for this reaction is:

$$\text{rate} = K[C_2H_5Br][OH^-].$$

It is a second-order reaction. In fact, all nucleophilic substitution reactions of primary alkyl halides are second-order. These reactions are called S_N2 reactions. The "S" signifies a substitution reaction, while the "N" denotes the presence of a nucleophile, and the "2" indicates that they are bimolecular reactions. This means that there are two molecules involved in the transition state of an

S_N2 reaction. The S_N2 reaction also features a backside attack. That is, the nucleophile attacks the antibonding orbital on the opposite side of the molecule from which the leaving group departs. This causes a stereochemical inversion of the molecule's configuration if it is chiral. The three groups attached to the reacting carbon become inverted when a nucleophile attacks an alkyl halide through the S_N2 mechanism. Therefore when (R)-2-bromobutane reacts with Cl^-, it forms (S)-2-chlorobutane (Figure 6.2).

(R) 5-membered transition state (S)

$Cl^- + H\text{-}C(CH_3)(C_2H_5)\text{-}Br \longrightarrow [Cl\text{--}C(CH_3)(C_2H_5)\text{--}Br] \longrightarrow Cl\text{-}C(CH_3)(H)(C_2H_5) + Br^-$

inversion of configuration

Figure 6.2 Inversion of optically active bromide by S_N2 reaction.

Generally, the rate of an S_N2 reaction increases with increased basicity of the nucleophile. However, a basic compound may not always be nucleophilic. Stronger bases yield faster reactions. The rate also increases when the leaving group is a weaker base than the nucleophile.

Good nucleophile (bad leaving group)	Bad nucleophile (good leaving group)

$OH^- > F^- > Cl^- > Br^- > I^-$

This order of reactivity is dependent upon the type of solvent. In the gas phase (no solvent) and in polar aprotic solvents the order is as above. However the order is reversed if the reaction takes place in a protic solvent. In this case, the weaker base is the better nucleophile because it is weakly hydrogen-bonded to the solvent. A strong base is bogged down by multiple H-bonds with the solvent. Hydrogen-bonding greatly increases the free energy of activation and decreases the rate of the reaction.

Steric branching at the site of substitution decreases the rate of the S_N2 reaction. Subsequently, some secondary and all tertiary alkyl halides react very slowly by the mechanism. In fact, sometimes they react by a different mechanism, the elimination reaction. In the elimination reaction the secondary or tertiary alkyl halide loses a hydrogen from a beta carbon along with the halogen forming an alkene. This is called Beta Elimination or an E2 reaction. In order for an E2 reaction to occur, the proton to be removed and the leaving group must be in an anti-periplanar orientation (Figure 6.3).

$$CH_3O^- + H_3C-C(C_2H_5)_2Br \longrightarrow CH_3OH + H_2C{=}C(C_2H_5)_2 + Br^-$$

Anti-periplanar orientation

$$\text{rate} = k[CH_3O^-][\text{2-ethyl 2-bromobutane}]$$

Figure 6.3 Elimination of Alkyl Halide by E2 Reaction

E2 reactions are second-order.

The bases used as nucleophiles are usually alkoxide salts. Also, as with the S_N2 reaction, the E2 reaction is faster when the leaving group is a weak base.

S_N1 reactions Secondary and some tertiary halides undergo both S_N1 and E1. High temperatures lead preferentially to elimination reactions. These two mechanisms are in competition with each other. Relatively unbranched secondary alkyl halides reacting with unbranched bases tend to form more substitution products than elimination products. Branching in the structure of the alkyl halide or the base increases the percentage of elimination products.

Normally, in elimination the formation of the most highly substituted double bond will be favoured. However, if the nucleophile is highly branched, steric repulsions will force it to attack hydrogens from a less substituted carbon resulting in a less substituted alkene. Details of elimination reactions have been already discussed in chapter 3.

Tertiary alkyl halides containing good leaving groups undergo S_N1 substitution and E1 elimination reactions. These share a common rate-determining step—the formation of a carbocation.

$$(CH_3)_3C-Cl \longrightarrow (CH_{3+})_3C^+ + Cl^-$$

Since carbocations play an important role in S_N1 reactions here we present the relative stabilities of carbocations once again.

STABILITY OF CARBOCATIONS

The general stability order of simple alkyl carbocations is: (most stable) 3° > 2° > 1° > methyl (least stable).

$$\underset{3^\circ}{(H_3C)(CH_3)C^+(CH_3)} > \underset{2^\circ}{(H_3C)(H)C^+(CH_3)} > \underset{1^\circ}{(H_3C)(H)C^+(H)} > \underset{\text{methyl}}{(H)(H)C^+(H)}$$

As we have already discussed in chapter 1, alkyl groups are weakly electron-donating due to hyperconjugation and inductive effects. Resonance effects can further stabilize carbocations when present.

Tertiary halides undergo the S_N1 mechanism because of the stability of the tertiary carbocation. Primary halides do not undergo the S_N1 mechanism because of the instability of the primary carbocation.

The transition state of the rate-determining step contains only one molecule. Consequently, it is a first-order reaction.

After the carbocation is formed, a weak base can attack the positively charged carbon yielding a substitution product, or the carbocation can lose a beta hydrogen to form an elimination product. These are the product-determining steps of the reaction. Figure 6.4 gives the steps of the reaction.

$$(CH_3)_3C^+ + CH_3OH \xrightarrow{S_N1} (CH_3)_3C-O^+(H)-CH_3 \xrightarrow{Cl^-} (CH_3)_3C-O-CH_3$$

substitution product (ether)

$$(CH_3)_3C^+ \xrightarrow{E1} CH_2{=}C(CH_3)_2 + HCl$$

elimination product (alkene)

Figure 6.4 Steps of E1 Mechanism

The rate-determining step of the S_N1 and E1 reactions is the loss of the leaving group (e.g., the ionization of an alkyl halide). Therefore, solvents that promote ionization—polar, protic solvents—increase the overall reaction rate.

The tertiary carbocations which are achiral react to give optically inactive products. They can be attacked by a nucleophile on either side of the molecule with equal frequency, yielding two racemic products. This results from the fact that the positive charge on the electrophile is located in a p orbital whose lobes are above and below the plane of the molecule.

REARRANGEMENTS OF CARBOCATIONS

The carbocations formed from alcohols or alkyne halides are the same. Hence we use the carbon-cation formed from alcohols for illustration.

Carbocations are prone to rearrangement via 1,2-hyride or 1,2-alkyl shifts provided it generates a more stable carbocation. Figure 6.5 shows the formation of various alkenes on dehydration.

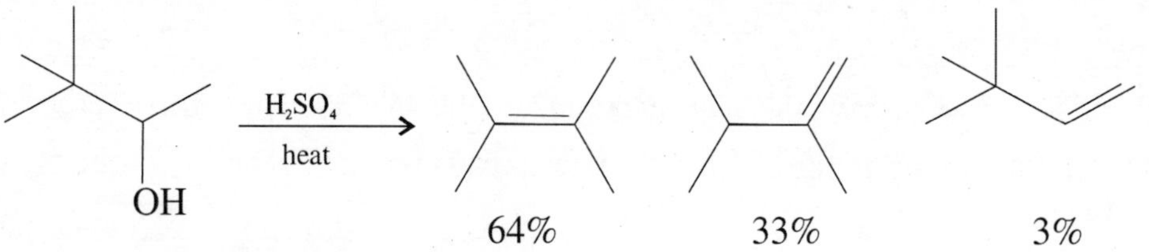

H_2SO_4, heat; OH; 64%; 33%; 3%

Notice that the "predicted" product is only formed in 3% yield, and that products with a different skeleton dominate. The reaction proceeds via protonation to give the better leaving group which departs to give the 2° carbocation shown. A methyl group rapidly migrates taking its bonding electrons along, giving a new skeleton and a more stable 3° carbocation which can then lose H^+ to give the more stable alkene as the major product.

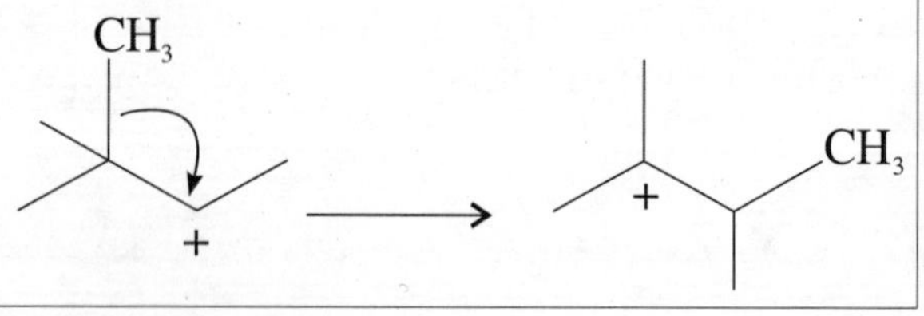

CH_3 + → + CH_3

Figure 6.5 Dehydration of alcohols

This is an example of a 1,2-alkyl shift. The numbers indicate that the alkyl group moves to an adjacent position. Similar migrations of H atoms, 1,2-hydride shifts are also known.

1,2-Hydride Shifts

If a molecule such as 1-bromopropane is being forced to react as shown in Equation (1), the carbocationic intermediate formed is an unstable species and it will do whatever it can to become more stable. Figure 6.6 shows an example of hydride shift.

$$CH_3CH_2CH_2—Br \underset{\Delta}{\overset{HCOH (O)}{\rightleftharpoons}} [CH_3CH_2\overset{\oplus}{C}H_2] + :\overset{\ominus}{Br} \quad (1)$$

Figure 6.6 Hydride shift in S_N1 or E1 elimination reactions

One option available to this cation is to react with a nucleophile to form a neutral product. While the carbocation generated in reaction 1 certainly recombines with bromide ion, this event is not of particular interest since it does not produce any new product. The other potential nucleophile in this mixture is the solvent, but as we have seen, formic acid is used to study S_N1 reactions because it is a highly ionizing but relatively non-nucleophilic solvent. The result of using a solvent of low nucleophilic reactivity is to increase the lifetime of the carbocationic intermediate. This, in turn, provides more time for other processes to occur. In the example shown in Figure 6.6, one such alternative involves the migration of a hydrogen atom with a pair of electrons from one carbon atom to the adjacent positively charged carbon atom. Such a molecular rearrangement is called a 1,2-hydride shift. It is called a 1,2-shift because the hydrogen atom moves from one atom to the adjacent atom. It is called a hydride shift because a hydrogen atom with a pair of electrons (H^-) shifts.

Note that the 1,2-hydride shift converts a 1° carbocation into a 2° carbocation. Once this more stable carbocation is formed, it can react with bromide ion to produce a new product as shown in Figure 6.7.

$$CH_3CH_2CH_2—Br \underset{\Delta}{\overset{HCOOH}{\rightleftharpoons}} \left[CH_3CH_2\overset{\oplus}{C}H_2\right] + :\overset{\ominus}{Br}$$

$$\downarrow \text{~1,2 hydride shift} \qquad (2)$$

$$\underset{\displaystyle Br}{CH_3\overset{|}{C}HCH_3} \xleftarrow{:\overset{\ominus}{Br}} \left[CH_3\overset{\oplus}{C}HCH_3\right]$$

Figure 6.7 Conversion of primary to secondary carbocation

Molecular rearrangements of this type frequently accompany S_N1 reactions, especially in highly ionizing, non-nucleophilic solvents such as formic acid and acetic acid. As a rule of thumb, rearrangements will occur whenever they can produce carbocations of equal or greater stability. Hydride shifts are not the only type of rearrangement seen under these conditions. Another common type of 1,2-shift is called a 1,2-methide shift.

1,2-Methide Shifts

The name methide refers to the H_3C^- group; in other words, a methyl anion. Figure 6.8 describes a reaction involving a 1,2-methide shift.

$$\underset{H_3C\quad H}{\overset{H_3C\quad Br}{CH_3C-CCH_3}} \xrightleftharpoons[\Delta]{HCOH} \underset{H_3C\quad H}{\overset{H_3C}{CH_3C-\overset{\oplus}{C}CH_3}} + :\overset{\ominus}{Br}$$

~1,2 CH_3^- shift

$$\underset{H_3C\quad H}{\overset{Br\quad CH_3}{CH_3C-CCH_3}} \xleftarrow{:\overset{\ominus}{Br}} \underset{H_3C\quad H}{\overset{CH_3}{CH_3\overset{\oplus}{C}-CCH_3}} \qquad (3)$$

Figure 6.8 Methide Shift

In this case solvolysis of 2-bromo-3,3-dimethylbutane generates a 2° carbocation. Rearrangement of one of the three methyl groups, with a pair of electrons, from C-3 to C-2 then generates a 3° carbocation which ultimately captures a bromide ion to yield the final product, 2-bromo-2,3-dimethylbutane.

1,2-Phenyl Shifts

In this shift, phenyl ring undergoes frequent rearrangement. When 3-methyl-3-phenyl-2-butanol is treated with aqueous HBr, the reaction outlined in Figure 6.9 occurs below.

$$C_6H_5-C(CH_3)_2-CH(OH)-CH_3 \xrightarrow[\Delta]{HBr/H_2O} C_6H_5-C(CH_3)_2-CH(Br)-CH_3 +$$

3-methyl 3-phenyl 2-butanol — 3-methyl 3-phenyl 2-bromobutane

$$CH_3-C(Br)(CH_3)-CH(C_6H_5)-CH_3 + H_2O \qquad (4)$$

2-methyl 3-phenyl 2-bromobutane

Figure 6.9 1,2 Phenyl shift

It seems likely that the formation of 3-methyl-3-phenyl-2-bromobutane involves simple replacement of the OH group of the starting material with a bromide ion but we have to account for the formation of the isomeric 2-methyl-3-phenyl-2-bromobutane. To understand the formation of this compound let us consider the role of aqueous HBr in the reaction. There are three points to be noted:

- the solvent is water, which means that it supports the formation of ions.
- HBr is a strong acid, which means that a significant percentage of the starting material will be protonated.
- HBr is a source of the nucleophilic bromide ions that are incorporated into the product.

With these points in mind, Figure 6.10 suggests a plausible pathway by which the 2-methyl-3-phenyl-2-bromobutane may be formed.

$$C_6H_5-C(CH_3)_2-CH(OH)-CH_3 \xrightleftharpoons{HBr\ /\ H_2O} C_6H_5-C(CH_3)_2-CH(\overset{\oplus}{O}H_2)-CH_3 \quad :\overset{\ominus}{Br}$$

$$\rightleftharpoons C_6H_5-C(CH_3)_2-\overset{\oplus}{C}H-CH_3$$

2° Carbocation

$$H_3C-C(CH_3)(C_6H_5)-\overset{\oplus}{C}H-CH_3 \rightleftharpoons H_3C-C(CH_3)(\overset{\oplus}{C_6H_5})-CH-CH_3 \rightleftharpoons$$

$$H_3C-\overset{\oplus}{C}(CH_3)-CH(C_6H_5)-CH_3 \rightleftharpoons H_3C-C(Br)(CH_3)-CH(C_6H_5)-CH_3$$

3° Carbocation

Figure 6.10 Formation of 2-methyl-3-phenyl-2-bromobutane

The process is initiated by an acid–base reaction in which the HBr protonates the alcohol. This converts the OH group to a better leaving group, water. Ionization of the C—O bond generates the 2° carbocation shown in Figure 6.10. When this ion captures a bromide ion, a molecule of 3-methyl-3-phenyl-2-bromobutane is formed. However, capture of a bromide ion competes with the intramolecular donation of electron density from the adjacent phenyl ring as shown by the black arrow. This change produces a "bridged" ion in which the aromatic ring is simultaneously bonded to two carbon atoms. The 1,2-shift of the phenyl ring is then completed by the electronic movement indicated by the grey arrow. This process generates a 3° carbocation, which ultimately traps a bromide ion to produce the second product, 2-methyl-3-phenyl-2-bromobutane.

The reactions that involve carbocations are:

1. Substitutions via the S_N1
2. Eliminations via the E1
3. Additions to alkenes and alkynes (HX, H_3O^+), additions to enol ethers.

As we have seen in these sections and earlier chapter on alkenes the terms nucleophiles, basicity, leaving groups have been extensively used. In the next section these concepts are explained.

NUCLEOPHILES

Nucleophile means "nucleus-loving" which describes the tendency of an electron-rich species to be attracted to the positive nuclear charge of an electron-poor species, the electrophile.

The nucleophilicity expresses the ability of the nucleophile to react in this fashion. A collection of nucleophiles is present in Figure 6.11.

In general terms this can be appreciated by considering the availability of the electrons in the nucleophile. The more available the electrons, the more nucleophilic the system. Hence the first step should be to locate the nucleophilic centre. At this point we will be considering "Nu" that contains lone pairs and may be anionic, however the high electron density of a C=C is also a nucleophile.

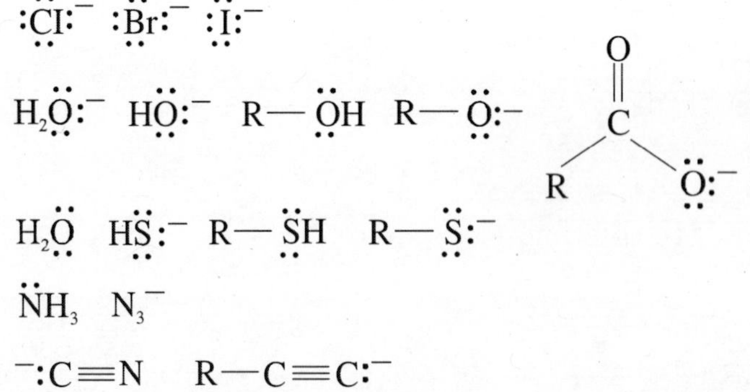

$:\ddot{Cl}:^-$ $:\ddot{Br}:^-$ $:\ddot{I}:^-$

$H_2\ddot{O}:^-$ $H\ddot{O}:^-$ $R—\ddot{O}H$ $R—\ddot{O}:^-$ $RC(=O)\ddot{O}:^-$

$H_2\ddot{O}$ $H\ddot{S}:^-$ $R—\ddot{S}H$ $R—\ddot{S}:^-$

$\ddot{N}H_3$ N_3^-

$^-:C\equiv N$ $R—C\equiv C:^-$

A collection of important nucleophiles are shown to the left.

Figure 6.11 Collection of nucleophiles

Nucleophilicity and Factors Related to it

1. Across a row in the periodic table nucleophilicity (lone pair donation) order is $C^- > N^- > O^- > F^-$, since increasing electronegativity decreases the lone pair availability. This is the same order as for basicity.
2. If one is comparing the same central atom, higher electron density will increase the nucleophilicity, e.g. an anion will be a better "Nu" (lone pair donor) than a neutral atom such as $HO^- > H_2O$. This is the same order as for basicity.
3. Within a group in the periodic table, increasing polarization of the nucleophile as we go down a group enhances the ability to form the new C—X bond and increases the nucleophilicity, so $I^- > Br^- > Cl^- > F^-$. The electron density of larger atoms is more readily distorted, i.e., polarized, since the electrons are further from the nucleus.

The following is a table of relative nucleophilicities as measured in methanol (CH_3OH).

Very Good	I^-, HS^-, RS^-
Good	Br^-, HO^-, RO^-, NC^-, N_3^-
Fair	NH_3, Cl^-, F^-, RCO_2^-
Weak	H_2O, ROH
Very Weak	RCO_2H

Nucleophilicity Versus Basicity

Nucleophilicity and basicity are *very similar properties* in that species that are nucleophiles are usually also bases (e.g. HO^-, RO^-).

$$Nu^- + {}^+C\!\!< \longrightarrow Nu{-}C\!\!<$$

$$Nu^- \quad {-}C{-}LG \longrightarrow {-}C{-}Nu \quad LG^-$$

Both these reactions depict a nucleophile reacting with an electrophilic C atom.

Kinetically controlled reactions of lone pair donor with an electrophilic carbon (usually) atom result in the formation a new C—X bond.

$$A^- + H^+ \rightleftharpoons H{-}A$$

$$B: \quad H{-}A \rightleftharpoons \overset{+}{B}{-}H + A^-$$

Both these reactions depict a base reacting with an electrophilic H atom, a proton

Equilibrium (thermodynamic) reaction of the lone pair donor with a proton, forming a new H—X bond.

The following general reaction mechanisms show why it is important to appreciate the differences, since an anion that is reacting as a nucleophile will result in a substitution, but if it reacts as a base, then elimination will result, be it on a carbocation (S_N1 vs E1) or with the loss of a leaving group (S_N2 vs E2).

Substitutions

$$Nu^- + C^+ \longrightarrow Nu—C \qquad S_N1$$

$$Nu^- \quad C^{\delta+}—LG^{\delta-} \longrightarrow —C—Nu \quad LG^- \qquad S_N2$$

Eliminations

$$B: \quad H—C—C^+ \longrightarrow B—H \quad C=C \qquad E1$$

$$\bar{B}: \quad H—C—C—LG \longrightarrow B—H \quad C=C \quad LG^- \qquad E2$$

LEAVING GROUPS

A leaving group, LG is an atom (or a group of atoms) that is displaced as stable species taking with it the bonding electrons. Typically the leaving group is an anion (e.g., Cl^-) or a neutral molecule (e.g., H_2O). The better the leaving group, the more likely it is to depart. A "good" leaving group can be recognized as being the conjugate base of a strong acid.

This can be understood by writing the chemical equations for the two processes of deprotonation and the removal of leaving group.

$$H—A \overset{K_3}{\rightleftharpoons} H^+ + A^-; \qquad —C—LG \longrightarrow —C^+ + LG^-$$

These two equations represent Bronsted acid dissociation and loss of a leaving group in an S_N1 type reaction. Note the similarity of the two equations: both show heterolytic cleavage of an s bond to create an anion and a cation.

For acidity, the more stable the A^-, the more the equilibrium will favour dissociation and release of protons meaning that HA is more acidic.

For the leaving group, the more stable the leaving group, the more it favours "leaving".

Hence factors that stabilize A^- also apply to the stabilization of a LG^-.

Here is a table classifying some common leaving groups.

Excellent	TsO^-, NH_3
Very Good	I^-, H_2O
Good	Br^-
Fair	Cl^-
Poor	F^-
Very Poor	HO^-, NH_2^-, RO^-

Note We know that HO^- is a poor leaving group. The conjugate base, water, is a good leaving group, since it is the conjugate base of H_3O^+, which is a strong acid.

The S_N1 and S_N2 reactions are affected in different ways by solvents. The details are given below.

SOLVENT EFFECTS

In general terms, the choice of solvent can have a significant effect on the performance of a reaction.

Factors when choosing a solvent are:

- reagents must be in same phase for better solubility
- usually solvent has to be unreactive with reagent. If it is a nucleophile as in the case of solvolysis reaction it may react.

S_N1 Reactions

For an S_N1 reaction, the polarity and ability of the solvent to stabilize the intermediate carbocation is of paramount importance, as shown by the relative rate data for the solvolysis of *t*-butyl chloride.

Solvent	**Dipole moment, μ**	**Dielectric constant, ε**	**Relative Rate**
CH_3CO_2H	1.68	6	1
CH_3OH	2.87	33	4
H_2O	1.84	78	150,000

Dipole moment, μ in debye

S_N2 Reactions

For an S_N2 reaction, the effect of solvent polarity is usually much less, but the ability S_N2 of the solvent to solvate the nucleophile is important, as shown by the relative rate data for the S_N2 reaction of n-butyl bromide with N_3^-.

Solvent	Dipole moment, μ	Dielectric constant, ε	Relative Rate	Type
CH_3OH	2.87	33	1	protic
H_2O	1.84	78	7	protic
DMSO	3.96	49	1,300	aprotic
DMF	3.82	37	2,800	aprotic
CH_3CN	3.92	38	5,000	aprotic

For S_N2 reactions the solvent should be polar aprotic so that the unsolvated anion is available for S_N2 reaction.

Polar aprotic solvents are polar but have no ability to be hydrogen bond. Examples are acetone, acetonitrile, DMSO and DMF. They have the following features.

- They have dipoles due to polar bonds.
- They do not have H atoms that can be donated into a H-bond.
- The anions are not solvated and are "naked" and the reaction is not inhibited.

Another aspect namely phase transfer catalysis is important in S_N2 reactions. This is discussed below.

Phase transfer catalysis We have already mentioned that polar aprotic solvents are important in S_N2 reactions so that substrate dissolves and the anion of the attacking reagent is unsolvated for reaction. Another procedure is to use tetralkylammonium salts as phase transfer catalysts. The tetralkylammonium ion is soluble in organic solvents and the anion Cl^- is unsolvated and reacts with CN^- by S_N2 process to form RCN from RCl. Another method is to use 18-crown-6 in benzene. The crown ether binds potassium ion from KF and the anion travels to organic phase to maintain electrical neutrality and reacts with RCl to give RF. More detailed discussion on phase transfer catalysts and crown ethers are presented later in this chapter.

Since nucleophilic substitution reactions and elimination reactions take place with the substrates, namely alkyl halides there is a competition between these reactions. This aspect is presented here.

Substitution Versus Elimination

When temperatures are high, elimination reactions take place preferentially. Substitution and elimination reactions are strongly influenced by many factors. Some of the more important factors are outlined in the following table. The significance of the effect is stated first, then the system that will favour the reaction is stated. This should help us to decide when the anion functions as a Nu and when it functions as a "Base".

Reaction	Solvent	Nu or Base	Leaving Group	Substrate (alkyl halide)	Typical Conditions
S_N1	Very Strong Polar solvents	Weak Good Nu and weak base	Strong Good LG	Tertiary or resonance stabilized	$AgNO_3$ / aq. EtOH
S_N2	Strong Polar aprotic solvents	Strong Good Nu and weak base	Strong Good LG	Strong Methyl or primary tertiary	NaI / Acetone
E1	Very Strong Polar solvents	Weak Weak base	Strong Good LG	Strong or resonance stabilised	H_2SO_4, heat
E2	Strong Polar aprotic solvents	Strong Poor Nu and strong base	Strong Good LG	Strong tertiary	KOH, heat

The E1 and E2 reactions have been discussed in the chapter on alkenes.

PHASE TRANSFER CATALYSTS

A **phase transfer catalyst** or **PTC** in chemistry is a catalyst which facilitates the migration of a reactant in a heterogeneous system from one phase into another phase where the reaction can take place. Ionic reactants are often soluble in an aqueous phase but insoluble in an organic phase unless the phase transfer catalyst is present. **Phase transfer catalysis** refers to the acceleration of the reaction by the phase transfer catalyst.

Phase transfer catalysts for anion reactants are often quaternary ammonium salts. The corresponding catalysts for cations are often crown ethers. A PTC works by encapsulating the ion. The PTC-ion system has a hydrophilic interior containing the ion, and a hydrophobic exterior.

For example, the nucleophilic aliphatic substitution reaction of an aqueous sodium cyanide solution with the alkyl halide, 1-bromooctane, does not ordinarily take place because 1-bromooctane does not readily dissolve in the aqueous solution. By the addition of small amounts of a phosphonium salt such as hexadecyltributylphosphonium bromide, cyanide ions can be freed from the water phase into an organic phase (e.g., bromooctane + some hydrocarbon). With the phase transfer catalyst, nonanenitrile (1-cyanooctane) forms quantitatively in 90 minutes at reflux.

$$C_8H_{17}Br(org) + NaCN(aq) \rightarrow C_8H_{17}CN(org) + NaBr(aq) \text{ [catalyzed by a } R_4P^+Cl^- \text{ (PTC)]}$$

Subsequent work by Herriott and Picker demonstrated that many such reactions can be performed rapidly at around room temperature by using catalysts such as tetra-n-butylammonium bromide or methyltrioctylammonium chloride in benzene/water systems.

By using a PTC process, one can achieve faster reactions, obtain higher conversions or yields, make fewer by-products, eliminate the need for expensive or dangerous solvents which dissolve all the reactants in one phase, eliminate the need for expensive raw materials and/or minimize waste problems. Phase transfer catalysts are especially useful in green chemistry; by allowing the use of water, the need for organic solvents is reduced.

Contrary to common perception, PTC is not limited to systems with aqueous and organic soluble reactants. PTC is sometimes employed in liquid/solid and liquid/gas reactions. As the name implies, one or more of the reactants are transported into a second phase, which contains the other reactants.

CROWN ETHERS

In 1967 C.J. Pedersen, a DuPont chemist, reported the results of a study of the chemistry of a group of cyclic polyethers that have since come to be called crown ethers. One of the simplest of these compounds contains an 18-membered ring in which 6 of the atoms are oxygens. It is called 18-crown-6 (Figure 6.12). This compound forms complexes with metal ions as shown below. Clearly this is a simple example of host–guest interactions; the crown ether contains a cavity into which a metal ion such as a potassium ion fits snugly.

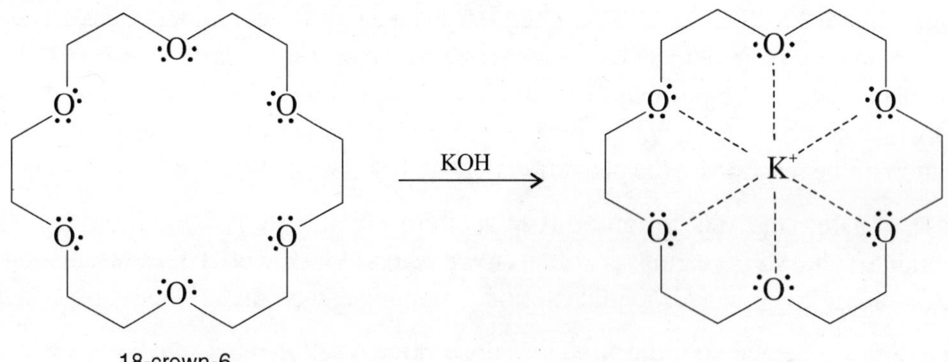

Figure 6.12 18-Crown-6 and the capture of K^+ in the cavity

Other crown ethers are 15-Crown-5 and 12-Crown-4 used with Na^+ and Li^+.

For S_N2 reaction generally we require polar aprotic solvent. Instead of using such solvents salts like KF, KCN and NaN_3 can be dissolved in organic solvents with the help of crown ethers. For potassium salts we use 18-crown-6 and for sodium salts we use 15-crown-5. The cavity size matches with size of metal ion. The metal ion is strongly solvated leaving an anion bar for S_N2 reaction.

$$K + CN^- + RCH_2X \text{ (18-crown-6 in benzene)} \rightarrow RCH_2CN + KX.$$

When KF is used in the presence of 18-crown-6 in benzene, 100% conversion of $C_6H_5CH_2Cl$ to $C_6H_5CH_2F$ takes place.

The crown ethers can be used for oxidation of alkenes by $KMnO_4$. The $KMnO_4$ solution can be dissolved in benzene in the presence of 18-crown-6 and then oxidation can easily be achieved. The cation is bound by Crown ether and anion follows it.

The next section discusses the Grignard reaction which is useful from synthetic point of view.

GRIGNARD REACTION

The Grignard reaction is a versatile and popular technique for making carbon–carbon bonds and building elaborate carbon frameworks. The Grignard reagent is an organometallic compound—a compound with a bond between carbon and a metal that has much covalent character. The metal is usually magnesium but lithium is also used and makes more reactive reagents. The Grignard reagent is made from the direct reaction of magnesium with an alkyl halide.

$$RX + Mg \xrightarrow{\text{ether}} RMgX$$

$$ArX + Mg \xrightarrow{\text{ether}} ArMgX$$

$$>C{=}C<^{X} + Mg \xrightarrow{\text{ether}} >C{=}C<^{MgX}$$

Iodides are more reactive than bromides which are more reactive than chlorides. The unshared pair of electrons on ether–oxygen complexing with Mg is believed to contribute to the stability of the reagent. Notice that aryl halides (ArX) and vinyl halides which are inert to nucleophilic substitution are reactive in this reaction.

Figure 6.13 shows the grignard reagent reacting with various C=O groups.

$$\mathrm{R{-}\overset{\displaystyle O}{\overset{\|}{C}}{-}H} \xrightarrow[(2)\ \mathrm{H_3O^+}]{(1)\ \mathrm{RMgX}} \mathrm{R{-}\underset{\displaystyle H}{\underset{|}{\overset{\displaystyle H}{\overset{|}{C}}}}{-}OH}$$

$$\mathrm{R'{-}\overset{\displaystyle O}{\overset{\|}{C}}{-}H} \xrightarrow[(2)\ \mathrm{H_3O^+}]{(1)\ \mathrm{RMgX}} \mathrm{R{-}\underset{\displaystyle H}{\underset{|}{\overset{\displaystyle R'}{\overset{|}{C}}}}{-}OH}$$

$$\mathrm{R'{-}\overset{\displaystyle O}{\overset{\|}{C}}{-}R''} \xrightarrow[(2)\ \mathrm{H_3O^+}]{(1)\ \mathrm{RMgX}} \mathrm{R{-}\underset{\displaystyle R''}{\underset{|}{\overset{\displaystyle R'}{\overset{|}{C}}}}{-}OH}$$

$$\mathrm{R'{-}\overset{\displaystyle O}{\overset{\|}{C}}{-}OR''} \xrightarrow[(2)\ \mathrm{H_3O^+}]{(1)\ 2\,\mathrm{RMgX}} \mathrm{R{-}\underset{\displaystyle R}{\underset{|}{\overset{\displaystyle R'}{\overset{|}{C}}}}{-}OH} + \mathrm{R''OH}$$

$$\mathrm{O{=}C{=}O} \xrightarrow[(2)\ \mathrm{H_3O^+}]{(1)\ \mathrm{RMgX}} \mathrm{R{-}C({=}O){-}OH}$$

Figure 6.13 Grignard reagent with various C=O groups

The greater the number of alkyl substituents on the carbonyl carbon the more substituted the product alcohol, from primary to tertiary. In all the cases a magnesium alkoxide intermediate is hydrolysed to complete the reaction.

$$\mathrm{RCH_2^{\delta-}{-}Mg^{\delta+}X + H_2O \longrightarrow RCH_3 + XMgOH}$$

The Grignard reagent cannot be used on substrates with unprotected acidic protons such as alcohols and amines.

The preparation of triphenylmethanol is described below. This compound is the parent to a family of dyes called, sensibly, triphenylmethane dyes and includes the familiar crystal violet and phenolphthalein. The Grignard reagent prepared is phenylmagnesium bromide.

$$C_6H_5Br \xrightarrow[\text{ether}]{Mg} C_6H_5MgBr$$

Besides the reaction with water, there are several predictable side-products that can be formed, as given below.

$$C_6H_5MgBr + O_2 \longrightarrow ROOMgX \xrightarrow[2)\ H_3O^+]{RMgX} 2\ C_6H_5OH$$

$$C_6H_5MgBr + CO_2 \xrightarrow[2)\ H_3O^+]{} C_6H_5CO_2H$$

$$C_6H_5MgBr + C_6H_5Br \longrightarrow C_6H_5{-}C_6H_5$$

In the preparation of triphenyl methane one mole of methyl benzoate is added for every two moles of the phenylmagnesium bromide, because esters react twice with the Grignard reagent (Figure 6.14).

$$C_6H_5MgBr + C_6H_5C(=O)OCH_3 \xrightarrow[(2)\ H_3O^+]{\text{ether}} (C_6H_5)_3C{-}OH + C_6H_5MgBr$$

Figure 6.14 Addition of ethylbenzoate to Grignard reagent

ARYL HALIDES

METHODS OF PREPARATION

The methods of preparation of chlorobenzene (benzene + Cl_2/Fe) and bromobenzene (benzene + Br_2/Fe) have already been discussed in aromatic hydrocarbons chapter. The next method is by Sandmayer reaction which follows. The HCl used in all reactions is at 5°C.

$$\text{Benzene} + NaNO_2 + HCl + CuCl \longrightarrow C_6H_5Cl$$

Similarly,

$$\text{CuBr Benzene} + NaNO_2 + HCl \longrightarrow C_6H_5Br$$

Fluorides and iodides Fluorine is very reactive and iodine is very slow in reaction and special methods are used to prepare C_6H_5F and C_6H_5I.

$$\text{Benzene} + NaNO_2 + HCl + HBF_4 \longrightarrow C_6H_5F$$

$$\text{Benzene} + NaNO_2 + HCl + KI \longrightarrow C_6H_5I$$

Benzene + I_2 in HNO_3 also gives C_6H_5I

Aryl halides do not undergo the conventional S_N1 and S_N2 reactions. This is because benzene ring in aryl halide prevents backside attack in S_N2 reactions.

Phenyl cations are very unstable and hence S_N1 reactions do not occur. The C—X bonds are shorter than those of allylic and benzylic halides in case of vinyl and aryl halides. Shorter bonds are stronger and difficult to break.

Nucleophilic aromatic substitution has already been discussed in chapter 5.

I. "Explain Why" Questions and Small Structure-Based Problems

1. The dehydrohalogenation of $C_6H_5CH_2CH_2Br$ is not preferred relative to $C_6H_5CHBrCH_3$ to synthesize styrene. Why?

   Answer

   The carbocation formed in the case of $C_6H_5CH_2CH_2Br$ is primary and that in case of $C_6H_5CHBrCH_3$ is secondary which is more stable.

2. If the rate of reaction of $RCH_2CH_2CH_2Cl$ with NaNu is 'X' and when a catalytic amount of NaI is added in acetone the rate of $RCH_2CH_2CH_2Cl$ with NaNu is 'Y' and 'Y' is much greater than X. Explain the increase in rate of the substitution reaction.

Answer

Iodide replaces chlorine and iodide is a better leaving group and hence rate becomes faster.

3. $CH_3CH(CH_3)CH_2CH_2$ (A) reacts slower than $CH_3C(CH_3){=}CHCH_2I$ in S_N1 reactions. Why?

Answer

When the cation is formed from (B)

$$CH_3C(CH_3){=}CH{-}\overset{+}{C}H_2$$

It is allyl cation and resonance-stabilized and hence more reactive in S_N1 reaction relative to the primary cation formed from $CH_3CH(CH_3)CH_2CH_2I$, and is less stable.

4. $CH_3CH(CH_3)C(CH_3)_2Br$ (A)

$$\begin{array}{c} CH_3 \\ | \\ CH_3CHCBr \\ | \\ CH_3 \end{array} \quad (A)$$

On dehydrohalogenation, A with ethanolic KOH produces more of tetramethylethylene (B) and less of

$$CH_3CH(CH_3)C(CH_3){=}CH_2 \ (C)$$

Why?

Answer

The carbocation formed is tertiary which produces tetramethyl ethylene. The terminal alkene (C) is produced from a primary carbocation which is been stable.

5. When benzyl bromide is added to a suspension of KF in benzene, no reaction occurs. But when a catalytic amount of 18-crown-6 when added to the solution, $C_6H_5CH_2F$ is obtained in high yield. Explain.

Answer

When benzyl bromide is added to a suspension of KF in benzene, no reaction takes place since KF does not dissolve, but addition of crown ether makes K^+ bind to crown ether cavity and anion is free to react with $C_6H_5CH_2Cl$.

6. When 1-chlorooctane is treated with NaCN in aqueous medium, there is no reaction. But when the substrate is treated with NaCN in decane in the presence of $R_4N^+Br^-$, the reaction is complete in less than 2 hours and the yield is 95%. Explain.

Answer

Phase transfer catalysis by the tetralkyl ammonium salt is responsible for the faster rate in $R_4N^+Br^-$.

7. Methyl iodide reacts with NaCN in ethanol to produce CH_3CN and does not undergo elimination. Why? If solvent is changed dimethyl formamide reaction is faster. Why?

Answer

CH_3I has no β hydrogen and cannot undergo elimination. Only substitution occurs leading to CH_3CN. Dimethyl formamide is a polar aprotic solvent and suitable for S_N2 process.

8. n-propyl bromide with methoxide ion in methanol at 323 K gives the major product as $CH_3CH_2CH_2OCH_3$. The minor product is propylene. Explain.

Answer

Temperature is low suitable for nucleophilic substitution and S_N2 reaction takes place yielding ether as major product. By E_2 pathway propylene is formed as minor product.

9. n-propyl bromide with $(CH_3)_3OK$ in *t*-butanol gives propylene as major product. Why?

Answer

Since base is very strong and sterically hindered S_N2 reaction is not permitted and elimination takes place.

10. *t*-butyl bromide in methanol at 298 K gives only $(CH_3)_3COCH_3$. Why?

Answer

This is solvolysis which is an S_N1 reaction leading to ether.

11. Can we prepare Ph-O-Ph using PhOH and PhI?

Answer

No. Aryl halides cannot undergo the S_N2 reaction.

12. 5-chloro-1,3-cyclopentadiene undergoes S_N1 solvolysis very slowly even though it is doubly allylic. Explain.

Answer

The cation formed is cyclopentadienyl with a positive charge which is unstable according to Hückel rule.

13. When *trans*-2-methylcyclohexanol dehydrates and the product is made to undergo ozonolysis in reducing conditions we get $CH_3COCH_2CH_2CH_2CH_2CHO$, but when *trans*-1-bromo-2-methylcyclohexane is dehydrohalogenated and the product is made to undergo ozonolysis in reducing conditions we get $CHOCH(CH_3)CH_2CH_2CH_2CHO$. Explain.

Answer

The alcohol dehydration follows E1 mechanism and the alkene formed is 1-methylcyclohexene. But dehydrohalogenation proceeds by E2 pathway and the alkene formed is 3-ethylcyclohexene.

II. Multiple Choice Questions (One out of four answers is correct)

1. One of the following cations reacts by S_N1 at fastest rate.

a) $CH_3\overset{+}{C}H_2$ b) $(CH_3)_2\overset{+}{C}H$ c) $(CH_3)_3\overset{+}{C}$ d) $CH_3O\overset{+}{C}H_2$

2. Arrange in increasing order the rate of RX with iodide ion.

I) CH_3Cl II) $CH_2{=}CHCH_2Cl$ III) CH_3OCH_2Cl IV) $C_6H_5COCH_2Cl$

a) $I<II<III<IV$ b) $IV<II<III<I$ c) $I<IV<II<IV$ d) $I<III<II<IV$

3. Compound existing as enantiomers will be

a) 1-bromohexane b) 1-bromo-1-methylpentane

c) 2-bromo-2-methylpentane d) 3-bromo-3-methylpentane

4. One of the following alkyl halides gives only one alkene on dehydrohalogenation.

a) *t*-butylchloride b) 2-bromo-2-methylpentane

c) 2, 2-dimethyl-4-bromopentane d) 2,2-dimethyl-5-bromohexane

5. Alkyl halide giving mixture of alkenes on dehydrohalogenation

a) 1-bromopentane b) 1-bromohexane

c) benzyl chloride d) 2,2-dimethyl-4-bromopentane

6. If I) is $CH_3CHClCH_2CH_3$ II) is $CH_3CHBrCH_2CH_3$ and III) is $CH_3CHICH_2CH_3$ then the rate of S_N2 reaction with NaCN in aprotic solvent will be $I<II<III$, because of

a) steric factor b) polarity factor

c) leaving group ability d) absence of beta hydrogen

7. 2-bromo-3-methylbutane yields only 2-methyl-2-butanol on hydrolysis. This is due to
   a) stability of tertiary carbonium ion relative to primary carbonium ion.
   b) stability of tertiary carbonium ion relative to secondary carbonium ion.
   c) stability of secondary carbonium ion relative to primary carbonium ion.
   d) methyl shift
8. *t*-butylfluoride is solvolysed only in very acidic solutions because
   a) fluoride is a very good leaving group
   b) fluoride is very small in size
   c) of nucleophilic catalysis
   d) fluoride is a poor leaving group and hydrogen-bonding in strong acids facilitates removal of F^- ion
9. When 2-chloro-1-hydroxycyclohexane is treated with NaOH/water at 298 K for 1 hour, the product obtained is
   a) 1-hydroxycyclohexene b) 3-hydroxycyclohexene
   c) oxirane d) cyclohexyl ether
10. 1-hydroxy-7-bromoheptane with KOH and *p*-xylene gives
    a) 7-hydroxycycloheptene b) 7-hydroxy-1-heptene
    c) 7-hydroxy-2-heptene d) cyclic ether with formula $C_6H_{12}O$
11. Arrange in increasing rate of S_N2 reaction of the following halides.
    I. isopropyl bromide II. cyclohexyl bromide III. cyclopentyl bromide IV. cyclobutyl bromide
    a) IV<III<I<II b) IV<III<I<II c) IV<II<I<III d) I<II<IV<III
12. When CH_3Br reacts with another halide ion in gas phase with no solvent, the rate decrease in order of X^- is
    a) F > Cl > Br b) Br > C l> F c) Br = F > Cl d) Br = Cl = F
13. Arrange in increasing order, the heat of formation of
    I. *t*-butyl carbocation II. isopropylcarbocation III. ethyl cation IV. methyl cation
    a) I<II<III<IV b) IV<II<III<I c) I<II<IV<III d) I<III<II<IV
14. In the case of $(CH_3)_2CHCHBrCH_3$, treatment with $(Bu)_4N^+Cl^-$ gives S_N2 and E2 product almost equally but in the case of $CH_3CH_2CH_2Br$ only S_N2 product is obtained. This is because
    a) primary carbocation is much unstable and substitution is preferred over elimination.
    b) of the steric factor of bulky groups in $(CH_3)_2CHCHBrCH_3$.
    c) secondary halides cannot easily undergo elimination.
    d) there is no beta hydrogen in the primary halide.

15. Arrange in the increasing order of % of 2-alkene formed in preference to 1-alkene for dehydrohalogenation of $CH_3CH_2CH_2CH_2CHXCH_3$.
    I X is iodide II X is bromide III X is chloride IV X is fluoride
    a) IV<III<II<I b) IV<III<I<II c) IV<II<III<I d) I<III<II<IV
16. When Friedel–Crafts reaction is conducted between benzene and RX, the alkyl halide which gives a single alkylated benzene is
    a) $CH_3CH_2CH_2X$ b) $CH_3CH_2CH_2CH_2X$ c) $(CH_3)_2CX$ d) $(CH_3)_3CCH_2X$
17. *m*-methoxyaniline can be obtained from
    a) *ortho*-methoxybromobenzene only
    b) *ortho*-methoxybromobenzene and *meta*-methoxybromobenzene
    c) *p*-methoxybromobenzene
    d) *p*-bromoaniline
18. The compound which cannot undergo nucleophilic substitution involving benzyne intermediate is
    a) o-methoxybromobenzene b) 2,6-dinitrobromobenzene
    c) chlorobenzene d) 2-nitrobromobenzene
19. When alcohols react with $SOCl_2$ or PCl_3 or PBr_3, triethylamine or pyridine is used as
    a) nucleophile b) electrophile
    c) catalyst d) base used to mop up the acidic product
20. Reaction of primary alcohols with HCl is carried out in the presence of $ZnCl_2$ which
    a) acts as a catalyst.
    b) acts as Lewis base.
    c) forms a complex with alcohol and provides a poorer leaving group than water.
    d) forms a complex with alcohol and provides a better leaving group than water.

KEY

1. d	2. a	3. c	4. a	5. d	6. c	7. b
8. d	9. c	10. d	11. c	12. a	13. a	14. a
15. a	16. c	17. b	18. b	19. d	20. d	

III. Multiple Choice Questions (More than one answer is correct)

1. S_N1 reaction is not possible for
   a) CH_3Br b) $(CH_3)_3CCH_2Br$ in nonionizing solvent
   c) $CH_3CH_2CH_2Br$ d) CH_3OCH_2Cl

2. The solvents which can be chosen for the reaction of $CH_3CH_2CH_2I$ with NaOH to give best yield are

   a) water b) acetone c) dimethylformamide d) methanol

3. Compounds existing as enantiomers will be

   a) 1-bromohexane b) 2-bromo-3-methylpentane
   c) 2-bromo-2-methylpentane d) 3-bromo-3-methylpentane

4. Some the following alkyl halides give more than one alkene on dehydrohalogenation

   a) *t*-butylchloride b) 2-bromo-2-methylpentane
   c) 2,2-dimethyl-4-bromopentane d) 2,2-dimethyl-5-bromohexane

5. Alkyl halides not undergoing elimination are

   a) benzyl bromide b) 1-bromohexane
   c) CH_3Br d) 2,2-dimethyl-4-bromopentane

6. Some of the following statements are true for 2-fluoropentane when it undergoes dehydrohalogenation.

   a) It obeys Hoffmann rule.
   b) The product is decided by the stable carbanion.
   c) Zaytzeff rule is obeyed.
   d) 2-pentene is the major product.

7. (1-methyl)cyclopentylchloroethane with alc.KOH undergoes dehydrohalogenation. Some statements about major product are true

   a) The major product is 1,2-dimethylcyclohexene.
   b) Ozonolysis in reducing conditions gives $CH_3COCH_2CH_2CH_2CH_2COCH_3$.
   c) The major product is (1-methyl) cyclopentylethylene.
   d) Ozonolysis in reducing conditions gives HCHO as one of the products.

8. The following conversions follow S_N2 pathway

   a) $CH_3CH_2CH_2Cl + OH^-$
   b) $CH_3OCH_2Cl + OH^-$
   c) $(CH_3)_3CCH_2I$ in nonionizing solvent with OH^-
   d) The solvolysis of $(C_2H_5)_3CBr$ in ethanol.

9. 2-bromo-2-methylbutane by elimination gives

   a) 2-methyl-2-butene b) 2-methyl-1-butene
   c) 2-methylpropene d) 1-pentene

10. HBr with (R) 3-methyl-3-hexanol leads to
    a) racemic 3-bromo-3-methylhexane
    b) (S)-3-bromo-3-methylhexane
    c) partial inversion and retention take place
    d) product is optically active
11. $CH_3CHBrC(CH_3)_3$ undergoes elimination to give alkene whose ozonolysis products in $(CH_3)_2S$ are
    a) HCHO b) $(CH_3)_3CHO$ c) CH_3CHO d) $(CH_3)_3CCOCH_3$
12. 1-methyl-1-chlorocyclohexane on elimination gives two alkenes and these on ozonolysis in dimethyl sulphide followed by reduction with $NaBH_4$ yield
    a) $CH_3CHOH(CH_2)_4OH$ b) CH_3OH
    c) cyclohexanol d) cyclopentanol
13. $CH_2{=}CH_2{-}\underset{\displaystyle CH_3}{\overset{\displaystyle Br}{C}}{=}CH_3$ on dehydrohalogenation followed by ozonolysis in reducing conditions gives
    a) HCHO and CHOCHO and acetone b) HCHO and $CHOCH_2COCH_3$
    c) HCHO and $CHOCH_2CHOHCH_3$ d) $CHOCH_2COCH_3$ only
14. The dehydrohalogenation of some of these alkyl halides do not produce *cis–trans* isomers.
    a) 2-chloropentane b) 1-chlorocyclohexane
    c) $CH_3CHCH_3CH_2Cl$ d) 2-bromobutane
15. Some of the following alkyl halides on dehydrohalogenation produce more than 3 isomers including geometric isomers.
    a) 2-bromobutane b) 2-chloropentane
    c) 3-chloro-3-methylhexane d) 3-bromohexane

KEY

1. a,b,c	2. b,c	3. b,c	4. b,c,d	5. a,c	6. a,b	7. a,b
8. a,c	9. a,b	10. a,c	11. a,b	12. a,b,c	13. a,b	14. b,c
15. c,d						

IV. ASSERTION–REASON QUESTIONS

The questions consist of assertion in column 1 and reason in column 2. Use the following key to choose the appropriate answer.

a) If both assertion and reason are correct and reason is the correct explanation of assertion
b) If both assertion and reason are correct but reason is not the correct explanation of assertion
c) If assertion is correct but reason is incorrect
d) If assertion is incorrect but reason is correct

No.	Assertion	Reason
1.	2-fluoropentane on dehydrohalogenation gives 2-pentene as major product.	Fluoride ion is a poor leaving group and elimination depends on more stable carbanion and hence 2-pentene should be a major product.
2.	The rate of CH_3Cl + NaNu is increased by catalytic amount of NaI acetone.	Iodide is a better leaving group and it replaces Cl^- and then quickly leaves to be substituted by Nu.
3.	Benzyl chloride cannot give rise to an alkene by elimination.	There is an aromatic ring in benzyl chloride.
4.	Neopentyl chloride does not undergo E2 elimination.	Beta hydrogen is absent in neopentyl chloride.
5.	Ortho disubstituted bromobenzene cannot undergo aromatic nucleophilic substitution.	Benzyne intermediate is possible only when at least one *ortho* position is vacant.
6.	For the reaction of NaOH with C_2H_5Cl, rate is increased if OH concentration is increased.	$C_2H_5Cl + OH^-$ reaction is unimolecular.
7.	Hydrolysis of 2-bromo 3-methylbutane gives only 2-methyl 2-butanol.	The more stable tertiary carbocation is formed from secondary cation to give 2-methyl 2-butanol.
8.	The rate of reaction of $(CH_3)_3CCl$ with nucleophiles is increased by the addition of $AgNO_3$.	Ag^+ has more affinity for Cl^- than the solvent and Cl^- leaves and is easily replaced by nucleophile.
9.	t-butylfluoride is solvolysed very fast in strong acid solution.	F^- is a poor leaving group and hydrogen-bonding in acid reduces the rate of solvolysis.
10.	t-butylchloride is more easily solvolysed than 2-chloro 1,1,1-trifluoro 2-methylpropane.	The carbocation in the fluorine compound is destabilized due to smaller size of fluoride ion.
11.	t-butylchloride is solvolysed more slowly in the mixture of 90% D_2O and 10% dioxane relative to 90% H_2O and 10% dioxane mixture.	Hydrogen-bonding is weaker in D_2O relative to H_2O.
12.	$ClCH_2OCH_2CH_3$ solvolyses in ethanol faster than $ClCH_2CH_2CH_2CH_3$.	The $\overset{+}{C}H_2OCH_2CH_3$ will be resonance-stabilized and hence the rate is higher.
13.	The dehydrohalogenation of 2-bromobutane yields the major product as *trans* 2-butene	According to Zaytzeff rule more substituted alkene is formed as a major product. *Trans* being more stable, it is formed in high %.

No.	Assertion	Reason
14.	When cyclohexene is dissolved in chloroform and 50% NaOH added there is no reaction but if a small amount of benzyltriethylammonium chloride is added, the dichlorocarbene is formed which adds on alkene to give the cyclohexane derivative.	Since water layer and organic layer are immiscible, in the first case there is no reaction. When the quaternary salt is added, it dissolves in inorganic solvent and goes to organic layer taking the anion for electroneutrality (phase transfer catalysis). Hence the reaction occurs.
15.	S_N2 reaction of $C_6H_5CH_2Cl$ with KF takes place easily in 18-crown-6 in benzene as solvent.	S_N2 reactions require polar aprotic solvents so that the anion remains unsolvated. When 18-crown-6 is added to benzylchloride and KF, the potassium binds to crown ether and the anion is available for substitution reaction and this is phase transfer catalysis.

KEY

1. a	2. a	3. b	4. a	5. a	6. c	7. a
8. a	9. d	10. c	11. a	12. a	13. a	14. a
15. a						

V. Comprehension Passages

Passage I

Alkyl halides can be prepared in many ways.

1. Addition of HX to ROH. If alcohol is tertiary due to the stability of tertiary carbocation, HX addition is fast to give the alkyl halide. If primary alcohol is used, it is necessary to use acid to convert to protonated alcohol in NaBr for the formation of bromide. $ZnCl_2$ catalyst is used for forming alkyl chloride. $SOCl_2$, PCl_3, PBr_3 are used to form the halides.

Answer the following questions

1. ROH to RBr does not take place in
   a) dry NaBr
   b) NaBr + H_2SO_4
   c) dry KBr
   d) Br_2
2. When $SOCl_2$ is added to ROH, pyridine is used to
   a) catalyse the reaction
   b) to mop up the acid formed by combining with it
   c) act as a nucleophile
   d) prevent elimination

3. In the reaction of ROH with PBr_3, Br^- displaces
   a) $POBr_3$ b) HPOBr c) POBr d) $POBr_4^-$

KEY

1. a 2. b 3. b

Passage 2

Alkyl halides undergo two sets of important reactions, substitution (S_N1 and S_N2 and E1 and E2) and elimination reactions. Substitution reaction takes place at low temperature and strong nucleophiles. If bases are strong and temperature is high elimination takes place. Elimination requires the presence of beta hydrogen on alkyl halide.

Answer the following questions

1. The alkyl halide which undergoes elimination in E2 conditions is
   a) $CH_3CH_2CH_2Cl$ b) benzyl bromide
   c) neopentyl chloride d) CH_3Cl
2. Arrange in increasing order of S_N1 solvolysis for
   I. isopropyl chloride II. *t*-butyl chloride III. $ClCH_2OCH_3$ IV. methyl chloride
   a) IV<I<II<III b) IV<II<I<III c) IV<III<II<I d) IV<I<III<II
3. Elimination produces product according to Hoffmann rule for
   a) $CH_3CH_2CH_2CH_2Cl$ b) $CH_3CH_2CH_2CH_2CH_2Br$
   c) 2-fluoropentane d) benzyl bromide

KEY

1. a 2. a 3. c

Passage 3

Aromatic nucleophilic substitution takes place in terms of intermediate known as benzyne. Aryl halide do not react by S_N1 or S_N2 pathway. Benzyne formation requires at least one ortho position free in the aryl halide.

Answer the following questions

1. Aromatic nucleophilic substitution is not possible for
   a) bromobenzene b) *m*-nitrobromobenzene
   c) *p*-nitrobromobenzene d) 2,6-dichlorobromobenzene

2. Aromatic nucleophilic substitution is not possible in
   a) *o*-chloronitrobenzene
   b) *p*-chloronitrobenzene
   c) *m*-chloronitrobenzene
   d) 2,6-dibromochlorobenzene
3. One of the following is not an evidence for benzyne mechanism in nucleophilic aromatic substitution.
   a) If carbon bearing halogen is labelled with ^{14}C and when NH_2 is added the product has ^{14}C at carbon bearing NH_2 and also carbon adjacent to that which is substituted.
   b) If both *ortho*-positions are occupied, no aromatic nucleophilic substitution takes place.
   c) If there are two substituents on the ring, more than one substituted product is observed.
   d) 2-nitro-4-nitrochlorobenzene undergoes substitution with OCH_3 group.

KEY

1. d 2. c 3. d

VI. Matrix Matching

Question I

Column 1	Column 2
a) optically active due to dissymmetry	p) 1,3-dichloroallene
b) 2-fluoropentane + CH_3OK	q) 1,2-*trans*-dichlorocyclopentane
c) 2-bromo 3-methylbutane with C_2H_5OH gives	r) major product is 1-pentene
d) 1-bromo 3-methylbutane with alc.KOH	s) Hofmann rule is followed
	t) 3-methyl 1-butene
	u) 2-methyl 2-butene
	v) 2-ethoxy 2-methylbutane
	x) 2-methyl 1-butene

$[a \rightarrow p,q]\,[b \rightarrow r,s]\,[c \rightarrow t,u,v,x]\,[d \rightarrow t]$

Question 2

Column 1	Column 2
a) NBS + propylene (500°C)	p) $BrCH_2CH{=}CH_2$
b) In reaction between $(CH_3)_3O$- and $CHCl_3$, intermediate is	q) free radical reaction
c) 2-chloropentane + CH_3OK	r) dichlorocarbene
d) 2-fluoropentane + CH_3OK	s) 2-pentene is major product
	t) 1-pentene is major product
	u) carbanion mechanism

Answer

$[a \rightarrow p,q]\,[b \rightarrow r]\,[c \rightarrow s]\,[d \rightarrow t,u]$

Question 3

Column 1	Column 2
a) Benzyl bromide is mixed with --- in catalytic amounts to KFwe getbenzyl fluoride	p) 18-Crown-6
b) *trans*-3-hexene + Br_2 in CCl_4	q) meso dibromide
c) *cis*-3-hexene + dil.$KMnO_4$	r) racemic mixture of diol
d) *trans*-3-hexene + dil.$KMnO_4$	s) Rate of substitution is increased by addition of KI
e) CH_3Br + NaNu (where Nu is nucleophle + KI in catalytic amounts	t) meso diol

Answer

$[a \rightarrow p]\,[b \rightarrow q]\,[d \rightarrow r]\,[e \rightarrow s]$

VII. Structure-Based and Numerical Problems

1. Compound A answers test with $NaHCO_3$ contains aromatic ring and when it is heated with CaO, a gives B. B reacts with C in $AlCl_3$ to give D. C is the only monohalogenation product of Z. Z is obtained by Wurtz reaction of CH_3I with Na. D with NBS gives E. E on hydrolysis gives F. B with CH_3COCl in $AlCl_3$ gives G. G undergoes haloform reaction to give CHI_3 and anion of compound A. B with Br_2 in $FeBr_3$ gives H. H with Mg in ether gives J. J with CO_2 in acid by Kolbe method gives A. Find A to J and also identify Z.

Answer

A is Benzoic acid

B is Benzene

C is C_2H_5Cl

D is Ethyl benzene

E is $C_6H_5CHBrCH_3$

F is $C_6H_5CHOHCH_3$

G is $CH_3COC_6H_5$

H is C_6H_5Br

J is C_6H_5MgBr

Z is C_2H_6

2. A and B are two chlorides and contain aromatic ring and differ by a CH_2 group. A does not undergo elimination but undergoes S_N1 or S_N2 reactions. B undergoes substitution and elimination. B on dehydrohalogenation gives the main product C. C on ozonolysis in reducing conditions gives C_6H_5CHO and HCHO. C on complete hydrogenation with Ni/H_2 gives D. How many monochlorinated products are possible for D? Find A to D.

Answer

A is $C_6H_5CH_2Cl$;

B is $C_6H_5CH_2\,CH_2Cl$;

C is C — $C_6H_5CH{=}CH_2$;

D is $C_6H_{11}CH_2\,CH_3$; D gives 6 monochlorinated products

3. A and B are two alkenes. A is obtained by decarboxylation of sodium salt of succinic acid. A with HOCl gives C. C with Na_2CO_3 and water gives D. B has molecular weight 2 times that of A. B is obtained by dehydration of E which gives rise to a very stable carbocation. B reacts with HBr to give the major product F. Find the isomers of F. What reagents are used to prepare these isomers?

Answer

A is ethylene;

B is 2-methylpropene;

C is $HOCH_2CH_2Cl$;

D is ethylene oxide;

E is $(CH_3)_3C—OH$

F is $(CH_3)_3C—Br$

Isomers of F are $CH_3—\underset{\underset{CH_3}{|}}{\overset{\overset{H}{|}}{C}}—CH_2Br$, $CH_3—CH_2—CH_2—CH_2Br$ and

$CH_3—\underset{\underset{Br}{|}}{\overset{\overset{H}{|}}{C}}—CH_2CH_3$.

Both are primary bromides and we can use NaBr and H_2SO_4 or PBr_3.

4. Compounds A and B are the isomers and B gives precipitate with ammoniacal $AgNO_3$ and they have molecular weight of 68 grams. B on hydrogenation in Pd/carbon/$BaSO_4$ gives C. D is obtained from isomer of B by Na/NH_3. D on ozonolysis gives CH_3CHO and $CHOCH_2CH_2CH_3$ in reducing conditions. E is a fluoro compound and with CH_3OK undergoes elimination to give mainly C. If fluorine is replaced by bromine in this halogen compound, major product is D. Complete hydrogenation of A or B gives F. How many monochlorinated products are possible for F? One of the ozonolysis products of A in reducing condition in HCHO. Find A to F.

Answer

A is $CH_2=CH—CH_2—CH_2—CH=CH_2$

B is $CH≡C—CH_2CH_2CH_2CH_3$

C is $CH_2=CHCH_2CH_2CH_2CH_3$

$CH_3C≡C—CH_2CH_2CH_3$ (isomer of B)

D is $CH_3CH=CHCH_2CH_2CH_3$

E is $CH_3CHFCH_2CH_2CH_2CH_3$

F is n-hexane and gives 3 monochlorides

5. A and B are the two isomers with molecular weight 82 grams. Both decolorize $KMnO_4$. A on complete hydrogenation gives C which gives only one monochlorinated product. A on ozonolysis in reducing conditions gives a dialdehyde. B on complete hydrogenation gives D which gives 3 monochlorinated products. E has a molecular weight 12 grams more than that of A or B and obtained by dehydration of F in acid. G is an isomer of E and is obtained by dehydrohalogenation of H. Find A to H. Account for the different type of products obtained in dehydration and dehydrohalogenation of F and H respectively.

Answer

Both A and B have double bonds as indicated by $KMnO_4$ colour test.

A is Cyclohexene;

B is 1,5 hexadiene;

C is Cyclohexane;

D is n-hexane;

E is 1-methylcyclohexene;

F is *trans*-2-methylcyclohexanol;

G is 3-methylcyclohexene;

H is *trans*-1-bromo-2-methylcyclohexane.

The alcohol F undergoes dehydration by E1 pathway but the cyclohalogen compound. H undergoes dehydrohalogenation by E2 pathway. Hence different products are obtained.

6. Compound A with Cl_2 at 773 K gives B. A with Cl_2 in CCl_4 gives C. A with Cl_2 and water gives D. D with Na_2CO_3/H_2O gives E. A on hydrogenation gives F. F is also obtained when G is hydrogenated with Ni/H_2. G can be prepared by the reaction of diazomethane with ethylene. F gives two monochlorides in the % ratio 45:55. H is obtained by hydrolysis of Mg_2C_3. H on partial reduction gives A with Lindlar Pd. Find A to H.

Answer

A is propylene;

B is $CH_2{=}CHCH_2Cl$;

C is $CH_3CHClCH_2Cl$;

D is $OHCH_2CH_2Cl$;

E is ethylene oxide;

F is propane;

G is cyclopropane;

H is propyne;

7. A is always one of the products in the reaction of methylketones, C_2H_5OH or CH_3CHOHR when treated with NaOH and I_2. A reacts with Ag to give B. B reacts with $NaNH_2$ to give C. C with n-propyl iodide gives D. D on partial reduction with Lindlar Pd gives E. E with HBr in peroxide gives F. E with HBr in the absence of peroxide gives G which is optically active. What is the name given for E related to formation of G? Find A to G.

Answer

A is CHI_3;

B is acetylene;

C is sodium acetylide;

D is 1-butyne;

E is 1-butene;

F is $CH_3\ CH_2\ CH_2\ CH_2Br$;

G is $CH_3\ CH_2\ CHBr\ CH_3$.

E is prochiral towards reaction with HBr to give G.

8. Glycerol reacts with 3 moles of HI to give A. A with Mg in ether gives B. B and A react to give C and MgI_2. C on ozonolysis in reducing conditions gives HCHO and $CHOCH_2CH_2CHO$. Write all the isomers of C which have two double bonds.

Answer

A is allyl iodide.

B is $CH_2{=}CHCH_2MgI$.

C is $CH_2{=}CHCH_2CH_2CH{=}CH_2$.

Isomers are $CH_2{=}CH{-}CH{=}CH{-}CH_2CH_3$; $CH_2{=}CH{-}CH_2{-}CH{=}CHCH_3$; $CH_2{=}C(CH_3){-}C(CH_3{=}CH_2$.

9. Benzene reacts with Br_2/Fe to give A. A with $NaNH_2$ gives B. C is aromatic and has a molecular weight of 28 grams more than A. C does not react with $NaNH_2$ to give substitution product. When C is oxidized with $KMnO_4$, two COOH groups are introduced *ortho* to the bromine atom and is called D. Benzene reacts with $CH_3COCl/AlCl_3$ to give E. E with Cl_2/Fe gives F. F by Clemmenson reduction gives G. G with $NaNH_2$ gives H, I and J. Find A to J. Why are three products obtained from G with $NaNH_2$?

Answer

A is bromobenzene;

B is aniline;

C is 2,6-dimethylbromobenzene;

D is 2,6-dicarboxylbromobenzene;

E is $C_6H_5COCH_3$;

F is *m*-chloroacetophenone.;

G is *m*-chloroethylbenzene;

H is *o*-aminoethylbenzene;

I is *m*-aminoethylbenzene;

J is *p*-aminoethylbenzene;

Two benzyne intermediates give rise to three products on substitution.

10. Compound A can give rise to three monochlorides on chlorination in sunlight. The ratio of chlorinated products is 22:52:26 in percentage.

One of these chlorinated products (52%) is found to be in S-configuration. Find the number of dichlorides when this is further chlorinated. Find their configurations if they are optically active. Otherwise find which have diastereoisomers, and which are achiral, which occur in meso form. If the halogen is replaced by fluorine in this hydrocarbon and elimination is done with CH_3OK what is the major product? What are the ozonolysis products from this alkene in reducing conditions?

Answer

A is n-pentane. The chloro compound 2-chloropentane is in S-configuration.

The Fischer Projection diagram is

```
      H
      |
Cl ---+--- CH3 (S form of 2-chloropentane).
      |
     C3H7
```

There are 5 dichlorides.

```
        H
        |
I. Cl --+--- CH2Cl   (R form for the dichloride since priority for CH2Cl is higher than
        |
       C3H7
```

that for C_3H_7).

```
         Cl
         |
II. Cl --+--- CH3 Not chiral since two atoms are the same.
         |
        C3H7
```

III. $CH_3CH_2CHClCHClCH_3$

There are two asymmetric carbons and we have diastereoisomers namely (2S, 3R) and (2R, 3S) compounds.

IV. $CH_3CHClCH_2CHClCH_3$ (There are two enantiomers (2R,4R) and (2S,4S) and a meso form).

H
|
V. Cl ----------- CH_3 (s-form)
|
$CH_2CH_2CH_2Cl$

The dehydrohalogenation of 2-fluoropentane gives according to Hoffmann rule 1-pentene as a major product. This by ozonolysis in reducing conditions gives HCHO and $CHOCH_2CH_2CH_3$.

7

ALCOHOLS

NOMENCLATURE

In the common system, alcohols are named as alkyl alcohols. The word alcohol is added after the name of the alkyl group to which the hydroxyl group is attached. For example CH_3OH is methyl alcohol.

In the IUPAC system, the names of saturated alcohols are derived from corresponding alkanes by replacing 'e' of alkanes by 'ol'. Some examples are shown below.

$$CH_3—CH_2—OH$$

Ethanol (IUPAC name)

$$\overset{3}{C}H_3—\overset{2}{C}H_2—\overset{1}{C}H_2—OH$$

Propan-1-ol n-Propyl alcohol
(IUPAC name common name)

$$\overset{4}{C}H_3—\overset{3}{C}H_2—\overset{2}{C}H_2—\overset{1}{C}H_2OH$$

Butan-1-ol (IUPAC name)

The numbering is done such that the carbon atom attached to the —OH group gets the lowest number.

$$\overset{1}{C}H_3—\underset{\displaystyle OH}{\underset{|}{\overset{2}{C}H_2}}—\overset{3}{C}H_2—\overset{4}{C}H_3$$

Butan-2-ol

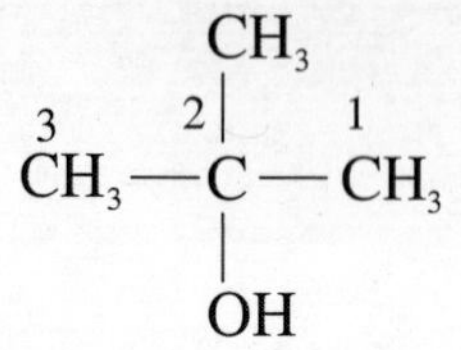

$$CH_3-\overset{CH_3}{\underset{OH}{C}}-CH_3$$

(carbon numbering: 3, 2, 1)

2-methylpropane-2-ol

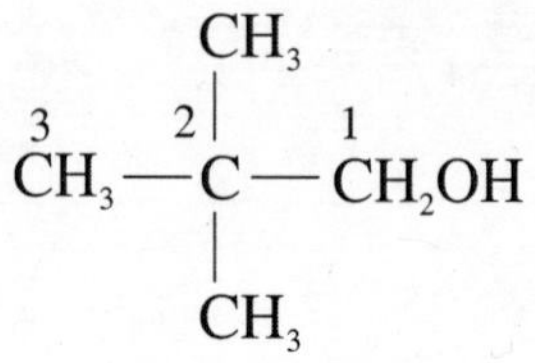

$$CH_3-\overset{CH_3}{\underset{CH_3}{C}}-CH_2OH$$

(carbon numbering: 3, 2, 1)

2, 2-dimethylpropane-1-ol

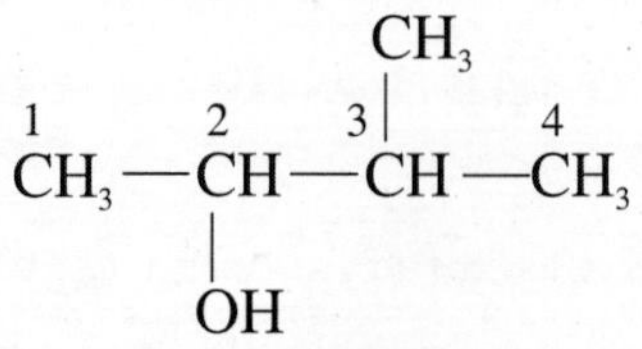

$$CH_3-\underset{OH}{CH}-\overset{CH_3}{CH}-CH_3$$

(carbon numbering: 1, 2, 3, 4)

3-methylbutan-2-ol

For naming polyhydric alcohols, the name of the alkane is retained and the ending "e" is not dropped. Thus dihydric alcohols are named as alkane diols and trihydric alcohols are named as alkane triols (Figure 7.1).

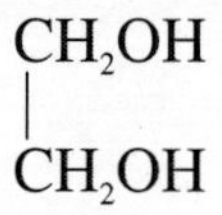

$$\begin{array}{l} CH_2OH \\ | \\ CH_2OH \end{array}$$

Ethane -1,2 diol

$$\begin{array}{l} CH_2OH \\ | \\ CHOH \\ | \\ CH_2OH \end{array}$$

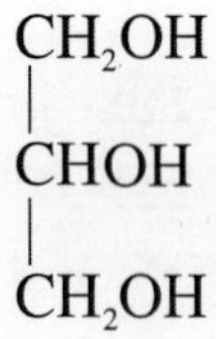

Propane -1,2,3, triol

Figure 7.1 Structures of ethane-1,2-diol and propane-1,2,3-triol

ISOMERISM IN ALCOHOLS

Alcohols exhibit the following types of isomerism.

1. Chain Isomerism

Alcohols with four or more carbon atoms exhibit this type of isomerism in which the carbon skeleton is different.

$$CH_3—CH_2—CH_2—CH_2OH$$

Butan-1-ol

$$\begin{array}{c} \quad CH_3 \\ \quad | \\ CH_3—CH—CH_2OH \end{array}$$

2-Methylbutan-1-ol

2. Position Isomerism

Alcohols with three or more carbon atoms can exhibit position isomerism. In this type of isomerism the position of the functional group, i.e., the —OH group varies. In other words the carbon atoms to which the —OH group is attached is different (Figure 7.2).

$$CH_3—CH_2—CH_2OH$$

Propane-1-ol

$$\begin{array}{c} CH_3—CH—CH_3 \\ | \\ OH \end{array}$$

Propane-2-ol

Figure 7.2 Two isomers of propyl alcohol

3. Functional Isomerism

Alcohols with two or more carbon atoms can exhibit functional isomerism with ethers. Thus ethers and alcohols have the same molecular formula but have different functional groups, hence they are called functional isomers.

$$CH_3CH_2CH_2CH_2OH$$

Butan-1-ol

$$CH_3—CH_2—O—CH_2CH_3$$

Ethoxyethane

Figure 7.3 Functional isomers of formula $C_4H_{10}O$

4. Optical Isomerism

Alcohols containing chiral centre can exhibit enantiomerism or optical isomerism. The optical isomers can rotate the plane of plane polarized angles in different directions (Figure 7.4).

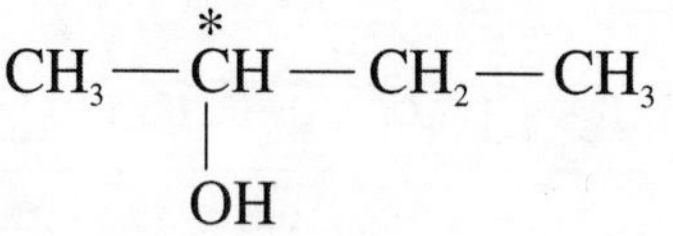

$$CH_3-\overset{*}{C}H(OH)-CH_2-CH_3$$

Butan-2-ol

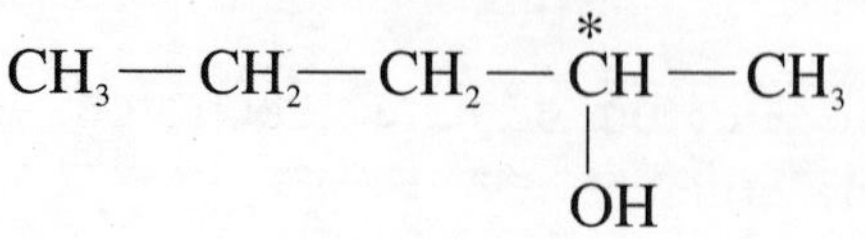

$$CH_3-CH_2-CH_2-\overset{*}{C}H(OH)-CH_3$$

Pentan-2-ol

$$CH_3-CH(CH_3)-\overset{*}{C}H(OH)-CH_3$$

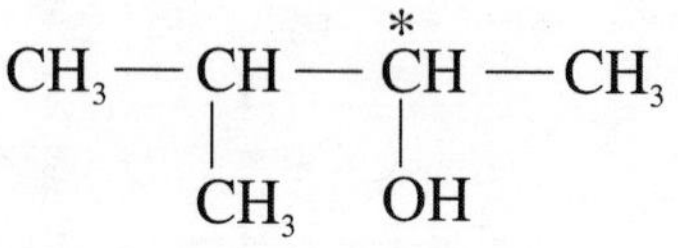

3-Methylbutan-2-ol

(* represents an asymmetric carbon atom.)

Figure 7.4 Examples of optically active alcohols

PHYSICAL PROPERTIES

Alcohols and phenols consist of two parts, an alkyl/aryl group and a hydroxyl group. The properties of alcohols and phenols are due to the —OH group.

The alkyl and aryl groups modify these properties.

1. The lower members of alcohols are colourless, volatile liquids with a characteristic alcoholic smell and burning taste whereas higher alcohols are odourless and tasteless.

   Higher alcohols having 12 or more carbon atoms are colourless waxy solids. Phenols are colourless, crystalline solids or liquids. (They may become coloured due to slow oxidation with air).

2. *Solubility of alcohols and phenols* The first three members are completely miscible with water. The solubility rapidly decreases with increase in molecular mass. The higher members are almost insoluble in water but are soluble in organic solvents like benzene, ether, etc.

   The solubility of lower alcohols is due to the existence of hydrogen bonds between water and polar —OH group of alcohol molecules. Phenols too, are sparingly soluble in water.

   The —OH group in alcohols and phenols contain a hydrogen bonded to an electronegative oxygen atom. Thus they form hydrogen bonds with water molecules.

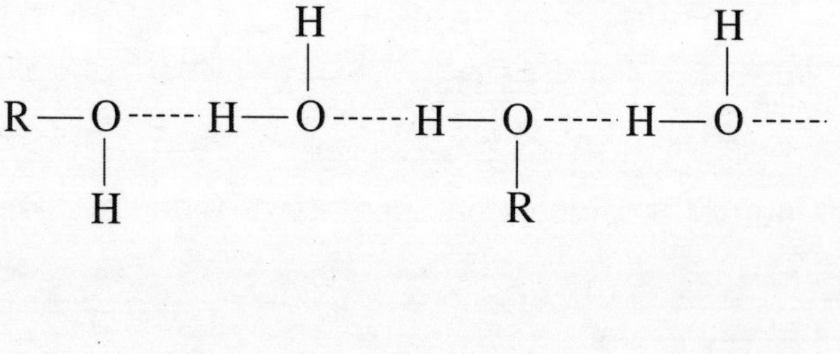

$$R-\underset{H}{\underset{|}{O}}\cdots H-\overset{H}{\overset{|}{O}}\cdots H-\underset{R}{\underset{|}{O}}\cdots H-\overset{H}{\overset{|}{O}}\cdots$$

The solubility of alcohols in water decreases with increase in molecular mass. Due to the increase in molecular mass, the nonpolar alkyl group becomes predominant and masks the effect of polar —OH group.

In addition, among the isomeric alcohols, the solubility increases with branching of chain. As the surface area of the non-polar part in the molecule decreases, the solubility increases.

Phenols are sparingly soluble in water but readily soluble in organic solvents such as alcohol and ether.

HYDROGEN BONDS

For example in alcohols a positively polarized hydrogen atom from one molecule is attracted to a negatively polarized oxygen of another alcohol molecule. This is intermolecular hydrogen bonding which increases the molecular weight and leads to higher boiling point. Generally an electronegative atom is in between two electropositive atoms or vice versa in a hydrogen-bonded system. HF and HCN also have intermolecular hydrogen bonding. In HCN the carbon is sp-hybridized and makes carbon electronegative and correspondingly hydrogen becomes more electropositive and a partial positive charge is created. Nitrogen gets a partial negative charge and NCH—NCH hydrogen bonding is possible. Sometimes the hydrogen bond is formed within a molecule (intramolecular) as in the case of *o*-nitrophenol or *o*-nitrobenzoic acid. A six-membered ring is formed in the *o*-nitrophenol.

Such molecules are steam-volatile and have low boiling points and are easily separated by steam distillation. 1,3-diketones are stable in enol form mainly because of the intramolecular hydrogen bonding.

If we compare diethyl ether and n-butyl alcohol (functional isomers), the alcohol has a boiling point of 117.7°C but ether has a boiling point of 34.6°C. Ethers do not form hydrogen bonds with themselves like alcohols. But their solubilities are similar to that of alcohols in water. This is because they form hydrogen bonds with water. In addition to carboxylic acids and alcohols, amides also form intermolecular hydrogen-bonding and have high boiling points. Boiling point of acetamide is 21.2°C. Primary and secondary amines have high boiling points due to intermolecular hydrogen bonding. When molecules are intermolecularly hydrogen-bonded we call them associated molecules. Peptides also have hydrogen bonds. Trouton's rule namely Δ S(vap) is not 21 cal/mole for alcohols since they are associated by hydrogen-bonding. Molecules like acetone which have no hydrogens bonding have Δ S(vap) = 21 cal/mole.

SOLVED EXAMPLE 1

1,2-propanediol and 1,3-propanediol have higher boiling points than any of butyl alcohols of almost same molecular weight. Explain.

Answer

The two OH groups in these diols give rise to more hydrogen-bonding than the monohydric alcohols.

Alkylation of alcohols Primary alcohols can be converted to *tert*. butyl ethers by dissolving them in a strong acid such as H_2SO_4 by adding isobutylene to the mixture.

$$RCH_2OH + CH_2{=}C(CH_3)_2 \xrightarrow{(H_2SO_4)} RCH_2O{-}\underset{CH_3}{\overset{CH_3}{\underset{|}{\overset{|}{C}}}}{-}CH_3$$

PROTECTION OF ALCOHOLS

Such ethers are used to protect OH groups when a reaction is carried on some other part of the reactant. For example let us convert $OHCH_2CH_2CH_2Br$ to 4-pentyne-1-ol. If we add sodium acetylide to $OHCH_2CH_2CH_2Br$ we get $NaOCH_2CH_2CH_2Br$ + acetylene. Sodium replaces hydrogen of hydroxyl group. The derived acetylenic alcohol is not obtained. The OH group has to be protected. Protection is done as follows. If we add strong acid such as H_2SO_4 along with isobutylene to the mixture, at first OH gets protected and we get

$$BrCH_2CH_2CH_2O{-}\underset{CH_3}{\overset{CH_3}{\underset{|}{\overset{|}{C}}}}{-}CH_3$$

Now addition of sodium acetylide leads to $(CH_3)_3COCH_2CH_2CH_2C{\equiv}C{-}H$

Finally hydrolysis removes the ether group and OH is reintroduced as shown below.

$$BrCH_2CH_2CH_2O{-}\underset{CH_3}{\overset{CH_3}{\underset{|}{\overset{|}{C}}}}{-}CH_3 + NaC{\equiv}C{-}H \rightarrow (CH_3)_3\,COCH_2CH_2CH_2C{\equiv}C{-}H$$

$$(CH_3)_3COCH_2CH_2CH_2C{\equiv}C{-}H + H_2O \rightarrow HO\ CH_2CH_2CH_2C{\equiv}C{-}H$$

The alcohol is 4-pentyne-1-ol.

METHODS OF PREPARATION

From Alkenes

i. *Hydration* Alkenes undergo hydration (addition of water across C=C bond) in the presence of dilute H_2SO_4 to produce alcohols. The alkyl hydrogen sulphate is formed which on hydrolysis with hot water gives alcohol (Figure 7.5).

$$>C=C< \xrightarrow{HOH} >\underset{H}{C}-\underset{OH}{C}<$$

Figure 7.5 Preparation from alkenes

The addition of water to the double bond is in accordance with Markovnikov rule. The alkene is obtained by cracking of hydrocarbons. The alkene is then absorbed by passing it into sulphuric acid at 353 K and 30 atmosphere pressure. The acid is diluted and treated with steam to release the alcohol.

The preparation of ethyl alcohol is done starting with ethene.

$$\underset{\text{Ethene}}{CH_2{=}CH_2} + H_2SO_4 \longrightarrow \underset{\text{Ethyl hydrogen sulphate}}{CH_3CH_2H_2SO_4} \xrightarrow[-H_2SO_4]{H_2O} \underset{\text{Ethanol}}{CH_3CH_2OH}$$

Similarly isopropyl alcohol is prepared from propene.

$$CH_3-CH{=}CH_2 + H_2SO_4 \longrightarrow \underset{\text{Iso-propyl hydrogen sulphate}}{CH_3-\overset{OSO_3H}{\overset{|}{CH}}-CH_3} \xrightarrow[-H_2SO_4]{H_2O} \underset{\text{Iso-propyl alcohol}}{CH_3-\overset{OH}{\overset{|}{CH}}-CH_3}$$

ii. *Oxymercuration–demercuration* Alkenes react with mercuric acetate in the presence of water to yield hydroxy mercurial compounds. These are reduced to alcohols by sodium borohydride (Figure 7.6).

$$>C=C< + Hg(OAc)_2 + H_2O \longrightarrow >C=C<(HgO) \xrightarrow{NaBH_4} -\underset{OH}{C}-\overset{H}{\underset{H}{C}}-H$$

Figure 7.6 Oxymercuration–demercuration

- This reaction gives a good yield of alcohol.

The alcohol obtained corresponds to Markovnikov's addition of water to an alkene.

$$CH_3-CH_2-CH=CH_2 \xrightarrow[H_2O]{Hg(OAc)_2} CH_3 \quad CH_2-\underset{OH}{CH}-CH_3=CH_3$$

iii. *Hydroboration* Alkenes react with diborane (B_2H_6), which is an electron-deficient molecule to yield alkylboranes (R_3B). These are oxidized to alcohols on reaction with hydrogen peroxide in the presence of alkali (Figure 7.7).

$$CH_3-CH=CH_2 + H-BH_2 \longrightarrow CH_3-\underset{H}{CH}-\underset{BH_2}{CH_2}$$

n-Propyl borane

$$\downarrow CH_3CH=CH_2$$

$$(CH_3-CH_2-CH_2)_3B \xleftarrow{CH_3CH=CH_2} (CH_3CH_2CH_2)_2\ BH$$

Tri-n-Propyl borane — Di-n-Propyl borane

$$(CH_3CH_2CH_2)_2\ B \xrightarrow{3H_2O_2/OH^-} 3CH_3-CH_2CH_2OH + B(OH)_3$$

Propan-1-ol — (i.e.,H_3BO_3)

Figure 7.7 Hydroboration oxidation of alkenes

In each addition step, the boron atom is attached to the sp^2 carbon atom that is bonded to greater number of hydrogen atoms. The hydrogen atom of the boron atom attaches to the other carbon of the double bond. Thus this is anti-Markovnikov's addition.

During the oxidation of trialkyl borane, boron is replaced by—OH group. The yield of alcohol in this method is good and the product is easy to isolate.

From Grignard's Reagent

Grignard reagents (R MgX) are alkyl or aryl magnesium halides.

The C $\leftarrow$ Mg bond in Grignard reagent is a highly polar bond as carbon is electronegative relative to electropositive magnesium. Due to this polar nature of C—Mg bond, Grignard reagents are very versatile magnets in organic synthesis.

Grignard reagents react with aldehydes and ketones to form products, which decompose with dil. HCl or dil H_2SO_4 to give primary, secondary and tertiary alcohols (Figure 7.8).

$$>C{=}O + \overset{\delta-}{R}{\leftarrow}\overset{\delta+}{MgX} \longrightarrow >C(R){-}\bar{O}{-}MgX \text{ Adduct}$$

$$\xrightarrow{HOH} -C(R){-}OH + Mg(OH)X$$

Alcohol

Figure 7.8 Preparation from Grignard's reagent

The overall result is to bind the alkyl group of Grignard reagent to carbon of the carbonyl group and hydrogen to oxygen. Formaldehyde gives primary alcohol whereas all other aldehydes give secondary alcohols and ketones furnish tertiary alcohols.

$$>C{=}O \xrightarrow{CH_3MgBr} \left[H{-}C(H)(CH_3){-}OMgBr\right] \xrightarrow{H_2O/H^+} CH_3CH_2OH + Mg(OH)Br$$

Methanol (formaldehyde) — Ethanol — Hydroxy magnesium bromide

$$CH_3{-}C(H){=}O \xrightarrow{CH_3MgBr} \left[CH_3{-}C(CH_3)(H){-}OMgBr\right] \xrightarrow{H_2O/H^+} CH_3{-}CH(CH_3){-}OH + Mg(OH)Br$$

Ethanol (Acetaldehyde) — Propan-2-ol

$$CH_3-\underset{\underset{CH_3}{|}}{C}=O \xrightarrow{CH_3MgBr} \left[CH_3-\overset{\overset{CH_3}{|}}{\underset{\underset{CH_3}{|}}{C}}-OMgBr \right] \xrightarrow{H_2O/H^+} CH_3-\overset{\overset{CH_3}{|}}{\underset{\underset{CH_3}{|}}{C}}-OH + Mg\begin{matrix} Br \\ OH \end{matrix}$$

Propanone (Acetone) — 2-methylpropan-2-ol

From Aldehydes and Ketones

Aldehydes and ketones are reduced to the corresponding alcohols by

a) Addition of hydrogen in the presence of catalysts like finely divided platinum, palladium, nickel and ruthenium.

$$R-CHO + H_2 \xrightarrow{Pd} RCH_2OH$$

This method is called catalytic hydrogenation.

b) Treatment with chemical reagents such as sodium borohydride ($NaBH_4$) or lithium aluminium hydride ($LiAlH_4$).

$$\underset{\text{Ethanal}}{CH_3CHO} + 2[H] \xrightarrow[Li\,Al\,H_4]{H_2/Ni} \underset{\text{Ethanol}}{CH_3CH_2OH}$$

Aldehydes yield primary alcohols while ketones give secondary alcohols (Figure 7.9).

$$\underset{\text{Propanone}}{\begin{matrix} CH_3 \\ \quad \diagdown \\ \quad C=O \\ \quad \diagup \\ CH_3 \end{matrix}} + 2[H] \xrightarrow[Li\,Al\,H_4]{H_2/Ni} \underset{\text{Propan-2-ol}}{CH_3-\underset{\underset{OH}{|}}{CH}-CH_3}$$

Figure 7.9 Alcohols from ketones

From Carboxylic Acids and Esters

Carboxylic acids are reduced to primary alcohols in the presence of strong reducing agent like lithium aluminium hydride.

$$RCOOH \xrightarrow[(ii)\ H_2O]{(i)\ Li\,Al\,H_4} RCH_2OH$$

The yield of alcohol here is high but $LiAlH_4$ being an expensive reagent, this method is not commonly used.

Commercially acids are reduced to alcohols by converting them to the esters followed by their reduction using either

i. Hydrogen in the presence of a catalyst (Catalytic hydrogenation) or

ii. Sodium and alcohol.

$$R-COOH \xrightarrow{R'OH} RCOOR' \xrightarrow[\text{Catalyst}]{H_2} RCH_2OH + R'OH$$

$$\underset{\text{Ethyl ethanoate}}{CH_3-\overset{\overset{\Large O}{\|}}{C}-OC_2H_5} + 4[H] \xrightarrow{Li\,Al\,H_4} \underset{\text{Ethanol}}{2CH_3CH_2OH}$$

REACTIONS OF ALCOHOLS

Alcohols are one of the most important functional groups in organic chemistry and an appreciation of their reactions is critical to success, particularly in synthesis questions.

Alcohols, R—OH, are able to undergo several different types of reactions:

Nucleophilic Substitution	"–OH" acts as the leaving group, or –OH or –O^- can function as the Nu.
Nucleophilic Addition	-OH acts as the Nu forming acetals with aldehydes or ketones.
Nucleophilic Acyl Substitution	-OH acts as the Nu forming esters with carboxylic acids and derivatives.
Elimination	-OH is removed (as water), C=C formed.
Oxidation	-OH is converted to C=O.

A more detailed summary of their specific reactions is given in Figure 7.10.

Substitution with HX $H{-}X + {-}\overset{|}{\underset{|}{C}}{-}OH \xrightarrow[X=Br,\ Cl]{} {-}\overset{|}{\underset{|}{C}}{-}X + H{-}OH$

Substitution by $SOCl_2$, etc. $SOCl_2\ {-}\overset{|}{\underset{|}{C}}{-}OH + \xrightarrow[Et_3N \text{ or pyridine}]{} {-}\overset{|}{\underset{|}{C}}{-}Cl + SO_2 + H{-}Cl$

Conversion to Tosylates (for further substitution) ${-}\overset{|}{\underset{|}{C}}{-}OH + TSCl \xrightarrow[\text{pyridine}]{} {-}\overset{|}{\underset{|}{C}}{-}O{-}Ts + H{-}Cl$

Conversion to Ethers ${-}\overset{|}{\underset{|}{C}}{-}OH \overset{B:^-}{\rightleftharpoons} {-}\overset{|}{\underset{|}{C}}{-}O^- \xrightarrow{R'{-}X} {-}\overset{|}{\underset{|}{C}}{-}O{-}R'$

Conversion to Esters $RC(=O)OH + R'OH \xrightarrow[heat]{H^+} RC(=O)OR' + H_2O$

Acid-catalysed Dehydration $H{-}\overset{|}{\underset{|}{C}}{-}\overset{|}{\underset{|}{C}}{-}OH \xrightarrow{H^+} {>}C{=}C{<} + H{-}OH$

Oxidation to carbonyls ${-}\overset{OH}{\underset{|}{C}}{-} \overset{[O]}{\rightleftharpoons} {>}C{=}O \overset{[O]}{\rightleftharpoons} {-}C(=O)OH$

Figure 7.10 Summary of the specific reactions of alcohols

The reactions namely, nucleophilic substitution, dehydration, reaction with other chlorinating agents have been already discussed in alkenes chapter and alkyl halides chapters. The esterification reaction will be presented in the section on carboxylic acids. One of the important reactions of alcohols is with halogen acids and this is described in the next section.

Reaction of Alcohols With HX

When alcohols react with HX, a substitution takes place producing an alkyl halide and water.

$$ROH + HX \longrightarrow RX + H_2O$$

The order of reactivity of hydrogen halides is HI > HBr > HCl (HF is generally unreactive). The order of reactivity of alcohols is tertiary > secondary > primary. The reaction is acid-catalysed. They react with HCl, HBr and HI. No reaction takes place in nonacidic NaCl, NaBr, NaI. Primary and secondary alcohols can be converted to halides by treatment with NaX and H_2SO_4.

All the reactions are acid-catalysed. The role of acid for secondary and tertiary alcohols is to generate the carbocations. With methanol and primary alcohols, the function of acid is to produce a substrate in which the leaving group is a weakly basic water molecule rather than the strongly basic hydroxide ion.

Since chloride ion is a weaker nucleophile than bromide or iodide, HCl does not react with primary or secondary alcohols unless zinc chloride or some Lewis acid is added. $ZnCl_2$ forms a complex with the alcohol through association with an unshared pair of electrons on the oxygen atom. This provides a better leaving group than water.

$$R{-}\ddot{O}(H){:} + ZnCl_2 \longrightarrow R{-}O^{+}(H){-}\bar{Z}nCl_2$$

$$Cl^{-} + R{-}O^{+}(H){-}\bar{Z}nCl_2 \longrightarrow R{-}Cl + [Zn(OH)_2Cl_2]^{-}$$

$$[Zn(OH)_2Cl_2^{-}]^{-} + H^{+} \longrightarrow ZnCl_2 + H_2O$$

Another important product of alcohols used in synthetic organic chemistry is the alkoxides. The details of alkoxide preparation is given in the next section.

Reaction of Alcohols with Na

Alcohols react with Na or K to form alkoxides. Sodium and potassium alkoxides are much used in organic chemistry.

$$CH_3CH_2OH + 2Na \longrightarrow CH_3CH_2ONa + H_2$$

The reactivity of alcohols towards metals like sodium is $R_3COH < R_2CHOH < RCH_2OH$.

When tertiary alcohol is primary or secondary, the reactivity of Na is sufficient in concentrated solutions of alkoxide in alcohol to be easily prepared. For tertiary alcohols, potassium is used. Alternatively RONa are prepared by reacting ROH with NaH.

$$ROH + NaH \longrightarrow RONa + H_2$$

When tertiary alcohols react with HX, the reaction may go through carbocations and rearranged products are possible. Hence from the synthetic point of view the reaction with PBr_3 or $SOCl_2$ leading to the bromide or chloride, is more useful. These methods are given below.

Reaction of Alcohols with PBr_3

PBr_3 reacts with alcohols to give RBr and H_3PO_3. Phosphorous acid is water-soluble and may be removed by washing the alkyl bromide with water in dilute aqueous base. These reactions do not involve rearrangement whereas reactions of alcohols with HCl may involve carbocation formation leading to rearranged products.

The next section discusses oxidation reactions of alcohols.

Conversion of Alcohols into Aldehydes and Ketones (Oxidation of Alcohols)

$$\mathrm{-\overset{OH}{\underset{|}{\overset{|}{C}}}-} \underset{[R]}{\overset{[O]}{\rightleftharpoons}} \mathrm{\overset{O}{\overset{\|}{C}}} \; (\diagup\;\diagdown) \underset{[R]}{\overset{[O]}{\rightleftharpoons}} \mathrm{\diagup\overset{O}{\overset{\|}{C}}\diagdown OH}$$

The type of the reaction involves oxidation–reduction. The common reagents used are given below.

Aqueous oxidants		Anhydrous oxidants
Chromic acid	H_2CrO_4	Pyridinium chlorochromate (PCC)
Chromate salts	CrO_4^{2-}	Pyridinium cation ($C_5H_5N^+$–H) $ClCrO_3^-$
		Pyridinium dichromate (PDC)
Dichromate salts	$Cr_2O_7^{2-}$	[Pyridinium cation ($C_5H_5N^+$–H)] $Cr_2O_7^{2-}$
Permanganate	MnO_4^-	

- The outcome of oxidation reactions of alcohols depends on the substituents on the carbinol carbon.
- In order for each oxidation step to occur, there must be a H on the carbinol carbon.
- Primary alcohols can be oxidized to aldehydes or further to carboxylic acids.
  - In aqueous media, the carboxylic acid is usually the major product.
  - PCC or PDC, which are used in dichloromethane, allow the oxidation to be stopped at the intermediate aldehyde.

$$R\text{—}CH_2\text{—}OH \underset{[R]}{\overset{[O]}{\rightleftharpoons}} RCHO \underset{[R]}{\overset{[O]}{\rightleftharpoons}} RCOOH$$

- Secondary alcohols can be oxidized to ketones but no further.

$$R_2CH\text{—}OH \xrightarrow{[O]} RCOR$$

- Tertiary alcohols cannot be oxidized (no carbinol C—H).

$$R_3COH \xrightarrow{[O]} \text{No reaction}$$

The next section describes mechanism of oxidation of alcohols by chromate salt.

Mechanism of Oxidation of Alcohols by Chromates

The mechanism of chromate oxidation of alcohols is given below.

The first step is the formation of chromate ester of the alcohol. This is shown in Figure 7.11.

The chromate ester is unstable and it transfers a proton to a base (water) and simultaneously eliminates as $HCrO_3^-$ ion. The overall result of the second step is the reduction of $HCrO_4^{2-}$ to $HCrO_3^-$, a 2-electron change in the oxidation state of chromium from Cr(VI) to Cr(IV). Simultaneously alcohol is oxidized to the ketone.

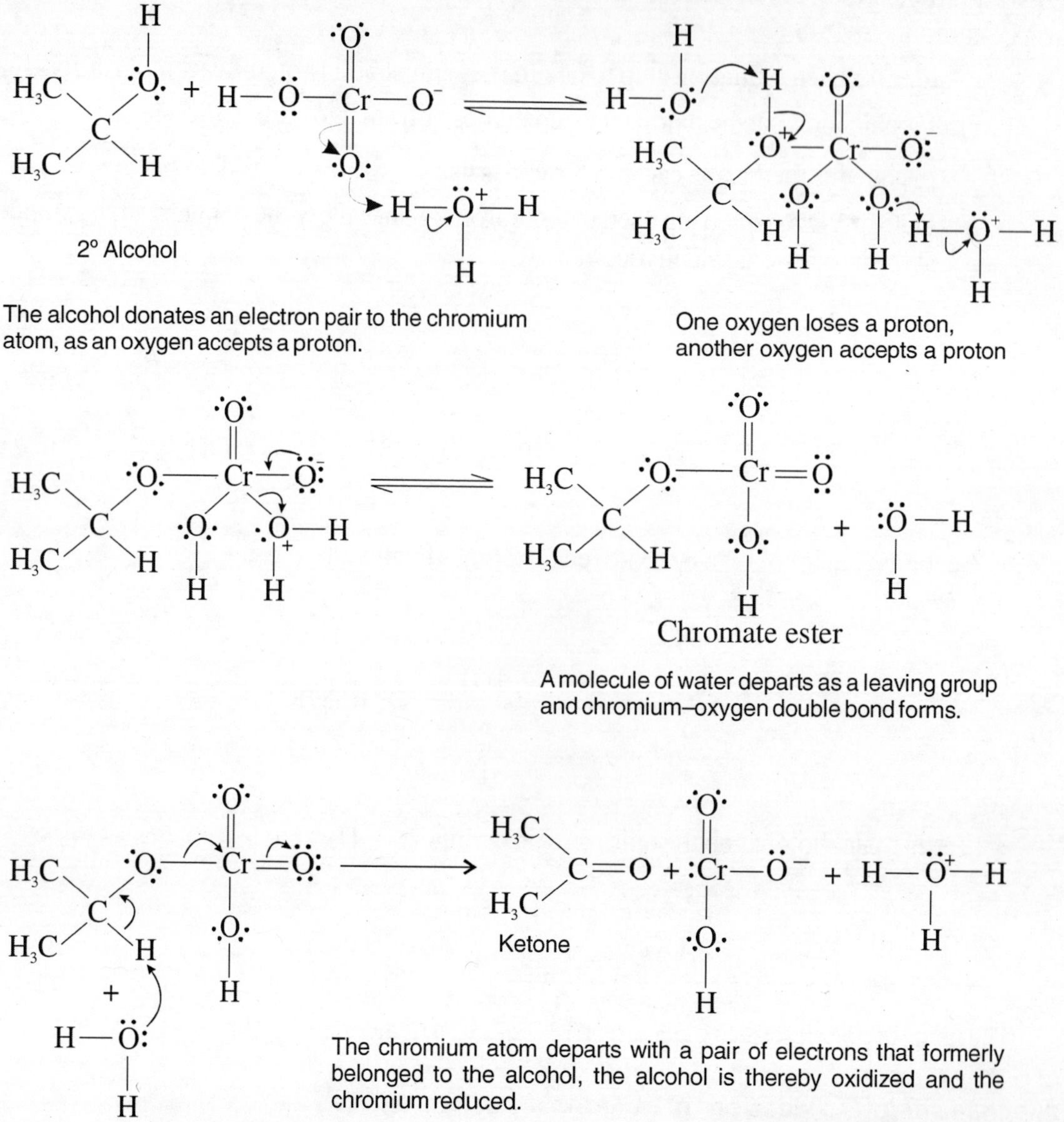

2° Alcohol

The alcohol donates an electron pair to the chromium atom, as an oxygen accepts a proton.

One oxygen loses a proton, another oxygen accepts a proton

Chromate ester

A molecule of water departs as a leaving group and chromium–oxygen double bond forms.

Ketone

The chromium atom departs with a pair of electrons that formerly belonged to the alcohol, the alcohol is thereby oxidized and the chromium reduced.

Figure 7.11 Mechanism of oxidation of alcohols by chromates

Alcohols and Haloform Reaction

Those alcohols which get oxidized to methyl ketones (CH_3COR) or CH_3CHO give the haloform reaction which is the conversion of methyl ketones by NaOH and I_2 or Br_2 or Cl_2 to CHX_3 and RCOONa. NaOH and halogen oxidize alcohols to ketones or carbonyl compounds and those having CH_3CO group answer haloform reaction. The iodoform formed is a yellow solid.

TESTS TO DISTINGUISH THREE TYPES OF ALCOHOLS

Three test tubes contain a solution of zinc chloride ($ZnCl_2$) in concentrated aqueous hydrochloric acid (HCl). The primary alcohol, 1-butanol, the secondary alcohol, 2-butanol, and the tertiary alcohol, 2-methyl-2-propanol, are added to separate test tubes. Immediately the tertiary alcohol forms a cloudy solution as the alcohol is converted to the water-insoluble alkyl chloride by a S_N1 reaction. The primary alcohol, 1-butanol, does not react. The secondary alcohol, 2-butanol, reacts after about 6.5 minutes.

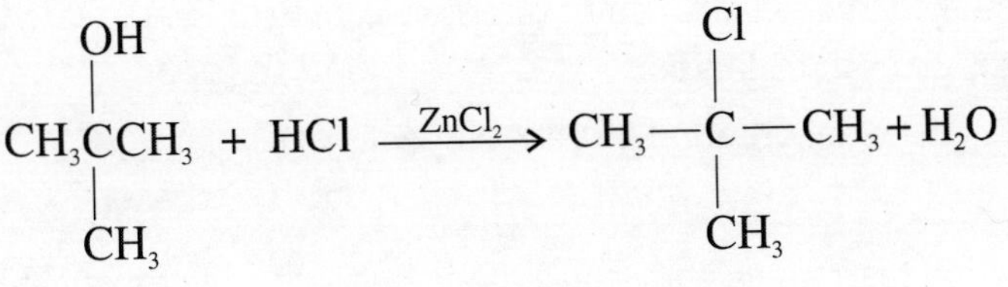

$$CH_3\underset{\underset{CH_3}{|}}{\overset{\overset{OH}{|}}{C}}CH_3 + HCl \xrightarrow{ZnCl_2} CH_3-\underset{\underset{CH_3}{|}}{\overset{\overset{Cl}{|}}{C}}-CH_3 + H_2O$$

PROBLEMS

I. "EXPLAIN WHY" QUESTIONS AND SMALL STRUCTURE-BASED PROBLEMS

1. Ethanol with dry NaBr has no reaction. But when ethanol is mixed with ($NaBr + H_2SO_4$), ethyl bromide is formed. Why?

   Answer

   —OH is a poor leaving group but when $NaBr + H_2SO_4$ is there alcohol is protonated $C_2H_5OH_2^+$ and H_2O is a good leaving group and Br^- attaches to C_2H_5 group to give bromide.

2. Neopentyl alcohol on dehydration gives the main product as 2-methyl-2-butene. Explain.

   Answer

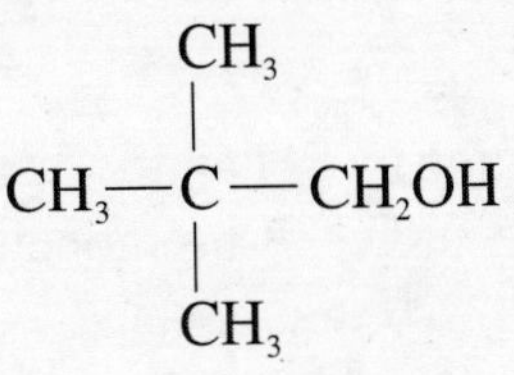

   $$CH_3-\underset{\underset{CH_3}{|}}{\overset{\overset{CH_3}{|}}{C}}-CH_2OH$$ gives first the primary carbocation.

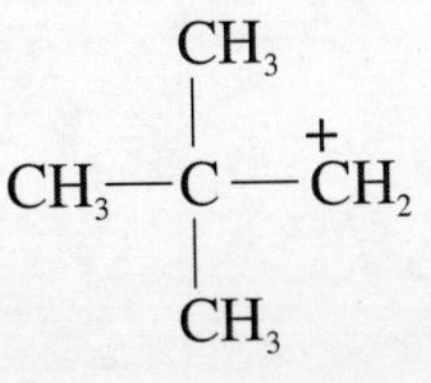

   $$CH_3-\underset{\underset{CH_3}{|}}{\overset{\overset{CH_3}{|}}{C}}-\overset{+}{C}H_2$$ Which rearranges to more stable tertiary cation.

$$CH_3-\overset{\overset{\displaystyle CH_3}{|}}{\underset{+}{C}}-CH_2CH_3$$ Which loses a proton to give 2-methyl-2-butene.

3. (1-methyl)cyclobutyl ethanol on dehydration gives a product which by ozonolysis in reducing conditions gives $CH_3COCH_2CH_2CH_2COCH_3$. Explain.

   Answer

   The carbocation rearranges with ring expansion leading to 1,2-dimethyl cyclopentene which by ozonolysis gives the diketone.

4. Cyclobutylmethanol on dehydration gives cyclopentene. Explain.

   Answer

   The carbocation rearranges with ring expansion to give cyclopentene.

5. $(C_6H_5)_2CHCH_2OH$ with HBr gives mainly $C_6H_5CHBrCH_2C_6H_5$. Explain.

   Answer

   Ph group migrates preferentially compared to methyl when alcohol is protonated to give the compound $C_6H_5CHBrCH_2C_6H_5$.

6. Neopentyl alcohol cannot be converted to neopentyl chloride by HCl but $SOCl_2$ or PCl_3 can bring about this conversion. Explain.

   Answer

   In $SOCl_2$ or PCl_3 no rearrangement takes place but in HCl carbocation rearrangement takes place.

7. Hydration of 3-phenyl-1-butene is not a good method to form 3-phenyl-2-butanol. Why?

   Answer

   The carbocation which is secondary changes to tertiary by hydride shift to yield 2-phenyl-2-butanol.

8. 3-pentanol with HBr gives a mixture of 3- and 2-bromopentanes. Explain.

   Answer

   A secondary alcohol like 3-pentanol can undergo substitution by S_N1 and S_N2 pathways. By S_N2 pathway it gives 3-bromopentane. By S_N1 pathway it gives 3-bromopentane due to secondary carbocation. In addition hydride shift leads to 2-bromopentane.

9. Lucas test is suitable for distinguishing alcohols that contain 6 or lower number of carbons only. Why?

   Answer

   The higher member alcohol are not soluble.

10. Oxymercuration–demercuration in $NaBH_4$ gives 3,3-dimethyl-2-butanol but acid-catalysed hydration yields 2,3-dimethyl-2-butanol. Explain.

Answer

Hydration in acid involves carbocation and rearrangement is possible, but oxymercuration–demercuration does not involve rearrangement.

II. Multiple Choice Questions (One out of four answers is correct)

1. The alcohol which does not undergo any rearrangement to give alkene is
   a) Ethanol　b) Cyclobutylmethanol
   c) (1methyl)cyclopentyl ethanol　d) 3,3-dimethyl-2-butanol
2. Which one of the following will be most readily dehydrated in acidic condition?
   a) $CH_3COCH_2CHOHCH_3$　b) $CH_3CHOHCH_2CH_3$
   c) $CH_3COCHOHCH_2CH_3$　d) $CH_3COCH_2CH_2CHOHCH_3$
3. One of the following alcohols does not undergo dehydration
   a) $C_6H_5CH_2OH$　b) $(CH_3)_3CCH_2OH$
   c) C_2H_5OH　d) 3,3-dimethyl-2-butanol
4. An alcohol cannot be converted to RX by
   a) $POCl_3$　b) PCl_3　c) PBr_3　d) $SOCl_2$
5. The alcohol which does not undergo oxidation is
   a) $(C_2H_5)_3COH$　b) Ethanol
   c) $(CH_3)_3CCH_2OH$　d) Isopropyl alcohol
6. A mixture of benzaldehyde and formaldehyde on heating with aq. NaOH solution gives
   a) benzyl alcohol and sodium formate　b) sodium benzoate and methyl alcohol
   c) sodium benzoate and sodium formate　d) benzyl alcohol and methyl alcohol
7. 1-propanol and 2-propanol can be distinguished by
   a) oxidation with alkaline $KMnO_4$ followed by reaction with Fehling solution
   b) oxidation with acidic dichromate followed by reaction with Fehling solution
   c) oxidation by heating with Cu followed by reaction with Fehling solution
   d) oxidation with conc. H_2SO_4 followed by reaction with Fehling solution
8. An alkene on ozonolysis in reducing conditions gives acetone and HCHO. This alkene on hydration with H_2SO_4 gives
   a) isopropyl alcohol　b) 1-butanol　c) 2-butanol　d) *t*-butyl alcohol
9. Acetic acid is obtained by oxidation of
   a) methanol　b) *t*-butyl alcohol　c) isopropyl alcohol　d) ethanol

10. One of the following does not undergo haloform reaction

a) acetone b) ethanol c) $C_6H_5CHOHCH_3$ d) CH_3COOH

11. When ethanol is oxidized with chromate ion, the oxidation requires ________ Faradays.

a) 1 b) 2 c) 3 d) 4

12. When alcohols react with $SOCl_2$ or PCl_3 or PBr_3 triethylamine or pyridine is used as a

a) nucleophile b) electrophile

c) catalyst d) base used to mop up the acidic product

13. Reaction of primary alcohols with HCl is carried out in the presence of $ZnCl_2$ which

a) acts as a catalyst

b) acts as a Lewis Base

c) forms a complex with alcohol and provides poorer leaving group than water

d) forms a complex with alcohol and provides a better leaving group than water

14. One of the following statements is wrong.

a) The reaction of alcohols with HBr will involve carbocations and rearranged products are possible.

b) In the conversion of alcohols to RCl, $ZnCl_2$ forms complex with alcohol and provides better leaving group than water.

c) The reaction of ROH with $SOCl_2$ gives RCl without any rearrangement.

d) Pyridine is used as a catalyst in the reaction of ROH with $SOCl_2$.

15. A seven-carbon tertiary alcohol on dehydration gives only one alkene.

a) $CH_3-C(CH_3)_2-C(CH_3)_2-OH$ (structure printed as: CH_3-C-CH_3-OH with CH_3 above and below each of the two central carbons)

b) $CH_3-C(CH_3)(OH)-CH_2-CH(CH_3)-CH_3$

c) $CH_3-C(CH_3)(OH)-CH(CH_3)-CH_2CH_3$

d) $CH_3-C(CH_3)(OH)-CH_2-CH_2-CH_2CH_3$

16. The role of acids in the case of tertiary and secondary alcohols is

a) to produce stable free radicals b) stable carbocations

c) to create a better leaving group d) no suitable answer

17. The leaving group in the S_N2 reaction of alcohols with $SOCl_2$ in pyridine is

a) pyridine b) SO_2 c) Cl^- d) $ClSO_2^-$

18. When a chiral alcohol is treated with HCl, the product

a) exhibits geometric isomerism b) is optically active

c) is the meso form d) is racemic

19. An alcohol by acid-catalysed dehydration gives alkene A which by ozonolysis in dimethyl sulphide gives C_6H_5CHO only. A is

a) $PhCHOHCH_2Ph$ b) $PhCH_2CH_2OH$

c) $PhCH_2CH_2CH_2CH_2OH$ d) $(Ph)_3COH$

20. The optically active alcohol among the following is

a) $(C_2H_5)_3COH$ b) CH_3CH_2OH

c) n-butanol d) $CH_3CHOHCH_2CH_3$

21. If ROH has an optically active alkyl group with R configuration, after reaction with PBr_3 the product

a) has S-configuration b) consists of racemic mixture

c) has R configuration d) loses the asymmetric centre

22. 2-phenyl-2-propanol can be obtained by the following combination of reactants

a) $C_6H_5COCH_3 + CH_3MgBr + H^+$ b) $C_6H_5MgBr + CH_3COCH_3 + H^+$

c) $C_6H_5MgBr + CH_3CHO + H^+$ d) $C_6H_5CHO + 2C_3H_7MgBr + H^+$

KEY

1. c	2. c,d	3. b	4. c	5. d	6. d	7. d
8. d	9. c	10. d	11. c	12. d	13. d	14. d
15. a	16. b	17. d	18. d	19. a	20. d	21. a
22. c						

III. MULTIPLE CHOICE QUESTIONS (MORE THAN ONE ANSWER IS CORRECT)

1. The alcohols which give haloform reaction are

a) C_2H_5OH b) isopropyl alcohol

c) $C_6H_5CHOHCH_3$ d) CH_3OH

2. Optically active alcohols among the following are

a) 2-butanol b) 1-butanol

c) *trans*-2-methylcyclopropanol d) ethanol

3. The hydration of some of the following alkenes in sulphuric acid give more than one alcohol
   a) ethylene b) 2-butene c) $(CH_3)_2C{=}CHCH_3$ d) $(CH_3)_3C{-}CH{=}CH_2$
4. Two different ozonolysis products are obtained for each major alkene obtained from dehydration of some of these alcohols.
   a) *t*-butanol b) cyclobutylmethanol
   c) (1-methyl)cyclobutylethanol d) propanol
5. ROH to RX conversion is done using
   a) $ZnCl_2 + HCl$ b) PBr_3 c) $SOCl_2$ d) $POCl_3$
6. Grignard reaction of the following give alcohols.
   a) $RMgX{=}H_2CO + H_2O \rightarrow$ b) $RMgX + CH_3CHO + H_2O \rightarrow$
   c) $RMgX + RCOR + H_2O \rightarrow$ d) $RMgX +$ t-butanol $\rightarrow$
7. 1-pentanol is the product from the reactions of
   a) 1-pentene + $BH_3/THF(H_2O_2/OH^-) \rightarrow$
   b) 1-pentyl bromide + $OH^- \rightarrow$
   c) 1-bromobutane + Mg + ethylene oxide + $H^+ \rightarrow$
   d) 1-bromobutane + Mg + $H_2CO + H_2O \rightarrow$
8. Alcohols produced from some of these reactions do not answer haloform reaction.
   a) 1-pentene + $BH_3/THF(H_2O_2/OH^-) \rightarrow$ b) 1-pentyl bromide + $OH^- \rightarrow$
   c) $CH_3MgX + H_2CO + H_2O \rightarrow$ d) $CH_3COC_6H_5 + NaBH_4 \rightarrow$
9. Some of the following alkenes by ozonolysis in oxidizing conditions followed by addition of $NaBH_4$ give alcohols or diols.
   a) Ethylene b) tetramethylethylene
   c) 1,2-dimethylcyclopentene d) propylene
10. Some of the following compounds react with CH_2N_2 in the absence of catalyst to give —CH_2-inserted products.
    a) phenol b) carboxylic acids c) ethylene d) alcohols
11. The following combinations of RMgX and R′COR′ give the same alcohol.

$$C_6H_5CH_2{-}\underset{\displaystyle OH}{\overset{\displaystyle CH_3}{\underset{|}{\overset{|}{C}}}}{-}CH_2CH_3$$

    a) $C_6H_5CH_2MgCl$ + 2-butanone b) $CH_3CH_2MgBr + C_6H_5CH_2COCH_3$
    c) $CH_3MgI + C_6H_5CH_2COCH_2CH_3$ d) C_6H_5MgCl + 2-butanone

12. Some statements about *t*-butyl alcohol are true.
   a) It is obtained by hydration of 2-methylpropene.
   b) It gives very stable carbocation.
   c) It does not undergo oxidation.
   d) t-butyl alcohol + $POCl_3 \rightarrow$ *t*-butyl chloride.
13. Alcohols obtained in some of these reactions answer haloform reaction
   a) $C_6H_5CHO + CH_3MgBr + H_2O \rightarrow$ b) Acetone + $C_6H_5CH_2MgBr + H_2O \rightarrow$
   c) $CH_3CHO + C_6H_5MgBr + H_2O \rightarrow$ d) $H_2CO + CH_3MgBr + H_2O \rightarrow$
14. The alcohols which undergo oxidation with $KMnO_4$ are
   a) t-butyl alcohol b) isopropyl alcohol
   c) ethanol d) neopentyl alcohol
15. Some of the following reactions lead to the formation of ether
   a) $C_6H_5OH + CH_3l$ b) $CH_3OH + C_6H_5l$
   c) $C_2H_5OH + CH_3l$ d) $C_6H_5OH + C_6H_5l$
16. Some of the statements about ether formed from the following are true.
   a) $C_6H_5CH_2CHOHCH_3$ + KOH$\rightarrow$ followed by addition of C_2H_5Br gives ether with retention of configuration
   b) $C_6H_5CH_2CHOHCH_3$ + *p*-toluenesulphonyl chloride followed by addition of $C_2H_5OH(K_2CO_3)$ gives ether with inverted configuration.
   c) $C_6H_5CH_2CHOHCH_3$ + *p*-toluenesulphonylchloride followed by addition of $C_2H_5OH(K_2CO_3)$ gives ether with retention of configuration.
   d) $C_6H_5CH_2CHOHCH_3$ + KOH followed by addition of C_2H_5Br gives ether with inversion of configuration.
17. *t*-butyl methyl ether is prepared by
   a) t-butyl alcohol + CH_3I b) isobutylene + $H_2SO_4 + CH_3I$
   c) CH_3OH + *t*-butyl iodide d) C_2H_5OH + *t*-butyl iodide
18. The products from some of these reactions do not undergo elimination.
   a) $C_6H_5CH_2OH + PBr_3$ b) $CH_3OH + SOCl_2$
   c) $C_6H_5CH_2CH_2OH + PCl_3$ d) $C_2H_5OH + PBr_3$
19. Some of the following reactions yield 1,2-dimethyl cyclohexene as a product
   a) (1methyl)cyclopentyl ethanol on dehydration
   b) 1-hydroxy-2,2-dimethylcyclohexane on dehydration
   c) (1-methyl)cyclobutyl ethanol on dehydration
   d) 1-methyl-1-hydroxycyclohexane on dehydration

20. Primary alcohols are only obtained in some of the following reactions.

a) $RCHO + Na/C_2H_5OH$ b) $RCOOR + Na/C_2H_5OH \rightarrow$

c) $R_2CO + NaBH_4 \rightarrow$ d) $RMgX + \text{Ethylene oxide} + H^+ \rightarrow$

KEY

1. a,b,c	2. a,c	3. c,d	4. a,d	5. a,b,c	6. a,b,c	7. a,b,d
8. a,b	9. b,c	10. a,b,c	11. a,b,c	12. a,b,c	13. a,b,d	14. b,c,d
15. a,c	16. a,b	17. a,b	18. a,b	19. a,b	20. a,b,d	

IV. Assertion–Reason Questions

The questions consist of assertion in column 1 and reason in column 2. Use the following key to choose the appropriate answer.

a) If both assertion and reason are correct and reason is the correct explanation of assertion

b) If both assertion and reason are correct but reason is not the correct explanation of assertion

c) If assertion is correct but reason is incorrect

d) If assertion is incorrect but reason is correct

No.	Assertion	Reason
1.	CH_3CH_2OH has pKa 18 but CF_3CH_2OH has pKa 12.4.	The –I effect of fluorine increases acidity for the fluoroethanol.
2.	di-tertiary butyl ether can be prepared by Williamson ether synthesis.	Williamson synthesis involves S_N2 pathway and tertiary halides cannot react by S_N2 pathway.
3.	A conc. aq. HBr reacts with ethanol but dry NaBr does not react with ethanol.	The alcohol needs to be protonated to produce $C_2H_5OH + H_2O$ so that H_2O can leave relative to Br^- which will replace it. This requires the presence of acid.
4.	The reaction of CH_3Li with cyclohexanone followed by addition of water and acidification produces methylene cyclohexane only.	The alcohol formed dehydrates giving methylene cyclohexane and 1-methylcyclohexene.
5.	If ROH is heated with NaH we get the alkoxide but if we treat ROH with NaOH we have an equilibrium involving reactants and products as $ROH + NaOH \leftrightarrow RONa + H_2O$.	The base NaH is strong and hence alkoxide formation is favoured.

(Contd.)

No.	Assertion	Reason
6.	CH_3MgBr reacts with H_2CO and H^+ to give a product which answers haloform reaction but product of reaction CH_3MgBr + ethylene oxide + H^+ does not answer haloform reaction.	Both reactions lead to primary alcohols.
7.	$(C_2H_5)_3COH$ with CrO_3 +H_2SO_4 has no reaction.	Tertiary alcohol cannot undergo oxidation like secondary and primary alcohols.
8.	When ketone is reduced with $NaBH_4$, we use aq.NH_4Cl to reduce to alcohol.	Aq. NH_4Cl is a stronger acid than water and hydrolyses the B—O bond.
9.	5-chloropentanal with $NaBH_4$ produces 5-chloropentanol only.	The alkoxide can react with terminal chloride to give cyclic ether by Williamson intramolecular reaction.
10.	$(CH_3)_3C\text{-}CHOHC(CH_3)_3$ reacts much slower than $(CH_3CH_2)_2CHOH$ in oxidation.	Molecular weight of the first compound is high.

KEY

1. a	2. d	3. a	4. d	5. a	6. b	7. a
8. a	9. d	10. b				

V. Matrix Matching

Question I

Column 1	Column 2
a) Ethanol	p) does not oxidize with $KMnO_4$
b) t-butyl alcohol	q) gives C_2H_5Br with aq.HBr.
c) Benzyl alcohol	r) dehydration gives ethylene.
d) di-tertiary butyl ether	s) elimination is not possible.
	t) pKa higher than that of CF_3CH_2OH.
	u) not possible by Williamson synthesis.

Answer

$[a \to q,r,t], [b \to p], [c \to s], [d \to u]$

Question 2

Column 1	Column 2
a) Neopentyl alcohol	p) pyridine is there to mop up HCl.
b) ROH + $SOCl_2$ + pyridine	q) beta hydrogen is absent.
c) 2-methylpropene	r) no ether is possible.
d) $C_6H_5ONa + C_6H_5Cl$	s) dehydration of t-butyl alcohol.
	t) elimination in ionizing solvent produces $(CH_3)_2C{=}CHCH_3$.

Answer

$[a \rightarrow q,t], [b \rightarrow p], [c \rightarrow s], [d \rightarrow r]$

Question3

Column 1	Column 2
a) Ethanol + CH_2N_2 (no catalyst)	p) anisole is produced.
b) Phenol + CH_2N_2	q) methylacetate is produced.
c) $CH_3COOH + CH_2N_2$	r) no reaction.
d) Acetone + $NaBH_4$	s) the product answers haloform reaction.

Answer

$[a \rightarrow r], [b \rightarrow p], [c \rightarrow q], [d \rightarrow s]$

VI. Structure-Based and Numerical Problems

1. Compound A is obtained by decarboxylation of potassium salt of succinic acid. A with Br_2/H_2O gives B. B is reacted with C in H_2SO_4 to get D. C is obtained by dehydration of E whose molecular weight is 74 grams. E gives rise to a very stable carbocation. D reacts with sodium salt of F to give G. F on partial hydrogenation in Lindlar Pd gives A. G on hydrolysis gives H. Find A to H.

Answer

A is ethylene;

B is $OHCH_2CH_2Br$;

C is 2-methylpropene;

$$\text{D is } BrCH_2CH_2O-\overset{\overset{\displaystyle CH_3}{|}}{\underset{\underset{\displaystyle CH_3}{|}}{C}}-CH_3;$$

E is *t*-butyl alcohol;

F is acetylene;

G is $(CH_3)_3-C-O-CH_2CH_2-C{=}CH$;

H is $CH \equiv CCH_2CH_2OH$.

2. Alcohol A on dehydration gives B. A and its oxidation product both answer haloform reaction. B with Br_2 in H_2O gives C. C with NaOH gives D. D on acidification gives E. E with HIO_4 undergoes cleavage to give the same products F and G as the ozonolysis of B in reducing conditions. Ethyl benzene with ZnO gives B. Find A to G.

Answer

A is $C_6H_5CHOHCH_3$;

B is styrene;

C is $C_6H_5CHOHCH_2Br$;

D is phenyloxirane;

E is $C_6H_5CHOHCH_2OH$;

F is HCHO;

G is C_6H_5CHO;

3. Compound A is an epoxide and reacts with $LiAlH_4$ to give a compound B which can make hydrogen bond. A with PCl_3 gives C. C with Li, followed by addition of acetone followed by addition of water gives D. D does not undergo oxidation. D on treatment with acid gives E which by ozonolysis gives cyclohexanone and acetone. Find A to E.

Answer

A is cyclohexene oxide;

B is cyclohexanol;

C is Chlorocyclohexane;

$$\text{D is } C_6H_{11}-\overset{\overset{\displaystyle CH_3}{|}}{\underset{\underset{\displaystyle CH_3}{|}}{C}}-OH;$$

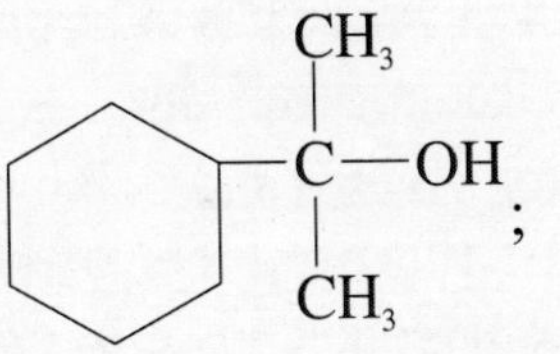

E is isopropylidene cyclohexane;

4. Compound A has a benzene ring and does not undergo elimination but undergoes nucleophilic substitution by S_N1 and S_N2 pathways. A on treatment with NaOH gives B. A and B undergo oxidation to give the same acid C. A and B in the presence of NaOH give rise to E. (Grignard reagent of A with H_2CO in acid gives F.) F also on oxidation gives the same acid similar to the oxidation of A or B. B with HCl gives G. G is treated with Mg in ether and then mixed with D_2O to get H. B on treatment with MnO_2 gives D which is one of the oxidation products of ozonolysis in reducing conditions of J. J is obtained by treatment of ethylbenzene with ZnO. Find A to J.

Answer

A is benzyl bromide;

B is benzyl alcohol;

C is benzoic acid;

D is C_6H_5CHO;

E is $C_6H_5CH_2OCH_2C_6H_5$;

F is $C_6H_5CH_2CH_2OH$;

G is $C_6H_5CH_2Cl$;

H is $C_6H_5CH_2D$;

J is styrene;

5. An alcohol A is obtained by hydration of 2-methylpropylene. A was reacted with Na metal and when the metal was consumed ethyl bromide is added to get B. In another experiment Na metal is made to react with C, an alcohol obtained by reaction of C_2H_5Br with OH^-. Then *t*-butyl bromide is added to the above mixture, and a gas D evolved. Work-up of the resulting mixture yielded only ethanol and an inorganic salt. Identify A to D and account for the difference in behaviour of the two experiments.

Answer

A is *t*-butyl alcohol

B is $(CH_3)_3COC_2H_5$

C is C_2H_5OH C_2H_5ONa + t-butylbromide $\rightarrow (CH_3)_2C{=}CH_2 + CH_3CH_2OH + NaBr$

D is $(CH_3)_2C{=}CH_2$, a gas which escaped.

When an alkoxide reacts with alkyl halide, the alkyl halide has to be primary for S_N2 reaction. If halide is tertiary elimination takes place leading to alkene.

6. A cyclic alcohol A with formula $C_8H_{16}O$ on dehydration gives B and C. B by ozonolysis in dimethyl sulphide gives D which is also one of the products of ozonolysis of $CH_2{=}C(CH_3){-}(CH_2)_4{-}C(CH_3){=}CH_2$ under same conditions. C by ozonolysis under these conditions gives cyclopentanone, acetone and water. B is also produced when (1-

methyl)cyclopentyl ethanol is dehydrated. Find the hydrocarbons E and F when B and C are reduced. How many monochloro products are possible for E and F. Find A to F. Can A undergo oxidation? What is the product?

Answer

A is 2,2-dimethylcyclohexanol;

B is 1,2-dimethylcyclohexene;

C is isopropylidene cyclopentane;

D is $CH_3C(=O)-(CH_2)_4-C(=O)-CH_3$;

E is 1,2-dimethylcyclohexane (cyclohexane ring bearing CH_3 on two adjacent carbons);

F is isopropylcyclopentane;

E gives 4 monochloro derivatives and F gives 5 monochloro derivatives. Since A is a secondary alcohol it undergoes oxidation to give a ketone which is 2,2-dimethylcyclohexanone.

7. Identify the reagents in the following reaction scheme:

$$\text{Cyclohexanone} \xrightarrow{A} \text{cyclohexanol} \xrightarrow{B} \text{cyclohexylbromide} \xrightarrow{\text{C and D}}$$

$$\text{cyclohexylmethanol} \xrightarrow{E} \text{cyclohexyl} - \text{CHO} \xrightarrow{F} \text{cyclohexylbenzylmethanol}$$

$$\xrightarrow{G} \text{cyclohexylphenylethylene}$$

Answer

A is $NaBH_4$;

B is PBr_3;

C is Mg/ether;

D is HCHO;

E is PCC in CH_2Cl_2;

F is $C_6H_5CH_2MgBr$;

G is $POCl_3$ and pyridine

8. A and B are two hydrocarbons differing in molecular weight by 2 grams. A is obtained by dehydration of C which has 4 carbons. C cannot undergo oxidation. Sodium salt of succinic acid by Kolbe electrolysis gives D. D with Br_2/H_2O gives E. E reacts with A and H_2SO_4 to give F. F with sodium acetylide gives G. G on hydrolysis gives H. H gives a precipitate with amm. $AgNO_3$ and also answers test for alcohols. B gives only one brominated compound in high yield. On chlorination it gives the primary and tertiary chlorides in ratio 64:36 %. Find A to H.

Answer

A is 2-methylpropylene;

B is isobutane;

C is tertiary butyl alcohol;

D is ethylene;

E is $BrCH_2CH_2OH$;

F is $(CH_3)_3CO—CH_2CH_2Br$;

G is $(CH_3)_3CO—CH_2CH_2C{\equiv}CH$;

H is 4-butyne-1-ol;

9. Compound A gives precipitate with ammoniacal $AgNO_3$. 3 moles of it on heating in a red hot tube gives an aromatic compound. A with Na gives B. B with C gives D. C is obtained from E by chlorination as one of the three monochlorinated products. This chloro compound undergoes S_N2 reaction and by elimination gives 1-pentene. D on partial reduction with Lindlar /Pd or Na/NH_3 gives F. F with BH_3, THF, H_2O_2/OH^- gives G. G with CrO_3, pyridine and CH_2Cl_2 gives H. H is one of the ozonolysis products of alkene J under reducing conditions. The other ozonolysis product is HCHO. Find A to J.

Answer

A is acetylene

B is sodium acetylide

C is $CH_3CH_2CH_2CH_2CH_2Cl$

D is 1-heptyne

E is n-pentane

F is 1-heptene

G is $CH_3(CH_2)_5CH_2OH$

H is heptanol

I is $CH_3(CH_2)_5CH{=}CH_2$

10. Compound A has molecular weight 98 grams and contains carbon and hydrogen. It decolorizes bromine water and reacts with BH_3/THF, H_2O_2/OH to give B. When heated with $KMnO_4$, B is oxidized to a carboxylic acid which is optically active. A by ozonolysis and work-up with H_2O_2 gives the same compound D which is formed by oxidation of 3-hexanol with chromic acid along with HCOOH. Give 10 isomers of A without cyclic compounds. Which are the isomers that exhibit geometrical isomerism? If A is hydrogenated find the hydrocarbon and give the number of monochlorides formed by it.

Answer

A is $CH_3CH_2—C=CH_2$;
with $CH_2CH_2CH_3$ on the C

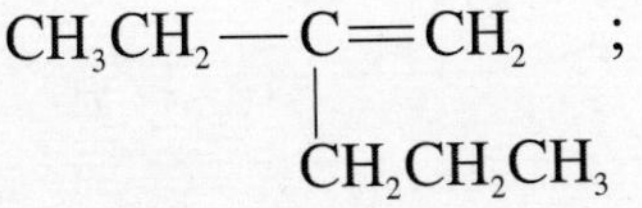

B is $CH_3CH_2—C(H)—CH_2OH$
with $CH_2CH_2CH_3$ on the C

C is $CH_3CH_2—C(H)—COOH$;
with $CH_2CH_2CH_3$ on the C

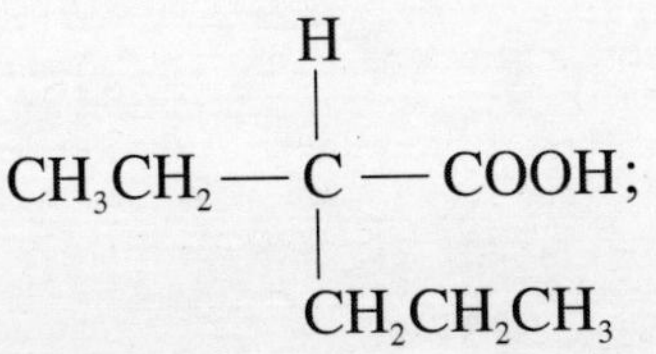

(optically active)

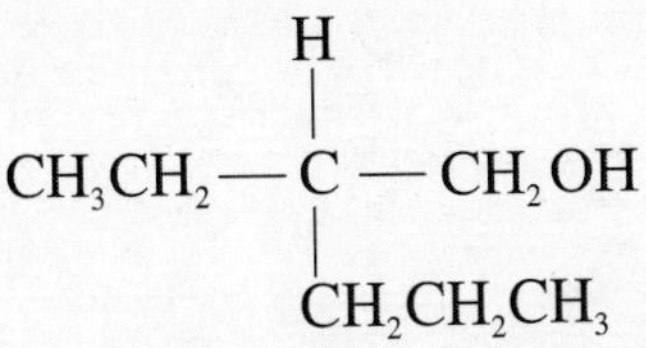

D is $CH_3CH_2CHOHCH_2CH_2CH_3 + CrO_3 \rightarrow CH_3CH_2COCH_2CH_2CH_3$

A on ozonolysis in oxidizing conditions gives D along with HCHO.

Isomers of A are

i. 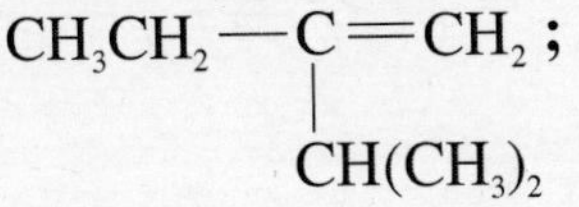 $CH_3CH_2—C=CH_2$;
with $CH(CH_3)_2$ on the C

ii. 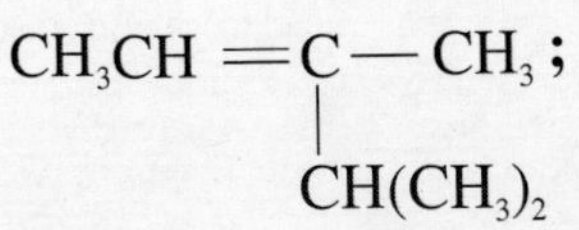 $CH_3CH=C—CH_3$;
with $CH(CH_3)_2$ on the C

iii. $CH_3CH=C(CH_3)—CH(CH_3)_2$;

iv. $$\begin{array}{l} \qquad\qquad\quad H \\ \qquad\qquad\quad | \\ CH_3CH-C=CH_2; \\ \quad\;\; | \\ \quad CH_2CH_2CH_3 \end{array}$$

v. $$\begin{array}{l} \qquad\qquad\quad H \\ \qquad\qquad\quad | \\ CH_2CH_2-C=CH_2; \\ | \\ CH_2CH_2CH_3 \end{array}$$

vi. $CH_3CH=CHCH_2-CH_2CH_2CH_3;$

vii. $(CH_3)_3C-CH=CHCH_3;$

viii. $(CH_3)_3C-CH_2-CH=CH_2;$

ix. $$\begin{array}{l} \qquad\quad CH_3 \\ \qquad\quad\;\; | \\ CH_3-C-C=CH_2 \\ \qquad\quad\;\; |\quad\;\; | \\ \qquad\quad CH_3\; CH_3 \end{array}$$

x. $(CH_3)_3C-CH_2-CH=CH_2$

A on hydrogenation gives $$\begin{array}{l} CH_3CH_2-CH-CH_3 \\ \qquad\qquad\;\; | \\ \qquad\qquad CH_2CH_2CH_3 \end{array}$$

It gives 7 monochlorinated products.

The isomers which exhibit geometrical isomerism are ii,iii,vi and vii

8

ETHERS AND PHENOLS

ETHERS

Ethers are alkoxy (RO—)-substituted alkanes, alkenes, and alkynes. As with alcohols, only saturated carbon atoms may be substituted in alkenes and alkynes.

$$CH_3—O—CH_3$$

Dimethyl ether

$$C_6H_5—O—CH_3$$

Anisole

NOMENCLATURE

Ethers are commonly named by listing the names of the groups attached to the oxygen atom and adding the word ether. Some examples include

$$CH_3—CH_2—\ddot{\underset{..}{O}}—CH_3$$

Ethyl methyl ether

$$CH_3—\ddot{\underset{..}{O}}—CH{=}CH_2$$

Methyl vinyl ether

$$C_6H_5—O—CH_3$$

Methyl phenyl ether

The IUPAC nomenclature names ethers as alkoxy alkanes, alkoxy alkenes, or alkoxy alkynes. The group in the chain that has the greatest number of carbon atoms is designated as the parent compound. In the case of aromatic ethers, the benzene ring is the parent compound.

$CH_3-CH_2-\ddot{O}-CH_3$ — Methoxyethane

$C_6H_5-\ddot{O}CH_3$ — Methoxybenzene

$H_2C=CH-\ddot{O}-CH_3$ — Methoxyethene

Cyclic ethers, oxygen-containing ring systems, are normally called by their common names.

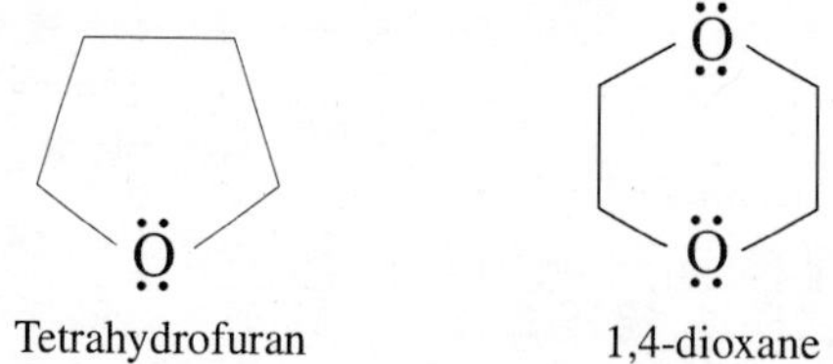

Tetrahydrofuran   1,4-dioxane

PHYSICAL PROPERTIES

The bonds between the oxygen atom and the carbon atoms of the alkyl groups in an ether molecule are polarized due to a difference in electronegativities between carbon and oxygen. In addition, the bond angle between the alkyl groups on the oxygen is 110°. These facts show that ether molecules must be dipoles (molecules having both a centre of positive and negative charge) with weak polarities. Thus, the structure of ether is similar to that of water.

However, in water the hydrogen atoms have a greater partial positive charge than the hydrogen atoms on ether. In water, the charge is localized only on the hydrogens and not delocalized (spread throughout) as with the alkyl groups, so the charge is stronger in water than in ethers.

Like water, ether is capable of forming hydrogen bonds. However, because of the delocalized nature of the positive charge on the ether molecule's hydrogen atoms, the hydrogens cannot take part in hydrogen-bonding. Thus, ethers only form hydrogen bonds to other molecules that have hydrogen atoms with strong partial positive charges. Therefore, ether molecules cannot form hydrogen bonds with other ether molecules. This leads to the high volatility of ethers. Ethers are capable, however, of forming hydrogen bonds to water, which accounts for the good solubility of low molecular weight ethers in water.

Table 8.1 shows boiling points for some simple ethers and the boiling points of alcohols with the same number of carbon atoms. Notice that due to the hydrogen-bonding between alcohol molecules, all alcohols have appreciably higher boiling points than their isomeric ethers.

Table 8.1 Boiling points of simple ethers and alcohols

Ether	Boiling point (°C)	Alcohol	Boiling point (°C)
CH_3OCH_3 Dimethyl ether	25	CH_3CH_2OH Ethanol	79
$CH_3OCH_2CH_3$ Methylethy ether	11	$CH_3CH_2CH_2OH$ Propanol	82
$CH_3CH_2OCH_2CH_3$ Diethyl ether	35	$CH_3CH_2CH_2CH_2OH$ Butanol	117

METHODS OF PREPARATION OF ETHERS

Williamson Synthesis

This method (Figure 8.1) is suitable for the preparation of a wide variety of unsymmetric ethers. The nucleophilic substitution of halides with alkoxides leads to the desired products.

OH + K —(− ½ H_2)→ O^- K^+

$\bar{O}K^+$ + Br (−K Br)→ O

Figure 8.1 Williamson synthesis

Figure 8.2 gives the mechanism of this synthesis.

If the halides are sterically demanding and there are accessible protons in the β-position, the alkoxide will act as a base, and the side products derived from elimination are isolated instead (Figure 8.3).

OH + K —(− ½ H_2)→ O^- K^+ Br (−K Br)→ O

Figure 8.2 Mechanism of Williamson synthesis

Figure 8.3 Elimination products

Limitations of this reaction This is an S_N2 reaction. Hence all types of alkyl halides cannot be used. Only primary alkyl halides can be used. Other alkyl halides will lead to elimination products as shown above in Figure 8.3.

For example di-tertiary butyl ether cannot be prepared by this method. Similarly diphenyl ether cannot be prepared by this method. If we want to prepare *t*-butyl methyl ether, the only combination of alkoxide and alkyl halide are potassium *t*-butoxide and CH_3I. Phenyl methyl ether is obtained by the combination of sodium phenoxide and CH_3I.

Other ethers which cannot be prepared by Williamson synthesis are

1. $R_2CHOCHR_2$ (both groups are secondary)
2. $R_2CH—O—CR_3$ (one group is tertiary and another one is secondary)
3. RCH$=$CHOCH$=$CHR′ (vinyl groups)
4. $R_3CCH_2OCH_2CR'_3$ (neopentyl groups are also unreactive to S_N2 pathway).

Di-tertiary butyl ether can be prepared by treating tertiary butyl chloride with Ag_2CO_3 to give $(CH_3)_3COC(CH_3)_3$. Ag^+ pulls off Cl^- and tertiary carbocation reacts with carbonate ion to give $(CH_3)_3COCO_2^-$ which loses CO_2 to give $(CH_3)_3CO^-$ which reacts with $(CH_3)_3C^+$ to give the ether.

Other Methods of Preparation of Ethers

i. Ethers may be prepared by addition of alcohols to alkenes in the presence of acid. Ethanol reacts with 2-ethylpropene in acid to give *t*-butylmethyl ether.

ii. Grignard reagent reacts with halogenoether to give ethers.

$$ROCH_2Cl + R'MgX \longrightarrow ROCH_2R' + MgXCl$$

iii. Alcohols with diazomethane in the presence of catalyst give ether.

$$ROH + CH_2N_2 \longrightarrow ROCH_3$$

The catalyst may be aluminium alkoxide. If primary alcohols are used, fluoroboric acid serves as a good catalyst. If a phenol is used to react with diazomethane, no catalyst is required since it is acidic.

VINYL ETHERS OR ENOL ETHERS

Ethers of the type $CH_2{=}CHOCH_3$ are called enol ethers. This is methyl vinyl ether.

Preparation of Enol Ethers

i. When ketones react with alkoxides of the type potassium tertiary butoxide in DMSO, enol ether is formed along with a C-alkylated by-product. For example acetophenone with $(CH_3)_3COK$ in DMSO gives $C_6H_5C(OMe){=}CH_2$ and $C_6H_5COCH_2Me$.

ii. Another method is to use aldehyde or keto acetals by reacting carbonyl compounds with ROH and then decompose the acetal in acid catalysis in the absence of water. Thus,

$$(CH_3CH_2CH)_2CHO + 2C_2H_5OLi\ (\text{in acid}) \longrightarrow \text{acetal.}$$

This acetal with catalytic acid without water reacts to give enol ether. The ethanol is distilled out. In the absence of water, elimination takes place from acetal to give enol ether.

iii. Acetylene reacts with methanol at 160–220°C in the presence of 1–2% potassium methoxide under pressure to give methyl vinyl ether.

iv. Divinyl ether is prepared by reacting ethylene chlorohydrin with sulphuric acid and passing the product over KOH at 200–240°C.

$$2ClCH_2CH_2OH \rightarrow (H_2SO_4) \rightarrow ClCH_2CH_2OCH_2CH_2Cl \rightarrow KOH \rightarrow CH_2{=}CHOCH{=}CH_2$$

This ether is used as an anaesthetic.

v. Acetylene and $CH_3CH_2CH_2ONa$ in OH^- gives $CH_3CH_2CH_2OCH{=}CH_2$.

vi. Ethylene + Br_2 + propanol$\rightarrow BrCH_2CH_2OCH_2CH_2CH_3 \rightarrow (\text{alc.KOH}) \rightarrow$ $CH_3CH_2CH_2OCH_2{=}CH_2$.

These ethers can involve in resonance structures when protonated and this aspect is used in the hydration reaction of these ethers. Since the protonated form is resonance-stabilized, the reactions of these enol ethers with dil. H_2SO_4 are about 10^{15} times faster than the reaction of ethylene with dil. H_2SO_4 (the protonated form has hyperconjugation).

Hydrolysis of Enol Ethers

The mechanism is given below:

$$-C{=}C-OR + H^+ \text{ (slow)} \longrightarrow -\underset{|}{CH}-\underset{|}{C^+}-OR \xrightarrow{(+H_2O)} -CH\overset{\overset{|}{O^+H_2}}{\underset{|}{C}}-OR$$

$$\downarrow (-H^+)$$

$$ROH + -\overset{\overset{H}{|}}{\underset{|}{C}}-\overset{\overset{OH}{|}}{\underset{|}{C^+}}- \xleftarrow{-ROH} -\overset{\overset{H}{|}}{\underset{|}{C}}-\overset{\overset{OH}{|}}{\underset{|}{C}}-\overset{+}{O}RH \longleftarrow -\overset{\overset{H}{|}}{\underset{|}{C}}-\overset{\overset{OH}{|}}{\underset{|}{C}}-OR + H^+$$

alcohol

$$\downarrow$$

$$-\overset{\overset{H}{|}}{\underset{|}{C}}-\underset{|}{C}{=}O$$

Carbonyl compound

Carbonyl compound is formed. The other product is ROH.

The following example given in the IIT-JEE 2004 examination indicates how enol ether gives rise to a ketone and alcohol and that the reaction proceeds 10^{15} times faster than hydration of ethylene.

SOLVED EXAMPLE I

An organic compound P having molecular formula $C_5H_{10}O$ is treated with dil. H_2SO_4 to give two compounds Q and R both of which answer positive the iodoform test. The reaction of $C_5H_{10}O$ with dil. H_2SO_4 is 10^{15} times faster than the reaction of sulphuric acid with ethylene. Identify P, Q and R and give the reason for the faster rate of $C_5H_{10}O$ with sulphuric acid. (IT-JEE 2004)

Answer

After we study aldehydes, ketones and ethers we will know that the possible compounds with formula $C_5H_{10}O$ are a ketone, an aldehyde, an enol ether, an unsaturated alcohol and cyclic ether. Among these structures only the enol ether given in Figure 8.4 can react with dil. H_2SO_4 to give a heteroatom cation which is stable due to resonance. The mechanism of addition of water is given in Figure 8.4. The products 4 and 5 both give positive haloform reaction (carbonyl to be discussed in Aldehydes and Ketones chapter.

CH_3
C CH_2 (1)
CH_2 Ö CH_3

CH_3
C CH_2
CH_2 Ö CH_3
(1)

$\xrightarrow{\cdot\cdot}$

CH_3
C^+ CH_2
CH_3 Ö CH_3
(2)

$\longleftrightarrow$

(3)
CH_3
C + CH_2
CH_3 O CH_3

$\downarrow H_2O$

CH_3
C + CH_3CH_2OH
CH_3 Ö
(4) (5)

Figure 8.4 Mechanism of reaction of enol ether with H_2SO_4

(1) is the enol ether which reacts with hydrogen ion from acid. The ether reacts with H^+ to give the oxygen-stabilized carbocation (2). (2) is delocalized by resonance and gives (3). (3) with H_2O gives (4) acetone and (5) ethanol both of which answer positive the haloform reaction. Haloform reaction is discussed in aldehydes, ketones chapter.

REACTIONS OF ETHERS

1. Ethers dissolve in conc. solution of acids to form oxonium salts.

$$R_2O + H_2SO_4 \longrightarrow [R_2\overset{+}{O}H]^{+} SO_4^{2-}$$

2. $R_2O + BF_3 \longrightarrow [R_2O + BF_3 + R_2O] \longrightarrow [R_3OR]^+ BF_4^-$
3. R_2O with dil. sulphuric acid under pressure form alcohols.
4. Ethers are readily attacked by conc. HI or HBr and the products depend on temperature of the reaction.

   In cold, we have

   $$R_2O + HI \longrightarrow RI + H_2O$$

   The reaction may go by S_N1 or S_N2 pathway.

   If mixed ethers are used, the alkyl iodide produced depends on the nature of alkyl groups. If one group is methyl and other group is primary or secondary, methyl iodide is produced. The S_N2 pathway is followed. I^- attacks smaller CH_3 group due to steric factor. If optically active secondary butylmethyl ether is cleaved with HI, optically active $CH_3-\underset{\displaystyle OH}{\underset{|}{CH}}-CH_2CH_3$ is formed and this indicate S that s-BuO bond is not broken in the reaction.

On heating,

$$R_2O + 2HI \longrightarrow 2RI + H_2O$$

$$ROH + HI \longrightarrow RI + H_2O$$

Zeisels method of estimation of methoxy groups

$$ArOCH_3 + HI \xrightarrow{(373K)} ArOH + CH_3I$$

This is the basis of quantitative estimation of methoxy groups.

Cleavage of ethers HI or HBr may be used. HCl does not cleave ethers. Ethyl isopropyl ether with HI gives isopropyl alcohol and ethyl iodide. The products isopropyl ether and ethyl iodide do not result since iodide attacks less hindered side. If tertiary butyl cyclohexyl ether is treated with trifluoroacetic acid the products are cyclohexanol and 2-methylpropene. The tertiary carbocation leads to elimination product.

Sometimes HBr cleavage happens by S_N1 pathway if alkyl group is tertiary. Tertiary butyl propyl ether with HBr gives by S_N1 pathway tertiary butyl bromide and propanol. When $(CH_3)_3COCH_3$ is treated with anhydrous HI in ether, the products are CH_3I and tertiary butyl alcohol. But when treated with conc. HI, the products are $(CH_3)_3Cl$ and CH_3OH. This is because when anhydrous HI is used, S_N2 pathway is followed and in conc. HI, S_N1 pathway is followed.

CYCLIC ETHERS

Three-membered rings containing oxygen are called epoxides or oxiranes. Others are referred to as cyclic ethers. Two important examples of cyclic ethers are given in Figure 8.5.

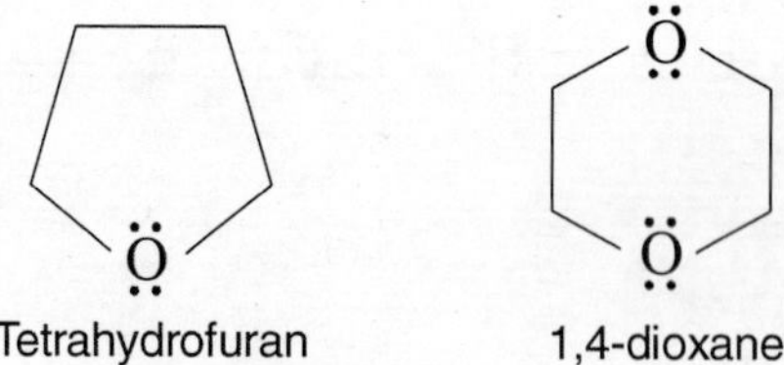

Figure 8.5 Cyclic ethers

Both are solvents since they are inert. But they are cleaved by acids. Three-membered epoxides behave differently from open-chain ethers due to strain of cyclopropane ring.

Naming of Cyclic Ethers

They have a common name. The IUPAC name can be of two kinds. For tetrahydrofuran we call as tetramethylene oxide or oxacyclopentane according to the IUPAC format. The total number of atoms including oxygen is 5 and hence this is called as oxacyclopentane taking into oxygen as another member in addition to 4 carbons.

Preparation of Cyclic Ethers

Cyclic ethers are prepared by intramolecular Williamson synthesis. Examples are given below:

i. Ethylene chlorohydrin ($ClCH_2CH_2OH$) with NaOH gives ethylene oxide (CH_2—CH_2 bridged by O)

ii. 6-bromo-1-hydroxyhexane with KOH in *p*-xylene at 413 K heated for 7 hours gives hexamethylene oxide or oxacycloheptane or oxepane.

PHENOLS

NOMENCLATURE

Functional group suffix—common: phenol, systematic: benzenol. Functional group prefix = *hydroxy*. Numbering of the ring begins at the hydroxyl-substituted carbon and proceeds in the direction of the next substituted carbon that possesses the lower number.

Ortho, meta or para Monosubstituted phenols are characterized using the prefix ortho (*o*-), meta (*m*-) or para (*p*-) depending on the placement of the substituent from the hydroxyl group or the hydroxyl group from a higher priority functional group

2-chlorophenol or *o*-chlorophenol	3-chlorophenol or *m*-chlorophenol	4-chlorophenol or *p*-chlorophenol

PHYSICAL PROPERTIES

- The polar nature of the O—H bond (due to the electronegativity difference of the atoms) results in the formation of hydrogen bonds with other phenol molecules or other H-bonding systems (e.g., water). The implications of this are:
  - high melting and boiling points compared to analogous arenes
  - high solubility in aqueous media

- The presence of *intramolecular* hydrogen bonding is the reason for the significantly lower boiling points of certain *ortho*-substituted phenols *vs* the *meta*- and *para*- analogs. For example *ortho*-nitrophenol has a lower boiling point than *meta* or *para* nitrophenol. Steam distillation is helpful to distill *o*-nitrophenol due to its volatility. *m*- and *p*-nitrophenols cannot be steam-distilled.

STRUCTURE

- The alcohol functional group consists of an O atom bonded to an sp^2-hybridized aromatic C atom and a H atom via bonds.
- Both the C—O and the O—H bonds are polar due to the high electronegativity of the O atom.
- Conjugation exists between an unshared electron pair on the oxygen and the aromatic ring (Figure 8.6).
- Compared to simple alcohols, phenols have:
    - a shorter carbon–oxygen bond distance.
    - a more basic hydroxyl oxygen.
    - a more acidic hydroxyl proton (—OH).

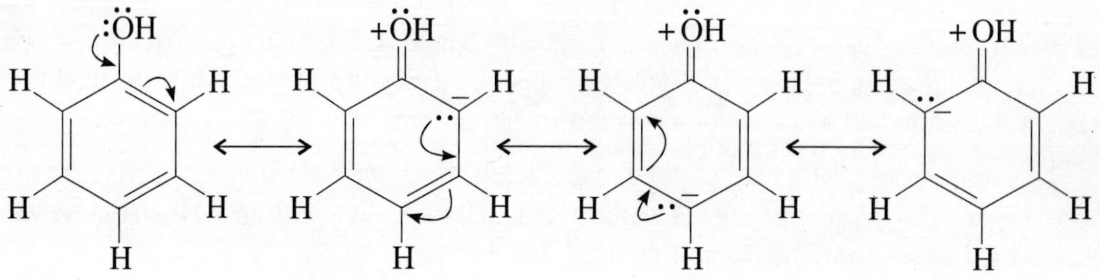

:ÖH H H H H H ⟷ +ÖH H H H H H ⟷ +ÖH H H H H H ⟷ +OH H H H H H

Figure 8.6 Resonance structures of phenol

ACIDITY

- Phenols are more acidic ($pK_a \approx 10$) than alcohols ($pK_a \approx 16$–20), but less acidic than carboxylic acids ($pK_a \approx 5$).
- The negative charge of the phenolate ion is stabilized by resonance due to electron delocalization on to the ring as shown in Figure 8.7.

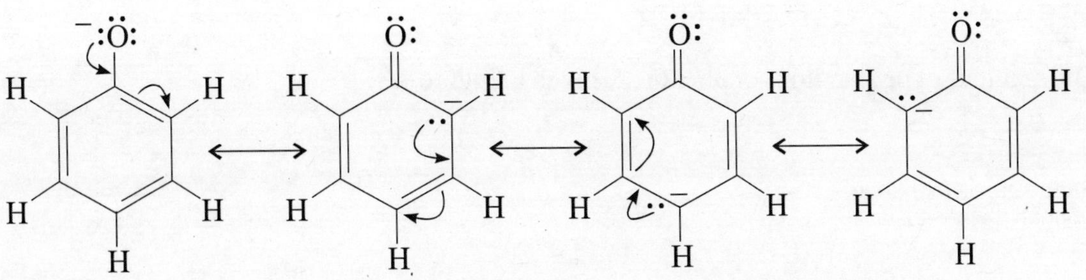

Figure 8.7 The delocalization of phenoxide ion

- The acidity difference means that it is possible to separate phenols from alcohols and/or carboxylic acids.
  - Mixing an ether solution of either phenol and alcohol or phenol and carboxylic acid, with dilute base (sodium hydroxide and sodium bicarbonate respectively), results in the stronger acid being converted to its alkali salt, which is then extracted to the aqueous phase and can be separated from the organic phase.
- Nucleophilic substitution reactions of phenols are generally carried out under basic conditions as the **phenolate ion** is a better nucleophile.

Acidity of Substituted Phenols

Substituents, particularly those located *ortho* or *para* to the —OH group, can dramatically influence the acidity of the phenol due to resonance and / or inductive effects. Electron withdrawing groups enhance the acidity, electron donating substituents decrease the acidity. The resonance stabilization of *o*-nitrophenol is shown below:

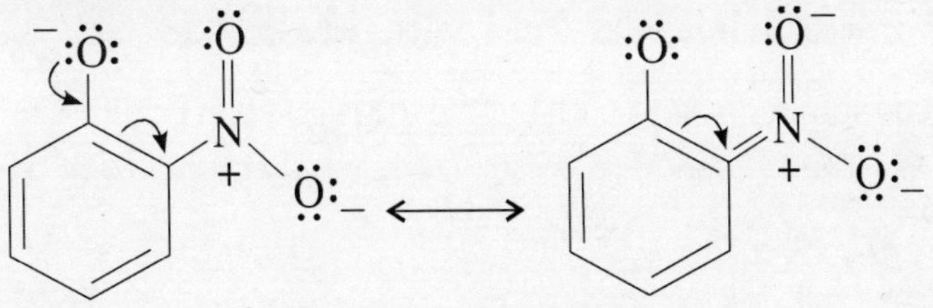

Compound	pK_a	Compound	pK_a
Phenol	10.0		
o-Methoxyphenol	10.0	*p*-Methoxyphenol	10.2
o-Methylphenol	10.3	*p*-Methylphenol	10.3
o-Chlorophenol	8.6	*p*-Chlorophenol	9.4
o-Nitrophenol	7.2	*p*-Nitrophenol	7.2
m-Nitrophenol	8.4		

PREPARATION OF PHENOL

The methods of preparations of phenol are given in Figure 8.8.

Method	Reagents	Conditions
Reaction of benzene sulphonic acid with base	NaOH + $C_6H_5SO_3H$	heat
Reaction of chlorobenzene with base	NaOH + C_6H_5Cl	heat
Acidic oxidation of cumene	O_2 + cumene	H_2O, H_2SO_4
Hydrolysis of diazonium salts	H_2O + $C_6H_5N_2^+$	H_2SO_4, 100°C

Product: phenol (C_6H_5OH)

Figure 8.8 Methods of preparation of phenol

Note The first three methods are primarily industrial methods while the hydrolysis of diazonium salts is the most important laboratory method.

Reaction of Benzene Sulphonic Acid with Hydroxide

The reaction of bezene sulphonic acid with hydroxide is given in Figure 8.9.

$$NaOH + C_6H_5SO_3H \xrightarrow{heat} C_6H_5OH$$

Figure 8.9 Preparation of phenol from benzene sulphonic acid

Reaction type Nucleophilic aromatic substitution

Details

- This is a useful industrial method for the synthesis of phenol.
- The reagents use are usually NaOH, at 300–350°C followed by an acidic work-up to neutralize the phenoxide.

- It also represents the oldest method for the industrial preparation of phenol.
- The reaction occurs via the addition–elimination mechanism with SO_3^{2-} functioning as the leaving group.

The mechanism for nucleophilic aromatic substitution in nitro-substituted aryl halides is shown in Figure 8.10.

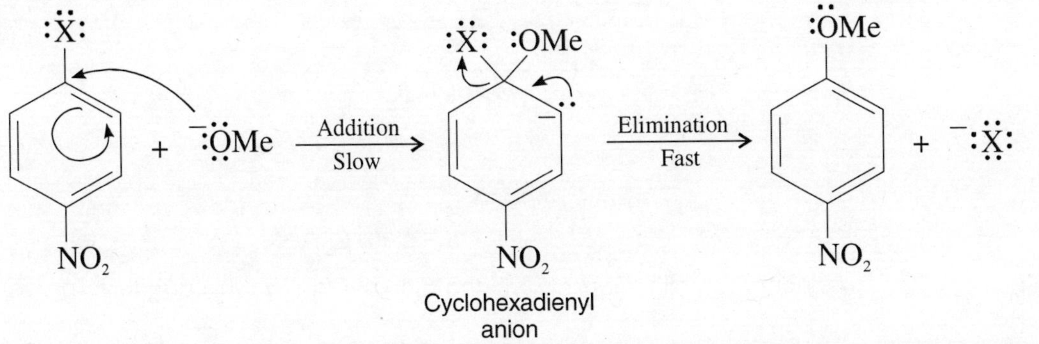

Figure 8.10 The mechanism for the conversion of *p*-nitrohalobenzene to *p*-nitroanisole.

- Attack of the strong nucleophile on the halogen-substituted aromatic carbon forming an anionic intermediate.
- Loss of the leaving group, the halide ion restores the aromaticity.
- Kinetics of the reaction are observed to be second-order.
- The addition step is the rate-determining step (loss of aromaticity).
- Nucleophilic substitution, and therefore reaction rate, is facilitated by the presence of a strong electron-withdrawing group (esp. NO_2) *ortho* or *para* to the site of substitution, which stabilize the cyclohexadienyl **anion** through resonance (Figure 8.11).

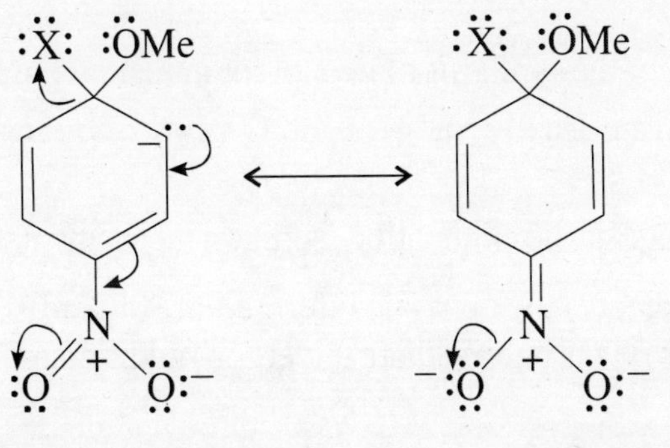

Figure 8.11 Resonance stabilization of cyclohexadienyl anion by nitro group

- Aryl halide reactivity : —F>—Cl>—Br>—I (note the contrast to simple nucleophilic substitution).

- The more electronegative the group, the greater the ability to attract electrons which increases the rate of formation of the cyclohexadienyl anion. Figure 8.12 indicates resonance stabilization when fluorine is present.

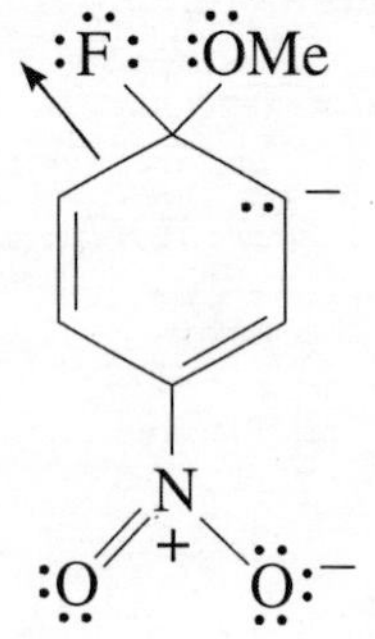

Figure 8.12 Resonance stabilization of cyclohexadienyl ion when fluorine is present.

Base Hydrolysis of Chlorobenzene

The reaction sequence is given in Figure 8.13.

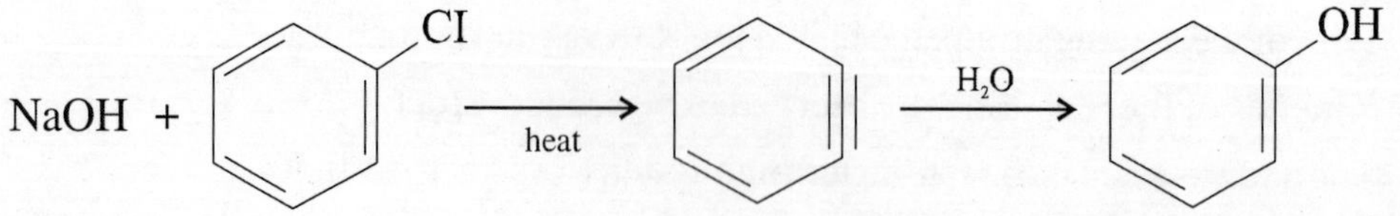

Figure 8.13 The reaction sequence of base hydrolysis of chlorobenzene

Reaction type Nucleophilic aromatic substitution

Details

- This is a method for the synthesis of phenol for industrial purposes..
- The reagents used are usually NaOH at 350°C followed by an acidic work-up to neutralize the phenoxide.
- The reaction occurs via the elimination–addition mechanism via a benzyne intermediate.

Elimination of HCl creates the benzyne intermediate (already discussed in the chapter on Benzene and its reactions). This intermediate undergoes addition of H_2O to produce the phenol.

Acidic Oxidation of Cumene

The next method of preparation of phenol involves acidic oxidation of cumene. Figure 8.14 shows the reaction sequence.

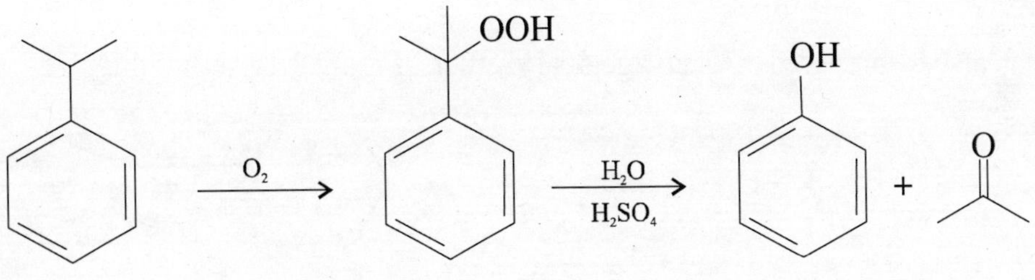

Figure 8.14 Oxidation of cumene

Reaction type Oxidation reaction

Details

- This industrial method is the primary source of phenol.
- Oxidation at the benzylic position gives the hydroperoxide which cleaves to give phenol and acetone.

Cheap reagents and important products make the process attractive.

Figure 8.15 shows the preparation of isopropyl benzene, where R is $CH_3CH_2CH_2Cl$

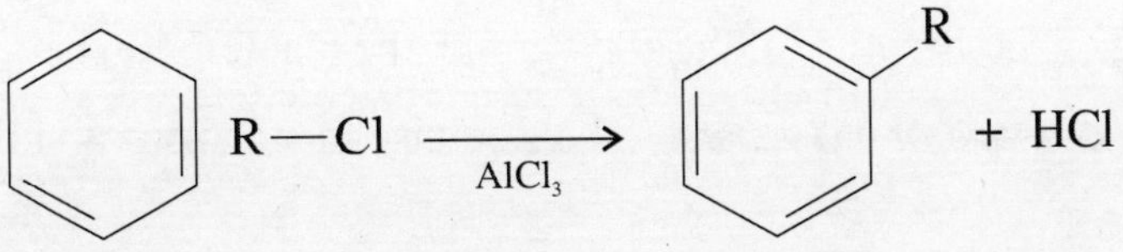

Figure 8.15 Friedel–Crafts alkylation of benzene

Hydrolysis of Aryl Diazonium Salts

Hydrolysis of aryl diazonium salt is given in Figure 8.16.

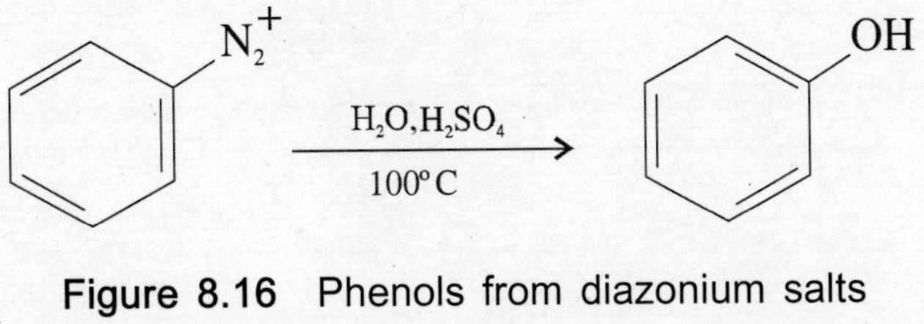

Figure 8.16 Phenols from diazonium salts

Details

- Aryl diazonium salts can be converted into phenols using H_2O / H_2SO_4 / heat.
- HCl should not be used in diazotization since Cl^- will get substituted instead of OH group.
- Aryl diazonium salts are prepared by reaction of aryl amines with nitrous acid, HNO_2.

Cleavage of Aryl Ethers

Phenols can also be prepared by cleavage of aryl ethers. Figure 8.17 shows this.

H—X + Ph—O—C— $\xrightarrow[X=Br,I]{}$ Ph—OH + X—C—

Figure 8.17 Cleavage of aryl ethers by HI or HBr.

Reaction type Nucleophilic substitution

Details

- Like alkyl ethers, alkyl aryl ethers, **R-O-Ar**, are cleaved by the strong acids HI or HBr in a nucleophilic substitution reaction.
- The products are phenol and the appropriate alkyl halide.
- Since phenols **do not** react with hydrogen halides to give aryl halides, there can be no further reaction (unlike alkyl ethers and simple alcohols).
- Protonation of the ethereal oxygen creates a good leaving group, a neutral phenol molecule.
- The halide ion, bromide or iodide are both good nucleophiles.
- Depending on the structure of the alkyl group, the reaction can be S_N1 or S_N2.

Mechanism for acid cleavage of aryl ethers The stepwise mechanism is shown in Figure 8.18.

CH_3CH_2—Ö—Ph

Fast H^+

CH_3CH_2—Ö$^+$(H)—Ph

Slow :Br:$^-$

CH_3CH_2—Br: + HÖ—Ph

Figure 8.18 Mechanism for acid cleavage of aryl ethers

Step 1 The first step is an acid/base reaction where there is protonation of the ether oxygen to make a better leaving group. This step is very fast and reversible. The lone pairs on the oxygen make it a Lewis base.

Step 2 The bromide ion functions as the nucleophile and attacks to displace the good leaving group, neutral phenol molecule, by cleaving the C—O bond. This results in the formation of an alkyl bromide and a phenol.

REACTIONS OF PHENOLS

Electrophilic Aromatic Substitution Reactions of Phenols

When an electrophile E^+ attacks phenol, the *ortho* and *para* compounds are formed as shown in Figure 8.19.

OH (phenol) —E^+→ OH, E (*para-*) + OH, E (*ortho-*) + H^+

Figure 8.19 Electrophile attacking phenol

Reaction type Electrophilic aromatic substitution

Details

- *Phenols* are potentially very reactive towards electrophilic aromatic substitution. This is because the hydroxy group, OH, is a strongly activating, *ortho-*/*para-*directing substituent.
- Substitution typically occurs *para* to the hydroxyl group unless the *para* position is blocked, then *ortho* substitution occurs.
- The strong activation often means that milder reaction conditions than those used for benzene itself can be used (see table below for a comparison).
- Phenols are so activated that polysubstitution can be a problem.

Table 8.2 Electrophilic substitution in phenol

Reaction	Phenol	Benzene
Nitration	Dil. HNO_3 in H_2O or CH_3CO_2H	HNO_3 / H_2SO_4
Sulphonation	Conc. H_2SO_4	H_2SO_4 or SO_3 / H_2SO_4
Halogenation	X_2	X_2 / Fe or FeX_3
Alkylation	ROH / H^+ or RCl / $AlCl_3$	RCl / $AlCl_3$
Acylation	RCOCl / $AlCl_3$	RCOCl / $AlCl_3$
Nitrosation	Aq. $NaNO_2$ / H^+	

Acylation of Phenols

There are two types of acylation (Figure 8.20) for phenols. These are

C-acylation = electrophilic aromatic substitution

O-acylation = nucleophilic acyl substitution

C-acylation

O-acylation

Figure 8.20 C-acylation and O-acylation in phenols

Details

- Phenols are examples of bidentate nucleophiles, meaning that they can react at two positions:
  - on the aromatic ring giving an aryl ketone via **C-acylation**—a Friedel–Crafts reaction or,
  - on the phenolic oxygen giving an ester via **O-acylation**—an Esterification.
- Reagents :

  C-acylation : acylating agent (acyl chloride or anhydride) and $AlCl_3$.

  O-acylation : acylating agent (acyl chloride or anhydride).

- The product of **C-acylation** is more stable and predominates under conditions of **thermodynamic control** (i.e., when $AlCl_3$ is present).
- The product of **O-acylation** forms faster and predominates under conditions of **kinetic control**.
- **O-acylation** can be promoted by either acid catalysis via protonation of the acylating agent, increasing its electrophilicity or base catalysis via deprotonation of the phenol, increasing its nucleophilicity.
- It is also known that aryl esters readily rearrange to aryl ketones in the presence of $AlCl_3$, a reaction known as the **Fries rearrangement** (Figure 8.21).

Figure 8.21 Fries Rearrangement

Carboxylation of Phenols

One of the most important reactions of phenol is carboxylation. Carboxylation of phenols is called Kolbe–Schmitt reaction (Figure 8.22).

Figure 8.22 Kolbe–Schmitt reaction

Reaction type Electrophilic aromatic substitution

Details

- Heating the nucleophilic phenolate salt with carbon dioxide under high pressure / temperature results in regioselective *ortho*-substitution.
- This process is also known as the **Kolbe–Schmitt synthesis**.
- *o*-hydroxybenzoic acid is more commonly known as salicyclic acid.

- Generally the *p*-substituted compound is formed in more % due to steric factor. In the present case intramolecular hydrogen bonding leads to preferential formation of the *ortho*-substituted compound.
- Figure 8.23 shows intramolecular hydrogen-bonding in salicylic acid.

The geometry allows for an intramolecular hydrogen-bond.

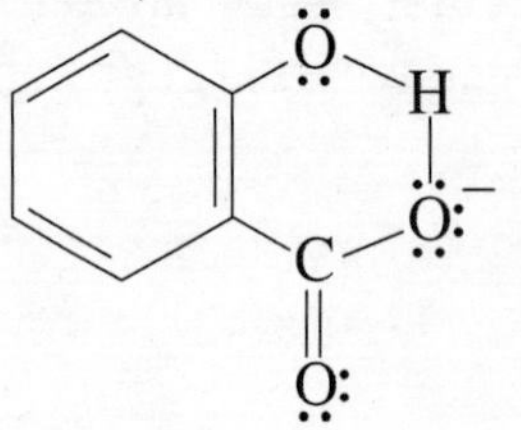

Figure 8.23 Intramolecular hydrogen-bonding in salicylic acid.

The mechanism for carboxylation of phenols is discussed in Figure 8.24.

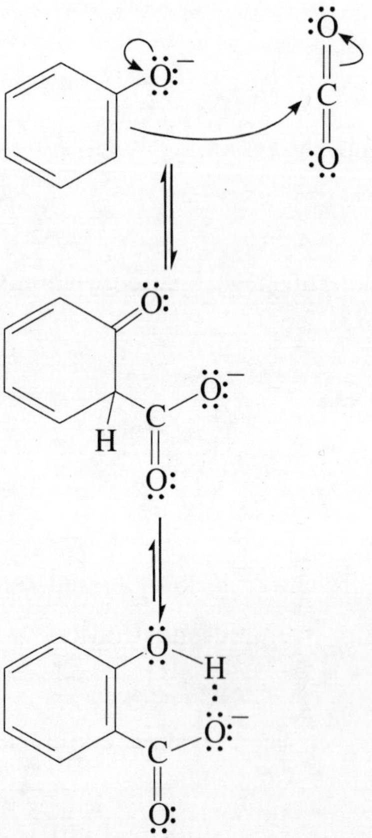

Figure 8.24 Mechanism for Carboxylation of Phenols

Step 1 The nucleophilic phenolate (reacting like an enolate) reacts with the electrophilic carbon of carbon dioxide in the ortho position (compare this with an Aldol reaction)

Step 2 The non-aromatic cyclohexadienonecarboxylate intermediate tautomerizes to the more stable aromatic enol which is further stabilized by an intramolecular hydrogen bond. An acidic work-up with HCl will give salicylic acid.

Preparation of Aryl Ethers from Phenols

Preparation of aryl ethers from phenols is given in Figure 8.25.

Figure 8.25 PhOR from phenols

Reaction type Electrophilic substitution

Details

- The reagents used include a base such as Na_2CO_3 to prepare the phenoxide; then add the alkyl halide.
- Since the reaction is S_N2, the halide should be methyl or primary alkyl halides (or tosylates).

This reaction is the Williamson ether synthesis applied to aromatic alcohols.

Oxidation of Phenols

Details

- In general, phenols are more easily oxidized than simple alcohols.
- Oxidation can achieved by reaction with silver oxide (Ag_2O) or chromic acid ($Na_2Cr_2O_7$), or other oxidizing agents.
- Particularly important are the oxidation of 1,2- and 1,4-benzenediol (catechol and hydroquinone, respectively) and their derivatives (Figure 8.26)
  - These types of systems are important in biological redox systems such as coenzyme Q and vitamin K.
  - Figure 8.27 gives a closer look at the two one-electron transfers that are believed to take place when hydroquinone is oxidized to benzoquinone.
  - Loss of a proton and an electron generates a phenoxy radical.
  - Loss of a second proton and a second electron completes the oxidation.

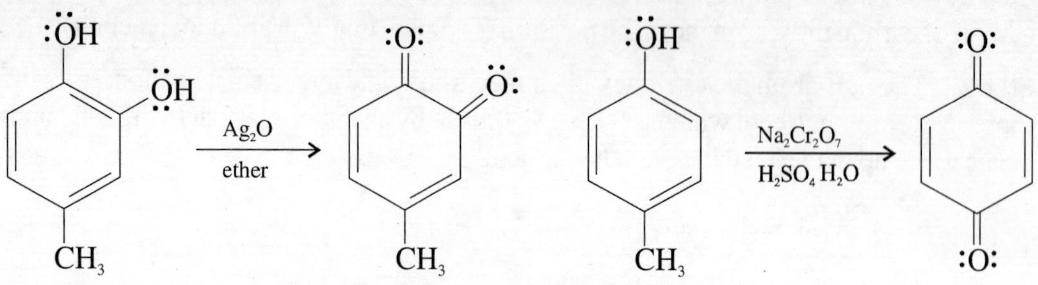

Figure 8.26 Oxidation of phenols

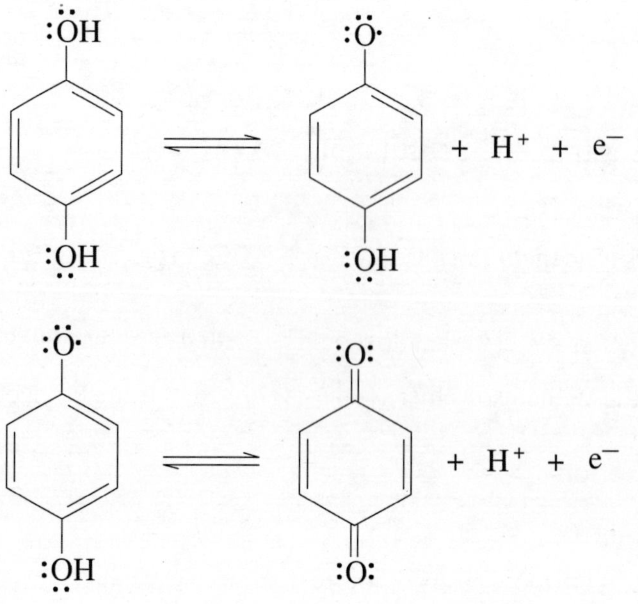

Figure 8.27 Mechanism of oxidation of hydroquinone

Claisen Rearrangement of Aryl Allyl Ethers

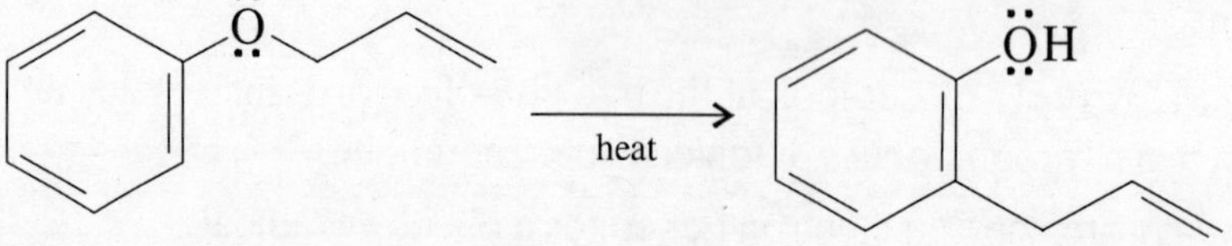

Figure 8.28 Claisen rearrangement of aryl allyl ethers with mechanism

Details

- Aryl allyl ethers undergo a thermal rearrangement to give *ortho*-allylphenols (Figure 8.28).
- This reaction is an intramolecular process.
- The mechanism of this rearrangement is given in Figure 8.29.

Step 1 This is an electrocyclic process. Push those electrons around the 6-membered ring. Best appreciated by starting the electron flow by having the aromatic C=C attacking the allyl C=C and displacing an O leaving group.

Step 2 Tautomerization of the ketone (actually a dienone) to the more stable aromatic enol, the phenol.

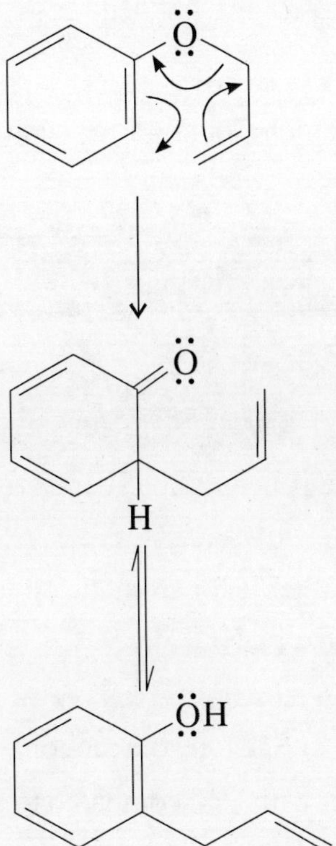

Figure 8.29 Mechanism of Claisen Rearrangement

Reimer–Teimann Reaction

When $CHCl_3$ in the presence of KOH is added to phenol, salicylaldehyde is produced. This is Reimer–Tiemann reaction. The mechanism is given in the Figure 8.30.

Figure 8.30 Reimer–Teimann reaction with mechanism

I. "Explain Why" Questions and Small Structure-Based Problems

1. Why is there no reaction between phenols and HBr to give arylbromides?

   Answer

   S_N1 or S_N2 pathway is not suitable for aromatic systems.

2. A diaryl ether cannot be cleaved with HI. Why?

   Answer

   This cleavage requires S_N1 or S_N2 type of reaction.

3. S_N2 cleavage of ethers is slower with HCl compared to HI. Why?

   Answer

   HI is a stronger acid than HCl and transfer of H^+ to ROR is greater with HI.

4. When ArOR ethers are cleaved, the only products are ArOH and RI. But when ROR′ is cleaved, R and R′ being primary alkyl groups, both sets of products ROH + R′I and R′OH and RI are formed. Why?

   Answer

   For aromatic rings S_N2 attack is not possible and hence ArOH and RI are only formed.

5. $ClCH_2CH_2OH$ cannot be used to prepare $CH_3CH_2CH_2OCH{=}CH_2$ by adding $ClCH{=}CH_2$. Why?

Answer

The chlorohydrin will prefer to form the epoxide. Also vinyl halide cannot react.

6. $R_3CCH_2O\,CH_2CR_3$ cannot be formed by Williamson synthesis. Why?

Answer

Neopentyl halides do not react by S_N2 pathway.

7. Phenol boils at a higher temperature than toluene although molecular weights are nearly same. Why?

Answer

Phenol has intermolecular hydrogen-bonding and hence the boiling point is high.

8. *o*-nitrophenol can be steam-distilled but *p*-nitrophenol cannot be steam-distilled. Why?

Answer

o-nitrophenol has intramolecular hydrogen-bonding and has low boiling point near that of water and has appreciable vapour pressure and hence can be steam-distilled. *p*-nitrophenol is intermolecularly hydrogen-bonded with high boiling point. In case of the *p*-isomer, attraction to water reduces vapour pressure.

9. Phenol has pKa equal to 10 but ethanol has pKa equal to18. Why?

Answer

The phenoxide ion is resonance-stabilized and hence phenol is much more acidic.

10. *p*-nitrophenol is more acidic than *p*-chlorophenol. Why?

Answer

The nitrogen of NO_2 has + ve charge and is electron-withdrawing, and its inductive effect is greater than that of Cl. It has also electron-withdrawing resonance effect.

II. Multiple Choice Questions (One out of four answers is correct)

1. The most acidic compound among the following is
   a) ethylene b) acetylene c) ethanol d) phenol
2. One of the following compounds is easily steam-distilled.
   a) *m*-nitrophenol b) *p*-nitrophenol c) *p*-nitrobenzoic acid d) *o*-nitrophenol
3. Among the following phenols the least acidic is
   a) phenol b) *o*-chlorophenol c) *o*-methoxyphenol d) *o*-nitrophenol
4. The pKa is maximum for one of the following compounds
   a) *p*-methoxyphenol b) *o*-chlorophenol c) *o*-nitrophenol d) *p*-nitrophenol

5. One of the following reactions involves benzyne intermediate.
   a) NaOH with benzyl bromide
   b) NaOH with chlorobenzene
   c) Cl_2 + Fe with benzene
   d) *t*-butyl chloride with NaOH
6. The conversion of benzene sulphonic acid to phenol by NaOH involves benzyne intermediate with —— as leaving group.
   a) SO_3H
   b) H_2O
   c) OH^-
   d) SO_3^{2-}
7. When nitro-substituted aromatic halides react with OMe^- to obtain *p*-methoxynitrobenzene the leaving group is
   a) nitro
   b) chloro
   c) methoxy
   d) cyclohexadienyl ion
8. In aromatic nucleophilic substitution the order of reactivity of halides is
   a) $F^- > Cl^- > Br^- > I^-$
   b) $F^- > I^- > Br^- > Cl^-$
   c) $F^- > Br^- > I^- > Cl^-$
   d) $Cl^- > Br^- > F^- > I^-$
9. The hydrolysis of chlorobenzene involves addition of ________ and elimination of ________
   a) water, HCl
   b) OH^-, HCl
   c) water, OH^-
   d) Cl^-, water
10. When isopropyl benzene is oxidized by air, the product is
    a) cumene hydroperoxide
    b) benzoic acid
    c) benzaldehyde
    d) acetophenone
11. When cumene is oxidized in air followed by reaction with water (sulphuric acid), the products are
    a) benzaldehyde, acetone
    b) benzoic acid, acetone
    c) phenol, acetone
    d) acetic acid, phenol
12. When Friedel–Crafts alkylation is carried out with n-propyl chloride the main product is
    a) n-propylbenzene
    b) ethylbenzene
    c) cumene
    d) n-butylbenzene
13. Friedel–Crafts alkylation involves
    a) carbocation mechanism
    b) benzyne formation
    c) free radical formation
    d) no suitable answer
14. The conversion of diazonium salts to phenol involves
    a) S_N1 mechanism
    b) carbanium
    c) free radicals
    d) S_N2 mechanism
15. Phenol is nitrated to give more of *p*-nitrophenol and less of *o*-nitrophenol by reacting phenol with
    a) nitric acid in acetic acid
    b) nitric acid and sulphuric acid
    c) oleum
    d) nitrous acid
16. Phenols are________ nucleophiles.
    a) active
    b) deactivating
    c) bidendate
    d) poor

17. When aliphatic amines are treated with nitrous acid, the diazonium salts are unstable but aromatic diazonium salts are stable. Choose the wrong statement among the following.
    a) Aryl carbocations are less stable due to sp^2 hybridization.
    b) Loss of nitrogen from aromatic diazonium salts is much slower.
    c) Aliphatic diazonium salts lead to more stable carbocations.
    d) Loss of nitrogen from aromatic diazonium salts is easy.
18. O-acylation of phenols is promoted by
    a) acid catalysis via protonation of the acylating agent, increasing its electrophilicity
    b) base catalysis via protonation of the phenol, increasing its' nucleophilicity.
    c) Presence of aluminium chloride
    d) high temperature
19. The rearrangement of ________ to aryl ketones in the presence of $AlCl_3$ is called Fries rearrangement.
    a) phenols b) alcohols c) aryl esters d) benzyl alcohol
20. One of the following statements is wrong.
    a) O-acylation is kinetically controlled.
    b) C-acylation is done using $AlCl_3$.
    c) Phenol is less acidic than carboxylic acids.
    d) Salicylic acid has intermolecular hydrogen bonding.
21. One of the following reactions is regioselective.
    a) bromination of benzene b) HCl addition to alkenes
    c) hydration of unsymmetrical alkenes d) carboxylation of phenols
22. The Kolbe–Schmitt reaction gives the *ortho*-substituted acid because
    a) The *ortho* acid is intermolecularly hydrogen-bonded
    b) The *ortho* acid is intramolecularly hydrogen-bonded
    c) There is no steric factor operating.
    d) *Ortho* compound is much resonance-stabilized.
23. The halides suitable for preparation of ethers from phenols are
    a) primary halides b) tertiary halide c) vinyl halides d) aryl halides
24. The formation of aryl ethers from phenols involves ________ mechanism.
    a) benzyne b) S_N1 c) S_N2 d) free radical

25. One of the following statements is true
    a) Carboxylation of phenols produces *p*-hydroxybenzoic acid.
    b) Carboxylation of phenols is not regiospecific.
    c) Oxidation of hydroquinone with dichromate gives *p*-carboxyl benzoic acid.
    d) Oxidation of hydroquinone with dichromate gives quinine.
26. Cleavage of aryl ethers by HI follows ________ mechanism.
    a) S_N1 b) S_N2
    c) S_N1 or S_N2 depending on alkyl group d) free radical
27. One of the following statements is true
    a) Fluoride is a good nucleophile for cleavage of aryl ethers.
    b) The aryl ether cleavage reaction is electrophilic substitution reaction.
    c) S_N1 path is always followed in cleavage of aryl ethers.
    d) Diaryl ethers cannot be cleaved because the reaction is S_N1 or S_N2 type.
28. Williamson ether synthesis makes use of
    a) phenols, base and allyl halide b) phenols, base and vinyl halide
    c) phenols, base and aryl halide d) phenols, acid and allyl halide
29. One of the following statements is wrong.
    a) When a mixture of phenol and alcohol is shaken with NaOH, phenol separates in aqueous layer.
    b) When a mixture of phenol and carboxylic acid is shaken with Na_2CO_3, phenol separates in organic layer.
    c) Salicylic acid is intramolecularly hydrogen-bonded.
    d) Williamson ether synthesis makes of tertiary halides.
30. The compound with maximum boiling point among the following is
    a) phenol b) toluene c) fluorobenzene d) dimethyl ether

KEY

1. d	2. d	3. c	4. a	5. b	6. d	7. b
8. a	9. a	10. a	11. c	12. c	13. a	14. a
15. a	16. c	17. d	18. a	19. c	20. d	21. d
22. b	23. a	24. c	25. d	26. c	27. d	28. a
29. d	30. a					

III. Screening Test Questions of Previous IIT-JEE Question Papers

1. A new C—C bond is possible in
   a) Cannizarro reaction
   b) Friedel–Crafts alkylation
   c) Clemmenson reduction
   d) Reimer–Teimann reaction (1998)
2. Benzene diazonium chloride reacts with phenol in weakly basic solution to give
   a) diphenylether
   b) *p*-hydroxyazobenzene
   c) chlorobenzene
   d) benzene (1998)
3. In the following compounds the order of acidity is I. phenol II. *p*-cresol III. *m*-nitrophenol IV. *p*-nitrophenol
   a) III>IV>I>II b) I>IV>III>II c) II>I>III>IV d) IV>III>I>II (1996)
4. The products formed in the reaction $C_6H_5OCH_3 + HI$ are
   a) C_6H_5OH, CH_3I b) C_6H_5I, CH_3OH c) $C_6H_5CH_3$, H_2O d) C_6H_6, CH_3OI (1995)

KEY

1. c 2. b 3. d 4. a

IV. Multiple Choice Questions (More than One Answer is Correct)

1. Some of these observations are true.
   a) Phenol has 100% enol form.
   b) Phenol is associated by intermolecular hydrogen-bonding.
   c) The freezing point depression of water produced by phenol of 1 molal concentration is different from that of same concentration of sucrose.
   d) Phenol is less acidic than alcohols.
2. The order of increasing acidity is correctly represented for
   a) phenol < *m*-nitrophenol < *p*-nitrophenol
   b) ethane < ethylene < acetylene
   c) ethanol < acetylene < phenol
   d) acetic acid < benzoic acid < peracetic acid
3. Some of the following reactions are true
   a) phenol + $Br_2(CCl_4) \rightarrow$ *p*-bromophenol
   b) phenol + $3Br_2(H_2O) \rightarrow$ 2,4,6-tribromophenol
   c) phenol + $3Br_2 \rightarrow$ 2,4-dibromophenol + HBr
   d) Phenol + $HNO_3(CHCl_3) \rightarrow$ *o*-nitro and *p*-nitrophenols

4. Phenol is produced in some of these reactions.
   a) $C_6H_5N_2 + HSO_4^- \rightarrow$ (hydrolysis) phenol
   b) Cumene + O_2 followed by treatment with $H_2SO_4 \rightarrow$ phenol + acetone
   c) $C_6H_5NH_2$ + NaOH $\rightarrow$ phenol
   d) $C_6H_6 + H_2O \rightarrow C_6H_5OH$
5. Anisole is produced from
   a) phenol + CH_2N_2
   b) sodium phenoxide + CH_3I
   c) phenol + CH_3I
   d) phenol + C_2H_5I
6. Some of the following compounds do not exhibit chelation by hydrogen-bonding
   a) *o*-cresol b) methyl salicylate c) *o*-cyanophenol d) *o*-fluorophenol
7. Among the following sets, some have decreasing order of acidity for I relative to II
   a) I. *p*-chlorophenol II. *p*-nitrophenol
   b) I. 2,4-dinitrophenol II. 2,4,6-trinitrophenol
   c) I. aminophenol II. *m*-aminophenol
   d) I. 2,4,6-trinitrophenol II. phenol
8. The ethers which cleave with HI to give products of two kinds are
   a) $CH_3OC_2H_5$ b) $CH_3OCH_2—CH{=}CH_2$ c) $PhOCH_3$ d) PhOPh
9. Some of the following ethers cannot be synthesized using Williamson synthesis.
   a) PhOPh
   b) di-tertiary butyl ether
   c) $(CH_3)_3CH_2OCH_2(CH_3)_3$
   d) $CH_3OC_2H_5$
10. When phenol is substituted by groups by electrophilic aromatic substitution mainly more *p*-product result in case of
    a) Kolbe–Schmitt reaction b) alkylation c) bromination d) nitrosophenol
11. One of the following methods does not yield phenol.
    a) Cumene + O_2 and heat
    b) benzene sulphonic acid + NaOH at 573 K
    c) diazonium salt obtained from aniline and $NaNO_2$ and H_2SO_4 and steam-distilled
    d) oxidation of toluene
12. Some of these reactions do not give rise to *o*-bromophenol.
    a) phenol + Br_2 in acetic acid at 298 K
    b) phenol + excess bromine in CCl_4 at 273 K
    c) Phenol + H_2SO_4 at 373 K for 3 hours followed by bromination and adding acidic steam
    d) phenol + Br_2 in Fe

13. Some of the following reactions do not involve benzyne mechanism

    a) chlorobenzene with NaOH at high pressure, and high temperatutre followed by acidification

    b) phenol + excess bromine in CCl_4 at 273 K

    c) phenol + H_2SO_4 at 373 K for 3 hours followed by bromination and adding acidic steam

    d) phenol + Br_2 in Fe

14. 100% enol isomer is not possible in

    a) phenol   b) $CH_3COCH_2COCH_3$ c) acetone   d) acetaldehyde

15. Some of these methods are used to prepare ethers.

    a) *t*-butyl alcohol + excess ethanol as solvent in H_2SO_4

    b) 2-methylpropene + H_2SO_4 + CH_3OH

    c) ArONa + CH_3I

    d) *t*-butyl alcohol + *t*-butyl iodide

16. Some of the following are the methods to prepare enol ethers

    a) acetylene + CH_3OH (160–200°C) in presence of CH_3OK

    b) acetophenone with $(CH_3)_3COK$ in DMSO

    c) $ROCH{=}CH_2 + R'OH \xrightarrow{\text{(Mercuric acetate as catalyst)}} R'OCH{=}CH_2 + ROH$

    d) $ClCH{=}CH_2 + RONa \rightarrow$

KEY

1. a,b,c	2. a,b	3. a,b,d	4. a,b	5. a,b	6. a,c	7. a,b,c
8. a,b	9. a,b,c	10. b,c,d	11. a,b,c	12. a,b,d	13. b,c,d	14. b,c,d
15. a,b,c	16. a,b,c					

V. Assertion–Reason Questions

The questions 1 to 20 consist of an *assertion* in column 1 and *reason* in column 2. Use the following key to choose an appropriate answer

a) If both assertion and reason are correct and reason is the correct explanation of assertion

b) If both assertion and reason are correct but reason is not the correct explanation of assertion

c) If assertion is correct but reason is incorrect

d) If assertion is incorrect but reason is correct

No.	Assertion	Reason
1.	Phenol is more acidic than alcohols.	Phenoxide ion is resonance-stabilized.
2.	Carboxylation of phenols produces p-substituted acid.	Intramolecular hydrogen bonding is present in salicylic acid.
3.	18-Crown-6 is used in the reaction of RCl with LiF.	Only K^+ has the correct size to fit into cavity in the crown ether.
4.	$ClCH_2CH_2ONa$ with $ClCH{=}CH_2$ does not produce ether.	Vinyl halides do not undergo S_N2 reactions.
5.	*o*-nitrophenol can be steam-distilled.	It is volatile due to intramolecular hydrogen-bonding.
6.	In the mixture of phenol and alcohol when separation is done with NaOH, the phenol is extracted in aqueous layer.	Phenol is a stronger acid than alcohols.
7.	Nucleophilic substitution reaction of phenol is conducted under basic conditions.	Phenoxide ion is a good electrophile.
8.	Ethers invoving three-membered rings are different from other cyclic ethers in nucleophilic addition with OH.	These three-membered rings have strain and hence they react differently.
9.	PhOPh will not undergo cleavage with HI.	Aryl halides do not make use of S_N2 pathway.
10.	In phenol C-acylation predominates at high temperatures.	C-acylation is kinetically controlled.
11.	Williamson synthesis is not used to get diaryl ethers.	Aryl halides cannot undergo nucleophilic substitution.
12.	Phenols do not react with HBr to give aryl bromides.	S_N1 or S_N2 pathway is not suitable for aromatic systems.
13.	Phenol and toluene have same molecular weight but phenol boils at 182°C but toluene boils at 111°C.	Phenol is intramolecularly hydrogen-bonded.
14.	2,4-dinitrophenol has pKa= 3.96 but phenol has pKa = 9.89.	The two electron-withdrawing nitro groups increase acidity.
15.	Phenol reacts with NaOH to give phenoxide ion, but cyclohexanol does not give significant amount of base in NaOH.	Cyclohexanol is more acidic than phenol.
16.	When alkyl ethers react with aq. HBr they give alcohols which further react with HBr, but aromatic ethers produce phenols which do not react further with HBr.	The carbon–oxygen bond in phenol is strong.

(*Contd.*)

No.	Assertion	Reason
17.	Di-tertiary butyl ether can be prepared by Williamson synthesis.	Tertiary halides generally undergo elimination to give alkenes.
18.	Chlorobenzene reacts with NaOH under high pressure. But *o*-nitrochlorobenzene reacts with aq.$NaHCO_3$ at 150° to produce phenol.	In the second case carbanion is formed which is stabilized by electron-withdrawing groups.
19.	When sodium phenoxide is treated with allyl bromide a mixture of allylphenyl ether and *o*-allyl phenol are produced.	Allylphenyl ether undergoes Claisen rearrangement to give *o*-allyl phenol
20.	Normally phenol is acylated with CH_3COCl but resorcinol is acylated with RCOOH and $ZnCl_2$.	The two OH groups reinforce each other in electrophilic substitution to permit Friedel–Crafts acylation.

KEY

1. a	2. d	3. d	4. a	5. a	6. a	7. c
8. a	9. a	10. c	11. a	12. a	13. c	14. a
15. c	16. a	17. d	18. a	19. a	20. a	

VI. Comprehension Passages

Passage I

There are some important methods of preparation of ethers and among these the most important one is Williamson synthesis. RONa and RI react to give ethers. Diaryl ethers, di-tertiary butyl ethers cannot be prepared by this method since aryl halides do not choose S_N2 pathway and tertiary halides undergo elimination. Cleavage of ethers produces ROH and RI with HI. Here also cleavage of diaryl ethers does not take place. Enol ethers undergo important reaction of hydrolysis to give carbonyl compounds and alcohols.

Answer the following questions

1. Williamson synthesis is not possible for one of the following ethers.

   a) $PhOCH_3$ b) $PhOCH_2Ph$
   c) $(C_2H_5)_3COC(C_2H_5)$ d) $C_2H_5OC_2H_5$

2. One of the following methods is not a correct method of preparing t-butyl methyl ether.

   a) $(CH_3)_3CONa + CH_3I$ b) *t*-butyl alcohol + CH_3OH (excess as solvent)
   c) $CH_3ONa + (CH_3)_3CI$ d) 2-methyl propene + H_2SO_4 + CH_3OH

3. One of the following reactions does not happen.
   a) $PhOCH_3 + HI \rightarrow PhOH + CH_3I$
   b) $CH_3OCH_3 + HI \rightarrow CH_3OH + CH_3I$
   c) $PhOPh + HI \rightarrow PhOH + PhI$
   d) $(CH_3)_3COCH_3 + HI$ (anhydrous in ether $\rightarrow CH_3I + (CH_3)_3COH$
4. One of the following reactions is not a proper reaction of enol ethers.
   a) $CH_3OCH{=}CH_2 + Dil.H_2SO_4 \rightarrow CH_3CHO + CH_3OH$
   b) $CH_2{=}C(CH_3){-}OCH_3 + Dil.H_2SO_4 \rightarrow$ acetone $+ CH_3OH$
   c) $CH_2{=}C(CH_3){-}OCH_3 + Dil.H_2SO_4 \rightarrow CH_3CH_2CHO + CH_3OH$
   d) $CH_2{=}CHOCH_3 + CH_3OH$ (with a trace of HCl) $\rightarrow CH_3CH(OCH_3)_2$

KEY

1. c 2. c 3. c 4. c

Passage 2

Phenols can be prepared by some important reactions. One reaction involves oxidation cumene in oxygen. Aromatic nucleophilic substitution in chlorobenzene with OH^- is another method. Phenols are acidic because of many resonance structures of phenoxide ion.

Answer the following questions

1. When cumene is oxidized with oxygen, the product other than phenol answers
   a) Tollens test
   b) $KMnO_4$ unsaturation test
   c) Bromine water test
   d) Iodoform test
2. One of the following reactions does not produce phenols.
   a) Sodium salt of benzene sulphonic acid with NaOH gives phenol.
   b) $ArMgBr + O_2$ followed by addition of ArMgBr gives phenol.
   c) $ArN_2^- + HSO_4^-$ when steam-distilled, gives phenol.
   d) C_6H_5Cl on hydrolysis in acid gives C_6H_5OH.

3. One of the following statements relate to acidity of phenols

   a) Phenol gives violet colour with neutral $FeCl_3$.

   b) Phenol with diazomethane without a catalyst gives anisole but alcohols require a catalyst to react with CH_2N_2.

   c) Phenols with CH_3Cl gives *p*-methyl phenol in $AlCl_3$.

   d) Phenols with NI catalyst and hydrogen give cyclohexanol.

KEY

1. d **2. d** **3. b**

Passage 3

Phenols undergo electrophilic aromatic substitution. The OH group in phenol is highly activating.

Answer the following questions

1. *p*-bromophenol is obtained when

   a) Phenol + Br_2/Fe

   b) Phenol + bromine water

   c) Phenol + Br_2 in CS_2 at 0°C

   d) Phenol + NaBr

2. *o*-bromophenol is produced when

   a) Phenol + Br_2/Fe

   b) Phenol + bromine water

   c) Phenol + H_2SO_4 followed by Br_2/NaOH followed by boiling in acid

   d) Phenol + NaBr

3. 1,3,5-tribromophenol is obtained from phenol using

   a) Phenol + HBr

   b) Phenol + bromine water

   c) Phenol + Br_2 in CS_2 at 0°C

   d) Phenol + H_2SO_4 followed by Br_2/NaOH followed by boiling in acid

KEY

1. c **2. c** **3. b**

VII. Matrix Matching

Question 1

Column 1	Column 2
a) oxapane	p) oxacycloheptane
b) $ClCH_2CH_2OH$ + NaOH	q) hexamethylene oxide
c) oxatane	r) ethylene oxide
d) oxalane	s) intramolecular Williamson synthesis
	t) trimethylene oxide
	u) tetramethylene oxide
	v) tetrahydrofuran

Answer

$[a \rightarrow p, q], [b \rightarrow r, s], [c \rightarrow t], [d \rightarrow u, v]$

Question 2

Column 1	Column 2
a) $C_2H_5OCH_3$	p) $C_2H_5ONa + CH_3I$
b) $PhOCH_3$	q) $C_2H_5I + CH_3ONa$
c) 2-methyl propene + $CH_3OH + H_2SO_4$	r) $PhOH + CH_3I$
d) $C_2H_2 + CH_3OH$	s) t-butyl methyl ether
	t) $CH_2{=}CHOCH_3$

Answer

$[a \rightarrow p, q], [b \rightarrow r], [c \rightarrow s], [d \rightarrow t]$

Question 3

Column 1	Column 2
a) benzene + isopropylchloride ($AlCl_3$) followed by addition of oxygen and heat	p) *o*-allylphenol
b) $C_6H_5ONa + CH_2ClCH{=}CH_2$ in K_2CO_3	q) Claisen rearrangement
c) phenol + CH_2N_2	r) phenol + acetone
d) $C_2H_5OH + CH_2N_2$ (with catalyst)	s) anisole
	t) $C_2H_5OCH_3$

Answer

$[a \rightarrow r], [b \rightarrow p, q], [c \rightarrow s], [d \rightarrow t]$

VIII. Structure-Based and Numerical Problems

1. Compound A on dehydration gives B. B on ozonolysis gives C which is also one of the ozonolysis products of $CH_2{=}CH{-}(CH_2)_4{-}CH{=}CH_2$. D on complete hydrogenation with Ni/H_2 gives A. D reacts with NaOH to give E. E with CO_2 in acid gives F. F is intramolecularly hydrogen bonded. E reacts with $CHCl_3$ and NaOH at 343 K and on acidification gives G. D on condensation with phthalic anhydride gives H. H is used as indicator in acid–base titrations.

Answer

A is cyclohexanol;

B is cyclohexene;

C is $CHO{-}(CH_2)_4CHO$;

D is phenol;

E is C_6H_5ONa;

F is salicylic acid;

G is salicyl aldehyde;

H is phenolphthalein.

2. A is obtained by treating B with CaO on heating. A reacts with $HNO_3 + H_2SO_4$ to give C. C on reduction with Sn/HCl gives D. D with $NaNO_2/H_2SO_4$ gives E. E on hydrolysis gives F. F with benzoyl chloride gives G. F with Allyl iodide in K_2CO_3 gives H. H undergoes rearrangement at 200°C to give *o*-allyl phenol. Find A to G. What is the name of the rearrangement reaction?

Answer

A is benzoic acid;

B is benzene;

C is nitrobenzene;

D is aniline;

E is diazonium salt;

F is phenol;

G is phenyl benzoate;

H is phenyl allyl ether. The rearrangement is called Claisen rearrangement.

3. Compound A undergoes dehydration to give B. B on ozonolysis in Zn/H_2O gives C and D. C is also obtained as the ozonolysis product of 2-butene in reducing conditions. D is obtained by the HIO_4 cleavage of the diol produced from ethylene. Benzene reacts with C in sulphuric acid to give E. E on oxidation in air gives F. E on oxidation with $KMnO_4$ gives benzoic acid. F in acid gives G and H. H gives violet colouration with neutral $FeCl_3$. G is a solvent for S_N2 reactions. Find A to H.

Answer

A is isopropyl alcohol;

B is propylene;

C is acetaldehyde;

D is formaldehyde;

E is cumene;

F is cumene hydroperoxide;

G is acetone;

H is phenol.

4. Benzene with Cl_2 + Fe gives A. A with NaOH under high pressure gives B. What is the intermediate in the formation of B? B with bromine water gives C. B with Br_2 in CS_2 at 0°C gives CC and D. B with nitrous acid at 7.5°C gives E which exhibits tautomerism and the tautomeric form is F. Find A to F.

Answer

A is chlorobenzene;

B is phenol (benzyne intermediate);

C is tribromophenol;

CC is *o*-bromophenol;

D is *p*-bromophenol

E is *p*-nitrosophenol

F is quinone monoxime

5. A is a hydrocarbon which by monochlorination gives two products in the ratio 45:55 %. One of these chlorides B is a product obtained by addition of HCl to isopropyl alcohol (C). B with benzene in $AlCl_3$ gives D. D on oxidation in air gives E and F. F is the oxidation product of C. E reacts with CH_2N_2 to give G. G has the formula C_7H_8O and it reacts with Li/NH_3 and ethanol to give H. H in dil. H_2SO_4 gives J and CH_3OH. What is the special structural feature in J? G can also be prepared by reaction of C_6H_5ONa and CH_3I. Why does the method $C_6H_5I + CH_3ONa$ fail in the case of preparation of G? Find A to J and give necessary explanation for the questions given.

Answer

A is propane;

B is isopropylchloride;

C is isopropyl alcohol;

D is isopropylbenzene;

E is phenol;

F is acetone;

G is anisole;

H is 1-methoxy-1,4-cyclohexadiene;

J is 2-cyclohexenone. It is a conjugated ketone. Aryl halide C_6H_5I cannot be used in S_N2 reaction to produce anisole.

If we choose C_6H_5I and CH_3ONa to prepare G, then aryl halide has to react by S_N2 pathway which is not possible.

6. Benzene reacts with Br_2/Fe to give A. A with Mg in ether gives B. B in oxygen along with B in acid gives 2 moles of C. C with bromine water gives D. C with Br_2 in CS_2 at 273 K gives E. C with H_2SO_4 gives F. F with $Br_2/NaOH$ gives G. G in acid on heating gives H. Find the relationship between E and H. C with mercuric acetate gives J. J in aq. NaCl gives K. K with I_2 in $CHCl_3$ gives L. *p*-aminophenol with $NaNO_2$ and H_2SO_4 gives M. M with KI gives N which is an isomer of L. Find A to N.

Answer

A is bromobenzene;

B is C_6H_5MgBr;

C is C_6H_5OH;

D is 1,3,5-tribromophenol;

E is *p*-bromophenol;

OH

SO_3H

F is 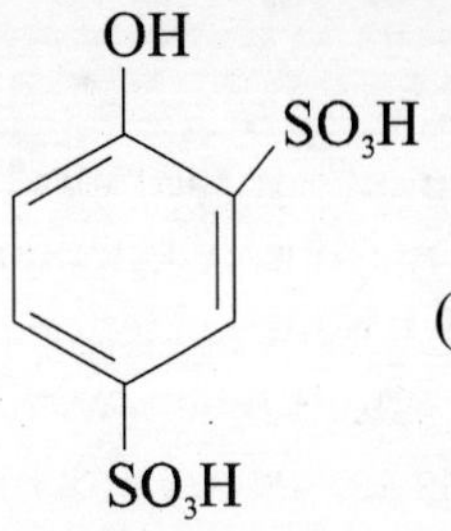 (phenol di-sulphonic acid);

SO_3H

OH

Br SO_3Na

G is 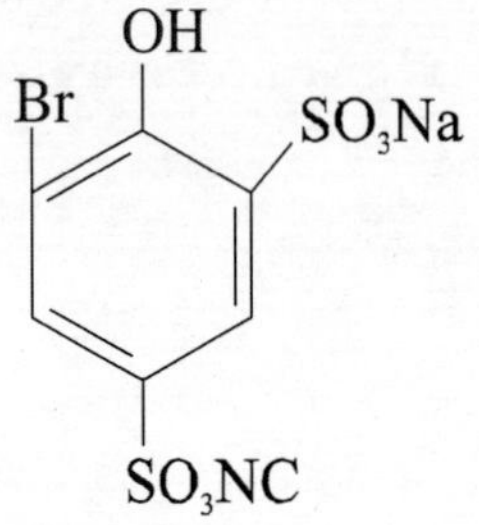 (1-SO_3Na-4-SO_3Na-6-bromophenol);

SO_3NC

OH

Br

H is 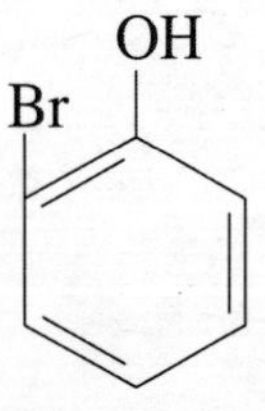 (*o*-bromophenol);

J is *o*-($HgOCOCH_3$) phenol;

K is *o*-(HgCl)phenol;

L is *o*-iodophenol;

OH

M is ;

$N_2^+ HSO_4^-$

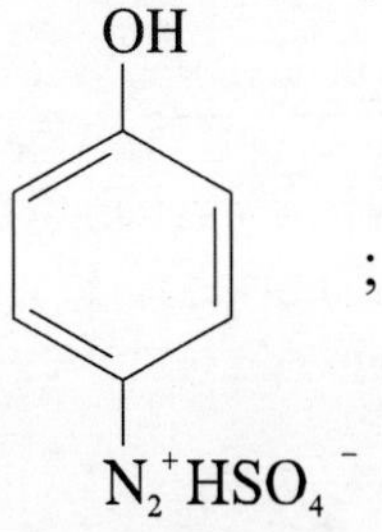

N is *p*-iodophenol;

L and N are the isomers.

7. Compound A on ozonolysis in $(CH_3)_2S$ gives B and C. C is the ozonolysis product of ethylene under reducing conditions. B is a product when isopropyl benzene is oxidized in air along with D. A with H_2SO_4 and CH_3I gives E. E can also be obtained from F which is a sodium salt of G and CH_3I. E with HI gives G and H. Find A to H.

Answer

A is 2-methylpropene;

B is acetone;

C is HCHOD;

D is Phenol;

E is *t*-butylmethyl ether;

F is sodium *t*-butyl oxide;

G is *t*-butyl alcohol;

H is CH_3I.

8. A is one of the products of reaction of methyl ketones with NaOH and I_2. A with Ag powder gives B. B with CH_3OH gives C. C on hydrolysis gives D and E. E also reacts with NaOH/I_2 to give A. F is another compound whose molecular weight is 28 grams more than that of C. Hydrolysis of F with dil. H_2SO_4 gives G and H. Both G and H react with NaOH/I_2 to give A. Find A to H.

Answer

A is CHI_3;

B is acetylene;

C is $CH_2{=}CHOCH_3$;

D is CH_3OH;

E is CH_3CHO;

F is $C_2H_5O-\overset{\overset{\large CH_3}{|}}{C}{=}CH_2$;

G is C_2H_5OH;

H is acetone.

9. Compound A with formula $C_5H_{10}O$ on hydrolysis in dil. H_2SO_4 gives B and C. A does not show test for alcohols. It decolorizes $KMnO_4$. A does not answer Tollens Test and does not react with sodium bisulphate. A is not cyclic. B on dehydration gives D. D with Cl_2/H_2O gives E. E with H_2SO_4 followed by addition of KOH gives F. E with *t*-butyl alcohol gives G. G with sodium acetylide gives H. H on hydrolysis in acid gives 3-butyne-1-ol. Find A to H.

A is — $C_2H_5O-\overset{\overset{\large CH_3}{|}}{C}{=}CH_2$;

B is ethanol;

C is acetone;

D is ethylene;

E is $OHCH_2CH_2Cl$;

F is $CH_2{=}CHO - CH{=}CH_2$;

G is $ClCH_2CH_2O\text{-}(C(CH_3)_3$;

H is $(CH_3)_3C - OCH_2CH_2C{\equiv}CH$. H on hydrolysis gives 3-butyne-1-ol.

10. A and B (isomers. A is insoluble in aq. $NaHCO_3$ and is soluble in NaOH. With bromine and $FeBr_3$, A rapidly forms C. C contains three bromine atoms. B does not dissolve in NaOH but shows reactions of A. B is obtained from E by reaction with CH_2N_2. E is obtained from D by treating with NaOH under high pressure and at high temperature. B reacts with Li/NH_3 to give F. F with dil. H_2SO_4 gives G and H. G is the oxidation product of ethylene by ozonolysis in oxidizing conditions. Find A to G.

A is *p*-cresol;

B is anisole. A and B are isomers;

C is 1,3,5-tribromo-2-methylphenol;

D is C_6H_5Cl;

E is phenol;

F is 1-methoxy-1,4-cyclohexadiene;

G is CH_3OH;

H is 2-cyclohexenone.

9

ALDEHYDES AND KETONES

NOMENCLATURE OF ALDEHYDES AND KETONES

The common names of aldehydes are derived from the names of the corresponding carboxylic acids.

$$\overset{O}{\overset{\|}{HCOH}}$$ Formic acid $\qquad$ $$\overset{O}{\overset{\|}{HCH}}$$ Formaldehyde

$$CH_3\overset{O}{\overset{\|}{C}}OH$$ Acetic Acid $\qquad$ $$CH_3\overset{O}{\overset{\|}{C}}H$$ Acetaldehyde

The systematic names for aldehydes are obtained by adding -al to the name of the parent alkane.

$$\overset{O}{\overset{\|}{HCH}}$$ Methanal $\qquad$ $$CH_3\overset{O}{\overset{\|}{C}}H$$ Ethanal

The presence of substituents is indicated by numbering the parent alkane chain from the end of the molecule that carries the —CHO functional group. For example

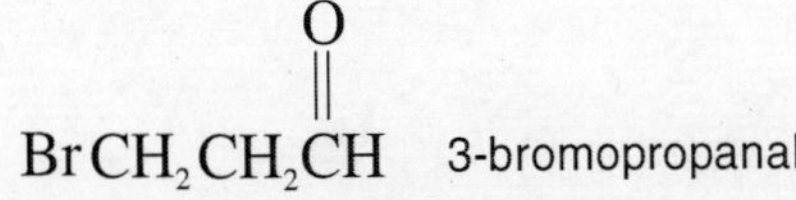
$$Br\,CH_2\,CH_2\overset{O}{\overset{\|}{C}}H$$ 3-bromopropanal

The common name for a ketone is constructed by adding *ketone* to the names of the two alkyl groups on the C=O double bond, listed in alphabetical order.

$$CH_3\overset{\overset{\displaystyle O}{\|}}{C}CH_2CH_3 \quad \text{Ethyl methyl ketone}$$

The systematic name is obtained by adding *-one* to the name of the parent alkane and using numbers to indicate the location of the C=O group.

$$CH_3\overset{\overset{\displaystyle O}{\|}}{C}CH_2CH_3 \quad \text{2-butanone}$$

COMMON ALDEHYDES AND KETONES

Formaldehyde has a sharp, somewhat unpleasant odour. The aromatic aldehydes in Figure 9.1 on the other hand, have a very pleasant flavour. Benzaldehyde has the characteristic odour of almonds, vanillin is responsible for the flavour of vanilla, and cinnamaldehyde makes an important contribution to the flavour of cinnamon.

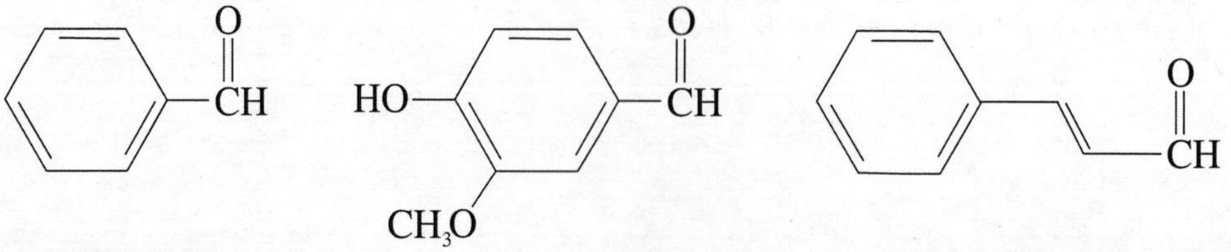

Benzaldehyde Vanillin Cinnamaldehyde

Figure 9.1 Some examples of aldehydes

Aldehydes and ketones play an important role in the chemistry of carbohydrates. The term *carbohydrate* literally means a "hydrate" of carbon, and was introduced to describe a family of compounds with the empirical formula CH_2O. Glucose and fructose, for example, are the carbohydrates with the formula $C_6H_{12}O_6$. These sugars differ in the location of the C=O double bond on the six-carbon chain, as shown in Figure 9.2. Glucose is an aldehyde; fructose is a ketone.

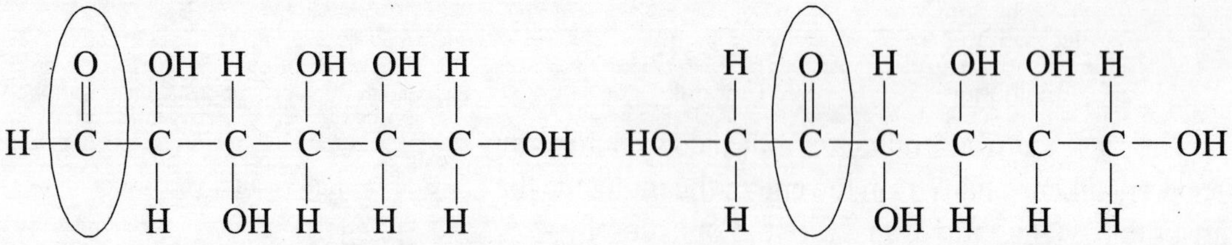

Glucose Fructose

Figure 9.2 Structures of glucose and fructose

METHODS OF PREPARATION OF ALDEHYDES AND KETONES

Many methods of preparation of aldehydes and ketones have already been discussed.

1. Ozonolysis of alkenes under reducing conditions.
2. Treatment of alkenes with dil. $KMnO_4$ followed by cleavage with HIO_4. The method of preparation of aldehydes and ketones from alcohols is presented once again here.
3. Conversion of alcohols into aldehydes and ketones (oxidation of alcohols) (Figure 9.3).

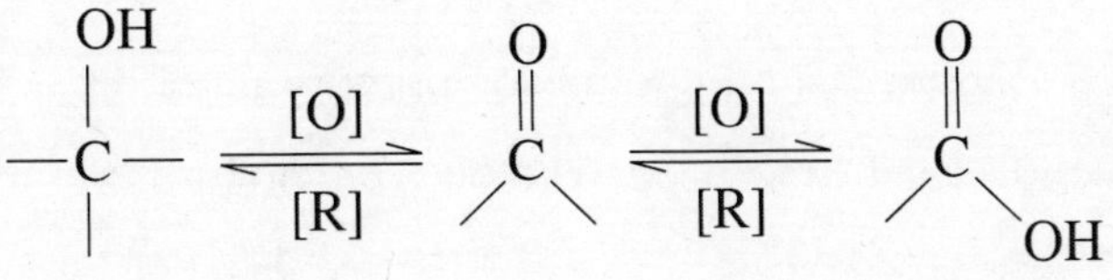

$$-\overset{OH}{\underset{|}{\overset{|}{C}}}- \underset{[R]}{\overset{[O]}{\rightleftharpoons}} \overset{O}{\overset{\|}{C}} \underset{[R]}{\overset{[O]}{\rightleftharpoons}} \overset{O}{\overset{\|}{C}}\text{—OH}$$

Figure 9.3 Alcohols to carbonyl compounds

Reaction type Oxidation–reduction.

Reagents used The reagents used in oxidation of alcohols are listed in the following table.

Aqueous Oxidants	Anhydrous Oxidants
Chromic acid H_2CrO_4	Pyridinium chlorochromate (PCC)
Chromate salts CrO_4^{2-}	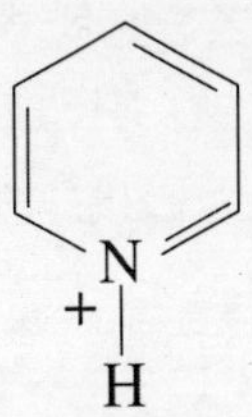 (pyridinium, $C_5H_5N^+$–H) $ClCrO_3^-$
Dichromate salts $Cr_2O_7^{2-}$	Pyridinium dichromate (PDC)
Permanganate MnO_4^-	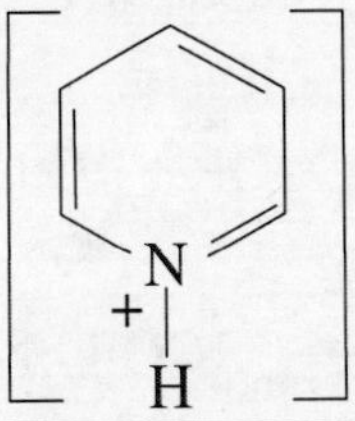 [(pyridinium, $C_5H_5N^+$–H)] $Cr_2O_7^{2-}$

- The outcome of oxidation reactions of alcohols depends on the substituents on the carbinol carbon.
- In order for each oxidation step to occur, there must be a H on the carbinol carbon.
- Primary alcohols can be oxidized to aldehydes or further to carboxylic acids (Figure 9.4).

- In aqueous media, the carboxylic acid is usually the major product.
- PCC or PDC, which are used in dichloromethane, allow the oxidation to be stopped at the intermediate aldehyde.

$$R-\underset{H}{\overset{OH}{\underset{|}{\overset{|}{C}}}}-H \underset{[R]}{\overset{[O]}{\rightleftharpoons}} R-\overset{O}{\overset{\|}{C}}-H \underset{[R]}{\overset{[O]}{\rightleftharpoons}} R-\overset{O}{\overset{\|}{C}}-OH$$

Figure 9.4 Scheme for primary alcohols

- Secondary alcohols can be oxidized to ketones *but* no further.

$$R-\underset{R}{\overset{H}{\underset{|}{\overset{|}{C}}}}-OH \xrightarrow{[O]} R-\overset{O}{\overset{\|}{C}}-R$$

- Tertiary alcohols cannot be oxidized (no carbinol C—H)

  $3^\circ\ R_3COH \xrightarrow{(O)}$ No reaction

4. Aldehydes from RCOCl

   $RCOOH + SOCl_2 \longrightarrow RCOCl \xrightarrow[\text{(ii) } H_2O]{\text{(i) LiAlH}(t\text{-bu})_3\text{, ether}} RCHO$

   Lithium tritertiary butyl hydride is called DIBAL-H and its formulae is LiAlH(t-bu)$_3$
5. Aldehydes from esters and nitriles

   $RCOOR' + \text{DIBAL-H} \longrightarrow H_2O \longrightarrow RCHO$

   $RCN + \text{DIBAL-H} \longrightarrow H_2O \longrightarrow RCHO$
6. Ketones from lithium dialkyl cuprates (R_2CuLi)

   $2RLi + CuI \longrightarrow R_2CuLi + LiI$

   $R'COCl + R_2CuLi \longrightarrow R'COR + RCu + LiCl$
7. Ketones from nitriles.

   $RCN + R'MgX + H_3O^+ \longrightarrow RCOR' + NH_4X + Mg^{+2}$

   $RCN + R'Li + H_3O^+ \longrightarrow RCOR' + NH_4^+ + Li^+$

REACTIONS OF ALDEHYDES AND KETONES

One of the important reactions of aldehydes in synthesis is the Wittig reaction. This reaction produces alkenes with exocyclic double bond if we use cyclic ketones.

Addition of Ylides (The Wittig Reaction)

The reaction of aldehydes or ketones with phosphorus ylides produces alkenes of unambiguous double-bond locations (Figure 9.5). Phosphorus ylides are prepared by reacting a phosphine with an alkyl halide, followed by treatment with a base. Ylides have positive and negative charges on adjacent atoms.

$$(C_6H_5)_3P: \; + \; CH_3-\underset{|}{\overset{Cl}{CH}}-CH_3 \xrightarrow[THF]{BuLi} (C_6H_5)_3\overset{+}{P}-\overset{CH_3}{\underset{CH_3}{C}}:^- + Cl^-$$

Triphenylphosphine 2-chloropropane

$$CH_3-\overset{O}{\overset{\|}{C}}H + (C_6H_5)_3\overset{+}{P}-\overset{CH_3}{\underset{CH_3}{C}}:^- \longrightarrow CH_3-CH_2-\overset{CH_3}{\overset{|}{C}}=CH_2 + O=P(C_6H_5)_3$$

acetaldehyde Dimethyltriphenyl phosphonium ylide 2-methyl butene triphenyl phosphine oxide

Figure 9.5 Wittig reaction

Nucleophilic Addition Reactions of Aldehydes and Ketones

In general, aldehydes are more reactive than ketones in nucleophilic addition reaction. The steric factors favour addition to small atoms like hydrogen of aldehyde. The alkyl groups are electron-releasing, and aldehydes have only one electron-releasing group and the positive charge of carbon in carbonyl is partially neutralized. For ketones, the additional alkyl groups increases the neutralization of the positive charge, and the equilibrium constant for the tetrahedral product is less.

Hydrate Formation Aldehydes and ketones react with water to form gemdiols. The reaction is reversible and involves equilibrium. If we consider formaldehyde and hydrate, there will be 99.9% of hydrate and 0.1% of aldehyde. If it is acetaldehyde then the % of hydrate is reduced. The % of hydrate becomes 0.1% for acetone. The reaction is acid- or base-catalysed. The equilibrium constant for hydrate formation for formaldehyde is 2300 but for acetaldehyde it is 1.06 and for acetone it is 0.002. The greater the substitution in the carbon of carbonyl, lesser in hydrate formation. The steric factor plays an important role. Also the greater the number of electronegative groups, greater is the hydration. Hexafluoroacetone has very high equilibrium constant for hydration compared to acetone. Benzene rings which interact with double bond of carbonyl by resonance decrease hydrate formation.

SOLVED EXAMPLE 1

Arrange the following compounds in increasing order of hydrate formation—*m*-chlorobenzaldedhyde, chloroacetaldehyde, trichloroacetaldehyde, *tert.* butyl carboxyaldehyde, trifluoroacetaldehyde, hexafluoroacetone.

Answer

m-chlorobenzaldehyde < *tert.* butyl carboxyaldehyde < chloroacetaldehyde < trichloroacetaldehyde < trifluoroacetaldehyde < hexafluoroacetone

SOLVED EXAMPLE 2

The equilibrium constant for hydrate formation is low for propanone but high for cyclopropanone. Explain.

Answer

A cyclic ketone suffers increased steric hindrance since bond angle changes from 120 to 109° on moving from sp^2 to sp^3 hybridization. In cyclopropanone, change of angle is from 60 to 120 (angle in the 3-membered ring) (changes to sp^2 hybridization angle). But in the hydrate, angle has to change from 60 to 109 (changes to sp^2 hybridization angle). Hence hydrate, is favoured for cyclopropanone.

Addition of alcohols When alcohols are added to aldehydes or ketones, the addition of one mole of alcohol leads to hemiacetal formation.

The mechanism for addition of alcohol to an aldehyde is given in Figure 9.6.

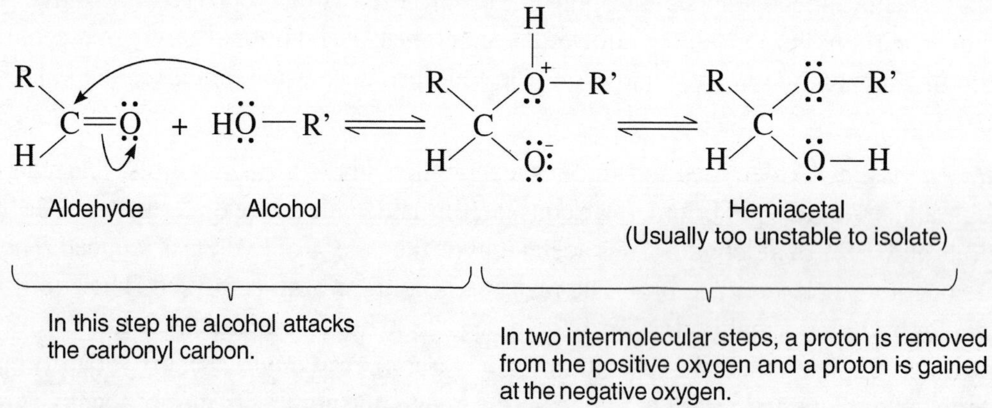

Figure 9.6 Addition of alcohol to an aldehyde

A hemiacetal is

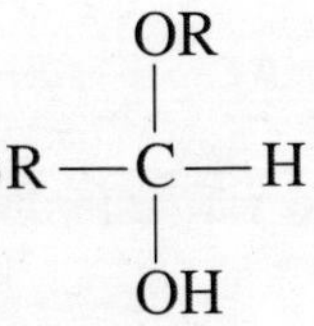

$$R-\underset{\displaystyle OH}{\overset{\displaystyle OR}{\underset{|}{\overset{|}{C}}}}-H$$

The reaction can be acid-catalysed or base-catalysed.

Acetal formation If alcohol solution of aldehyde (ketone) is reacted with gaseous HCl, two moles of alcohol add to give acetal which is written as

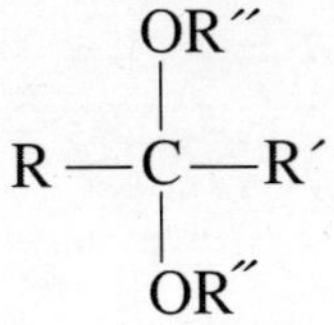

$$R-\underset{\displaystyle OR''}{\overset{\displaystyle OR''}{\underset{|}{\overset{|}{C}}}}-R'$$

R′ can be hydrogen (for aldehyde)

The detailed mechanism of acetal formation is as follows: Reactions of aldehydes with alcohols produce either **hemiacetals** (a functional group consisting of one —OH group and one —OR group bonded to the same carbon) or **acetals** (a functional group consisting of two —OR groups bonded to the same carbon), depending upon conditions. Mixing the two reactants together produces the hemiacetal. Mixing the two reactants with hydrochloric acid produces an acetal. For example, the reaction of methanol with ethanal produces the following results as given in Figure 9.7.

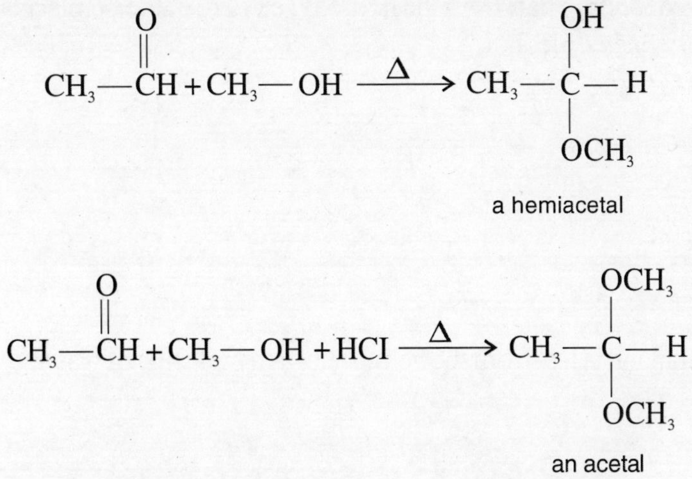

$$CH_3-\overset{\displaystyle O}{\overset{\|}{C}}H + CH_3-OH \xrightarrow{\Delta} CH_3-\underset{\displaystyle OCH_3}{\overset{\displaystyle OH}{\underset{|}{\overset{|}{C}}}}-H$$

a hemiacetal

$$CH_3-\overset{\displaystyle O}{\overset{\|}{C}}H + CH_3-OH + HCl \xrightarrow{\Delta} CH_3-\underset{\displaystyle OCH_3}{\overset{\displaystyle OCH_3}{\underset{|}{\overset{|}{C}}}}-H$$

an acetal

Figure 9.7 Detailed steps of acetal formation

A nucleophilic substitution of an OH group for the double bond of the carbonyl group forms the hemiacetal through the following mechanism:

1. An unshared electron pair on the alcohol's oxygen atom attacks the carbonyl group.

$$CH_3-\overset{O}{\overset{\|}{C}}-H + H\ddot{O}-CH_3 \longrightarrow CH_3-\underset{H}{\underset{|}{\overset{:\ddot{O}:^-}{\overset{|}{C}}}}-\underset{H}{\underset{|}{\overset{\cdot\cdot +}{O}}}-CH_3$$

2. The loss of a hydrogen ion to the oxygen anion stabilizes the oxonium ion formed in Step 1.

$$CH_3-\underset{H}{\underset{|}{\overset{:\ddot{O}:^-}{\overset{|}{C}}}}-\underset{H}{\underset{|}{\overset{\cdot\cdot +}{O}}}-CH_3 \rightleftharpoons CH_3-\underset{H}{\underset{|}{\overset{:\ddot{O}H}{\overset{|}{C}}}}-\ddot{O}-CH_3$$

The addition of acid to the hemiacetal creates an acetal through the following mechanism:

1. The proton produced by the dissociation of hydrochloric acid protonates the alcohol molecule in an acid–base reaction.

$$CH_3\ddot{O}H + H^+ \longrightarrow CH_3\underset{+}{\overset{H}{\overset{|}{O}}}H$$

2. An unshared electron pair from the hydroxyl oxygen of the hemiacetal removes a proton from the protonated alcohol.

$$CH_3-\underset{H}{\underset{|}{\overset{:\ddot{O}H}{\overset{|}{C}}}}-\ddot{O}CH_3 + H\underset{+}{\overset{H}{O}}-CH_3 \longrightarrow CH_3-\underset{H}{\underset{|}{\overset{\overset{H}{|}}{\overset{+:\ddot{O}H}{\overset{|}{C}}}}}-\ddot{O}CH_3 + CH_3\ddot{O}H$$

3. The oxonium ion is lost from the hemiacetal as a molecule of water.

$$CH_3-\underset{H}{\underset{|}{\overset{\overset{H}{^+OH}}{\overset{|}{C}}}}-\ddot{O}CH_3 \longrightarrow CH_3-\underset{H}{\underset{|}{C}}=\overset{+}{\ddot{O}}CH_3 + H_2\ddot{O}:$$

4. A second molecule of alcohol attacks the carbonyl carbon that is forming the protonated acetal.

$$CH_3-\overset{H}{C}=\overset{+}{\ddot{O}}CH_3 + CH_3\ddot{O}H \longrightarrow CH_3-\underset{H}{\overset{CH_3-\overset{+}{\ddot{O}}H}{C}}-\ddot{O}CH_3$$

5. The oxonium ion loses a proton to an alcohol molecule, liberating the acetal.

$$CH_3-\underset{H}{\overset{CH_3-\overset{+}{\ddot{O}}H}{C}}-\ddot{O}CH_3 + CH_3\ddot{O}H \longrightarrow CH_3-\underset{H}{\overset{:\ddot{O}CH_3}{C}}-\ddot{O}CH_3 + CH_3\overset{H}{\underset{+}{\ddot{O}}}H$$

Stability of acetals Acetal formation reactions are reversible under acidic conditions but not under alkaline conditions. This characteristic makes an acetal an ideal protecting group for aldehyde molecules or ketones that must undergo further reactions. A protecting group is a group that is introduced into a molecule to prevent the reaction of a sensitive group while a reaction is carried out at some other site in the molecule. The protecting group must have the ability to easily revert to the original group from which it was formed. An example indicating the protecting action of acetals is given below. Acetals are stable in basic solutions. They function as protecting groups for ketones. If we want to convert the following keto ester to keto alcohol, $LiAlH_4$ reduces both ester and keto group. To prevent the keto group from reduction the cyclic acetal is formed using ethylene glycol and the product is reduced with $LiAlH_4$ and acetal is then hydrolysed. Figure 9.8 illustrates this point.

O, $\overset{O}{\parallel}$ COC_2H_5 — HA / $HOCH_2CH_2OH$ → O, O, $\overset{O}{\parallel}$ COC_2H_5

(1) $LiAlH_4$ / (2) H_2O → O, O, CH_2OH — H_3O^+ / H_2O → O, CH_2OH

Figure 9.8 Conversion of keto ester to keto alcohol

Cyanohydrin Formation

$$RCHO + HCN \rightarrow R{-}\underset{\underset{CN}{|}}{\overset{\overset{H}{|}}{C}}{-}OH$$

The aldehyde is prochiral since the achiral C=O compound creates a chiral compound on addition of HCN. The reaction is very slow with HCN since it is a poor source of nucleophile. The addition of KCN increases the rate of cyanohydrin dramatically.

The mechanism is given in Figure 9.9.

$$R\overset{\overset{O}{||}}{C}H + HCN \rightleftharpoons (R)(H)C(OH)(CN)$$

$$RC(=O)R' + HCN \rightleftharpoons (R)(R')C(OH)(CN)$$

Cyanohydrins

Figure 9.9 Cyanohydrin formation.

A base stronger than CN^- converts HCN to cyanide ion which reacts with aldehyde.

Cyanohydrins are important intermediates since acidic hydrolysis may convert them to α hydroxy acids or α,β unsaturated acids as illustrated below.

$$CH_3CH_2COCH_3 + HCN \rightarrow CH_3CH_2{-}\underset{\underset{CH_3}{|}}{\overset{\overset{OH}{|}}{C}}{-}CN + HCl\ (\text{Water heat}) \rightarrow C_2H_5{-}\underset{\underset{CH_3}{|}}{\overset{\overset{OH}{|}}{C}}{-}COOH$$

In the presence of 95% sulphuric acid, the product is the unsaturated acid shown below.

$$CH_3CH{=}\underset{\underset{CH_3}{|}}{C}{-}COOH$$

With $LiAlH_4$ the cyanohydrin gives β-amino alcohol. Figure 9.10 illustrates the conversion of cyclohexanone to the amino alcohol.

$$\text{Cyclohexanone} \xrightarrow{\text{HCN}} \text{1-hydroxycyclohexane-1-carbonitrile (OH, CN)} \xrightarrow[\text{(2) } H_2O]{\text{(1) } LiAlH_4} \text{1-(aminomethyl)cyclohexanol (OH, } CH_2NH_2\text{)}$$

Figure 9.10 The conversion of cyclohexanone to the amino alcohol

Reaction with Derivatives of Ammonia

Aldehydes and ketones react with primary amines (RNH_2) to form imines ($RCH{=}NR$ or $R_2C{=}NR$). The reaction is acid-catalysed and the product is a mixture of Z and E isomers.

$$R{-}\underset{\underset{R'}{|}}{C}{=}O + H_2N{-}R(H_3O^+) \longleftrightarrow R{-}\underset{\underset{R'}{|}}{C}{=}N{-}R \text{ (Z and E forms)}$$

R′ can be H. Imine formation is fastest between pH 4 and 5. Acid catalyst is important. The main step is one in which protonated amino alcohol loses a molecule of water to become the iminium ion. By protonating the alcohol group acid converts a poor leaving group OH to a better one OH_2^+. The mechanism is given in Figure 9.11.

$$>C{=}\ddot{O}: + H_2NR \longleftrightarrow >C(NH_2{-}R)(:\ddot{O}^-) \longleftrightarrow >C(NH_2{-}R)(\ddot{O}H) \overset{H_3O^+}{\rightleftarrows}$$

Aldehyde or ketone; 1° amine; Dipolar intermediate; Amino alcohol

The amine adds to the carbonyl group a dipolar tetrahedral intermediate.

Intermolecular proton transfer from nitrogen form; then amino alcohol is formed.

$$>C(\ddot{N}H{-}R)(\ddot{O}H_2^+) \underset{}{\overset{-H_2O}{\longleftrightarrow}} >C{=}\overset{+}{N}(H)(R) \quad :\ddot{O}H_2 \longleftrightarrow >C{=}\ddot{N}{-}R + H_3O^+$$

Protonated amino alcohol; Iminium ion; Imine [(E) and (Z) isomers]

Protonation of the oxygen produces a good leaving group. Loss of a molecule of water yields an iminium ion.

Transfer of a proton to water produces the imine and regenerates the catalytic hydronium ion.

Figure 9.11 Mechanism of carbonyl compound reacting with amines

Imines from compounds like hydroxylamine (NH_2OH), hydrazine (NH_2NH_2), semicarbazide ($NH_2NHCONH_2$) are useful derivatives of aldehydes and ketones. The products of reactions of aldehydes and ketones with 2,4-dinitrophenylhydrazine, semicarbazide and hydroxylamine are used to identify aldehydes and ketones. The compounds 2,4-dinitrophenylhydrazones, semicarbazones and oximes are insoluble solids with sharp melting points.

$$RCHO\ (RCOR) + NH_2OH \rightarrow \underset{\text{(Oxime)}}{RCH{=}NOH}\ (\text{Z and E isomers})\ \text{or}\ RCR{=}NOH$$

$$RCHO + H_2NNHC_6H_5 \rightarrow \underset{\text{(Phenylhydrazone)}}{RCH{=}NNH{-}C_6H_5}$$

$$RCHO + H_2NNHCONH_2 \rightarrow \underset{\text{(semicarbazone)}}{RCH{=}NNHCONH_2}$$

Simple hydrazones have low melting points.

Wolf–Kishner Reduction

The reaction of hydrazine with ketones forms hydrazone intermediates which are easily converted to alkenes. This is the basis of Wolf–Kishner reduction.

$$RCOR'NH_2NH_2 \longrightarrow RR'C{=}NNH_2 \longrightarrow RCH_2R' + N_2$$

The reaction is conducted in NaOH in the presence of triethylene glycol at 473 K.

$$C_6H_5COCH_2CH_3 + NH_2NH_2 \xrightarrow{\text{(NaOH, 473 K, triethylene glycol)}} C_6H_5CH_2CH_2CH_3 + N_2.$$

The Wolf Kishner reduction can be accomplished at much lower temperatures using dimethylsulphoxide as solvent. This reduction converts (C=O) to (C—CH_2-) groups. It takes place in strongly basic solution and is used for compounds sensitive to acids. In contrast, Clemmenson reduction of (C=O) to (—CH_2^-) is conducted in acidic solutions and used for compounds sensitive to bases. The mechanism of Wolf–Kishner reduction is given in Figure 9.12.

The strong base brings about tautomerization of hydrazone to a derivative RCHR—N=NH. This undergoes base-catalysed elimination of nitrogen. The elimination of stable N_2 is the driving force for the reaction. More reduction and oxidation reactions are given in the sections that follow.

Step 1 Formation of a hydazone.

$$>C{=}O + H_2\ddot{N}{-}\ddot{N}H_2 \rightleftharpoons >C{=}\ddot{N}{-}\ddot{N}H_2 + H_2O$$

Step 2 Base-catalysed tautomerization and elimination of a molecule of nitrogen.

$$>C{=}\ddot{N}{-}\ddot{N}H_2 \underset{H_2O}{\overset{OH^-}{\rightleftharpoons}} \left[>C{=}\ddot{N}{-}\ddot{N}H^- \longleftrightarrow -\overset{..}{\underset{|}{C}}{-}\ddot{N}{=}\underset{..}{N}H \right] \underset{OH}{\overset{H_2O}{\rightleftharpoons}}$$

$$-\underset{|}{\overset{|}{C}}{-}\ddot{N}{=}\ddot{N}{-}H \underset{H_2O}{\overset{OH^-}{\rightleftharpoons}} -\underset{|}{\overset{H}{\overset{|}{C}}}{-}\ddot{N}{=}\ddot{N}{:}^- \xrightarrow{-N_2} -\underset{|}{\overset{H}{\overset{|}{C}}}{:}^- \xrightarrow{H_2O} -\underset{|}{\overset{H}{\overset{|}{C}}}{-}H$$

Figure 9.12 Wolf–Kishner reduction

Oxidation of Aldehydes and Ketones

Aldehydes are readily oxidized to carboxylic acids but ketones are generally inert towards oxidation. This is because aldehydes have a CHO proton that can be abstracted during oxidation but ketones do not. The structures of aldehydes and ketones reveal this point as indicated below.

$$R{-}\overset{H}{\overset{|}{C}}{=}O \quad \text{and} \quad R{-}\overset{R'}{\overset{|}{C}}{=}O$$

Many oxidizing agents like hot HNO_3 and $KMnO_4$ convert aldehydes into carboxylic acids but Jones reagent (aqueous sulphuric acid) is a more common choice for laboratory purposes. The oxidation occurs rapidly at room temperature and the yield is good.

Hexanol gives hexanoic acid in 96% yield but one drawback is that side reactions take place since the reaction is carried out in acidic conditions. In such cases the oxidation is carried out using a solution of silver oxide in aqueous ammonia (called Tollens' reagent). The aldehyde is oxidized to acid without affecting carbon–carbon double bonds or other functional groups in the molecule. For example benzaldehyde is converted to benzoic acid and a shining silver mirror is deposited on the of walls of the reaction flask and this forms the qualitative test for aldehydes in practical organic chemistry.

The aldehyde oxidation occurs through intermediate 1,1-diols (hydrates) which form by reversible nucleophilic addition to C=O group. Even though the hydrate is formed in small amounts at equilibrium it reacts like typical primary or secondary alcohols and is oxidized to the acid. The scheme is shown below.

$$R-\overset{\displaystyle H}{\overset{|}{C}}=O + H_2O \longleftrightarrow R-\underset{\displaystyle OH}{\underset{|}{\overset{\displaystyle OH}{\overset{|}{C}}}}-H \rightarrow (CrO_3, H_3O^+)\ RCOOH$$

Ketones are inert to most of the oxidizing agents but react with hot alkaline $KMnO_4$. The C—C bond next to carbonyl group is broken and the carboxylic acid is produced. This reaction is useful for symmetrical ketones only since mixture of products result for unsymmetrical ketones. Cyclohexanone gives hexanedioic acid.

Reduction of Aldehydes and Ketones

Aldehydes and ketones may be reduced to alcohols by the use of 1) metal catalyst (Na in alcohol) and hydrogen or (2) $LiAlH_4$. The more common reagent is $NaBH_4$. The balanced reaction is

$$4RCHO + NaBH_4 + 3H_2O \longrightarrow 4RCH_2OH + NaH_2BO_3$$

Butanal gives 1-butanol and butanone gives 2-butanol. The key step in the reduction reaction by $LiAlH_4$ or $NaBH_4$ is the transfer of hydride ion from the metal to the C=O carbon. The hydride ion is the nucleophile. The mechanism for reduction of a ketone to secondary alcohol is shown in Figure 9.13.

$$\underset{\delta+\ \delta^-}{\overset{H-\bar{B}H_3}{R_2C=O}} \longrightarrow \overset{H\qquad \bar{B}H_3}{R_2C-O} \xrightarrow{R_2C=O} \underset{\text{Tetraalkoxyborate}}{(R_2CHO)_4\bar{B}}$$

$$\underset{H-OH}{R_2CHO-\bar{B}(OCHR_2)_3} \longrightarrow R_2CHOH + HO\bar{B}(OCHR_2)_3 \xrightarrow{3H_2O} 3R_2CHOH + (OH)_4\bar{B}$$

Figure 9.13 Ketone to secondary alcohol

The hydrogen is transferred from BH_4^- and the steps are repeated until hydrogens are transferred. $NaBH_4$ is less powerful than $LiAlH_4$ and reduces only aldehydes and ketones. $LiAlH_4$ can reduce esters, aldehydes, ketones and acids. Since $LiAlH_4$ reacts violently with water, the reaction must be carried out in an anhydrous solution (anhydrous ether). Ethyl acetate is added cautiously after the reaction is complete to decompose excess $LiAlH_4$. Water is then added to decompose lithium complex.

$NaBH_4$ reductions can be carried out in water or alcohol solutions. The reaction is safe and handling $NaBH_4$ is easy. It can be easily weighed in open atmosphere and it is a white crystalline solid. $LiAlH_4$ and $NaBH_4$ reduce carbonyl and C=C. If the C=C has to remain unaffected, we can use $CeCl_3$ which is a hard Lewis acid metal salt and the reduction of carbonyl to alcoholic group takes place without disturbing the C=C. Cyclopenten-2-one is reduced by $NaBH_4$ and $CeCl_3$ to cyclopentanol.

$LiAlH_4$ or $NaBH_4$ reductions cannot be used to produce tertiary alcohols. One of the most important reactions of aldehydes and ketones is the nucleophilic addition reaction. A nucleophile attacks the electrophilic C=O atom from a direction perpendicular to the plane of the C=O group. Rehybridization of the carbonyl carbon from sp^2 to sp^3 occurs. The attacking nucleophile can be negatively charged (OH^-, H^-, carbanion, OR^-, CN^-) or neutral species like (H_2O, ROH, NH_3 or RNH_2). Some of these reactions are discussed below:

Halogenation Reactions

There are many reaction which involve the formation of enols if α hydrogens are present. Some of them involve enolate anions.

Enolate anions An important characteristic of aldehydes and ketones is the acidity of the hydrogens on the carbon atom adjacent to the C=O group. These are called α hydrogens. If we have the compound $C_6H_5COCH_3$ the hydrogen of methyl next to carbonyl is called alpha hydrogen. In the compound acetone there are two alpha hydrogens. In $C_6H_5COCH_2CH_3$ the hydrogen of methyl group is β hydrogen. In formaldehyde there is no alpha hydrogen. In CCl_3CHO there is no alpha hydrogen. The pK_a values of alpha hydrogens of C=O are of the order of 19–20. These hydrogens are more acidic than hydrogen of acetylene (pK_a=25). The reason for the acidity of alpha hydrogens is due to resonance stabilization of the anion of carbonyl compounds as indicated below.

$$:\ddot{O}=\overset{|}{\underset{|}{C}}-\overset{|}{\underset{|}{C}}-H \;\; :\bar{B} \Leftrightarrow O=\underset{|}{C}-\underset{|}{\ddot{C}}- \longleftrightarrow :\ddot{\bar{O}}:-\underset{|}{C}=\underset{|}{C}- + H-B$$

The resonance-stabilized anion can accept a proton at carbonyl carbon or at oxygen to form the *enol*. The resonance-stabilized anion is called the *enolate* anion. The enolate anions are involved in reactions like aldol addition and haloform reaction.

Aldol condensation or addition is a very important reaction for carbonyl compounds containing α. Hydrogen and is discussed in detail.

Aldol Addition

When acetaldehyde reacts with dilute NaOH at room temperature or below a dimerization takes place forming 3-hydroxybutanal. Since this contains OH and CHO groups it is called *aldol*. The reaction is called aldol addition reaction.

$$2CH_3CHO \xrightarrow{(10\%NaOH,\ H_2O\ 5°C)} CH_3\underset{}{\overset{OH}{\overset{|}{C}}}HCH_2CHO$$

(3-hydroxybutanal or aldol)

The mechanism of aldol reaction is given in Figure 9.14.

Step 1 In this step the base (a hydroxide ion) removes aproton from the a carbon of one molecule of acetaldehyde to give a resonace - stabilised enolate anion

$$H-\ddot{\underset{..}{O}}:^- + H-CH_2-\overset{O}{\overset{\|}{C}}-H \rightleftharpoons \left[\overset{-}{:}CH_2-\overset{:\ddot{O}:}{\overset{\|}{C}}-H \longleftrightarrow CH_2=\overset{:\ddot{O}:^-}{\overset{|}{C}}-H \right] + H-\ddot{O}:-H$$

Enolate anion

Step 2 The enolate anion then acts as a nucleophile (as a carbanion) and attacks the carbonyl carbon of a second molecule of acetaldehyde, producing an alkoxide anion.

$$CH_3-\overset{:\ddot{O}:}{\overset{\|}{C}}-H + \overset{-}{:}CH_2-\overset{:\ddot{O}:}{\overset{\|}{C}}-H \rightleftharpoons CH_3-\overset{:\ddot{O}:^-}{\overset{|}{C}H}-CH_2-\overset{:\ddot{O}:}{\overset{\|}{C}}-H$$

$$\updownarrow$$

$$CH_2=\overset{:\ddot{O}:^-}{\overset{|}{C}}-H$$

An alkoxide anion

Step 3 The alkoxide anion now removes a proton from a molecule of water to form the aldol.

$$CH_3-\overset{:\ddot{O}:^-}{\overset{|}{C}H}-CH_2-\overset{:\ddot{O}:}{\overset{\|}{C}}-H + H-\ddot{O}-H \rightarrow CH_3-\overset{:\ddot{O}-H}{\overset{|}{C}H}-CH_2-\overset{:\ddot{O}:}{\overset{\|}{C}}-H + \bar{O}-H$$

Stronger base — Aldol — Weaker base

Figure 9.14 The aldol reaction

The mechanism shows two characteristics:

1. acidity of alpha hydrogens
2. the tendency of C=O to undergo nucleophilic addition

Dehydration of aldol addition product If the aldol addition product is heated, dehydration takes place to form 2-butenal. Dehydration takes place readily due to the acidic hydrogens.

The product is stabilized by conjugation of C=O with C=C. In some aldol reactions dehydration takes place as soon as aldol is formed.

Aldol condensation Condensation refers to the process when two molecules are joined together through intermolecular elimination of water or alcohol. The mechanism is given in Figure 9.15.

These hydrogens are acidic

The double bonds of the alkene and carbonyl are conjugated

$$CH_3-\underset{|}{\overset{\overset{:\ddot{O}-H}{|}}{C}H}-\underset{\underset{H}{|}}{C}H-\overset{\overset{\ddot{O}}{\|}}{C}-H + H-\ddot{O}:^- \rightarrow CH_3-CH=CH-\overset{\overset{\ddot{O}}{\|}}{C}-H + H-\underset{\underset{H}{|}}{\ddot{O}}: + H-\ddot{O}:^-$$

2-Butenal (crotonaldehyde)

Figure 9.15 Dehydration process in aldol condensation

Synthetic applications of aldol condensation The reaction is of great importance in synthesis since it gives us a method of linking two similar molecules with carbon–carbon bond between them. Some examples of synthesis are:

1. Aldehyde to 1,3-diol.

$$2RCH_2CHO \xrightarrow{(OH^-, H_2O)} RCH_2CHOHCHRCHO \longrightarrow (NaBH_4)$$

$$\downarrow$$

$$RCH_2CHOHCHRCH_2OH(1,3\text{-diol})$$

2. Aldehyde to another aldehyde

$$2RCH_2CHO \xrightarrow{(OH^-, H_2O)} RCH_2CHOHCHRCHO \xrightarrow[H_2O]{HA} RCH_2CH=CRCHO$$

$$\downarrow H_2/Pd/C$$

$$RCH_2CH_2CHRCHO$$

3. Aldehyde to allylic alcohol

$$2RCH_2CHO \rightarrow RCH_2CHOHCHRCHO \xrightarrow[H_2O]{HA} RCH_2CH=CRCHO \longrightarrow LiAlH_4$$

$$\downarrow$$

$$RCH_2CH=CRCH_2OH$$

4. Aldehyde to alcohol

$$2RCH_2CHO \rightarrow RCH_2CHOHCHRCHO \rightarrow RCH_2CH{=}CRCHO \xrightarrow{H_2/Ni(\text{high pressure})} RCH_2CH_2CHRCH_2OH$$

SOLVED EXAMPLE 3

Synthesize butanol from acetaldehyde.

Answer

$$2CH_3CHO \rightarrow CH_3CHOHCH_2CHO \xrightarrow[H_2O]{HA} CH_3CH{=}CHCHO \xrightarrow{H_2/Ni\ (\text{high pressure})} CH_3CH_2CH_2CH_2OH$$

Ketones undergo aldol addition but equilibrium is unfavourable. The reaction is carried out in a special apparatus which allows the product to be removed from contact with base as it is formed. The product is diacetone alcohol if starting ketone is acetone. The structure of diacetone alcohol is as follows.

$$CH_3COCH_2COH(CH_3)_2$$

Soxhlet extraction is used. The ketone is boiled in a flask and aluminium tritertiary butoxide is used as a catalyst at elevated temperatures to promote the condensation. Ketone distills in the condensor. It drips to come in contact with the catalyst. As the ketone boils, the level of liquid in trap rises and the whole solution is siphoned from the chamber into the flask containing ketone. There is no catalyst in the flask and the aldol product cannot revert to acetone. The mixture continues to boil and the remaining acetone which has a low boiling point compared to the product comes in contact with the catalyst and the aldol amount increases. This leads to equilibrium shifted to right.

The advantage of dehydration in case of ketones is that, this step drives the reaction towards the right even though the aldol formation step (equilibrium) is unfavourable.

Intramolecular Aldol Condensation

Aldol condensation offers a convenient way of synthesizing 5 or 6-membered rings. This can be done by intramolecular aldol condensation using a dialdehyde or ketoaldehyde or diketone as the substrate (Figure 9.16). For example the following ketoaldehyde yields 1-cyclopentenyl methylketone.

$$CH_3COCH_2CH_2CH_2CH_2CHO \rightarrow \text{cyclopentenylmethylketone}$$

$$CH_3\overset{O}{\overset{\|}{C}}CH_2CH_2CH_2CH_2\overset{O}{\overset{\|}{C}}H \longrightarrow \text{(cyclopentenyl)}-\overset{O}{\overset{\|}{C}}CH_3$$

Figure 9.16 Intramolecular aldol condensation

The mechanism is given in Figure 9.17.

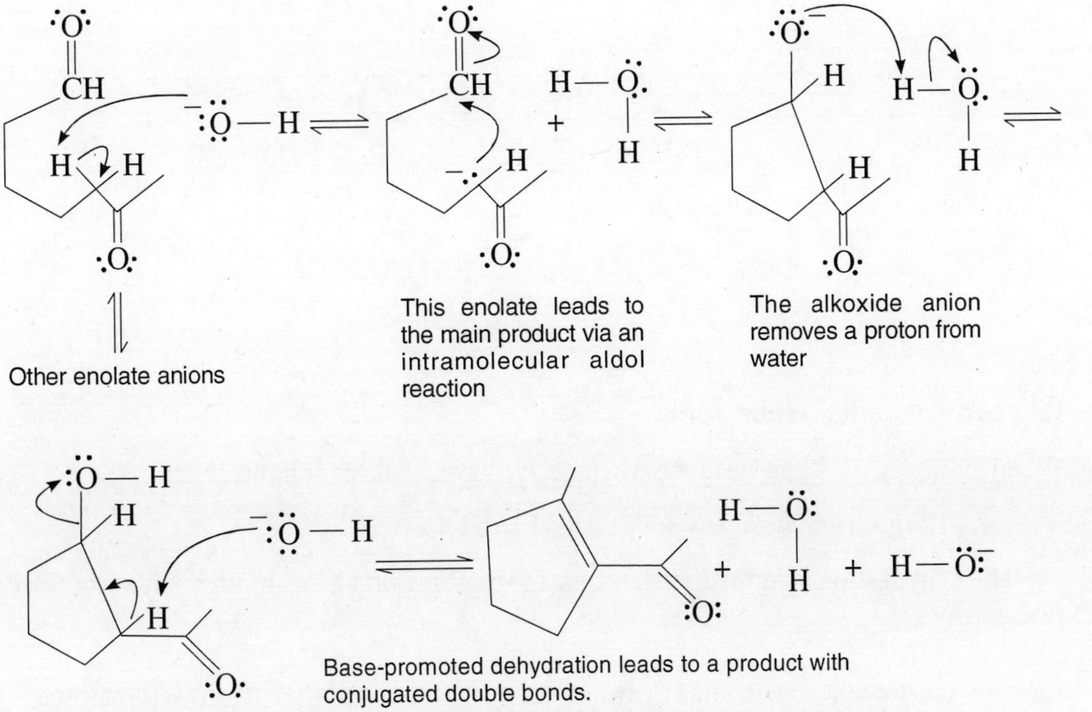

Other enolate anions

This enolate leads to the main product via an intramolecular aldol reaction

The alkoxide anion removes a proton from water

Base-promoted dehydration leads to a product with conjugated double bonds.

Figure 9.17 Mechanism of intramolecular aldol condensation

If there is a possibility of 5- and 7-membered rings, 5-membered ring is preferentially formed. The aldehyde group undergoes addition in preference to ketone since the carbon of C=O of aldehyde is more positive and less sterically hindered.

Crossed aldol reactions These reactions use NaOH as base and one of the two aldehydes not having alpha hydrogen.

1. benzaldehyde + $CH_3CH_2CHO \rightarrow C_6H_5CH{=}\underset{CH_3}{\underset{|}{C}}{-}CHO$

2. $C_6H_5CHO + CH_3CHO \rightarrow C_6H_5CHOHCH_2CHO \rightarrow C_6H_5CH{=}CHCHO$

Aldol reaction involving lithium enolates Lithium diisopropyl amide abbreviated as LDA is prepared by dissolving diisopropyl amine in diethyl ether or THF and reacted with alkyl lithium.

$$(i-C_3H_7)_2NH + C_4H_9Li \rightarrow (i-C_3H_7)_2N^- Li^+ C_4H_{10}$$

$$LDA + CH_3COCH_3 \rightarrow CH_3\underset{}{\overset{O-Li^+}{\overset{|}{C}}}=CH_2$$

$$CH_3\overset{O-Li^+}{\overset{|}{C}}=CH_2 + CH_3CH_2CHO \rightarrow CH_3COCH_2\underset{O-Li^+}{\underset{|}{C}H}CH_2CH_3$$

Addition of water gives the aldol which is

$$CH_3COCH_2\underset{OH}{\underset{|}{C}H}CH_2CH_3.$$

Claisen–Schmitt Reaction

When the alpha hydrogen system is a ketone, we have Claisen–Schmitt reaction

$$C_6H_5CHO + CH_3COCH_3 \longrightarrow C_6H_5CH=CHCOCH_3$$

The dehydration is immediate because the double bond is in conjugation with $C=O$ and benzene ring.

Other Carbonyl Condensation Reactions and Substitution Reactions

Both carbonyl condensation reactions and substitution reactions take place under basic conditions involving *enolate* ions. We have to suitably alter the procedure depending on whether we need aldol product or substitution product.

Substitution product Alpha substitution reactions require full equivalent of the strong base and are normally carried out so that carbonyl is rapidly and completely converted to enolate ion at low temperature. An electrophile is rapidly added (lithium diisopropyl amide (LDA) for alkylation) so that reactive *enolate* ion is quenched, solvent is THF at dry ice (–78°C).

There will be no reactant left for the condensation reaction. Immediately an alkyl halide is added to complete alkylation reaction.

Cyclohexanone + LDA, $\xrightarrow{\text{THF, }-78°\text{C}}$ lithium enolate $\rightarrow$ CH_3I $\rightarrow$ 2-methyl cyclohexanone

In Aldol reaction we need only a catalytic amount of weak base. Once condensation occurs, the catalyst is regenerated. For propanal we dissolve aldehyde in methanol and add 0.05 equivalent of $NaOCH_3$ and warm the mixture to get aldol. Aldol reactions are also acid-catalysed. In acid-catalysed reactions, enol is formed and OH is protonated and water is expelled (Figure 9.18).

Enol

Figure 9.18 Acid-catalysed aldol reaction

Halogenation of Ketones

Ketones that have a hydrogen undergo acid- or base-catalysed halogenation. In acid-catalysed halogenation the reaction stops after stage 1. Monohalogenated compound is formed. The mechanism is given in Figure 9.19.

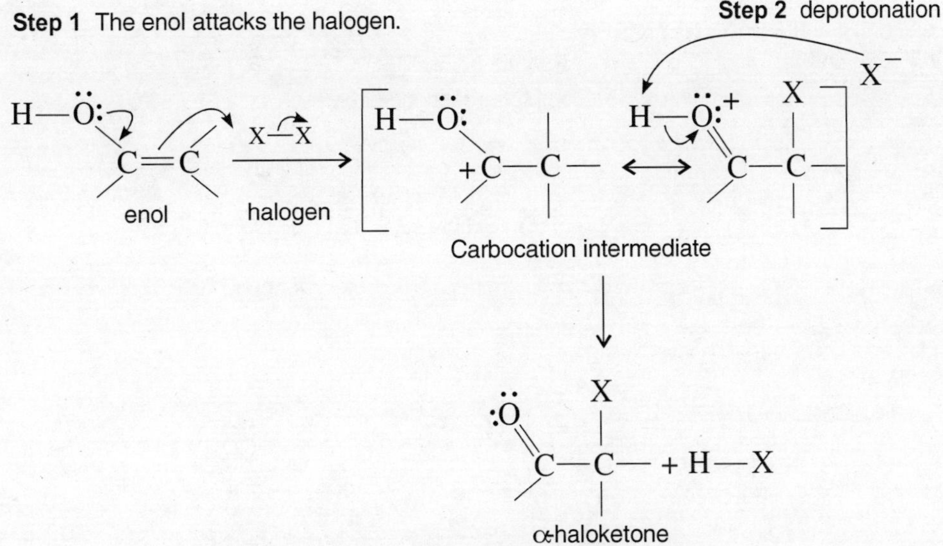

Figure 9.19 Halogenation of ketones

Haloform reaction When halogenation takes place in the presence of base, the enolate anion is formed slowly to form the halo compound. The inductive effect helps in replacing two more hydrogens of methyl ketones to give $RCOCX_3$. The $RCOX_3$ reacts with OH^- to give $RCOO^-$ and CHX_3, the haloform. The mechanisms are given in Figure 9.20.

$$C_6H_5-\overset{O}{\overset{\|}{C}}-CH_2-H + :\bar{B} \rightleftharpoons C_6H_5-\overset{O}{\overset{\|}{C}}-\bar{C}H-H \xrightarrow{X-X} C_6H_5-\overset{O}{\overset{\|}{C}}-CHX-H + X^-$$

$$\updownarrow$$

$$^-:\ddot{O}-C(C_6H_5)=CH_2$$

Enolate anion

(a) Halogenation step of the haloform reaction to form enolate anion

$$C_6H_5-\overset{O}{\overset{\|}{C}}-CHX-H + :\bar{B} \rightleftharpoons C_6H_5-\overset{O}{\overset{\|}{C}}-\bar{C}HX \xrightarrow{X-X} C_6H_5-\overset{O}{\overset{\|}{C}}-CHX-X + X^-$$

$$\updownarrow \qquad \downarrow :\bar{B} \text{ then } X_2^-$$

$$^-:\ddot{O}-C(C_6H_5)=CHX \qquad C_6H_5-\overset{O}{\overset{\|}{C}}-CX_3$$

Acidity is increased by the electron withdrawing halogen atom

Enolate anion

(b) Complete reaction scheme

$$C_6H_5-\overset{O}{\overset{\|}{C}}-CX_3 + ^-:\ddot{O}-H \rightleftharpoons C_6H_5-C(O^-)(\ddot{O}H)-CX_3$$

$$C_6H_5-C(O^-)(\ddot{O}H)-CX_3 \rightleftharpoons C_6H_5-\overset{O}{\overset{\|}{C}}-\ddot{O}-H + ^-:CX_3$$

$$\rightleftharpoons C_6H_5-\overset{O}{\overset{\|}{C}}-\ddot{O}:^- + H-CX_3$$

Carboxylate anion Haloform

(c) Cleavage step of the haloform reaction

Figure 9.20 Haloform reaction

$$R—COCH_3 + 3I_2 + 3OH^- \longrightarrow R—COCl_3 + 3I^- + 3H_2O.$$

$RCOCX_3$ ($RCOCBr_3$), CBr_3^- or CX_3^- is a better leaving group than OH^- since pK_a of the conjugate acid $CHBr_3$ is 9 (but for the conjugate acid of OH^-, namely water, $pK_a = 14$) The carbanion is much stabilized by the 3 halogen atoms. Hence the following reaction occurs.

$$R—COCl_3 + OH^- \longrightarrow RCOO^- + CHI_3.$$

This reaction is a rare example of nucleophilic substitution at carbonyl group resulting in cleavage of C—C single bond.

Iodoform is a yellow solid and can be used to characterize all methyl ketones. CH_3CHO, ethanol, isopropyl alcohol or RCHOHAr can also undergo this reaction. The oxidation of the secondary alcohol gives methylketone which will lead to the haloform.

If the compound contains alpha hydrogen on both sides of C=O, two acids may result. Hence such compounds are not useful in the synthesis of carboxylic acids by this reaction.

If halogenation has to stop in one stage we must carry out the halogenation in acid medium only. If we consider an unsymmetrical dialkyl ketone, halogenation takes place at the less highly substituted side of C=O group in base but in acid, halogenation takes place at the more substituted side of C=O group.

$$CH_3CH_2COCH_3 + I_2 + NaOH \longrightarrow CH_3CH_2COCI_3 \longrightarrow CH_3CH_2COO^- + CHI_3$$

$$CH_3CH_2COCH_3 + Br_2 + HOAC \longrightarrow CH_3CHBrCOCH_3$$

With a symmetric ketone if more than one equivalent of Br_2 is present in acetic acid, bromination takes place on both sides of the ketone to give 1,3-dibromoacetone.

Cannizarro reaction

Hydroxide ions will give products from non-enolizable carbonyl compounds. For example if we allow benzaldehyde to stand in the presence of a strong base, a redox reaction occurs and we get benzoate ion and benzyl alcohol. If one of the two carbonyl compounds is formaldehyde we always get the formate anion along with alcohol of the other C=O compound. This reaction is called Cannizarro reaction (Figure 9.21). If the reaction is carried out in deuterated solvent, no carbon–deuterium bonds appear in alcohol.

$$C_6H_5CHO + D_2O(NaOD) \longrightarrow C_6H_5CH_2OD + C_6H_5COO^-\ Na^+$$

The hydrogen reducing the aldehyde must come from another aldehyde and not the solvent. The detailed mechanism of Cannizarro reaction is given in Figure 9.22.

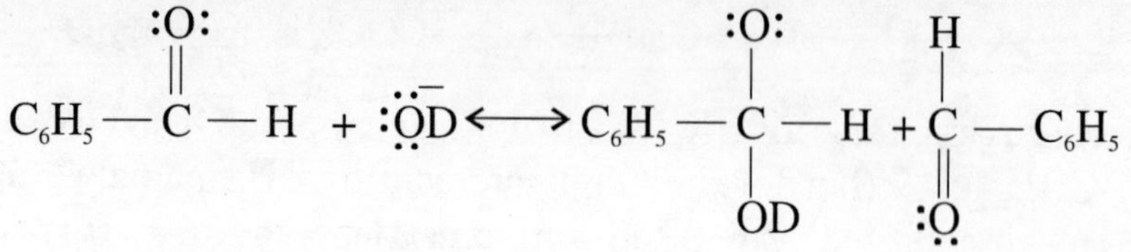

$$C_6H_5-\overset{\overset{:O:}{\|}}{C}-H + :\ddot{O}D^- \longleftrightarrow C_6H_5-\underset{OD}{\overset{:O:^-}{|}}{C}-H + \underset{:O}{\overset{H}{|}}{C}-C_6H_5$$

$$C_6H_5COOD + C_6H_5CH_2O^- \longleftrightarrow C_6H_5COO^- + C_6H_5CH_2OD$$

A hydride shift is at the heart of the mechanism of the Cannizarro reaction.

2 (2-methoxybenzaldehyde) $\xrightarrow{KOH}$ 2-methoxybenzoic acid + 2-methoxybenzyl alcohol

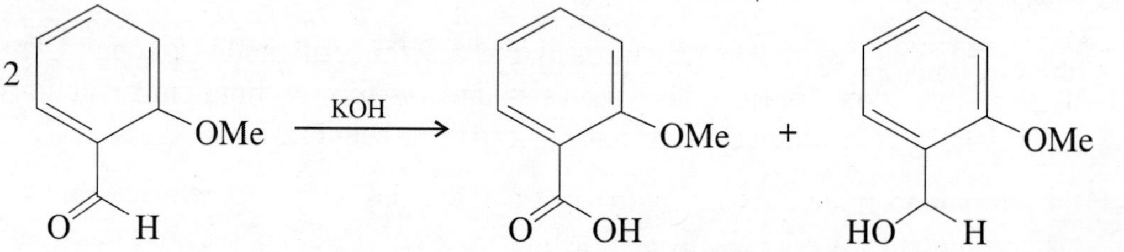

Figure 9.21 Examples of Cannizarro reaction

This redox disproportionation of non-enolizable aldehydes to carboxylic acids and alcohols is conducted in concentrated base.

α-keto aldehydes give the product of an intramolecular disproportionation in excellent yields.

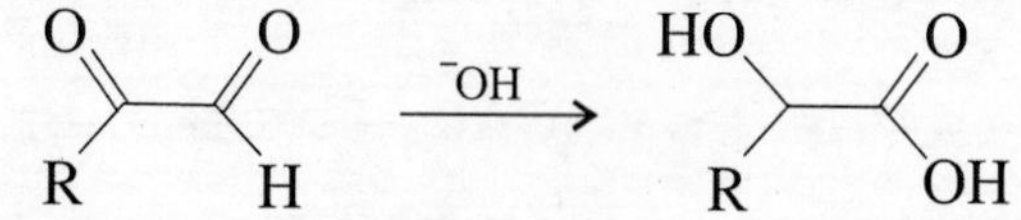

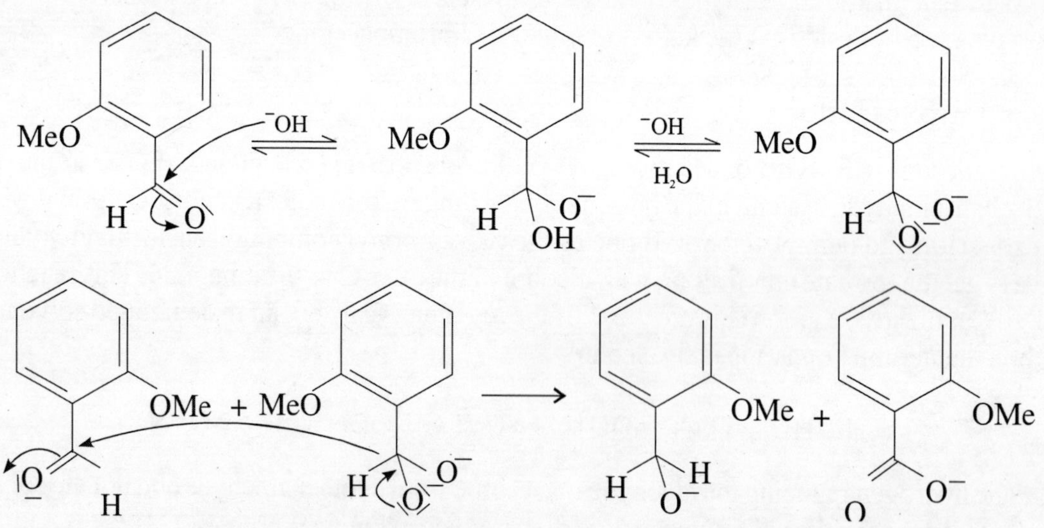

$$RCOCHO \xrightarrow{^-OH} RCH(OH)COOH$$

Figure 9.22 Mechanism of the Cannizarro reaction

An interesting variant, the crossed Cannizzaro reaction, uses formaldehyde as a reducing agent (Figure 9.23).

$$C_6H_5CHO + H_2C{=}O \xrightarrow[\text{2. Acidification}]{\text{1. 30\% NaOH}} C_6H_5CH_2OH + HCOOH$$

Figure 9.23 Cannizarro reaction of formaldehyde and benzaldehyde

At the present time, various oxidizing and reducing agents can be used to carry out such conversions (with higher yields). Hence the Cannizarro reaction has limited synthetic utility mainly for the above-mentioned conversion of α-keto aldehydes.

One important aspect of the Cannizarro reaction is the requirement of the Lewis acid acceptor of hydride ion. The reaction is applicable to formaldehyde and substituted benzaldehydes.

SOLVED EXAMPLE 4

When *o*-phthalaldehyde undergoes reaction with NaOH followed by addition of H_3O^+, what is the product?

Answer

This is an internal Cannizarro reaction. One aldehyde group becomes acid and another becomes alcohol and we get *o*-(hydroxymethyl) benzoic acid.

The following table shows the rates of Cannizarro reaction of substituted benzaldehydes relative to benzaldehyde at two different temperatures.

Aldehyde	**Rate relative to benzaldehyde**	
	298K	**373K**
Benzaldehyde	1	1
p-methylbenzaldehyde	0.2	0.2
p-methoxybenzaldehyde	0.05	0.1
p-dimethylaminobenzaldehyde	Very slow	0.0004
p-nitrobenzaldehyde	210	2200

The Cannizarro reaction goes much slower with electron-donating groups and goes much faster in the presence of electron-withdrawing groups in the ring.

PERKIN REACTION

In Perkin condensation acetic anhydride condenses with benzaldehyde (non-enolizable) giving an unsaturated intermediate which by hydrolysis leads to cinnamic acid. The salt of the acid acts as the base. The mechanism is given in the Figure 9.24.

$$(CH_3CO)_2O + PhCHO \xrightarrow{CH_3COO^{-}Na^{+}} Ph{-}CH{=}CH{-}CO_2H$$

$$CH_3{-}CO{-}O{-}CO{-}CH_3 + CH_3COO^{-} \rightleftharpoons {}^{-}CH_2{-}CO{-}O{-}CO{-}CH_3 + CH_3COOH$$

$$C_6H_5{-}CHO + {}^{-}CH_2{-}CO{-}O{-}CO{-}CH_3 \longrightarrow C_6H_5{-}CH(O^{-}){-}CH_2{-}CO{-}O{-}CO{-}CH_3$$

$$C_6H_5{-}CH(O^{-}){-}CH_2{-}CO{-}O{-}CO{-}CH_3 + CH_3COOH \longrightarrow C_6H_5{-}CH(O){-}CH_2{-}CO{-}O{-}CO{-}CH_3 + CH_3COO^{-}$$

$$C_6H_5{-}CH{=}CH{-}CO{-}O{-}CO{-}CH_3 \xrightarrow{H_3O^{+}} C_6H_5{-}CH{=}CHCOOH$$

Figure 9.24 Perkin condensation

Condensation of Carbonyl Compounds with Nitriles

The alpha hydrogens of nitriles are also appreciably acidic but less than that of aldehydes and ketones. pKa for CH_3CN is 25. Hence nitriles undergo aldol type condensation with aldehydes which do not have alpha hydrogens. For example benzaldehyde reacts with benzyl cyanide in ethoxide ion in ethanol to give

$$\begin{array}{l} C_6H_5CH{=}C{-}CN \\ | \\ C_6H_5 \end{array}$$

Condensation of Carbonyl Compounds with nitro compounds

The alpha hydrogens of nitroalkanes are acidic (pKa = 10). They have greater acidity than aldehydes and ketones due to resonance forms with electron-withdrawing NO_2 groups. Hence they condense with aldehydes in the presence of OH^-. For example, benzaldehyde condenses with nitromethane to give $C_6H_5CH{=}CHNO_2$. With H_2 and Ni both C=C and NO_2 reduce to give amines. The product is $C_6H_5CH_2CH_2NH_2$.

Grignard Addition to Carbonyl Compounds

Like oxygen and nitrogen centred nucleophiles organometallic reagents react with C=O compounds. Even though there is no free R- or H, the organometallic reagent can deliver R- group acting as nucleophile to the Lewis acidic carbon of carbonyl group. RMgX adds on RCOR as

$$RMgX + \begin{array}{l} R \\ \quad \diagdown \\ \qquad C{=}\ddot{O}: \\ \quad \diagup \\ R \end{array} \rightarrow R{-}\overset{|}{\underset{|}{C}}{-}O^-\overset{+}{Mg}X$$

Protonation happens next to give alcohol.

$$R{-}\overset{|}{\underset{|}{C}}{-}OH + HOMgX$$

With formaldehyde we get primary alcohols, with other aldehydes we get secondary alcohols and with ketones we get tertiary alcohols.

$$HCHO + C_2H_5MgBr)(H_3O^+) \longrightarrow C_2H_5\,CH_2OH$$

$$CH_2{=}CH{-}CHO + CH_3MgBr \longrightarrow H_3O^+ \rightarrow CH_2{=}CH{-}CH(OH){-}CH_3$$

$$(CH_3)_2CHCOCH(CH_3)_2 + CH_3MgBr\ (ether)\ (H_3\overset{+}{O})$$

$$(CH_3)_2CH-\underset{\displaystyle CH_3}{\overset{\displaystyle OH}{\underset{|}{\overset{|}{C}}}}-CH(CH_3)_2$$

The Grignard reaction between aldehydes and RMgX can be done in many alternative ways. For example let us consider the preparation of 3-phenyl 3-pentanol.

The routes are:

1. 3-pentanone + C_6H_5MgBr
2. $C_6H_5COCH_2CH_3$ + C_2H_5MgBr
3. $2CH_3CH_2MgBr$ + $C_6H_5COOCH_3$

Robinson Annulation

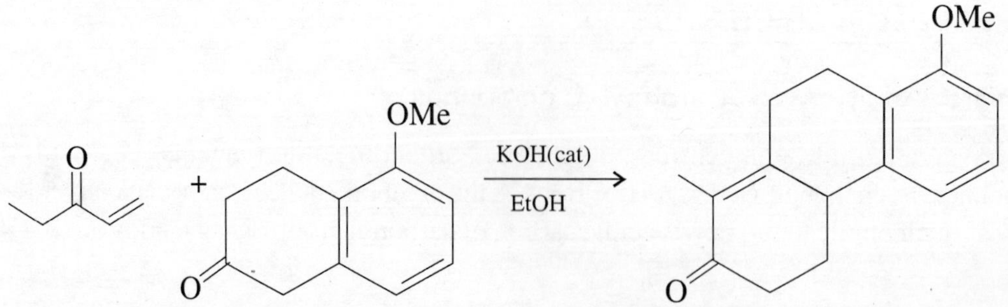

OMe OMe KOH(cat) EtOH

Figure 9.25 Robinson annulation

The Robinson annulation (Figure 9.25) is a useful reaction for the formation of six-membered rings in polycyclic compounds, such as steroids. It combines two reactions: the Michael addition and aldol-condensation. Figure 9.26 shows the mechanism of the Robinson annulation

OMe OMe OMe ^-OH H_2O

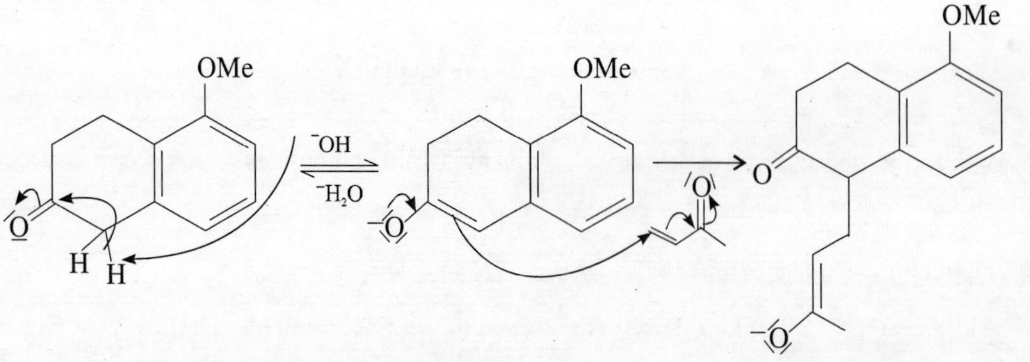

Figure 9.26 Mechanism of the Robinson Annulation

The newly formed enolate intermediate must first tautomerize for the conversion to continue.

The subsequent cyclization via Aldol addition which is followed by a condensation forms a six-membered ring enone.

I. "Explain Why" Questions and Small Structure-Based Problems

1. 2,4-pentenedione does not give intramolecular aldol condensation product. Why?

   Answer

   The 1,3-dione gives 2 enolate anions one of which is very stable and unreactive (A). The second enolate anion (B) gives 4-membered ring which is strained and hence not formed.

2. Cycloheptatrienone is very stable, but cyclopentadienone is reactive and undergoes rapidly Diels–Alder reaction. Why?

   Answer

   In the seven-membered ketone polarization of C=O bond produces positive charge on carbon and according to Hückel rule cycloheptatrienyl cation is very stable. In contrast the cyclopentadienyl cation has 4 pi electrons and does not follow Hückel rule.

3. Hexafluoroacetone has a high equilibrium constant for hydration reaction compared to acetone. Why?

   Answer

   The number of electronegative groups is high due to six fluorines and hence the hydration increases leading to high equilibrium constant.

4. Formaldehyde is much hydrated compared to acetone and has a better equilibrium constant for hydration than acetone. Why?

Answer

Unlike acetone, HCHO has only hydrogens and steric effect is not there and it readily moves from sp^2 to sp^3 state in hydrated form and hence has higher equilibrium constant for hydration.

5. Acetone undergoes very small % of hydration when water is added. Explain how it can be confirmed that hydration equilibrium exists in acetone.

Answer

If water containing ^{18}O is added to ketone, after hydration of the ketone is tested, it will have ^{18}O indicating that hydration reaction occurred.

6. Trichloroacetaldehyde exists as a crystalline substance in the form of chloral hydrate. Why?

Answer

The electronegative chlorines shift the equilibrium

$CCl_3CHO + H_2O \longleftrightarrow CCl_3CH(OH)_2$

completely to the right and hence the hydrate is isolated as a crystalline compound.

7. Phenol exists in 100% *enol* form. Why?

Answer

The aromatic ring is stable in the *enol* form and hence the *keto* form of phenol is not isolated.

8. Acetone has less than 1% *enol* form but $CH_3COCH_2COCH_3$ has 76% enol form. Explain.

Answer

In the *enol* form, the dione has intramolecular hydrogen bond and hence its percentage is high compared to *enol* of acetone.

9. The *enol* form is almost 100% in the case of $PhCH_2COCH_3$. Why?

Answer

The *enol* form will have double bond conjugated with benzene ring and hence will be very stable.

10. Hydrate of cyclopropanone has a higher equilibrium constant for the equilibrium compared to propanone. Why?

Answer

The hydrate formation in the case of cyclopropanone relieves the strain. Hence hydration of cyclopropanone has a higher equilibrium constant.

II. Multiple Choice Questions (One out of four answers is correct)

1. The reagent used to carry out the transformation of 4-methyl cyclohexane carboxylic acid to 4-methyl cyclohexyl methanol is ($CH_3COC_6H_{11}COOH$ to $CH_3CHOH\text{-}C_6H_{11}CH_2OH$)
   a) $LiAlH_4$ b) $NaBH_4$ C) H_2, Pd/C   d) Ni/H_2
2. The reagent which is ideal to carry out the transformation of $CHOC_6H_{11}COOCH_3$ to $CH_2OHC_6H_{11}COOCH_3$ is
   a) $LiAlH_4$ b) $NaBH_4$ c) H_2, Pd/C   d) Li/NH_3
3. One of the following is not an aldol condensation product
   a) 2,2,3-trimethyl-3-hydroxybutanal   b) $CH_3CH{=}CHCHO$
   c) 5-ethyl-4-methyl-4-hepten-3-one   d) $CH_2{=}CHCH_2OH$
4. Identify the ketone which can enolize in only one way to produce aldols
   a) acetone   b) $CH_3CH_2COCH_3$
   c) $(CH_3)_3CCOCH_3$ d) $(CH_3)_3CCOC(CH_3)_3$
5. The Lewis acid catalyst used for reduction of cyclopenten-2-one to cyclepenten-2-ol by $NaBH_4$ is
   a) $AlCl_3$ b) BF_3 c) $CeCl_3$ d) $FeCl_3$
6. $LiAlH_4$ cannot reduce
   a) C=C   b) C=O   c) ester C=O   d) C=O of acid
7. The alkyne not suitable for synthesis of ketones is
   a) 2-butyne   b) 2-pentyne   c) propyne   d) 3-hexyne
8. $LiAlH_4$ or $NaBH_4$ reduction of aldehydes and ketones cannot be used to produce________ alcohol.
   a) primary   b) secondary   c) tertiary   d) isopropyl
9. One of the following does not undergo haloform reaction
   a) acetone   b) ethanol   c) $C_6H_5CHOHCH_3$ d) CH_3COOH
10. One of the following has alpha hydrogen which is most acidic
    a) CH_3CHO b) CH_3COCH_3 c) $CH_3CH_2NO_2$ d) C_6H_5CHO
11. The hydrate which has maximum equilibrium constant for the equilibrium
    Carbonyl compound + water $\longleftrightarrow$ hydrate is
    a) chloral hydrate   b) formaldehyde hydrate
    c) hydrate of acetone   d) hydrate of acetaldehyde
12. The ozonolysis products of this alkene in dimethylsulphide do not answer haloform reaction.
    a) ethylene   b) tetramethyl ethylene   c) 2-butene   d) tetraethyl ethylene

13. Keto–enol tautomerism is not possible in

a) CH_3COCH_3 b) CH_3CHO c) $(CH_3)CCOC(CH_3)_3$ d) $C_6H_5COCH_3$

14. Cannizarro reaction is not possible for.

a) C_6H_5CHO b) $(CH_3)_3CCHO$ c) HCHO d) $C_6H_5CH_2CHO$

15. Benzoic acid is not the product in one of these reactions

a) $C_6H_5COCH_3 + NaOH/I_2$ followed by acidification

b) Bayer–Villiger oxidation of C_6H_5CHO

c) ozonolysis of $C_6H_5CH{=}CH_2$ in oxidizing conditions

d) ozonolysis of $C_6H_5CH_2C{\equiv}CH$

16. One of the following reactions does not produce a ketone.

a) ozonolysis of tetramethylethylene b) hydration of propyne in Hg^{2+}/H_2SO_4

c) RMgBr RCN followed by acidification d) hydroboration oxidation of propyne

17. The ketone obtained from the reaction scheme $CH_3(CH_2)_3COOH + PCl_5$ followed by reaction with $(CH_3)_2$ CuLi and acidification is also one of the products of ozonolysis of

a) $CH_3(CH_2)_3C(CH_3){=}CH_2$ b) $CH_3CH_2CH{=}CH_2$

c) $CH_3CH{=}CHCH_3$ d) cyclohexene

18. $C_6H_5CH_2CHO$ is the product of one of these reactions.

a) hydroboration oxidation of $C_6H_5C{\equiv}CH$

b) hydration in $Hg_2 + /H_2SO_4$

c) ozonolysis of styrene

d) KMnO4 oxidation of styrene

19. The ketone obtained by one of these reactions does not answer haloform reaction.

a) $C_2H_5COCl + (C_2H_5)_2CuLi \rightarrow$ b) $C_6H_5MgBr + CH_3CN +$ acidification

c) $CH_3COOH + 2C_2H_5Li \rightarrow$ d) ozonolysis of tetramethyl ethylene

20. The product of one of the following reactions answers Cannizarro reaction.

a) styrene on ozonolysis in reducing conditions

b) hydroboration oxidation of $C_6H_5CH_2C{\equiv}CH$

c) hydration of acetylene in Hg^{2+}/ H_2SO_4

d) Ozonolysis of $(CH_3)_2C{=}C{=}C(CH_3)_2$

KEY

1. a	2. b	3. a	4. c	5. c	6. a	7. b
8. c	9. d	10. c	11. a	12. d	13. c	14. d
15. d	16. d	17. a	18. a	19. a	20. a	

III. Screening Test Questions of Previous IIT-JEE Question Papers

1. When $C_6H_5C{\equiv}CCH_3$ is treated with $HgSO_4$ and dil. H_2SO_4 more than one product is formed. Predominant product is

   a) $C_6H_5COCH_2CH_3$ b) $CH_3COCH_2C_6H_5$

   c) $C_6H_5{-}C{=}CH(OH){-}CH_3$ d) $C_6H_5CH_2C(OH){=}CH_2$ (2003)

2. Identify the correct order of boiling points of the following compounds

   $CH_3CH_2CH_2CH_2OH$(1) $CH_3CH_2CH_2CHO$(2) $CH_3CH_2CH_2COOH$(3)

   a) 1>2>3 b) 3>1>2 c) 1>3>2 d) 3>2>1 (2002)

3. Compound A with molecular formula C_3H_8O is treated with acidified dichromate to form B (C_3H_6O). B forms a shining silver mirror on warming with ammoniacal silver nitrate. B when treated with aq. solution of $H_2NCONHNH_2$ and sodium acetate gives C. C is

   a) $CH_3CH_2CH{=}NNHCONH_2$ b) $(CH_3)_2C{=}NNHCONH_2$

   c) $(CH_3)_2C{=}NCONHNH_2$ d) $CH_3CH_2CH{=}NCONHNH_2$ (2002)

4. A mixture of benzaldehyde and formaldehyde on heating with aq. NaOH solution gives

   a) benzyl alcohol and sodium formate b) sodium benzoate and methyl alcohol

   c) sodium benzoate and sodium formate d) benzyl alcohol and methyl alcohol (1998)

5. Which of the following has the most acidic hydrogen?

   a) 3-hexanone b) 2,4-hexanedione

   c) 2,5-hexanedione d) 2,3-hexanedione (2000)

6. Which one of the following will be most readily dehydrated in acidic condition

   a) $CH_3COCH_2CHOHCH_3$ b) $CH_3CHOHCH_2CH_3$

   c) $CH_3COCHOHCH_2CH_3$ d) $CH_3COCH_2CH_2CHOHCH_3$

7. The *enol* form of acetone after treatment with D_2O gives

   a) CH_3CODCH_3 b) CD_3COCD_3 c) CH_3COHCH_2D d) $CD_2{=}COD{-}CD_3$

8. The product obtained by oxymercuration ($HgSO_4 + H_2SO_4$) of 1-butyne is
   a) $CH_3CH_2COCH_3$ b) $CH_3CH_2CH_2CHO$
   c) $CH_3CH_2CHO + HCHO$ d) $CH_3CH_2COOH + HCOOH$
9. A new carbon–carbon bond formation is possible in
   a) Cannizarro reaction b) Friedel–Crafts alkylation
   c) Clemmenson reduction d) conversion of ROH to RONa (1998)
10. Tautomerism is not exhibited by
   a) $C_6H_5CH{=}CHOH$ b) $O{=}C_6H_4{=}O$
   c) 2,4-cyclohexenedione d) 1,2-cyclohexanedione
11. Which of the following does not undergo aldol condensation?
   a) acetaldehyde b) propanaldehyde
   c) benzaldehyde d) trideuteroacetaldehyde
12. In Cannizarro reaction the slowest step is
   a) the attack of OH^- at carbonyl group
   b) the transfer of hydride to carbonyl group
   c) the abstraction of proton from carboxyl group
   d) the deprotonation of benzyl alcohol (1996)

KEY

1. a	2. b	3. a	4. c	5. b	6. a	7. a
8. a	9. b	10. b	11. c	12. b		

IV. Multiple Choice Questions (More than one answer is correct)

1. $LiAlH_4$ can be used for reduction of
   a) RCOR b) RCOOH c) RCOOR d) C=C
2. The following compounds answer haloform reaction
   a) CH_3CHO b) $CH_3CHOHCH_3$
   c) $CH_3COC(CH_3)_2COC_2H_5$ d) $CH_3COCH_2COCH_3$
3. The compounds which undergo Cannizarro reaction are
   a) CH_3CHO b) HCHO c) CHOCHO d) C_6H_5CHO
4. The ketones which can enolize in both ways to produce aldols are
   a) $CH_3COC_2H_5$ b) $CH_3CH_2COC_3H_7$
   c) $(CH_3)_3CCOCH_3$ d) $(CH_3)_3CCOC(CH_3)_3$

5. $PhCOCH_3$ is produced from
   a) ozonolysis of $C_6H_5C(CH_3){=}CH_2$
   b) $C_6H_5C{\equiv}CH$ on hydration in Hg^{2+}/H_2SO_4
   c) $C_6H_5CHOHCH_3$ on oxidation
   d) ozonolysis of cyclohexene
6. $LiAlH_4$ cannot reduce
   a) C=C b) C=O c) ester C=O d) C=C
7. The alkynes not suitable for synthesis of ketones are
   a) 2-butyne b) 2-pentyne c) propyne d) 3-heptyne
8. $LiAlH_4$ or $NaBH_4$ reduction of aldehydes and ketones can be used to produce—— alcohol.
   a) primary b) secondary c) tertiary d) *t*-butyl alcohol
9. Some of the following produce CHI_3 with NaOH and I_2
   a) acetone b) ethanol c) $C_6H_5CHOHCH_3$ d) CH_3COOH
10. Some of the following have more *enol* form relative to *keto* form,
    a) CH_3CHO
    b) CH_3COCH_3
    c) $CH_3COCH_2COCH_3$
    d) phenol
11. In the equilibrium Carbonyl compound + water $\longleftrightarrow$ hydrate, the reaction shifts more to right in case of
    a) chloral b) hexafluoroacetone c) acetone d) acetaldehyde
12. The ozonolysis products of these alkenes in dimethylsulphide answer haloform reaction
    a) ethylene
    b) tetramethyl ethylene
    c) 2-butene
    d) tetraethyl ethylene
13. Keto–enol tautomerism is possible in
    a) CH_3COCH_3
    b) CH_3CHO
    c) $(CH_3)_3CCOC(CH_3)_3$
    d) $C_6H_5COCH_3$
14. Cannizarro reaction produces only one product for
    a) C_6H_5CHO b) $(CH_3)_3CCHO$ c) CHOCHO d) phthalaldehyde
15. Benzoic acid is the product in some of these reactions
    a) $C_6H_5COCH_3 + NaOH/I_2$ followed by acidification
    b) Bayer–Villiger oxidation of C_6H_5CHO
    c) Ozonolysis of $C_6H_5CH{=}CH_2$ in oxidizing conditions
    d) Ozonolysis of $C_6H_5CH_2C{\equiv}CH$
16. Some of these reactions produce a ketone
    a) Ozonolysis of tetramethylethylene
    b) Hydration of 1-butyne in Hg^{2+}/H_2SO_4
    c) RMgBr RCN followed by acidification
    d) Hydroboration oxidation of 1-butyne

KEY

1. a,b,c	2. a,b,c	3. b,c,d	4. a,b	5. a,b,c	6. a,d	7. b,d
8. a,b	9. a,b,c	10. c,d	11. a,b	12. b,c	13. a,b,d	14. c,d
15. a,b,c	16. a,b,c					

V. Assertion–Reason Questions

The questions 1 to 10 consist of assertion in column 1 and reason in column 2.

Use the following key to choose appropriate answer

a) If both assertion and reason are correct and reason is the correct explanation of assertion
b) If both assertion and reason are correct but reason is not the correct explanation of assertion
c) If assertion is correct but reason is incorrect
d) If assertion is incorrect but reason is correct

No.	Assertion	Reason
1.	pKa of the hydrogens in 2,4- cyclopentadien one is higher than that of acetone	Intramolecular hydrogen bonding is present in the dienone.
2.	The crossed aldol reaction of benzophenone with acetaldehyde gives the same product as that of self condensation of acetaldehyde.	Benzophenone has no alpha hydrogens and cannot undergo aldol reaction.The steric effect of phenyl rings prevents crossed aldol reaction.
3.	Bicyclic ketones cannot undergo aldol condensation.	The ketone cannot enolize since double bonds at bridge head will be strained.
4.	Alkylation of enolate anions are carried out using methyl, primary, benzylic and allyl halides.	The reaction follows SN_2 pathway.
5.	Hydration of unsymmetrical alkynes is a good method of synthesis of ketones.	Hydration leads to mixture of ketones.
6.	NaOH is used with $CH_3COCH_2COCH_3$ to produce enolate anion.	The presence of active methylene group leads to high acidity and weak base is enough.
7.	Aldol condensation is generally followed by dehydration.	The enone formed is stable due to hyperconjugation.
8.	Even though two enolate anions are produced in intramolecular aldol condensation of $CH_3COCH_2CH_2COCH_3$ only the five- membered ring hydroxyketone is produced.	The other enolate ion produces strained three-membered ring.

(*Contd.*)

No.	Assertion	Reason
9.	The hydrate formed from propanone has low equilibrium constant but that from cyclopropanone has high equilibrium constant.	The cycloketone has to change from strained ring to tetrahedral carbon (60=109) which is more favourable.
10.	Alpha halogenation for ketones is easy but it is not so for carboxylic acids.	Carboxylic acids do not enolize easily.

KEY

1. a	2. a	3. a	4. a	5. d	6. a	7. a
8. a	9. a	10. a				

VI. Comprehension Passages

Passage I

There are several methods of preparing aldehydes and ketones. Ozonolysis of alkenes, hydration of alkynes, hydroboration oxidation of alkynes and RMgBr + RCN followed by hydrolysis.

Answer the following questions

1. The ketone obtained by ozonolysis of $C_6H_5C(CH_3){=}CH_2$ can be obtained also from the following reaction.
   a) oxidation of $C_6H_5CHOHCH_3$ b) hydration of $C_6H_5CH_2C{\equiv}CH$
   c) $CH_3MgBr + CH_3CN$ d) RMgBr + ethylene oxide
2. CH_3COCl on reduction with H_2 in boiling xylene with Pd catalyst in $BaSO_4$ gives the compound which is obtained also from
   a) hydration of acetylene with Hg^{2+} and H_2SO_4
   b) hydroboration oxidation of propyne
   c) hydration of propyne with Hg^{2+} and H_2SO_4
   d) oxidation of isopropyl alcohol
3. When 2 moles of acetic acid vapour is passed over manganous oxide at 573 K, the product formed
   a) answers haloform reaction
   b) answers Tollen's test
   c) is same as that obtained by ozonolysis of 2-butene in dimethyl sulphide
   d) is same as that obtained by oxidation of benzyl alcohol

KEY

1. a 2. a 3. a

Passage 2

Carbonyl compounds exhibit keto-enol tautomerism. Tautomerism is possible in carbonyl compounds containing enolizable hydrogens. Methyl ketones undergo haloform reaction. Ethanol, acetaldehyde and CH_3CHOHR also undergo this reaction.

Answer the following questions

1. The compound which cannot exhibit keto-enol tautomerism is
   a) acetone b) CH_3CHO
   c) CH_3CH_2OH d) di-*tert*. butylketone
2. Haloform reaction is possible for
   a) CH_3COCH_2COOEt b) CH_3COOH
   c) C_6H_5CHO d) $CH_3CO(CH_2)_3COC_2H_5$
3. The acid obtained by acidification after treatment with NaOH and I_2 of $C_6H_5COCH_3$ is same as that obtained by
   a) ozonolysis of $C_6H_5CH{=}CH_2$ in reducing conditions
   b) Ozonolysis of $C_6H_5CH{=}CH_2$ in oxidizing conditions
   c) oxidation of *t*-butyl benzene with $KMnO_4$
   d) ozonolysis of $C_6H_5CH_2C{\equiv}CH$

KEY

1. d 2. d 3. b

Passage 3

Aldehydes and ketones which have alpha hydrogen undergo aldol condensation. Those aldehydes not having alpha hydrogen undergo Cannizarro reaction. Diketones or dicarbonyl compounds undergo intramolecular aldol condensation

Answer the following questions

1. Cannizarro reaction is an example of
   a) double displacement reaction b) disproportionation reaction
   c) oxidation reaction d) reduction reaction
2. Aldol condensation is not possible for
   a) CH_3CHO b) C_6H_5CHO c) CH_3COCH_3 d) CH_3CH_2CHO
3. CHOCHO with NaOH gives
   a) CHOCOOH b) CH_2OHCHO c) $CH_2OHCOONa$ d) no product

KEY

1. b 2. b 3. c

VII. Matrix Matching

Question 1

Column 1	Column 2
a) Benzene by Friedel–Crafts acylation with CH_3COCl	p) gives $C_6H_5COCH_3$
b) Sodium formate is one of the products in	q) one of which answers haloform reaction and the other answers Cannizarro reaction
c) 2-methylpropene has ozonolysis products under reducing conditions	r) Cannizarro reaction of formaldehyde with any other aldehyde
d) Hydration of acetylene in Hg^{2+}, H_2SO_4 gives	s) CH_3CHO

Answer

$[a \rightarrow p], [b \rightarrow r], [c \rightarrow q], [d \rightarrow s]$

Question 2

Column 1	Column 2
a) $C_2H_5COCH_3$	p) $C_2H_5MgBr + CH_3CN$
b) $CHOCHO + NaOH$	q) ozonolysis of $C_2H_5C(CH_3){=}CH_2$
c) $C_6H_5COCH_3 + NaOH + I_2$	r) 1-butyne on hydration n Hg^{2+} /H_2SO_4
d) hydroboration oxidation of propyne	s) $CH_2OHCOONa$
	t) $C_6H_5COONa + CHI_3$
	u) CH_3CH_2CHO

Answer

$[a \rightarrow p,q,r], [b \rightarrow s], [c \rightarrow t], [d \rightarrow u]$

Question 3

Column 1	Column 2
a) Ethanal	p) oxidation product of ozonolysis of $C_2H_5CH = CHC_2H_5$ in reducing conditions
b) t-butyl alcohol	q) no enolization is possible
c) benzyl alcohol	r) CH_3MgBr with acetone and acidification
d) ditertiary butyl ketone	s) Cannizarro reaction of benzaldehyde and HCHO

Answer

$[a \rightarrow p], [b \rightarrow r], [c \rightarrow s], [d \rightarrow q]$

VIII. Structure-Based and Numerical Problems

1. A exhibits tautomerism and the enol tautomer is 100%. A shows violet colouration with neutral $FeCl_3$. A on catalytic reduction gives B. A with diazomethane gives C. C on cleavage with HI gives A and D which cannot undergo dehydrohalogenation. When E reacts with $LiAlH_4$ or $NaBH_4$ in acid, the compound formed is B, but when $NaBH_4$ reacts with E in $CeCl_3$ the product is F. Find A to F.

Answer

A is phenol;
B is cyclohexanol;
C is $C_6H_5OCH_3$;
D is CH_3I;
E is cyclohexen-2-one;
F is cyclohexene-2-ol.

2. CaC_2 is hydrolysed to compound A. 3 moles of A heated in a red-hot tube gives B. B with C in $AlCl_3$ gives D. C is the only monochlorinated product of E which is obtained by Wurtz reaction of CH_3I. B with CH_3COCl in $AlCl_3$ gives F. F by Wolf Kishner reduction gives D. F reacts with HCN to give G. G on hydrolysis in HCl gives H. There is no ($\alpha\beta$) unsaturated acid formed from reaction of G with HCN in 95% sulphuric acid. Find A to H. Why is the unsaturated acid not formed?

Answer

A is acetylene;
B is benzene;
C is C_2H_5Cl;
D is $C_6H_5CH_2CH_3$;

E is C_2H_6;

F is $C_6H_5COCH_3$;

G is $C_6H_5\overset{\overset{OH}{|}}{\underset{\underset{CH_3}{|}}{C}}CN$;

H is $C_6H_5\overset{\overset{OH}{|}}{\underset{\underset{CH_3}{|}}{C}}COOH$;

Since the group adjacent to $C(CH_3)OHCOOH$ is phenyl, no α,β-unsaturated acid is formed.

3. Aniline is treated with $NaNO_2$ and HCl at –5°C and then H_3PO_2 is added and heating is done to get A. A with CH_3COCl in $AlCl_3$ gives B. B with cold $KMnO_4$ gives C which on further oxidation gives D. D is obtained as oxidation product of $C_6H_5CH_2OH$, $C_6H_5CH_2Br$, C_6H_5CHO, $C_6H_5CHOHCH_3$ and so on. If tertiary butylbenzene is treated with $KMnO_4$, this acid is not produced. C can also be obtained from PhCOCl with CuCN followed by acidification. SeO_2 oxidizes B to a ketoaldehyde E. B reacts with NaOH and I_2 to give F and G. G is the sodium salt of D. Find A to G.

Answer

A is benzene;

B is $C_6H_5COCH_3$;

C is $C_6H_5COCOOH$;

D is C_6H_5COOH;

E is C_6H_5COCHO;

F is CHI_3;

G is C_6H_5COONa.

4. 1-methylcyclohexene (A) with ozone in dimethylsulphide gives B. B with KOH gives C. B is one of the products of ozonolysis of diene D along with 2 moles of HCHO. Under reducing conditions, A on reduction gives E. E can give 5 monochloro derivatives. Find A to E.

Answer

A is (1-methylcyclohexene ring with CH_3)

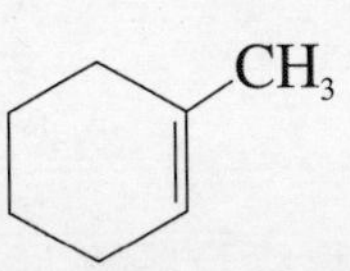

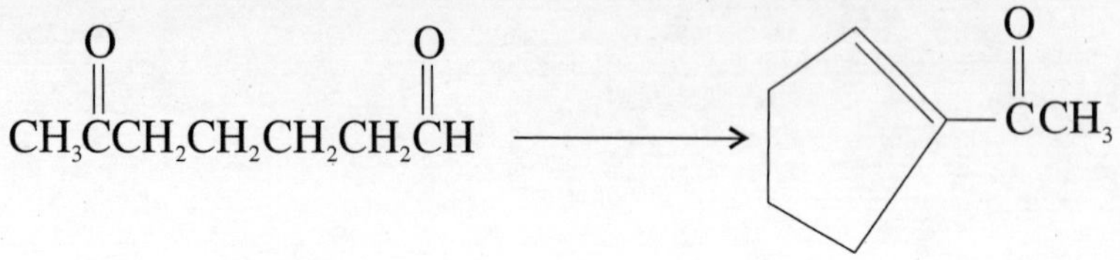

$$CH_3\overset{O}{\overset{\|}{C}}CH_2CH_2CH_2CH_2\overset{O}{\overset{\|}{C}}H \longrightarrow$$ (C): cyclopentene ring bearing $-\overset{O}{\overset{\|}{C}}CH_3$

(B) (C)

D is $CH_2{=}\underset{\displaystyle CH_3}{\underset{|}{C}}{-}(CH_2)_4{-}CH{=}CH_2$

E is (cyclohexane ring bearing CH_3)

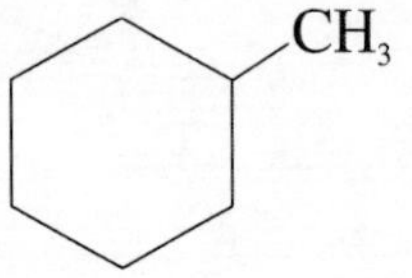

Figure Ex.9.1 Structures of compounds in intramolecular aldol condensation

5. Compound A decolorizes $KMnO_4$ and has the formula C_5H_{10}. On reduction, it gives B which can give 4 monochloroderivatives on chlorination. On hydration in the presence of BH_3, H_2O_2, OH^- it gives C. C on oxidation gives D. C and D both answer haloform reaction. The iodoform reaction yields sodium salt of isobutyric acid. Find A to D. If A reacts with H_2SO_4, find the product. Will it answer haloform reaction? Can it undergo oxidation?

Answer

A is $(CH_3)_2C{=}CHCH_3$;

B is $(CH_3)_2CHCH_2CH_3$;

C is $(CH_3)_2CHCHOHCH_3$;

D is $(CH_3)_2CHCOCH_3$;

The product with H_2SO_4 will be $(CH_3)_2COHCH_2CH_3$. This will not undergo haloform reaction since it does not have R—CHOH group. Since it is a tertiary alcohol it will not undergo oxidation.

6. Compound A is obtained by reaction of the product of isopropyl chloride and benzene in $AlCl_3$ in air along with acetone. A on complete hydrogenation gives B. B on oxidation gives C. C on treatment with CH_3MgBr followed by $POCl_3$ gives D and E, in the ratio 9:1. When C reacts with $(C_6H_5)_3P^+CH_2^-$ it gives 85% of E and triphenylphosphene oxide. E on hydrogenation gives F. F gives 5 monochloroderivatives on halogenation. Find A to F.

Answer

A is phenol;

B is cyclohexanol;

C is cyclohexanone;

D is 1-methylcyclohexene;

E is methylene cyclohexane;

F is 1-methylcyclohexane.

7. Compound A has formula C_6H_{12}. When it reacts with conc. H_2SO_4 it gives B. By hydroboration oxidation it gives C, an isomer of B. A on ozonolysis in reducing conditions gives D and E. D and E react in NaOH to give F and G. G on acidification gives H. H and E both answer Tollens Test. D on oxidation gives J. J does not undergo HVZ reaction. Find A to J

Answer

A is $(CH_3)_3CCH{=}CH_2$

B is $(CH_3)_3CCHOHCH_3$

C is $(CH_3)_3CCH_2CH_2OH$

D is $(CH_3)_3CCHO$

E is HCHO

F is $(CH_3)_3CCH_2OH$

G is HCOONa

H is HCOOH

J is $(CH_3)_3CCOOH$. J has no alpha hydrogen and hence does not answer HVZ reaction.

8. Compound X is inert to Br_2 in CCl_4. Vigorous oxidation of X with hot alkaline $KMnO_4$ gives benzoic acid. X gives precipitate with semicarbazide hydrochloride and with 2,4-dinitrophenylhydrazine. Find all possible isomers which are carbonyl compounds of X. Can all these isomers be distinguished by chemical tests? Find the alkenes which give on ozonolysis the different isomers of X given above under reducing conditions. Find the alkynes which by hydration or hydroboration oxidation give these isomers.

Answer

Since all give benzoic acid they contain the benzylic hydrogen and the possible structures of X are

1. $PhCOCH_2CH_3$ 2. $PhCH_2COCH_3$
3. $PhCH_2CH_2CHO$ 4. $PhCH(CH_3)CHO$

1 and 2 can be distinguished by haloform reaction since 2 contains $COCH_3$ group. 3 and 4 on mild oxidation give acids whereas 1 and 2 will not undergo oxidation under such conditions. 3 and 4 cannot be distinguished by simple methods.

$PhCOCH_2CH_3$ is obtained by ozonolysis in reducing conditions of $PhC(C_2H_5){=}CH_2$ along with HCHO. $PhCH_2COCH_3$ is obtained by ozonolysis under reducing conditions of $PhCH_2C(CH_3){=}CH_2$ along with HCHO. $PhCH_2CH_2CHO$ is obtained by ozonolysis in reducing conditions of $PhCH_2CH_2CH{=}CH_2$ along with HCHO. $PhCH(CH_3)CHO$ is obtained by ozonolysis under reducing conditions of $PhCH(CH_3)CH{=}CH_2$ along with HCHO. $PhC{\equiv}CCH_3$ gives by hydration ketone 1. $PhCH_2C{\equiv}CH$ gives on hydration ketone 2. $PhCH_2C{\equiv}CH$ by hydroboration oxidation gives 3. 4 cannot be obtained by hydroboration oxidation of any suitable alkyne.

9. Compound A with formula $C_5H_{10}O$ is used in Wittig reaction with the phosphonium ylide obtained from $(C_6H_5)_3P$ and C_2H_5Br to obtain B. B on ozonolysis in reducing conditions gives the ketone (A) and C. C is the hydration product of acetylene. A by Clemmenson reduction gives D. A gets reduced to $(C_2H_5)_2CHOH$. D gives 3 monochloroproducts on halogenation. Find A to D.

Answer

A is diethylketone or 3-pentanone;

B is 3-ethyl 2-pentene;

C is CH_3CHO;

D is n-pentane;

10. A is obtained by Wurtz reaction of CH_3I with Na. A with NBS gives B. B with Mg in ether gives C. C with CH_3CN and acid gives D. D with CH_3MgBr in water gives E. F is obtained by reaction of CHI_3 with Ag. G is a partial reduction product of F with Li/NH_3. G on reduction gives A. F on hydration in Hg^{2+}/H_2SO_4 gives H. H reacts with CH_3MgBr to give J. Find A to J.

Answer

A is C_2H_6;

B is C_2H_5Br;

C is C_2H_5MgBr;

D is $C_2H_5COCH_3$;

E is $C_2H_5{-}C(CH_3)_2{-}OH$;

F is acetylene;

G is ethylene;

H is CH_3CHO;

J is $CH_3CH(OH)CH_3$.

10

CARBOXYLIC ACIDS AND DERIVATIVES OF ACIDS

CARBOXYLIC ACIDS

INTRODUCTION

Carboxylic acids are the compounds which contain a —COOH group. Some examples of acids are given in Figure 10.1.

$H-C(=O)-OH$ Methanoic acid

$CH_3-C(=O)-OH$ Ethanoic acid

$CH_3-CH_2-C(=O)-OH$ Propanoic acid

$CH_3-CH_2-CH(CH_3)-C(=O)-OH$ 2-methylbutanoic acid

Figure 10.1 Examples of carboxylic acids

The name counts the total number of carbon atoms in the longest chain, including the one in the —COOH group. If side groups are attached to the chain, the numbering is done from the carbon atom in the —COOH group.

SALTS OF CARBOXYLIC ACIDS

Carboxylic acids are acidic because of the hydrogen in the –COOH group. When the acids form salts, this is lost and replaced by a metal. Sodium ethanoate, for example, has the structure as given in Figure 10.2.

$$CH_3-C(=O)-O^- Na^+$$

Sodium ethanoate

Figure 10.2 Salt of a carboxylic acid

Depending on the ionic nature of the compound, this would be simplified to $CH_3COO^- Na^+$ or just CH_3COONa.

NOMENCLATURE OF CARBOXYLIC ACIDS

As with aldehydes, the carboxyl group must be located at the end of a carbon chain. In the IUPAC system of nomenclature the carboxyl carbon is designated #1, and other substituents are located and named accordingly. The characteristic IUPAC suffix for a carboxyl group is "*oic acid*" and care must be taken not to confuse this systematic nomenclature with the similar common system. These two nomenclatures are illustrated in the Table 10.1, along with their melting and boiling points.

Table 10.1 Nomenclature of Carboxylic acids

Formula	Common Name	Source	IUPAC Name	Melting Point	Boiling Point
HCO_2H	Formic acid	Ants	Methanoic acid	8.4°C	101°C
CH_3CO_2H	Acetic acid	Vinegar	Ethanoic acid	16.6°C	118°C
$CH_3CH_2CO_2H$	Propanoic acid	Milk	Propanoic acid	–20.8°C	141°C
$CH_3(CH_2)_2CO_2H$	Butyric acid	Butter	Butanoic acid	–5.5°C	164°C
$CH_3(CH_2)_3CO_2H$	Valeric acid	Valerian root	Pentanoic acid	–34.5°C	186°C
$CH_3(CH_2)_4CO_2H$	Caproic acid	Goats	Hexanoic acid	–4.0°C	205°C
$CH_3(CH_2)_5CO_2H$	Enanthic acid	Vines	Heptanoic acid	–7.5°C	223°C
$CH_3(CH_2)_6CO_2H$	Caprylic acid	Goats	Octanoic acid	16.3°C	239°C
$CH_3(CH_2)_7CO_2H$	Pelargonic acid	Pelargonium	Nonanoic acid	12.0°C	253°C
$CH_3(CH_2)_8CO_2H$	Capric acid	Goats	Decanoic acid	31.0°C	219°C

Substituted carboxylic acids are named either by the IUPAC system or by common names. Some common names, the amino acid threonine for example, do not have any systematic origin and must simply be memorized. In other cases, common names make use of the Greek letter notation for carbon atoms near the carboxyl group. Some examples of both nomenclatures are provided in Figure 10.3.

CH_3 H O

$H_3^5C \overset{4}{—} C \overset{3}{—} CH_2 \overset{2}{—} C \overset{1}{—} C$

CH_3 C_2H_5 O—H

2-ethyl 4,4-dimethylpentanoic acid

CO_2H

$CH_3—CH(OH)—CH(NH_2)—CO_2H \equiv H_2N—|—H$

H—|—OH

CH_3

2-amino 3-hydroxybutanoic acid
The 2S, 3R isomer is threonine.

Br

H_3C β α CO_2H
γ

CH_3

2-bromo 3-methylbutanoic acid
(or) a-bromoisovaleric acid

CO_2H

CO_2H

cis-1,3-cyclohexanedicarboxylic acid

Figure 10.3 Nomenclature of carboxylic acids with examples

Simple dicarboxylic acids having the general formula $HO_2C–(CH_2)_n–CO_2H$ (where n = 0 to 5) are known by the common names: Oxalic (n = 0), Malonic (n = 1), Succinic (n = 2), Glutaric (n = 3), Adipic (n = 4) and Pimelic (n = 5) Acids.

Related Carbonyl Derivatives Other functional group combinations with the carbonyl group can be prepared from carboxylic acids, and are usually treated as related derivatives. Five common classes of these carboxylic acid derivatives are shown in Figure 10.4. Although nitriles do not have a carbonyl group, they are included here because the functional carbon atoms all have the same oxidation state. The top row (*not shaded*) shows the general formula for each class, and the bottom row (*shaded*) gives a specific example of each. As in the case of amines, amides are classified as 1°, 2° or 3°, depending on the number of alkyl groups bonded to the nitrogen.

$R-C(=O)-X$ X= F,Cl,Br or I	$R-C(=O)-O-C(=O)-R$	$R-C(=O)-O-R'$	$R-C(=O)-NR'_2$ R'=H or alkyl	$R-C\equiv N$
$C_2H_5-C(=O)-Cl$ Propanoyl chloride	$R-C(=O)-O-C(=O)-R$ Acetic anhydride	$H_3C-C(=O)-O-C_2H_5$ Ethyl acetate	$H-C(=O)-NH_2$ Formamide	$H_3C-C\equiv N$ Acetonitrile

Figure 10.4 Carboxylic acid derivatives

Functional groups of this kind are found in many kinds of natural products.

PROPERTIES OF CARBOXYLIC ACIDS

Physical Properties of Carboxylic Acids

Table 10.1 gave the melting and boiling points for a homologous group of carboxylic acids having from one to ten carbon atoms. The boiling points increased with size in a regular manner, but the melting points did not. Unbranched acids made up of an even number of carbon atoms have melting points higher than the odd-numbered homologues having one more or one less carbon. This reflects differences in intermolecular attractive forces in the crystalline state. In the case of fatty acids, the presence of a *cis*-double bond significantly lowers the melting point of a compound. Thus, palmitoleic acid melts over 60º lower than palmitic acid, and similar decreases occur for the C_{18} and C_{20} compounds. Again, changes in crystal packing and intermolecular forces are responsible.

The factors that influence the relative boiling points and water solubilities of various types of compounds were discussed earlier. In general, dipolar attractive forces between molecules act to increase the boiling point of a given compound, with hydrogen bonds being an extreme example. Hydrogen bonding is also a major factor in the water solubility of covalent compounds. The following table lists a few examples of these properties for some similar sized polar compounds (the non-polar hydrocarbon hexane is provided for comparison).

Table 10.2 Physical properties of some organic compounds

Formula	IUPAC Name	Molecular Weight	Boiling Point	Water Solubility
$CH_3(CH_2)_2CO_2H$	Butanoic acid	88	164 °C	very soluble
$CH_3(CH_2)_4OH$	1-pentanol	88	138 °C	slightly soluble
$CH_3(CH_2)_3CHO$	Pentanal	86	103 °C	slightly soluble
$CH_3CO_2C_2H_5$	Ethyl ethanoate	88	77 °C	moderately soluble
$CH_3CH_2CO_2CH_3$	Methyl propanoate	88	80 °C	slightly soluble
$CH_3(CH_2)_2CONH_2$	Butanamide	87	216 °C	soluble
$CH_3CON(CH_3)_2$	N,N-dimethylethanamide	87	165 °C	very soluble
$CH_3(CH_2)_4NH_2$	1-aminobutane	87	103 °C	very soluble
$CH_3(CH_2)_3CN$	Pentanenitrile	83	140 °C	slightly soluble
$CH_3(CH_2)_4CH_3$	Hexane	86	69 °C	insoluble

The first five entries in Table 10.2 all have functional groups with C═O or O—H and the relatively high boiling points of the first two is clearly due to hydrogen bonding. Carboxylic acids have exceptionally high boiling points, due in large part to dimeric associations involving two hydrogen bonds (Figure 10.5). Carboxylic acids have high boiling points due to intermolecular hydrogen bonding. The C═O bond distance is short (1.23 Å). The carboxylate anion is resonance-stabilized with equal C—O distances. The two carboxylate oxygens are equivalent.

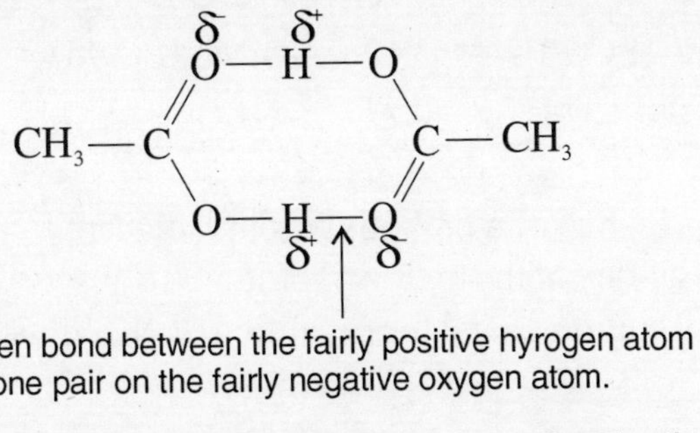

Figure 10.5 Intermolecular hydrogen bonding in the acid

This immediately doubles the size of the molecule and so increases the van der Waals dispersion forces between one of these dimers and its neighbours, resulting in a high boiling point.

Solubility in water In the presence of water, the carboxylic acids do not dimerize. Instead, hydrogen bonds are formed between water molecules and individual molecules of acid.

The carboxylic acids with up to four carbon atoms will mix with water in any proportion. When the two mix together, the energy released when the new hydrogen bonds form is much the same as is needed to break the hydrogen bonds in the pure liquids.

The solubility of the bigger acids decreases very rapidly with size. This is because the longer hydrocarbon "tails" of the molecules get between the water molecules and break the hydrogen bonds. In this case, these broken hydrogen bonds are only replaced by much weaker van der Waals dispersion forces.

The energetics of dissolving carboxylic acids in water is made more complicated because some of the acid molecules actually react with the water rather than just dissolving in it. This is the basis for the acidity of these compounds and is discussed on another page.

ACIDITY OF CARBOXYLIC ACIDS

The pK_a's of some typical carboxylic acids are listed in the Table 10.3. When we compare these values with those of comparable alcohols, such as ethanol ($pK_a = 16$) and 2-methyl 2-propanol ($pK_a = 19$), it is clear that carboxylic acids are stronger acids by over ten powers of ten! Furthermore, electronegative substituents near the carboxyl group act to increase the acidity.

Table 10.3 Acidity of carboxylic acids

Compound	pK_a	Compound	pK_a
HCO_2H	3.75	$CH_3CH_2CH_2CO_2H$	4.82
CH_3CO_2H	4.74	$ClCH_2CH_2CH_2CO_2H$	4.53
FCH_2CO_2H	2.65	$CH_3CHClCH_2CO_2H$	4.05
$ClCH_2CO_2H$	2.85	$CH_3CH_2CHClCO_2H$	2.89
$BrCH_2CO_2H$	2.90	$C_6H_5CO_2H$	4.20
ICH_2CO_2H	3.10	*p*-$O_2NC_6H_4CO_2H$	3.45
Cl_3CCO_2H	0.77	*p*-$CH_3OC_6H_4CO_2H$	4.45

Why should the presence of a carbonyl group adjacent to a hydroxyl group have such a profound effect on the acidity of the hydroxyl proton? To answer this question we must return to the nature of acid–base equilibria and the definition of pK_a, illustrated by the general equations given below.

Acid–base equilibria in acids reacting with water in an equilibrium

$$H{-}A + H_2O \rightleftharpoons H_3\overset{\oplus}{O} + \overset{\ominus}{A:} \qquad K_{eq} = \frac{[H_3\overset{\oplus}{O}][\overset{\ominus}{A:}]}{[HA][H_2O]}$$

$$K_a = \frac{[H_3\overset{\oplus}{O}][\overset{\ominus}{A:}]}{[HA]} \qquad pK_a = -\text{Log } K_a = \log\left(\frac{1}{K_a}\right)$$

We know that an equilibrium favours the thermodynamically more stable side, and that the magnitude of the equilibrium constant reflects the energy difference between the components of each side. In an acid–base equilibrium the equilibrium always favours the weaker acid and base (these are the more stable components). Water is the standard base used for pK_a measurements; consequently, anything that stabilizes the conjugate base ($A:^{(-)}$) of an acid will necessarily make that acid (H—A) stronger and shift the equilibrium to the right. Both the carboxyl group and the carboxylate anion are stabilized by resonance (Figure 10.6), but the stabilization of the anion is much greater than that of the neutral function, as shown in the following diagram. In the carboxylate anion the two contributing structures have equal weight in the hybrid, and the C—O bonds are of equal length (between a double and a single bond). This stabilization leads to a markedly increased acidity.

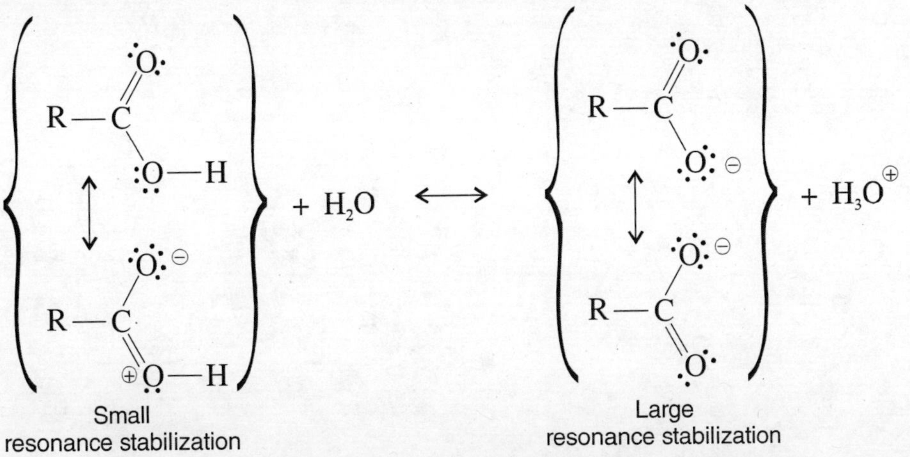

Figure 10.6 Resonance stabilization of carboxylate ion

The resonance effect described here is undoubtedly the major contributor to the exceptional acidity of carboxylic acids. However, inductive effects also play a role. For example, alcohols have pK_a's of 16 or greater but their acidity is increased by electron-withdrawing substituents on the alkyl group. Figure 10.7 illustrates this factor for several simple inorganic and organic compounds (row #1), and shows how inductive electron withdrawal may also increase the acidity of carboxylic acids (rows #2 & 3). The acidic hydrogen is shown in bold in all examples.

Water is less acidic than hydrogen peroxide because hydrogen is less electronegative than oxygen, and the covalent bond joining these atoms is polarized in the manner shown. Alcohols are slightly less acidic than water, due to the poor electronegativity of carbon, but chloral hydrate, $Cl_3CCH(OH)_2$, and 2,2,2,-trifluoroethanol are significantly more acidic than water, due to inductive electron withdrawal by the electronegative halogens (and the second oxygen in chloral hydrate). In the case of carboxylic acids, if the electrophilic character of the carbonyl carbon is decreased, the acidity of the carboxylic acid will also decrease. Similarly, an increase in its electrophilicity will increase the acidity of the acid. Acetic acid is ten times weaker than formic acid (first two entries in the second row), confirming the electron-donating character of an alkyl group relative to hydrogen. Electronegative substituents increase acidity by inductive electron withdrawal. As expected, the higher the electronegativity of the substituent the greater the increase in acidity (F > Cl > Br > I), and the closer the substituent is to the carboxyl group, the greater is its effect (isomers in the 3rd row). Substituents also influence the acidity of benzoic acid derivatives, but resonance effects compete with inductive effect. The methoxy group is electron-donating and the nitro group is electron-withdrawing (last three entries in the Table 10.3).

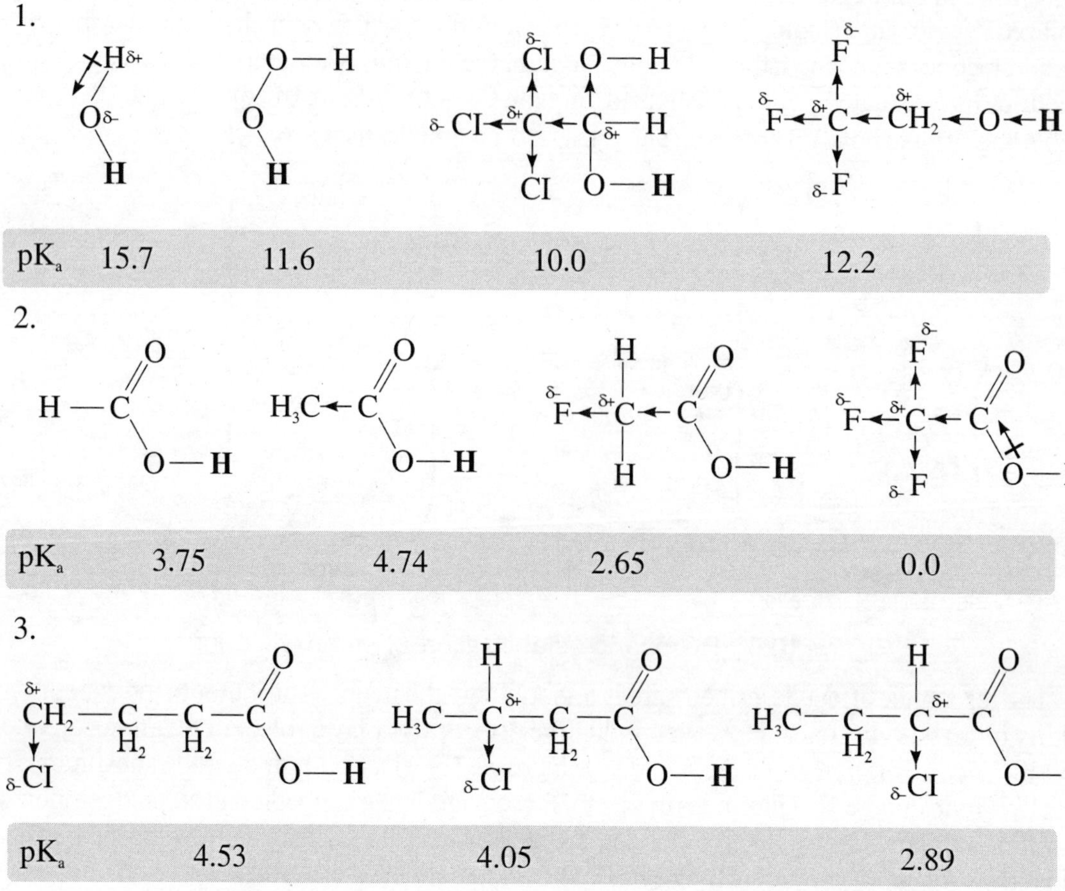

1.

pK_a	15.7	11.6	10.0	12.2

2.

pK_a	3.75	4.74	2.65	0.0

3.

pK_a	4.53	4.05	2.89

Figure 10.7 Inductive effect and acidity of carboxylic acids

If we compare $ClCH_2COOH$, $Cl_2CHCOOH$ and Cl_3CCOOH, the trichloro has maximum acidity. The electronegative chlorine withdraws electrons and greater the number of chlorines the more easily is the anion form atom. Thus inductive effect explains acidity of these substituted acids. The inductive effect depends on distance. 2-chlorobutanoic acid is stronger than 4 chlorobutanoic acid (chlorine is away from COOH group). If we compare acetic acid and peroxy acetic acid the peracid is less acidic. This is because resonance is not possible in the peracid anion. If we consider aromatic acids, benzoic acid is stronger than acetic acid due to resonance stabilization of $C_6H_5COO^-$. If we compare substituted benzoic acids, *p*-nitrobenzoic acid is stronger than benzoic acid but *p*-methoxy benzoic acid is weaker than benzoic acid. Nitro group is electron-withdrawing but methoxy is electron-releasing. The measure of acidity of substituted benzoic acids relative to benzoic acid indicates whether the groups are activating or deactivating. *p*-trifluoromethyl benzoic acid has $pK_a = 3.6$ relative to benzoic acid whose pKa is 4.19. Since the trifluoromethyl group increases acidity, it is a deactivating group.

PREPARATION OF CARBOXYLIC ACIDS

The carbon atom of a carboxyl group has a high oxidation state. It is not surprising, therefore, that many of the chemical reactions used for their preparation are oxidations.

1. Oxidative Methods in Preparing Carboxylic Acids

a. ***Oxidation of arene side-chains*** Oxidation of alkyl-substituted benzenes or these having benzylic hydrogen ($C_6H_5CH_2R$, $C_6H_5CH(CH_3)_2$, $C_6H_5CH_2OH$, $C_6H_5CH_2Cl$, etc.) oxidized by $KMnO_4$ to benzoic acid. *t*-butyl benzene is not oxidized to benzoic acid. Since it has no benzylic hydrogen.

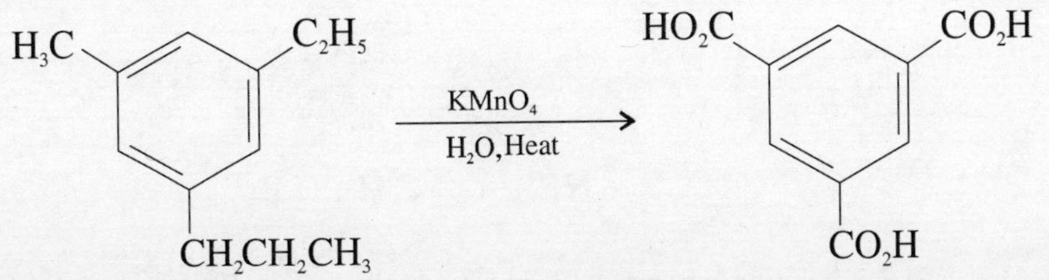

H_3C C_2H_5 $CH_2CH_2CH_3$ $\xrightarrow[H_2O,\text{Heat}]{KMnO_4}$ HO_2C CO_2H CO_2H

b. ***Oxidation of 1°-alcohols*** Oxidation of primary alcohols by CrO_3, H_2O and H_2SO_4 give acids.

$$H-\overset{CH_3}{\underset{CH_3}{C}}-CH_2-CH_2OH \xrightarrow[(H_2CrO_4)]{\text{Jones' Reagent}} H-\overset{CH_3}{\underset{CH_3}{C}}-CH_2-COOH$$

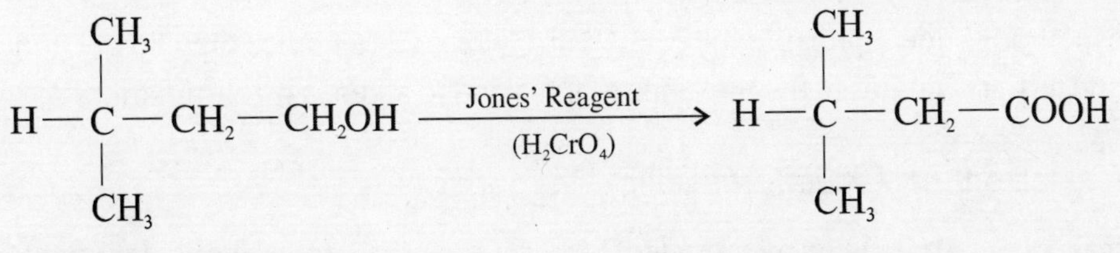

c. ***Oxidation of aldehydes*** Oxidation of aldehydes by Tollens' reagent, H_2O and H_2SO_4 give acids.

O C=O H $\xrightarrow[(Ag^+ \text{ in } NH_4OH)]{\text{Tollens' Reagaent}}$ O C=O OH

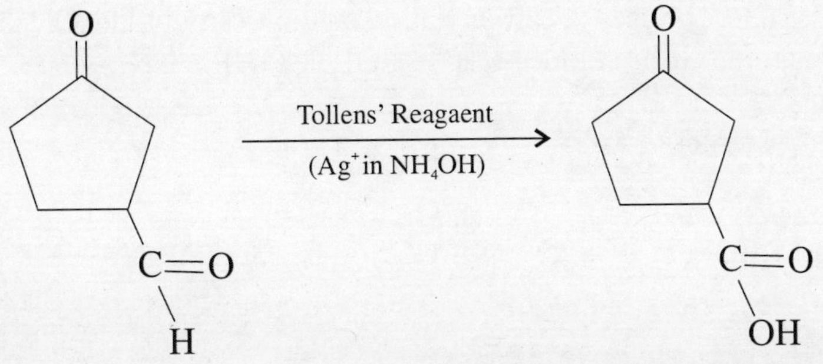

d. ***Oxidative cleavage of alkenes and alkynes*** Oxidative cleavage of the type given below give acids.

$$\underset{\text{(A)}}{(H)(R)C{=}C(R)(H)} \text{ or } \underset{\text{(B)}}{R-C\equiv C-H} \xrightarrow[H_2O,\ \text{Heat}]{KMnO_4} 2\,R-C(=O)OH$$

2. By Hydrolysis of Nitriles

Hydrolysis of nitriles by strong hot aqueous acid or base gives acid. This method begins with an organic halogen compound and the carboxyl group eventually replaces the halogen. In this method the nitrile intermediate is generated by S_N2 reaction, in which cyanide anion is a nucleophilic precursor of the carboxyl group.

$$R-CH_2-Br \xrightarrow[S_N2]{NaCN} R-CH_2-C\equiv N \xrightarrow[\text{heat}]{H_2O + H_3O^+} R-CH_2-C(=O)OH + NH_4^+$$

$$C_6H_5CH_3 + NBS \longrightarrow C_6H_5CH_2Br$$

$$C_6H_5CH_2Br + KCN \longrightarrow C_6H_5CH_2CN + H^+ \xrightarrow{H_2O} C_6H_5CH_2COOH$$

This method is mainly suited for primary halides and unhindered secondary halides which can react by S_N2 to give nitrile.

Nitrile method is not applicable for C_6H_5Cl or $CH_2{=}CHCl$ since they do not undergo S_N2 reactions and tertiary halides in the presence of CN^- lead to alkenes by elimination.

3. Carboxylation of Organometallic Reagents

Grignard reagent is carboxylated over dry ice (Solid CO_2) or dry CO_2 is bubbled through Grignard reagent solution to get the acid. The reaction is limited to alkyl halides which can form Grignard reagents. The compounds like $CNCH_2CH_2Br$, $OHCH_2CH_2Br$ cannot be used to prepare RMgX. Such compounds cannot be used in carboxylation method. Groups like OH, NH, COOH protonate the Grignard reagent. Groups like –CHO, COR, $CONR_2$, CN, —NO_2, —SO_3R add on to Grignard reagent.

The mechanism of CO_2 addition is as follows:

$$RX + Mg(\text{ether}) \longrightarrow RMgX + CO_2 \longrightarrow RCO_2MgX \xrightarrow{H_3O^+} RCOOH$$

The electrophilic halide is first transformed into a strongly nucleophilic method derivative, and this adds to carbon dioxide (an electrophilic). The initial product is a salt of the carboxylic acid, which must then be released by treatment with strong aqueous acid.

$$C_6H_5Br \xrightarrow[\text{Pentane or ether}]{2\ Li} C_6H_5Li + LiBr \xrightarrow[2)\ H_3O^+]{1)\ CO_2} C_6H_5COOH + LiBr$$

(Phenyl lithium)

4. By Phase Transfer Catalysis

18-crown-6 helps in dissolving $KMnO_4$ in the organic solvent benzene. For example $C_6H_5CH{=}CH{-}C_6H_5$ (*trans*) on oxidation with $KMnO_4$ in crown ether gives 2 moles of benzoic acid. The potassium of $KMnO_4$ is bound by crown ether and the anion also travels to the organic layer to maintain electroneutrality and helps in oxidizing the double-bonded compound.

5. By Haloform Reaction of Methyl Ketones

This reaction can be used to obtain acids; iodine and NaOH are used. It is one of the rare nucleophilic substitution reactions at carbonyl group leading to cleavage of C—C single bond. Iodination has to be base-catalysed for the methyl ketone to be trihalogenated.

REACTIONS OF CARBOXYLIC ACIDS

1. Salt Formation

Because of their enhanced acidity, carboxylic acids react with bases to form ionic salts, as shown in the following equations. In the case of alkali metal hydroxides and simple amines (or ammonia) the resulting salts have pronounced ionic character and are usually soluble in water. Heavy metals such as silver, mercury and lead form salts having more covalent character (3rd example), and the water solubility is reduced, especially for acids composed of four or more carbon atoms.

$$RCO_2H + NaHCO_3 \longrightarrow RCO_2^{(-)}Na^{(+)} + CO_2 + H_2O$$

$$RCO_2H + (CH_3)_3N: \longrightarrow RCO_2^{(-)}(CH_3)_3NH^{(+)}$$

$$RCO_2H + AgOH \longrightarrow RCO_2^{\delta(-)}\,Ag^{\delta(+)} + H_2O$$

Carboxylic acids and salts having alkyl chains longer than six carbons exhibit unusual behaviour in water due to the presence of both hydrophilic (CO_2) and hydrophobic (alkyl) regions in the same molecule. Such molecules are termed amphiphilic (Gk. *amphi* = both) or amphipathic. Depending on the nature of the hydrophilic portion, these compounds may form monolayers on the water surface or spherelike clusters, called micelles, in solution.

2. Substitution of the Hydroxyl Hydrogen

This reaction class could be termed electrophilic substitution at oxygen, and is defined as follows (E is an electrophile). Some examples of this substitution are provided in equations (1) through (4) of Figure 10.8.

If E is a strong electrophile, as in the first equation, it will attack the nucleophilic oxygen of the carboxylic acid directly, giving a positively charged intermediate which then loses a proton. If E is a weak electrophile, such as an alkyl halide, it is necessary to convert the carboxylic acid to the more nucleophilic carboxylate anion to facilitate the substitution. This is the procedure used in reactions 2 and 3. Equation 4 illustrates the use of the reagent diazomethane (CH_2N_2) for the preparation of methyl esters. This toxic and explosive gas is always used in an ether solution (bright yellow in colour). The reaction is easily followed by the evolution of nitrogen gas and the disappearance of the reagent's colour. This reaction is believed to proceed by the rapid bonding of a strong electrophile to a carboxylate anion.

1. Cyclopentane–COOH $\xrightarrow[H_2SO_4 \text{ Catalyst}]{(CH_3)_2C=CH_2}$ Cyclopentane–C(=O)–O–C(CH_3)$_3$

2. $(H_3C)_3C-C(=O)OH \xrightarrow{NaOH} (H_3C)_3C-C(=O)O^{\ominus}\,Na^{\oplus} \xrightarrow[S_N2]{CH_3-I} (H_3C)_3C-C(=O)O-CH_3 + NaI$

3. Benzene–COOH $\xrightarrow{AgOH}$ Benzene–C(=O)O$^{\delta-}$ Ag$^{\delta+}$ $\xrightarrow[S_N2]{\text{allyl-Br}}$ Benzene–C(=O)O–$CH_2CH=CH_2$ + AgBr

4. Cyclohexene–COOH $\xrightarrow[\text{ether}]{CH_2N_2}$ Cyclohexene–C(=O)O–CH_3 + N_2

Diazomethane, CH_2N_2 has the structure: $H_2C=\overset{\oplus}{N}=\ddot{\underset{..}{N}}^{\ominus}$

Figure 10.8 Reactions of acids (Electrophilic substitution)

Alkynes may also serve as electrophiles in substitution reactions of this kind, as illustrated by the synthesis of vinyl acetate from acetylene. Intramolecular carboxyl group additions to alkenes generate cyclic esters known as lactones. Five-membered (gamma) and six-membered (delta) lactones are most commonly formed. Electrophilic species such as acids or halogens are necessary initiators of lactonizations. Even the weak electrophile iodine initiates iodolactonization of λ, δ^{-} and δ, ε^{-} unsaturated acids.

3. Nucleophilic Acyl Substitution

This reaction converts RCOOH to RCOY by nucleophilic substitution. The conversion of RCOOH to RCOCl is done by reacting the acid with $SOCl_2$. The carboxylic acid is first converted to a reactive chlorosulphite intermediate which is then attached by nucleophilic chloride ion. Chloroform is used as a solvent.

The final products are RCOCl, SO_2 and Cl^-.

4. Substitution of the Hydroxyl Group

Reactions in which the hydroxyl group of a carboxylic acid is replaced by another nucleophilic group are important for preparing functional derivatives of carboxylic acids. The alcohols provide a useful reference in chemistry against which this class of transformations may be evaluated. In general, the hydroxyl group proves to be a poor leaving group, and virtually all alcohol reactions involved a prior conversion of –OH to a better leaving group. This has proven to be true for the carboxylic acids as well. Four examples of these hydroxyl substitution reactions are presented by the following equations (Figure 10.9). In each example, the new bond to the carbonyl group and the nucleophilic atom that has replaced the hydroxyl oxygen are shown in bold. The hydroxyl moiety is often lost as water, but in reaction 1 the hydrogen is lost as HCl and the oxygen as SO_2. This reaction parallels a similar transformation of alcohols to alkyl chlorides, although its mechanism is different. Other reagents that produce a similar conversion to acyl halides are PCl_5 and $SOBr_2$. The amide and anhydride formations shown in equations 2 and 3 require strong heating.

Carboxylic acids are Bronsted and Lewis acids. Unlike the carbonyl compounds where nucleophilic addition easily happens, here the possibility of removal of H from OH of COOH is higher. The anion formed is resistant to addition because this will introduce a second negative charge, but as an exception, RLi (organolithium reagent which is a strong nucleophile) adds to the acid to create dianion with 2 moles of RLi. There is no leaving group. When hydrolysis happens, hydrate is formed which leads to ketone since ketone hydrates are less stable than ketones (K for hydrate is small).

$$\text{Benzoic acid} + 2(CH_3)_3CLi + H_3O^+ \xrightarrow{[98\text{ K (pentane)}]} C_6H_5COC(CH_3)_3$$

The water quench also destroys any remaining RLi so that no further reaction of RLi will take place with ketone.

1. $$O_2N-C_6H_4-C(=O)OH \xrightarrow[\text{Heat}]{SOCl_2} O_2N-C_6H_4-C(=O)Cl + HCl + SO_2$$

2. $$H-C(=O)OH + C_6H_5-N(CH_3)H \xrightarrow[110^\circ C]{\text{Toluene}} H-C(=O)-N(CH_3)-C_6H_5 + H_2O$$

3. $$6F_3C-C(=O)OH + P_2O_5 \xrightarrow{\text{Heat}} 3F_3C-C(=O)-O-C(=O)-CF_3 + 2\,H_3PO_4$$

4. $$H_3C-C(=O)OH + C_2H_5OH \underset{H_2SO_4}{\rightleftharpoons} H_3C-C(=O)-O-C_2H_5 + H_2O$$

Figure 10.9 More reactions of acids

5. Fischer Esterification

The reaction between an acid and alcohol gives ester. Strong acid acts as a catalyst. The reaction was analysed by Fischer and therefore known as **Fischer esterification mechanism**, which as follows. The reaction is conducted in the presence of acids, and carboxylic acid becomes more reactive by the protonation of oxygen of C=O group. The carboxyl gets a positive charge and then loss of water molecule from the tetrahedral intermediate gives ester. All steps of the reaction are reversible. The detailed mechanism is given below in Figure 10.10.

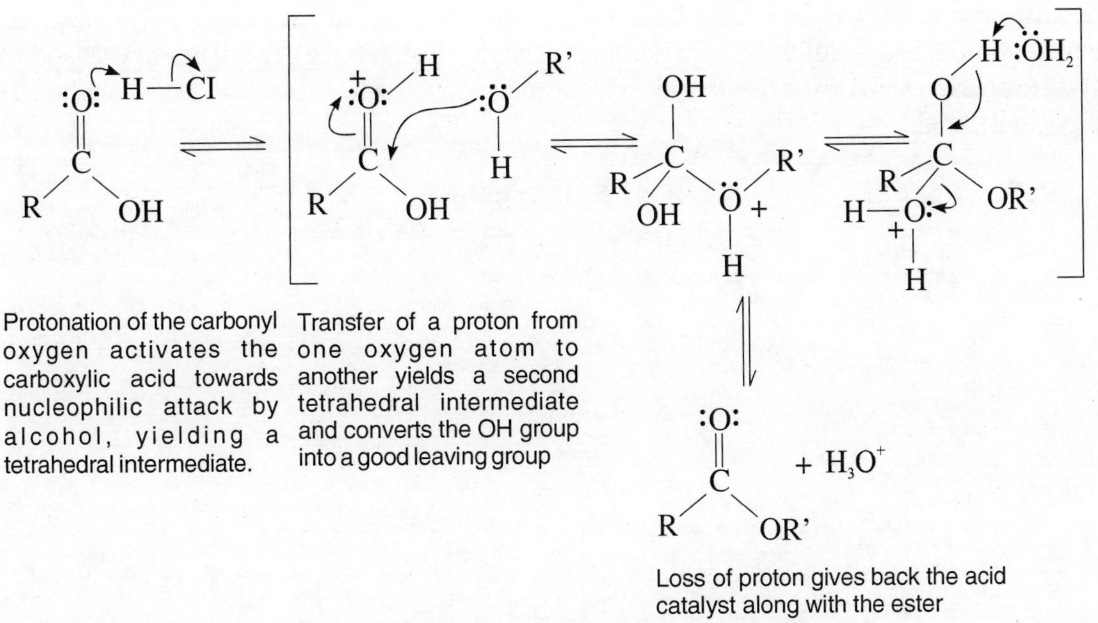

Protonation of the carbonyl oxygen activates the carboxylic acid towards nucleophilic attack by alcohol, yielding a tetrahedral intermediate.

Transfer of a proton from one oxygen atom to another yields a second tetrahedral intermediate and converts the OH group into a good leaving group

Loss of proton gives back the acid catalyst along with the ester

Figure 10.10 Mechanism of esterification

I. REDUCTIONS AND OXIDATIONS OF CARBOXYLIC ACIDS

1. ***Reduction*** The carbon atom of a carboxyl group is in a relatively high oxidation state. Reduction to a 1°-alcohol takes place rapidly on treatment with the powerful metal hydride reagent, lithium aluminium hydride, as shown by the following equation. One-third of the hydride is lost as hydrogen gas, and the initial product consists of metal salts which must be hydrolysed to generate the alcohol. These reductions take place by the addition of hydride to the carbonyl carbon, in the same manner noted earlier for aldehydes and ketones. The resulting salt of a carbonyl hydrate then breaks down to an aldehyde that undergoes further reduction.

$$3LiAlH_4 \xrightarrow{\text{Ether}} 4H_2 + 4RCH_2OLi + \text{Metal oxides} \xrightarrow{H_2O} 4RCH_2OH + \text{Metal hydroxides}$$

Diborane (B_2H_6) reduces the carboxyl group in a similar fashion. Sodium borohydride, ($NaBH_4$) does not reduce carboxylic acids; however, hydrogen gas is liberated and salts of the acid are formed. Partial reduction of carboxylic acids directly to aldehydes is not possible, but such conversions have been achieved in two steps by way of certain carboxyl derivatives.

2. ***Oxidation*** Because it is already in a high oxidation state, further oxidation removes the carboxyl carbon as carbon dioxide. Depending on the reaction conditions, the oxidation state of the remaining organic structure may be higher, lower or unchanged. The following reactions are all examples of decarboxylation (loss of CO_2). In the first, bromine replaces the carboxyl group, so both the carboxyl carbon atom and the remaining organic moiety are oxidized. Silver salts have also been used to initiate this transformation, which is known as the Hunsdiecker reaction. The second reaction is an interesting bis-decarboxylation, in which the atoms of the organic residue

retain their original oxidation states. Lead tetraacetate will also oxidize monocarboxylic acids in a manner similar to reaction 1. Finally, the third example illustrates the general decarboxylation of β-keto acids, which leaves the organic residue in a reduced state (note that the CO_2 carbon has increased its oxidation state).

1. $RCOO^{-}Ag^{+} + Br_2 \longrightarrow RBr + CO_2 + AgBr$

2. (cyclobutane-1,2-dicarboxylic acid) $+ Pb(OCOCH_3)_4 \xrightarrow{\text{Heat}}$ (cyclobutene) $+ 2CO_2 + 2CH_3CO_2H + Pb(OCOCH_3)_2$

3. (2-oxocyclohexanecarboxylic acid) $\xrightarrow{\text{Heat}} CO_2 +$ [enol, O—H] $\longrightarrow$ (cyclohexanone)

Figure 10.11 Reactions involving loss of CO_2

Hunsdiecker reaction The first reaction in Figure 10.11 is the Hunsdiecker reaction. The acid or the silver salt is used to initiate the reaction. If silver salt is used we get RBr and if acid is used, we get HBr.

The silver salt of a carboxylic acid is treated with bromine. The product is alkyl bromide. The mechanism is given below:

$$R-C(=O)-O-Ag^{+} + Br-Br \longrightarrow RCOOBr + AgBr$$

$$RCOOBr \longrightarrow RCOO^{\bullet} + Br^{\bullet}$$

$$RCOO^{\bullet} \longrightarrow R^{\bullet} + CO_2$$

$$R^{\bullet} + RCOOBr \longrightarrow RBr + RCOO^{\bullet}$$

When two molecules of acid react, water is eliminated to give anhydrides. With dicarboxylic acids at high temperatures, cyclic anhydrides are obtained.

Succinic acid gives succinic anhydride. If the acid is $COOHCH_2COOH$ (CH_2 flanked by two CO groups) then decarboxylation takes place. Acids like $COOHCH_2COOH$ and beta keto acids like $RCOCH_2COOH$ undergo decarboxylation readily. Reaction 3 in the Figure 10.12 describes decarboxylation of beta keto acids.

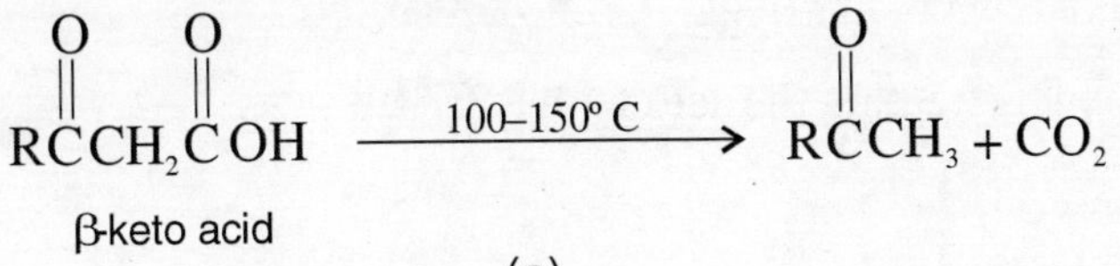

$$\underset{\beta\text{-keto acid}}{RCOCH_2COOH} \xrightarrow{100\text{–}150^\circ C} RCOCH_3 + CO_2$$

(a)

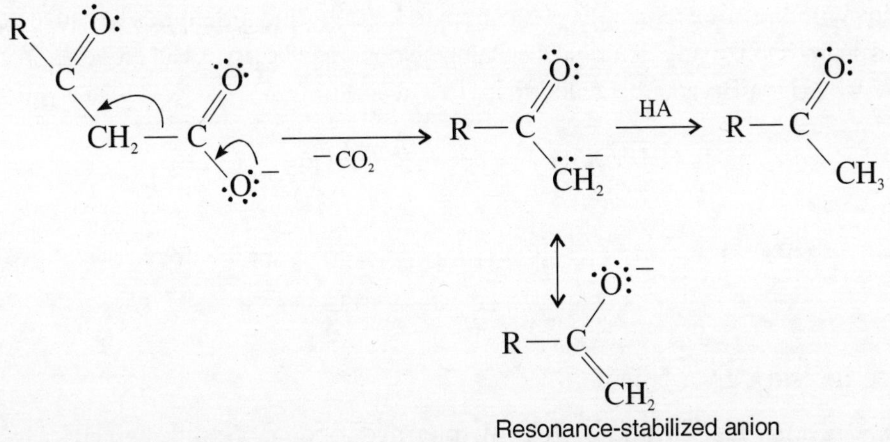

$$R\text{–}CO\text{–}CH_2\text{–}COO^- \xrightarrow{-CO_2} R\text{–}C(=O)\text{–}\ddot{C}H_2^- \xrightarrow{HA} R\text{–}C(=O)\text{–}CH_3$$

$$R\text{–}C(=O)\text{–}\ddot{C}H_2^- \longleftrightarrow R\text{–}C(\text{–}O^-)=CH_2$$

Resonance-stabilized anion

(b)

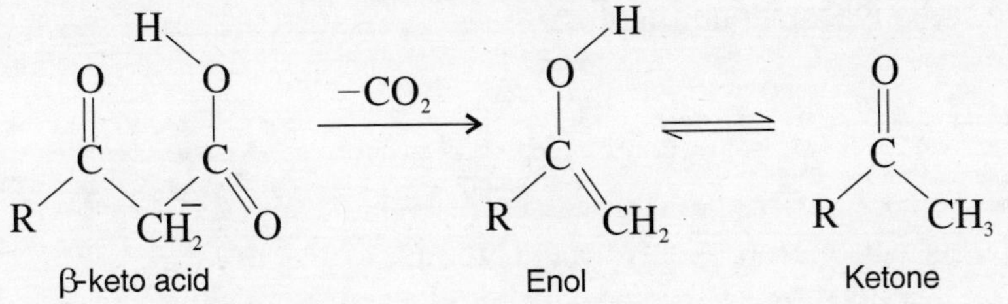

$$\underset{\beta\text{-keto acid}}{R\text{–}CO\text{–}CH_2\text{–}COOH} \xrightarrow{-CO_2} \underset{\text{Enol}}{R\text{–}C(OH)=CH_2} \rightleftharpoons \underset{\text{Ketone}}{R\text{–}CO\text{–}CH_3}$$

(c)

Figure 10.12 Decarboxylation reaction

DERIVATIVES OF CARBOXYLIC ACIDS

CARBOXYLIC ACID HALIDES

Acid halides are prepared by the reaction of carboxylic acid with sulphonyl chloride ($SOCl_2$) or treatment with PBr_3.

$$\text{RCOOH} + \text{SOCl}_2 \longrightarrow \underset{\text{Acid chloride}}{\text{RCOCl}}$$

RCOCl is used as a reagent in Friedel–Crafts acylation.

RCOBr is obtained by treating the acid with PBr_3.

HALO ACIDS

Hell–Volhard–Zelinsky (HVZ) reaction of acids Direct bromination of carbonyl compounds with bromine in acetic is restricted to ketones and aldehydes because acids, esters and amides do not enolize sufficiently so that halogenation can take place. But the acids are brominated alternatively by Br + PBr_3 and this reaction is HVZ reaction.

$$\text{R}_2\text{CHCOOH} + \text{Br}_2 \longrightarrow \underset{\text{(A)}}{\text{R}_2\text{CHCOBr}} + \text{HBr}$$

$$\text{(A) enolizes on } \text{R}_2\text{C}{=}\underset{\text{Br}}{\underset{|}{\text{C}}}{-}\text{OH}$$

The mechanism is given below.

The acid reacts with PBr_3 and Br_2 to form intermediate acid bromide and HBr.

$$\text{R}_2\text{CHCOOH}+\text{BR}_2 \longrightarrow \text{R}_2\text{CHCOBr} + \text{HBr}$$

The bromo carbonyl compound enolizes as

$$\text{R}_2\text{C}{=}\underset{\text{Br}}{\underset{|}{\text{C}}}{-}\text{OH}$$

This enol reacts with one more equivalent of bromine to give $BrCR_2$—COBr. This on hydrolysis gives $R_2CBrCOOH$. Since enol is formed which is planar, the optically active acid gets racemized.

Reactions of Acid Halides

1. *Friedal–Crafts acylation* (already discussed in earlier chapter)
2. *Acid halides to acids* The hydrolysis of acid chloride by water gives the acid
3. *Acid chlorides to esters* Acid chlorides react with alcohols in give esters. Pyridine is used to remove HCl. This reaction is affected by steric hindrance. The order of reactivity among alcohols is 1° > 2° > 3°. If we have cyclohexanol with CH_2OH at the 4th position, the CH_2OH is primary and less hindered. Esterification happens with CH_2OH only.

4. Acid halides react rapidly with ammonia to give amides. Since HCl is formed, two equivalents of NH_3 must be used.
5. Acid chlorides are converted to alcohols by $LiAlH_4$ in ether in acid to give primary alcohols.
6. RCOCl + RMgX in ether followed by addition of H^+ gives tertiary alcohol.

$$\text{Benzoyl chloride} + CH_3MgBr\,(\text{ether}) + H^+ \longrightarrow C_6H_5C(CH_3)_2OH$$

7. Reaction of acid chlorides with NBS in the presence of catalytic amount of acid gives alpha bromo acid chlorides.

$$CH_3CH_2CH_2CH_2CH_2COOH + SOCl_2 \xrightarrow{65°C} CH_3CH_2CH_2CH_2CH_2COCl \xrightarrow[85°C]{NBS,\,HBr} CH_3CH_2CH_2CH_2CHBrCOCl.$$

If NCS is used chloro compound is formed and I_2 at 130°C gives iodo compound. α-halo acids are useful in synthesis. They can be converted to hydroxy acids with OH^- and cyano compound is obtained by treatment with KCN. Then addition of K_2CO_3 and water to the cyano compound (heat) may be carried out. The next step is the treatment with NH_3 for long time followed by water addition in the presence of H^+. This leads to the alpha amino acid.

CARBOXYLIC ACID AMIDES

Amides are not easily prepared by the reaction of acids with amines because amines are strong bases and they form the carboxylate anions. These ions cannot react with nucleophiles except at very high temperatures. Sometimes the ammonium salt is formed from acid and amine and this is heated to give amides. The reaction is dehydration. For example, RCOOH is reacted with $C_6H_5CH_2NH_2$ in the presence of dicyclohexylcarbodiimide (DCC) to obtain amide as given below:

$$RCOOH + C_6H_5CH_2NH_2 \rightarrow R\text{—}CONHCH_2C_6H_5\text{(amide)}$$

The structure of DCC is given in Figure 10.13.

N=C=N

Figure 10.13 Structure of dicyclohexylcarbodiimide (DCC)

Basicity of Amides

Amides are much less basic than amines. Amides do not undergo protonation when treated with acids, indicating their lack of basicity. This is because amide is stabilized by delocalization of lone pair of nitrogen by orbital overlap with C=O group. Protonation is disfavoured since resonance stabilization will be lost.

Reactions of Amides

Hoffmann rearrangement Amides react with NaOH and bromine in the presence of water to give amines and CO_2.

The mechanism of the rearrangement is given in Figure 10.14.

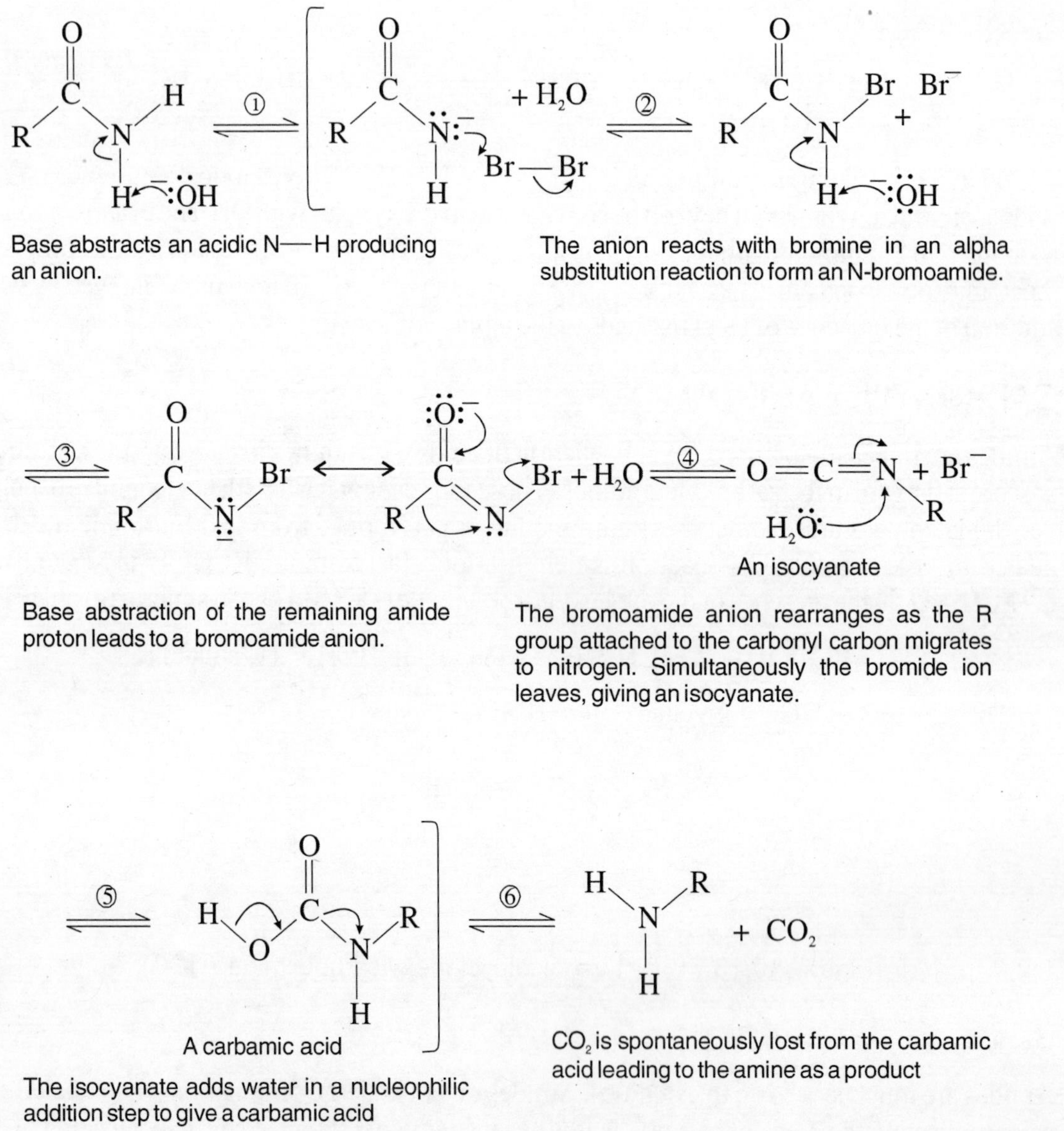

Base abstracts an acidic N—H producing an anion.

The anion reacts with bromine in an alpha substitution reaction to form an N-bromoamide.

An isocyanate

Base abstraction of the remaining amide proton leads to a bromoamide anion.

The bromoamide anion rearranges as the R group attached to the carbonyl carbon migrates to nitrogen. Simultaneously the bromide ion leaves, giving an isocyanate.

A carbamic acid

The isocyanate adds water in a nucleophilic addition step to give a carbamic acid

CO_2 is spontaneously lost from the carbamic acid leading to the amine as a product

Figure 10.14 Hoffmann rearrangement

If $RCON_3$ reacts with water on heating, it gives amines via the formation of R—N=C=O (isocyanate) and this is called Curtius rearrangement.

The formation of isocyanate (RNCO) from RCOCl is diagrammatically represented in Figure 10.15.

$$R-\overset{O}{\overset{\|}{C}}-Cl \xrightarrow{NaN_3} R-\overset{O}{\overset{\|}{C}}-\ddot{N}-\overset{+}{N}\equiv N \xrightarrow{\Delta} N_2 + O=C=N-R \xrightarrow{H_2O} R-NH_2 + CO_2$$

Acid chloride   Acyl azide   Isocyanate   Amine

Figure 10.15 Curtius rearrangement

NITRILES

$RCH_2Br + NaCN$ by S_N2 pathway gives RCH_2CN.

Reactions of Nitriles

1. *Hydrolysis*

$$RCN + H_2O \xrightarrow{\text{(in acidic or basic conditions)}} RCONH_2 + H_2O \rightarrow RCOOH$$

2. *Reduction*

$$RCN \xrightarrow[\text{(2) } H_2O]{\text{(1) } LiAlH_4 \text{ (ether)}} RCH_2NH_2$$

3. *Grignard reaction*

$$RCN \xrightarrow[\text{(2) } H_2O]{\text{(1) RMgX (ether)}} RCOR$$

OTHER METHODS OF PREPARATION OF CARBOXYLIC ACIDS USING THE ACID DERIVATIVES

Malonic Ester and Acetoacetic Ester Synthesis

Malonic ester synthesis Synthesis of mono- and disubstituted carboxylic acids is done using malonic ester synthesis.

The diester of malonic acid diethyl malonate is treated with $NaOC_2H_5$ followed by the reaction with an alkyl bromide to give a carboxylic acid by adding 50% KOH, and then dilute H_2SO_4 is added and refluxed.

$$C_2H_5OCOCH_2COOC_2H_5 \xrightarrow[\text{(2) } CH_3CH_2CH_2CH_2Br]{\text{(1) } (NaOC_2H_5)} \underset{\displaystyle CH_2CH_2CH_2CH_3}{C_2H_5OCOCHCOOC_2H_5}$$

$$\downarrow \text{(1) alc. KOH (50\%)} \quad \text{(2) dil. } H_2SO_4 \text{ reflux}$$

$$CH_3CH_2CH_2CH_2{-}CH_2COOH$$

Acetoacetic ester synthesis Ethylacetoacetate with $NaOC_2H_5$ in ethanol followed by addition of alkyl bromide gives a substituted acetoacetate. This on heating loses CO_2 and gives a ketone.

$$CH_3COCH_2COOC_2H_5 \xrightarrow[\text{(2) } CH_3CH_2CH_2CH_2Br]{\text{(1) } NaOC_2H_5/C_2H_5OH} \underset{\displaystyle CH_2CH_2CH_2CH_3}{CH_3COCHCOOC_2H_5}$$

$$\downarrow \text{(heat)}$$

$$CH_3COCH_2CH_2CH_2CH_2CH_3$$

2-heptanone

I. "EXPLAIN WHY" QUESTIONS AND SMALL STRUCTURE-BASED PROBLEMS

1. Amides are not basic like amines. Why?

Answer

Amides are much less basic than amines. Amides do not undergo protonation when treated with acids indicating their lack of basicity. This is because amide is stabilized by the delocalization of lone pair of nitrogen by orbital overlap with C=O group. Protonation is disfavoured since resonance stabilization will be lost.

2. CH_3COOH is stronger than CH_3CH_2OH. Why?

Answer

In carboxylic acid, there is resonance-stabilized carboxylate ion when proton is lost. Hence it is more acidic. In $CH_3CH_2O^-$, such resonance stabilization is not there.

3. F_3CCOOH is much stronger than CH_3COOH. Why?

Answer

The –I effect of fluorine pulls electrons towards fluorine and makes the acid much stronger.

4. Peracetic acid is much weaker than acetic acid. Why?

   Answer

   In CH_3COOH, the lone pair on oxygen of carboxyl ion $R-\overset{\overset{O}{\|}}{C}-O-$ can be shifted to next carbon which is sp^2-hybridized. Such shifting is not possible in $R—COOO^-$ since the bond next to oxygen with negative charge is sp^3-hybridized.

5. Trifluoromethyl group is a strong deactivating group than in electrophilic aromatic substitution. Why?

   Answer

   The acidity of trifluoromethylbenzoic acid is much higher than that of benzoic acid and hence trifluoromethyl group is a strong deactivating group.

6. Nitrile method is not suitable for converting bromobenzene to benzoic acid. Why?

   Answer

   Bromobenzene cannot react with NaCN to give C_6H_5CN since aryl halide does not undergo S_N2 reaction. Hence nitrile method is not possible.

7. $(CH_3)_3CCl$ cannot be converted to $(CH_3)_3CCOOH$ by nitrile method. Why?

   Answer

   Tertiary chlorides undergo elimination with CN^-.

8. Normally electrophilic aromatic substitution gives more para product than ortho product due to steric crowding. But if phenol is carboxylated, we get 100% *o*-hydroxybenzoic acid (salicylic acid). Why?

   Answer

   This is because the *ortho*-COOH group is hydrogen-bonded to the OH group intramolecularly.

9. *o*-hydroxybromobenzene cannot be converted to the corresponding *o*-hydroxy acid both by Grignard and nitrile methods. Why?

   Answer

   The compound does not undergo S_N2 reaction and hence nitrile method fails. Grignard reaction is not possible since OH group is present.

10. Grignard carboxylation is not the method to convert $OHCH_2CH_2CH_2Br$ to $OHCH_2CH_2CH_2COOH$. Why?

    Answer

    When OH group is there, Grignard reaction is not possible. Only nitrile method will work.

II. Multiple Choice Questions (One out of four answers is correct)

1. The acid which does not give anhydride on heating is
   a) adipic acid b) succinic acid c) $COOH(CH_2)_6COOH$ d) malonic acid
2. The acid which undergoes decarboxylation most easily is
   a) succinic acid b) acetic acid c) CH_3COCH_2COOH d) benzoic acid
3. The acid which does not undergo HVZ reaction is
   a) CH_3COOH b) $(C_2H_5)_3CCOOH$ c) $C_6H_5CH_2COOH$ d) $(CH_3)_2CHCOOH$
4. The compound obtained by treating acetic acid with P/Cl_2 is
   a) $ClCH_2COOH$ b) CH_3COCl c) $ClCH_2COCl$ d) $Cl_2CHCOCl$
5. The weakest acid among the following is
   a) CH_3COOOH b) C_6H_5COOH c) CH_3COOH d) F_3CCOOH
6. The ozonolysis product under oxidizing conditions of this unsaturated compound does not undergo HVZ reaction.
   a) propylene b) 2-butene c) $(CH_3)_3CCH{=}CH_2$ d) $CH_3C{\equiv}CH$
7. One of the following unsaturated compounds produces an acid different from that of others by ozonolysis in oxidizing conditions.
   a) 2-butene b) propyne c) 2-butyne d) 1-pentyne
8. A with PCl_5 followed by hydrolysis gives chloroacetic acid. A is
   a) CH_3COOH b) COOHCOOH c) $COOHCH_2COH$ d) $OHCH_2COOH$
9. CO_2 is not a product in one of the following reactions.
   a) 2-butene on ozonolysis in oxidizing conditions
   b) allene on ozonolysis
   c) malonic acid on heating
   d) $R_2C(OH)COOH + KMnO_4 \rightarrow$
10. An unsaturated carboxylic acid is produced on heating
    a) malonic acid b) $RCHOHCH_2COOH$
    c) $RCHOHCH_2CH_2COOH$ d) $RCHOHCH_2CH_2CH_2COOH$
11. The 5-carbon acid which is optically active is
    a) $CH_3CH_2CH_2CH_2COOH$ b) $(CH_3)_2CHCH_2COOH$
    c) $CH_3-\underset{\underset{COOH}{|}}{\overset{\overset{H}{|}}{C}}-CH_2CH_3$ d) $(CH_3)_3CCOOH$

12. The 5-carbon acid which cannot undergo HVZ reaction is

a) $CH_3\ CH_2CH_2CH_2COOH$ b) $(CH_3)_2CHCH_2COOH$

c) 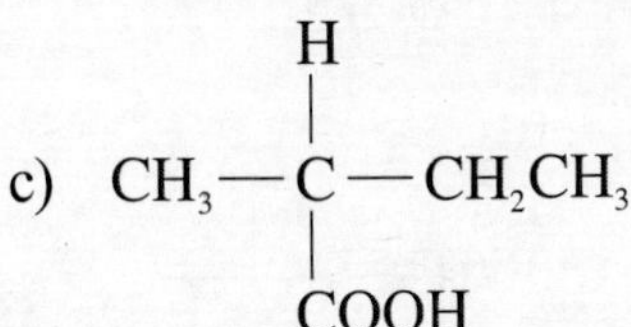

d) $(CH_3)_3CCOOH$

13. The reagent which cannot convert RCOOH to RCOCl is

a) P and Cl_2 b) PCl_3 c) PCl_5 d) $SOCl_2$

14. An acyl bromide can be converted to acyl fluoride by

a) F_2 b) PF_3 c) PF_5 d) NaF and anhydrous HF

15. One of the following reactions does not take place

a) 2-butyne + O_3 in $H_2O_2 \rightarrow 2CH_3COOH$

b) $CH_3COOH + CH_2{=}C{=}O \rightarrow CH_3COOCOCH_3$

c) $CH_3COOH + PF_5 \rightarrow CH_3COF$

d) $RCOOH + SOCI_2 \rightarrow RCOCI + SO_2 + HCI$

16. The optically inactive acid among the following is

a) COOHCHOHCHOHCOOH

b) 1,2-*trans*-cyclopentane dicarboxylic acid

c) $CH_3-\overset{\displaystyle H}{\underset{\displaystyle COOH}{C}}-CH_2CH_3$

d) *cis*-1,2-cyclopropane dicarboxylic acid

17. The optically active hydrocarbon which on $KMnO_4$ oxidation gives terephthalic acid (*p*-$HOOCC_6H_4COOH$) is

a) *p*-me $C_2H_5-\overset{\displaystyle H}{\underset{\displaystyle CH_3}{C}}-CH_2CH_3$ b) *p*-me-$C_6H_4-C(CH_3)_3$

c) *p*-ethyl$C_6H_4CH_2CH_2CH_3$ d) *p*-butyl$C_6H_4CH_3$

18. There are 4 optically active isomers of an acid with formula $C_4H_8O_3$ having two functional groups. Only one of them is reduced by $LiAlH_4$ to an optically inactive product. This is

a) $CH_3CH(CH_2OH)—COOH$
b) $CH_3CH(OH)—CH_2COOH$
c) $CH_3CH_2CH(OH)—COOH$
d) $CH_3CH(O—CH_3)—COOH$

19. The acid which answers Tollens test of aldehydes is

a) HCOOH b) CH_3COOH c) CH_3CH_2COOH d) C_6H_5COOH

20. The halo acid which gives rise to unsaturated carboxylic acid in base is

a) $CH_3CH_2CHBrCH_2COOH$
b) $CH_3CH_2CHBrCOOH$
c) $BrCH_2CHCH_2COOH$
d) $ClCH_2COOH$

KEY

1. d	2. c	3. b	4. a	5. a	6. c	7. d
8. a	9. a	10. b	11. c	12. d	13. a	14. d
15. c	16. d	17. a	18. a	19. a	20. a	

III. Multiple Choice Questions (More than one answer is correct)

1. The acids which give anhydride on heating are

a) adipic acid
b) succinic acid
c) $COOH(CH_2)_6COOH$
d) malonic acid

2. The acids which do not decarboxylate easily are

a) succinic acid
b) acetic acid
c) CH_3COCH_2COOH
d) CH_3COCH_2COOH

3. The acids which undergo HVZ reaction are

a) CH_3COOH
b) $(C_2H_5)_3CCOOH$
c) $C_6H_5CH_2COOH$
d) $(CH_3)_2CHCOOH$

4. Acyl chloride is obtained from the acid by adding

a) $SOCl_2$ b) PCl_3 c) PCl_5 d) Cl_2

5. The acids stabilized by resonance are

a) CH_3COOOH b) C_6H_5COOH c) CH_3COOH d) C_6H_5COOOH

6. The ozonolysis product under oxidizing conditions of these unsaturated compounds do not undergo HVZ reaction
   a) propylene
   b) 2-butene
   c) $(CH_3)_3CCH{=}CH_2$
   d) $C_6H_5CH{=}CH_2$
7. The ozonolysis under oxidizing conditions of some of these unsaturated compounds yield the same acid
   a) 2-butene
   b) propyne
   c) 2-butyne
   d) 1-hexyne
8. The compounds which on oxidation give $(CH_3)_2CHCOOH$ are
   a) $(CH_3)_2CHCH_2OH$
   b) $(CH_3)_2CHCHO$
   c) $(CH_3)_2CHCH{=}CHCH(CH_3)_2$
   d) $(C_2H_5)_2CHCOCH_3 + NaOH + I_2$
9. CO_2 is a product in some of the following reactions
   a) 2-butene on ozonolysis in oxidizing conditions
   b) Allene on ozonolysis
   c) Malonic acid on heating
   d) $R_2C(OH)COOH + KMnO_4 \rightarrow$
10. Some of these acids do not decolorize $KMnO_4$ when heated
    a) malonic acid
    b) $RCHOHCH_2COOH$
    c) $RCHOHCH_2CH_2COOH$
    d) $CHOHCH_2CH_2CH_2COOH$
11. The 5-carbon acids which are achiral are
    a) $CH_3CH_2CH_2CH_2COOH$
    b) $(CH_3)_2CHCH_2COOH$
    c) $CH_3-\underset{\substack{|\\COOH}}{\overset{\substack{H\\|}}{C}}-CH_2CH_3$
    d) $(CH_3)_3CCOOH$
12. The 4-carbon acids which undergo HVZ reaction are
    a) $CH_3CH_2CH_2COOH$
    b) $(CH_3)_2CHCOOH$
    c) $(CH_3)_2CClCOOH$
    d) $(CH_3)_3CCOOH$
13. Some of the following reactions take place
    a) 2-butyne $+ O_3$ in $H_2O_2 \rightarrow 2CH_3COOH$
    b) $CH_3COOH + CH_2{=}C{=}O \rightarrow CH_3COOCOCH_3$
    c) $CH_3COOH + PF_5 \rightarrow CH_3COF$
    d) $RCOOH + SOCl_2 \rightarrow RCOCl + SO_2 + HCl$

14. There are 4 optically active isomers of an acid with formula $C_4H_8O_3$ having two functional groups. Some of these get reduced by $LiAlH_4$ and are still optically active. These are

a) $CH_3CH(CH_2OH)—COOH$
b) $CH_3CH(OH)—CH_2COOH$
c) $CH_3CH_2CH(OH)—COOH$
d) $CH_3CH(O—CH_3)—COOH$

15. Some of these halo acids do not give rise to unsaturated carboxylic acid in base and these are

a) $CH_3CH_2CHBrCH_2COOH$
b) $CH_3CH_2CHBrCOOH$
c) $BrCH_2CH_2CH_2COOH$
d) $ClCH_2COOH$

KEY

1. a,b,c 2. a,b 3. a,c,d 4. a,b,c 5. b,c 6. c,d 7. a,b,c
8. a,b,c 9. b,c,d 10. a,c,d 11. a,b,d 12. a,b 13. a,b,d 14. b,c,d
15. b,c,d

IV. Assertion–Reason Questions

The questions 1 to 10 consist of assertion in column 1 and reason in column 2. Use the following key to choose appropriate answer

a) If both assertion and reason are correct and reason is the correct explanation of assertion
b) If both assertion and reason are correct but reason is not the correct explanation of assertion
c) If assertion is correct but reason is incorrect
d) If assertion is incorrect but reason is correct

No.	Assertion	Reason
1.	*o*-hydroxybenzyl bromide cannot be prepared by Grignard carboxylation reaction.	The hydroxyl group is acidic.
2.	The Grignard carboxylation fails in the conversation of $CH_3COCH_2CH_2CH_2I$ to $CH_3COCH_2CH_2CH_2COOH$.	The Grignard reagent RMgX will react with carbonyl group. It should be protected.
3.	2-chloro-2-methylpentane can be converted to acid via cyanide.	The alkyl halide is tertiary and CN^- is a base leading to elimination product.

(Contd.)

No.	Assertion	Reason
4.	Alpha halogenation for ketones is easy but it is not so for carboxylic acids.	Carboxylic acids do not enolize easily
5.	If an optically active acid undergoes HVZ reaction, the product is meso(D).	The intermediate is an acid bromide enol which is planar at C$=$C bond.
6.	Methylamine is more basic than benzamide.	The nitrogen lone pair in $CONH_2$ group is delocalized with C$=$O group and hence amides are less basic.
7.	Bromobenzene cannot be converted to benzoic acid by nitrile method.	Bromobenzene undergoes nucleophilic aromatic substitution.
8.	Peracetic acid has lower pKa than CH_3COOH.	Resonance is not possible in peracetic acid.
9.	Malonic acid undergoes decarboxylation much slower than acetic acid.	On decarboxylation, the dicarboxylate ion forms resonance-stabilized enolate anion.
10.	There are totally five isomers in a carboxylic acid with formula $C_5H_{10}O_2$.	2-methylbutanoic acid is optically giving two stereoisomers.

KEY

1. a　2. a　3. d　4. a　5. d　6. a　7. b
8. d　9. d　10. a

V. Comprehension Passages

Passage I

Carboxylic acids can be prepared by several methods. Ozonolysis of alkenes in oxidizing conditions, treatment with alkaline $KMnO_4$, oxidation of aromatic compounds, nitrile method, Grignard reaction and so on.

Answer the following questions

1. The alkene which gives propanoic acid by ozonolysis in oxidizing conditions is

a) $CH_3CH=CHCH_3$　　b) $CH_3CH_2CH=CHCHCH_2CH_3$

c) 2-methylpropene　　d) propylene

2. One of the following methods is suitable for preparation of
   a) C_6H_5COOH from C_6H_5Br by nitrile method
   b) $C_6H_5CH_2COOH$ from $C_6H_5CH_2Cl$ by nitrile method
   c) $(CH_3)_3CCOOH$ from $(CH_3)_3CCl$ by nitrile method
   d) $OHCH_2CH_2Cl$ to $OHCH_2CH_2COOH$ by Grignard reaction
3. The nitrile method is not suitable for converting $(C_2H_5)_3CCl$ to $(C_2H_5)_3CCOOH$ because
   a) Cl^- is a poor nucleophile
   b) the CN^- leads to elimination of halide
   c) CN^- does not undergo S_N2 reaction
   d) tertiary carbocation is very stable

KEY

1. b 2. b 3. b

Passage 2

Carboxylic acids are acidic. Acidity is controlled by inductive and resonance effects. The acidic property of acids is responsible for certain reactions.

Answer the following questions

1. The strongest acid among the following is
   a) CH_3COOH b) C_6H_5COOH c) CH_3COOOH d) C_2H_5COOH
2. When phenol is carboxylated, we get salicylic acid only because of
   a) steric factor b) intermolecular hydrogen bonding
   c) intramolecular hydrogen bonding d) –I effect
3. One of the following reactions is not true
   a) $CH_3COOH + CH_2N_2 \rightarrow CH_3COOCH_3$
   b) $C_6H_5OH + CH_2N_2 \rightarrow C_6H_5OCH_3$
   c) $RMgX + \text{weak acid} \rightarrow RH$
   d) $C_2H_5OH + CH_2N_2 \xrightarrow{\text{(no catalyst)}} C_2H_5OCH_3$

KEY

1. b 2. c 3. d

Passage 3

Carboxylic acids undergo several reactions. Some of these reactions involve free radical mechanism. Derivatives of acids like $RCONH_2$, RNC undergo rearrangements.

Answer the following questions

1. One of these reactions involve free radical mechanism
   a) $RCOOAg + Br_2 \rightarrow RBr$ b) $CH_3CH{=}CHCH_3 + HCl$
   c) $CH_3CH{=}CHCH_3 + H_2SO_4 \rightarrow$ d) $(CH_3)_3CCl + NaOEt \rightarrow$
2. When RCOCl reacts with NaN_3 and is heated, amine is produced. One statement about this reaction is wrong.
   a) N_2 is evolved.
   b) $O{=}C{=}N{-}R$ is an intermediate.
   c) $O{=}N{=}C{-}R$ is an intermediate.
   d) Hydrolysis of isocyanate gives the primary amine.
3. $RCONH_2$ in a base rearranges to give a primary amine. One statement about this reaction is not true.
   a) It is Hoffmann rearrangement.
   b) Bromoamide anion is an intermediate anion.
   c) Isocyanate is formed.
   d) It is called Curtius rearrangement.

KEY

1. a 2. c 3. d

VI. Matrix Matching

Question I

Column 1	**Column 2**
a) Free radical mechanism	p) salicylic acid
b) Intramolecular hydrogen bonding	q) peracetic acid is weaker than acetic acid
c) Resonance effect	r) Hunsdiecker reaction
d) Malonic acid	s) decarboxylates easily to give $CH_3COOH + CO_2$

Answer

$[a \rightarrow r], [b \rightarrow p], [c \rightarrow q], [d \rightarrow s]$

Question 2

Column 1	Column 2
a) C_6H_5COOH	p) prepared by oxidation of benzyl alcohol with $KMnO_4$
b) nitrile method not suitable for	q) Bayer Villiger oxidation of benzaldehyde
c) CH_3CONH_2	r) $C_6H_5MgBr + CO_2$ in acid
d) –I effect	s) $(CH_3)_3CCl$ to $(CH_3)_3CCOOH$ by nitrile method
	t) C_6H_5Cl to C_6H_5COOH
	u) less basic than CH_3NH_2
	v) responsible for higher acidity of $ClCH_2COOH$ relative to CH_3COOH

Answer

$$[a \rightarrow p,q,r], [b \rightarrow s,t], [c \rightarrow u], [d \rightarrow v]$$

Question 3

Column 1	Column 2
a) $C_2H_5COCH_3 + NaOH + I_2$	p) CHI_3 and C_2H_5COONa
b) $CH_3COOCH_3 + LiAlH_4, H_2O$	q) benzoic acid
c) $C_6H_5CH{=}CH_2$ with alkaline $KMnO_4$	r) $CH_3CH_2OH + CH_3OH$
d) 2-butene on ozonolysis in oxidizing conditions	s) CH_3COOH

Answer

$$[a \rightarrow p], [b \rightarrow r], [c \rightarrow q], [d \rightarrow s]$$

VII. Structure-Based Problems

1. A is an optically active compound with formula $C_6H_{12}O_2$ and answers Tollens test and by this reagent gets oxidized to $C_6H_{12}O_3$(B) which is optically active. Compound A with PCC in CH_2Cl_2 gives a compound optically inactive (C) which by Zn(Hg)HCl gives a hydrocarbon C_5H_{12}(CC) which gives 4 monochlorinated products, and on bromination gives a selectively brominated hydrocarbon (D). On vigorous oxidation, A gives E, an optically inactive dicarboxylic acid. Find A to E.

Answer

$$CH_3$$
|
A is $HO-CH_2-CH_2-CH-CH_2CHO$;

B is $HO-CH_2-CH_2-CH-CH_2COOH$;
|
$$CH_3$$

C is $CHOCH_2CH-CH_2CHO$;
|
$$CH_3$$

CC is $CH_3CH_2CHCH_2CH_3$;
|
$$CH_3$$

D is $CH_3CH_2CBrCH_2CH_3$;
|
$$CH_3$$

E is $COOHCH_2CHCH_2COOH$.
|
$$CH_3$$

2. A is an acid obtained by oxidation with $KMnO_4$ of C_6H_5R where R can be CH_3 or C_2H_5 or CH_2OH or CH_2Br or a compound with benzylic hydrogen. A reacts with 2 moles of CH_3Li in water to give B. B on Clemmenson reduction gives C which by oxidation gives A. B with NaOH and I_2 gives CHI_3 and the anion of A. C on treatment with ZnO gives D which by ozonolysis in Zn/H_2O) gives E and F. E on oxidation gives A. Find A to F.

Answer

A is benzoic acid;
B is $C_6H_5COCH_3$;
C is $C_6H_5CH_2CH_3$;
D is $C_6H_5CH=CH_2$;
E is C_6H_5CHO;
F is H_2CO.

3. A and B are two alkyl halides. A does not undergo elimination but undergoes substitution reactions. B undergoes elimination and substitution reactions. A is the monohalogen product of hydrocarbon obtained by hydrolysis of Al_4C_3. B is the monochlorinated product of hydrocarbon from Wurtz reaction of A. A and B react with CN^- to give C and D respectively.

C and D hydrolyse to give E and F. E and F react with $2CH_3Li$ individually and on hydrolysis give G and H. G is a polar aprotic solvent used in S_N2 reactions and H by Clemmenson reduction gives n-butane. Find A to H.

Answer

A is CH_3I;
B is C_2H_5I;
C is CH_3CN;
D is C_2H_5CN;
E is CH_3COOH;
F is C_2H_5COOH;
G is CH_3COCH_3;
H is $C_2H_5COCH_3$;

4. Compound A on ozonolysis in reducing conditions gives HCHO and $(CH_3)_2CHCHO$. A on hydroboration oxidation gives B. B with PBr_3 gives C. C with Mg and ether gives D. D on carboxylation gives E. E with $LiAlH_4$ gives B. B on oxidation with PCC gives F. F by Clemmenson reduction gives G. G gives 5 monochloro derivatives. G by selective bromination gives H. Find A to H.

Answer

A is $(CH_3)_2CHCH=CH_2$;
B is $(CH_3)_2CHCH_2CH_2OH$;
C is $(CH_3)_2CHCH_2CH_2Br$;
D is $(CH_3)_2CHCH_2CH_2MgBr$;
E is $(CH_3)_2CHCH_2CH_2COOH$;
F is $(CH_3)_2CHCH_2CHO$;
G is $(CH_3)_2CHCH_2CH_3$;
H is $(CH_3)_2CBr\,CH_2CH_3$.

5. A gives light yellow precipitate with $AgNO_3$ and with KCN gives B. B on hydrolysis gives C. C with Cl_2/P gives D. The silver salt of C with Br_2 gives. A. A with Mg and ether gives E. E with CH_3CHO and H^+ gives $C_6H_5CH_2CHOHCH_3$(F). F on oxidation gives G. F and G both answer haloform reaction. The haloform reaction of F gives CHI_3 and salt of acid C. Find A to G.

Answer

A is $C_6H_5CH_2Br$;
B is $C_6H_5CH_2CN$;
C is $C_6H_5CH_2COOH$;

D is $C_6H_5CHClCOOH$;
E is $C_6H_5CH_2MgBr$;
F is $C_6H_5CH_2CHOHCH_3$;
G is $C_6H_5CH_2COCH_3$.

6. Hydrocarbon A gives two monochloro derivatives (B and C) with the % ratio 28 : 72. B is obtained from D by $ZnCl_2$ + HCl. B with CN^- gives E. E on hydrolysis gives F. F is obtained by ozonolysis of alkene G in oxidizing conditions along with HCOOH. G is obtained by the partial reduction of H. H on hydration in Hg^{2+} / H_2SO_4 gives $CH_3CH_2CH_2CH_2COCH_3$. Find A to H.

Answer

A is $CH_3CH_2CH_2CH_3$;
B is $CH_3CH_2CH_2CH_2Cl$;
C is $CH_3CHClCH_2CH_3$;
D is $CH_3CH_2CH_2CH_2OH$;
E is $CH_3CH_2CH_2CH_2CN$;
F is $CH_3CH_2CH_2CH_2COOH$;
G is $CH_3CH_2CH_2CH_2CH{=}CH_2$;
H is $CH_3CH_2CH_2CH_2C{\equiv}CH$;

7. One of the 4 isomers (A to D) of acid with formula $C_5H_{10}O_2$ is optically active. A can be obtained by oxidation of aldehyde E. E by Clemmenson reduction gives F. Can F show optical activity? Give the structures of A, B, C and D and indicate the method of their preparation. Which of these acids undergo HVZ reaction. Find the terminal alkenes which can produce acid A to D by ozonolysis in oxidizing conditions.

Answer

A is $CH_3CH_2CH(CH_3)—COOH$

B is $CH_3CH_2CH_2CH_2COOH$

C is $CH_3—CH(CH_3)—CH_2COOH$

D is $(CH_3)_3—C—COOH$

E is $CH_3CH_2CH(CH_3)CHO$

F is $CH_3CH_2\overset{\displaystyle CH_3}{\overset{|}{C}}H—CH_3$

F cannot show optical activity since two identical groups are there around the third carbon.

A is produced from the corresponding aldehyde E by oxidation.

B can be obtained by Grignard reaction of $CH_3CH_2CH_2CH_2Cl$ with Mg, ether and carboxylation

C can be obtained from $CH_3\overset{\displaystyle CH}{\overset{|}{C}}H—CH_2—MgBr$ on carboxylation. Also nitrile method by hydrolysis is acceptable.

D is obtained from oxidation of $(CH_3)_3CCHO$.

A can be obtained from ozonolysis of $CH_3CH_2\overset{\displaystyle CH_3}{\overset{|}{C}}H—CH{=}CH_2$.

B can be obtained by ozonolysis of $CH_3CH_2CH_2CH_2CH{=}CH_2$.

C is obtained by ozonolysis of $CH_3\overset{\displaystyle CH_3}{\overset{|}{C}}H—CH_2—CH{=}CH_2$.

D is obtained by ozonolysis of $(CH_3)_3CCH{=}CH_2$.

HVZ reaction is not possible for D since it has no alpha hydrogen.

11

AMINES AND DIAZONIUM COMPOUNDS

AMINES

NOMENCLATURE AND STRUCTURE OF AMINES

Amines are derivatives of ammonia in which one or more of the hydrogens has been replaced by an alkyl or aryl group. The nomenclature of amines is complicated by the fact that several different nomenclature systems exist, and there is no clear preference for one over the others. Furthermore, the terms primary (1°), secondary (2°) and tertiary (3°) are used to classify amines in a completely different manner than they were used for alcohols or alkyl halides. When these terms are used for amines they refer to the number of alkyl (or aryl) substituents bonded to the nitrogen atom, whereas in other cases they refer to the nature of an alkyl group. The four compounds shown in Figure 11.1 are all $C_4H_{11}N$ isomers.

1. $CH_3CH_2CH_2CH_2NH_2$
2. $(CH_3)_3—C—NH_2$
3. $CH_3—NH—CH_2CH_2CH_3$
4. $CH_3—N(CH_3)—C_2H_5$

Figure 11.1 $C_4H_{11}N$ isomers

The first two are classified as 1°-amines, since only one alkyl group is bonded to the nitrogen; however, the alkyl group is primary in the first example and tertiary in the second. The third and fourth compounds in the row are 2° and 3°-amines respectively. A nitrogen bonded to four alkyl groups will necessarily be positively charged, and is called a quaternary ammonium cation. For example, $(CH_3)_4N^{(+)}Br^{(-)}$ is tetramethylammonium bromide. More examples are presented in Figure 11.2.

In Amines, nitrogen is sp^3-hybridized. Three different groups attached to nitrogen lead to optical activity. Unlike chiral hydrocarbons, chiral amine enantiomers cannot be resolved since they interconvert by inversion.

Methyl amine—CH_3NH_2

Ethyl amine—$C_2H_5NH_2$

Diphenyl amine—$C_6H_5NHC_6H_5$

Figure 11.2 Nomenclature of amines

If a group is attached to NH_2, it is primary. If groups are attached to NH the amine is secondary and if three groups are attached to nitrogen, it is a tertiary amine (Figure 11.3).

$(CH_3)_3C—NH_2$ is a primary amine

$(CH_3)_2NH$ is a secondary amine

$(CH_3)_3N$ is a tertiary amine

Figure 11.3 Primary, secondary and tertiary amines

PROPERTIES OF AMINES

1. Boiling Point and Water Solubility

It is instructive to compare the boiling points and water solubility of amines with those of corresponding alcohols and ethers. The dominant factor here is hydrogen bonding, and Table 11.1 demonstrates the powerful intermolecular attraction that results from —O—H---O— hydrogen-bonding in alcohols. Corresponding —N—H---N—hydrogen bonding is weaker, as the lower boiling points of similarly sized amines demonstrate. Alkanes provide reference compounds in which hydrogen-bonding is not possible, and the increase in boiling point for equivalent 1°-amines is roughly half the increase observed for equivalent alcohols.

Table 11.1 Boiling points of alkanes, alcohols and amines

Compound	CH_3CH_3	CH_3OH	CH_3NH_2	$CH_3CH_2CH_3$	CH_3CH_2OH	$CH_3CH_2NH_2$
Mol.Wt.	30	32	31	44	46	45
Boiling Point °C	–88.6°	65°	–6.0°	–42°	78.5°	16.6°

Table 11.2 illustrates differences associated with isomeric 1°, 2° & 3°-amines, as well as the influence of chain branching. Since 1°-amines have two hydrogens available for hydrogen-bonding, we expect them to have higher boiling points than isomeric 2°-amines, which in turn should boil higher than isomeric 3°-amines (no hydrogen bonding). Indeed, 3°-amines have boiling points similar to equivalent sized ethers; and in all but the smallest compounds, corresponding ethers, 3°-amines and alkanes have similar boiling points. In the examples shown here, it is further revealed that chain branching reduces boiling points by 10 to 15°C.

Table 11.2 Boiling points of isomeric 1°, 2° and 3°-amines

Compound	**Molecular weight**	**Boiling point °C**
$CH_3(CH_2)_2CH_3$	58	-0.5°
$CH_3(CH_2)_2OH$	60	97°
$CH_3(CH_2)_2NH_2$	59	48°
$CH_3CH_2NHCH_3$	59	37°
$(CH_3)_3CH$	58	-12°
$(CH_3)_2CHOH$	60	82°
$(CH_3)_2CHNH_2$	59	34°
$(CH_3)_3N$	59	3°

The water solubility of 1° and 2°-amines is similar to that of comparable alcohols. As expected, the water solubility of 3°-amines and ethers is also similar. These comparisons, however, are valid only for pure compounds in neutral water. The basicity of amines (next section) allows them to be dissolved in dilute mineral acid solutions, and this property facilitates their separation from neutral compounds such as alcohols and hydrocarbons by partitioning between the phases of non-miscible solvents.

2. Basicity of Amines

Like ammonia, most amines are Brønsted and Lewis bases, but their base strength can be changed enormously by substituents. It is common to compare basicities quantitatively by using the pK_a's of their conjugate acids rather than their pK_b's. Since $pK_a + pK_b = 14$, the higher the pK_a the stronger the base, in contrast to the usual inverse relationship of pK_a with acidity. Most simple alkyl amines have pK_a's in the range 9.5 to 11.0, and their water solutions are basic (have a pH of 11 to 12, depending on concentration). Aliphatic amines are stronger than aromatic amines since delocalization of the lone pair into the benzene ring leads to nonavailability of lone pair for protonation. If we consider pyridine and pyrrole, pyridine is basic since lone pair is not used for satisfying Hückel rule, but the lone pair on nitrogen is used to get sextet in pyrrole and hence it is nonbasic.

Figure 11.4 shows resonance structures of *p*-nitroaniline. This electron pair delocalization is accompanied by a degree of rehybridization of the amino nitrogen atom, but the electron pair delocalization is probably the major factor in the reduced basicity of these compounds.

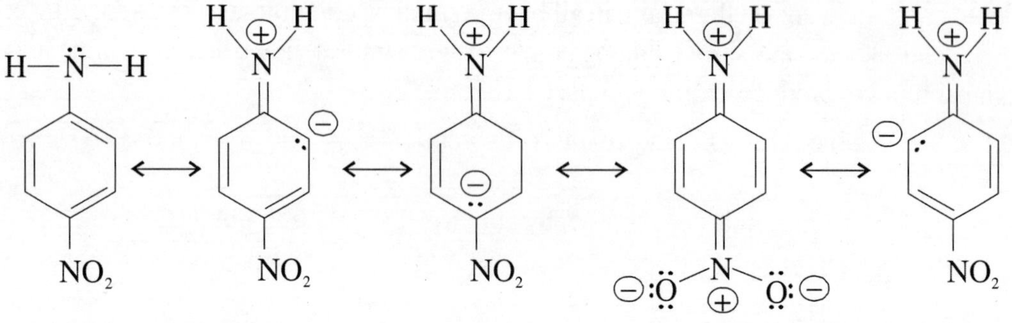

Reduced basicity of *para*-nitroaniline due to electron pair delocalization

Figure 11.4 Resonance structures of *p*-nitroaniline

More examples of basicity of amines The lone pair of electrons in amines are responsible for their basicity and nucleophilicity. In gaseous phase tertiary amines have highest basicity (aliphatic) compared to secondary and primary amines. In solution phase secondary amines have highest basicity. Compared to aliphatic amines, aromatic amines like aniline are less basic since lone pair is used for resonance in benzene ring. Cyclohexylamine is many times more basic than aniline. Among pyridine and pyrrole, pyridine is more basic since lone pair is not used for aromatic sextet but in pyrrole the lone pair is used for sextet of electrons (Hückel rule) and it is not basic.

If we compare *p*-methoxyaniline, *p*-nitroaniline and aniline, *p*-methoxyaniline is more basic relative to aniline since the electron-releasing group stabilizes anilinium ion. The *p*-NO_2 group is electron-withdrawing and reduces basicity. Guanidine is strongly basic and it has structure $HN{=}C(NH_2)_2$. On protonation the positive charge can be delocalized as shown in Figure 11.5.

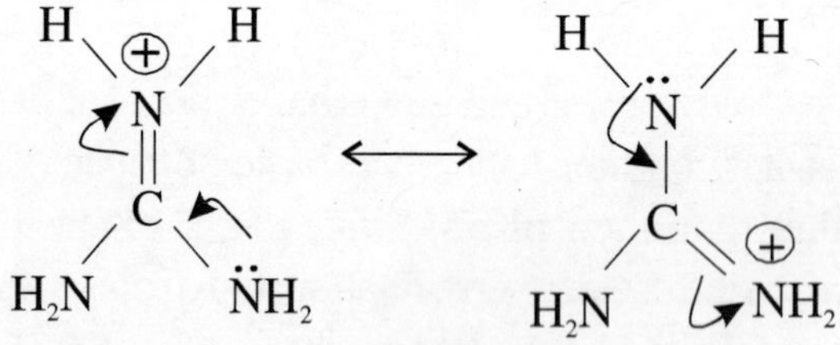

Figure 11.5 Basicity of guanidine

Imidazole is more basic than pyridine since in protonated form it has two nitrogens to stabilize positive charge but pyridine has one nitrogen only. The protonated resonance forms of imidazole are given in Figure 11.6

Figure 11.6 Resonance forms of Imidazole

3. Acidity of Amines

When we study amines the first thing which comes to our mind is the basicity of amines but it must be remembered that 1° and 2°-amines are also very weak acids (ammonia has a $pK_a = 34$). In this respect it should be noted that pK_a is being used as a measure of the acidity of the amine itself rather than its conjugate acid, as in the previous section. For ammonia this is expressed by the following hypothetical equation:

$$NH_3 + H_2O \longrightarrow NH_2^- + H_3O^+$$

The same factors that decreased the basicity of amines increase their acidity. This is illustrated by the examples given in Table 11.3, which are shown in the order of increasing acidity. The first compound is a typical 2°-amine, and the three next to it are characterized by varying degrees of nitrogen electron pair delocalization. The last two compounds show the influence of adjacent sulphonyl and carbonyl groups on N—H acidity. From previous discussion it should be clear that the basicity of these nitrogens is correspondingly reduced.

Table 11.3 Acidity of cyclic nitrogen compounds

Compound	N—H (piperidine)	NH_2 (aniline)	O_2N—C_6H_4—NH_2	N—H (pyrrole)	$C_6H_5SO_2NH_2$	N—H (succinimide, O / O)
pK_a	33	27	19	15	10	9.6

The acids shown here may be converted to their conjugate bases by reaction with bases derived from weaker acids (stronger bases). Three examples of such reactions are shown below, with the acidic hydrogen shown in bold in each case. For complete conversion to the conjugate base, as shown, a reagent base roughly a million times stronger is required.

$$C_6H_5SO_2\mathbf{NH}_2 + KOH \longrightarrow C_6H_5SO_2NH^{(-)}K^{(+)} + H_2O$$ a sulphonamide base

$$(CH_3)_3CO\mathbf{H} + NaH \longrightarrow (CH_3)_3CO^{(-)}Na^{(+)} + H_2$$ an alkoxide base

$$(C_6H_5)_2N\mathbf{H} + C_4H_9Li \longrightarrow (C_6H_5)_2N^{(-)}Li^{(+)} + C_4H_{10}$$ an amide base

Pyridine is commonly used as an acid scavenger in reactions that produce mineral acid co-products. For example in the reaction of alcohols with $SOCl_2$ we get HCl as a product in addition to RCl and SO_2. HCl is removed by pyridine and it acts as the scavenger.

The basicity and nucleophilicity of pyridine may be modified by steric hindrance, as in the case of 2,6-dimethylpyridine (pK_a=6.7), or resonance stabilization, as in the case of 4-dimethylaminopyridine (pK_a=9.7). The alkoxides are stronger bases that are often used in the corresponding alcohol as solvent, or for greater reactivity in DMSO.

One of the most important tests for distinguishing primary, secondary and tertiary amines is the Hinsberg test.

DISTINCTION OF 1°, 2°, 3°-AMINES

1. *Reaction with benzenesulphonyl chloride (The Hinsberg test)* An electrophilic reagent, benzenesulphonyl chloride, reacts with amines and this reaction provides a useful test for distinguishing primary, secondary and tertiary amines. As shown in Figure 11.7, 1° and 2°-amines react to give sulphonamide derivatives with loss of HCl, whereas 3°-amines do not give any isolable products other than the starting amine. In the latter case a quaternary "onium" salt may be formed as an intermediate, but this rapidly breaks down in water to liberate the original 3°-amine.

The Hinsberg test is conducted in aqueous base (NaOH or KOH), and the benzenesulphonyl chloride reagent is present as an insoluble oil. Because of the heterogeneous nature of this system, the rate at which the sulphonyl chloride reagent is hydrolysed to its sulphonate salt in the absence of amines is relatively slow. The amine dissolves in the reagent phase, and immediately reacts (if it is 1° or 2°), with the resulting HCl being neutralized by the base. The sulphonamide derivative from 2°-amines is usually an insoluble solid. However, the sulphonamide derivative from 1°-amines is acidic and dissolves in the aqueous base. Acidification of this solution then precipitates the sulphonamide of the 1°-amine.

Another method of distinction of the three types of amines is in their reaction with HNO_2.

2. *Reaction of Amines with Nitrous Acid* Nitrous acid (HNO_2 or HONO) reacts with aliphatic amines in a fashion that provides a useful test for distinguishing primary, secondary and tertiary amines.

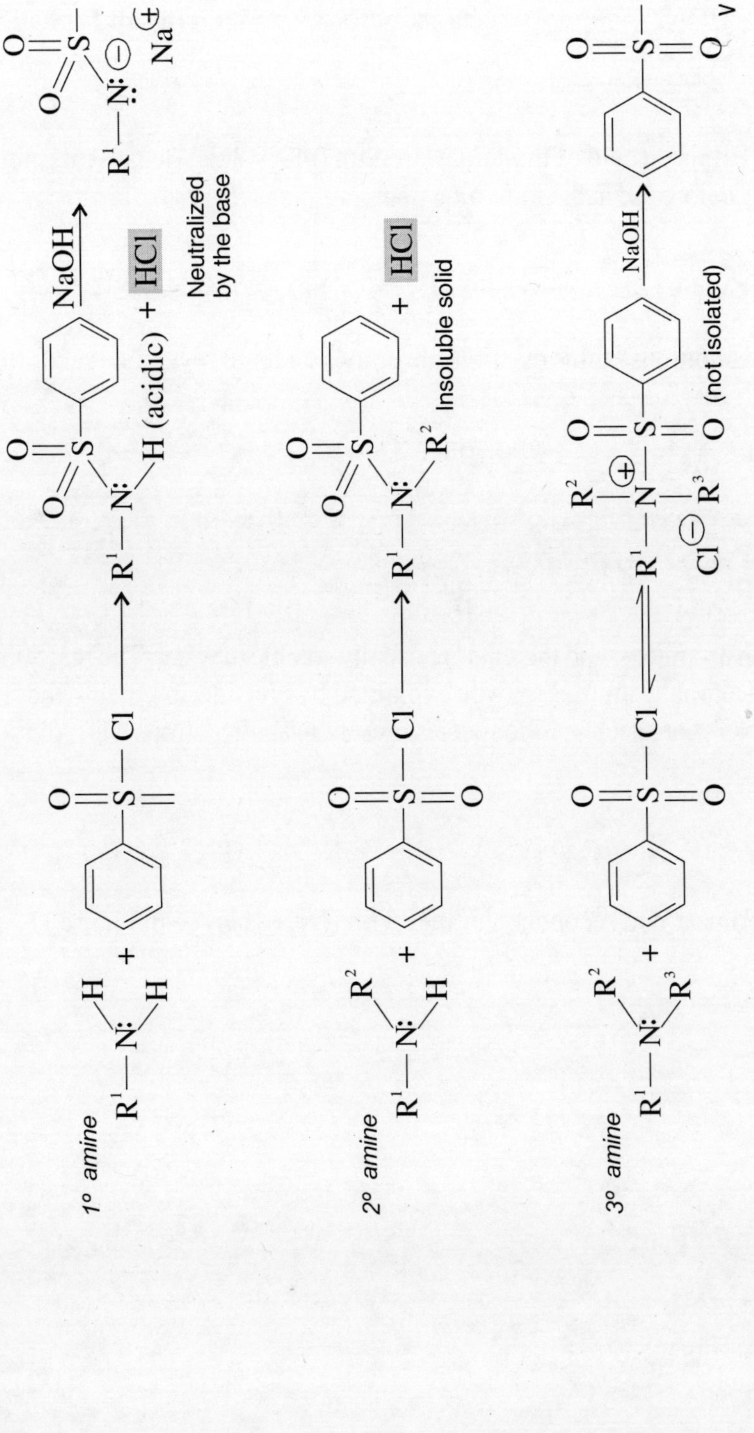

1º amine

$$R^1\text{—}\ddot{N}H_2 + C_6H_5SO_2Cl \longrightarrow R^1\text{—}\ddot{N}H\text{—}SO_2C_6H_5 \text{ (acidic)} + HCl \xrightarrow{NaOH} R^1\text{—}\ddot{N}^{\ominus}\text{—}SO_2C_6H_5\ Na^{\oplus}$$

HCl: Neutralized by the base

2º amine

$$R^1R^2\ddot{N}H + C_6H_5SO_2Cl \longrightarrow R^1R^2\ddot{N}\text{—}SO_2C_6H_5 + HCl$$

Insoluble solid

3º amine

$$R^1R^2R^3\ddot{N} + C_6H_5SO_2Cl \rightleftharpoons R^1R^2R^3N^{\oplus}\text{—}SO_2C_6H_5\ Cl^{\ominus} \xrightarrow{NaOH} C_6H_5SO_2\text{—}$$

(not isolated)

Figure 11.7 Hinsberg test for amines—the reactions

$$1^\circ\text{-Amines} + \text{HONO} \xrightarrow[\text{(Cold acidic solution)}]{} \text{Nitrogen gas evolution from a clear solution}$$

$$2^\circ\text{-Amines} + \text{HONO} \xrightarrow[\text{(Cold acidic solution)}]{} \text{An insoluble oil (N-nitrosoamine)}$$

$$3^\circ\text{-Amines} + \text{HONO} \xrightarrow[\text{(Cold acidic solution)}]{} \text{A clear solution (Ammonium salt formation)}$$

Nitrous acid is a Brønsted acid of moderate strength ($pK_a = 3.3$). Because it is unstable, it is prepared immediately before use in the following manner:

$$NaNO_2 + H_2SO_4 \xrightarrow{H_2O} \text{H}-\ddot{\underset{\cdot\cdot}{\text{O}}}-\ddot{\text{N}}=\ddot{\underset{\cdot\cdot}{\text{O}}} + NaHSO_4$$

Under the acidic conditions of this reaction, all amines undergo reversible salt formation

$$R^2-\underset{R^3}{\overset{R^1}{\underset{|}{\overset{|}{\text{N}}}}}\!: + HX \rightleftharpoons R^2-\underset{R^3}{\overset{R^1}{\underset{|}{\overset{|}{\text{N}}}}}\overset{\oplus}{-}H\ X^{\ominus} \quad \{X = HSO_4 \text{ or } NO_2\}$$

This happens with 3°-amines, and the salts are usually soluble in water. The reactions of nitrous acid with 1°- and 2°- aliphatic amines may be explained by considering their behavior with the nitrosonium cation, $NO^{(+)}$, an electrophilic species present in acidic nitrous acid solutions.

$$\text{H}-\ddot{\underset{\cdot\cdot}{\text{O}}}-\ddot{\text{N}}=\ddot{\underset{\cdot\cdot}{\text{O}}} + H_2SO_4 \longrightarrow \overset{\oplus}{\ddot{\text{N}}}=\ddot{\underset{\cdot\cdot}{\text{O}}}\ HSO_4^{\ominus} + H_2O$$

The reaction of primary and secondary amines with HNO_2 is given in Figure 11.8

Primary amines

$$R-\ddot{N}H_2 \xrightarrow[\oplus N=O]{HNO_2, 0^\circ} \left[R-\overset{H}{\underset{H}{\overset{|}{\underset{|}{N}}}}{}^{\oplus}-\ddot{N}=O\right] X^{\ominus} \xrightarrow{-H^{\oplus}} \left[\begin{matrix}R \\ H\end{matrix}\!\!>\!:N-\ddot{N}=O\right]$$

$$\Updownarrow \text{ tautomerism}$$

$$\left[R-\ddot{N}=\ddot{N}-OH\right]$$

$$\Updownarrow H^{\oplus}$$

$$\left.\begin{matrix}\text{Alcohols} \\ \text{and} \\ \text{Alkenes}\end{matrix}\right\} \xleftarrow{H_2O} [R\oplus] \xleftarrow{-N_2} [R-\overset{\oplus}{N}\equiv N:]X^{\ominus} \xrightleftharpoons{-H_2O} \left[R-\ddot{N}=\ddot{N}-\overset{\oplus}{O}H_2\right]$$

a diazonium cation

Secondary amines

$$\begin{matrix}R \\ R\end{matrix}\!\!>\!:N-H \xrightarrow[\oplus N=O]{HNO_2, 0^\circ} \left[R-\overset{H}{\underset{R}{\overset{|}{\underset{|}{N}}}}{}^{\oplus}-\ddot{N}=O\right] X^{\ominus} \xrightarrow{-HX} \begin{matrix}R \\ R\end{matrix}\!\!>\!:N-\ddot{N}=O$$

A N-nitrosoamine
(a yellow oily liquid)

Figure 11.8 Reaction of amines with nitrous acid

PREPARATION OF AMINES

Preparation of 1°-Amines

Method 1

$NaN_3 + CH_3CH_2CHBrCH_3 \rightarrow CH_3CH_2CHN_3CH_3 \xrightarrow{(H_2/Pd\ catalyst)\ or\ LiAPH_4} CH_3CH_2CHNH_2CH_3$

Method 2—Hydrolysis of isocyanides Isocyanides hydrolyse in water to give amines along with the carboxylic acid.

$$RNC + 2H_2O \longrightarrow RNH_2 + HCOOH$$

Method 3—Alkylation of ammonia This is an S_N2 reaction but further alkylation happens to give a mixture of primary, secondary and tertiary amines, and a quaternary salt.

Method 4—Gabriel phthalimide synthesis $C_6H_5CH_2Br + C_6H_4(CO)_2NK$ in DMF followed by addition of OH^- and water gives primary amines. $C_6H_5CH_2NH_2$ is the product.

Method 5—Reduction of nitriles

$$RBr \longrightarrow RCN \longrightarrow LiAlH_4(ether) + water \longrightarrow RCH_2NH_2$$

Method 6—Reductive amination Aldehydes or ketones by reductive amination with ammonia give amines.

$$C_6H_5CH_2COCH_3 \rightarrow NH_3, H_2/Ni \rightarrow C_6H_5CH_2\underset{\displaystyle CH_3}{\underset{|}{C}}HNH_2$$

Primary and secondary amines can also be used. The reducing agent used is $NaBH_4CN$ in methanol.

Preparation of Secondary Amines

$$\text{Aniline} + RX \longrightarrow C_6H_5NR_2 + HNO_2 \longrightarrow P\text{-nitroso}C_6H_5NR_2 + NaOH \downarrow p\text{-nitrosophenol} + R_2NH$$

$$CaNCN + NaOH \longrightarrow Na_2NCN \longrightarrow +2RX \longrightarrow R_2NCN \xrightarrow{\text{(acid or alkali)}} R_2NH + CO_2 + NH_3.$$

Preparation of Tertiary Amines

Reductive alkylation of a carbonyl compound in excess in the presence of a secondary amine, gives a tertiary amines.

$$R_2NH + R'CHO + H_2 \xrightarrow{Ni} \begin{matrix} R \\ & N - CH_2R' + H_2O \\ R \end{matrix}$$

REACTIONS OF AMINES

1. All amines combine with acids to form salts.

$$RNH_2 + HCl \rightarrow RNH_3 + Cl^-$$

2. Amines react with alkyl halides to form alkyl substituted ammonium halides.
3. If amine salt is heated at high temperature RX and NH_4X result.

$$R_2NH^+Cl^- \longrightarrow RCl + R_2NH \xrightarrow{HCl} RCl + RNH_2 \xrightarrow{HCl} RCl + NH_4Cl$$

4. Herzig–Meyer method of estimation of N-methyl groups in amines

$$ArNHCH_3 + HI \xrightarrow{423\text{ K}} ArNH_2 + CH_3I$$

5. Primary amines react with acetic anhydride to give CH_3CONHR.

$$RNH_2 + (CH_3CO)_2O \longrightarrow CH_3CONHR + CH_3COOH.$$

6. With acetyl chloride, R_2NH forms monoacyl derivative.

$$R_2NH + CH_3COCl \longrightarrow CH_3CONR_2 + HCl$$

7. $RNH_2 + Br_2 \xrightarrow{(aq.Na_2CO_3)} RNHBr + RNBr_2$.

8. The most important reaction of primary amines is the reaction with $CHCl_3$ and KOH to give isocyanide.

9. $RNH_2 + CHCl_3 + 3KOH \longrightarrow RNC + 3KCl + 3H_2O$.

If the amine used is aniline, phenylisocyanide is formed which has a very bad odour and this is a good test for primary amines. The reaction is known as carbylamine reaction.

10. Most important reaction of tertiary amines is the formation of amine oxide when oxidized with Caros acid or a peroxy acid or H_2O_2. Tertiary amine oxides of the type $RCH_2CH_2\overset{\overset{\displaystyle O}{|}}{\underset{\underset{\displaystyle O}{|}}{N}}—CH_3$ on heating give alkenes and the reaction is called COPE elimination.

$$RCH_2CH_2—\overset{\overset{\displaystyle O}{|}}{\underset{\underset{\displaystyle CH_3}{|}}{N}}—CH_3 \xrightarrow{150°C} RCH{=}CH_2 + (CH_3)_2NOH$$

11. The most important reaction of quatenary ammonium salts of amines which contain beta hydrogen is Hoffmann elimination.

Hoffmann elimination Amines are treated with CH_3I till we get the quaternary iodide. This is treated with Ag_2O to get the quaternary hydroxide where OH is more nucleophilic than I^-. The quaternary hydroxide on heating gives less substituted alkene (Hoffmann rule).

When secondary butylethyl ammonium hydroxide is heated in the presence of trimethyl amine, we get 95% of 1-butene and 5% of 2-butene.

The steric factor is responsible for E2 elimination leading to less substituted alkene. This aspect is shown in Figure 11.9.

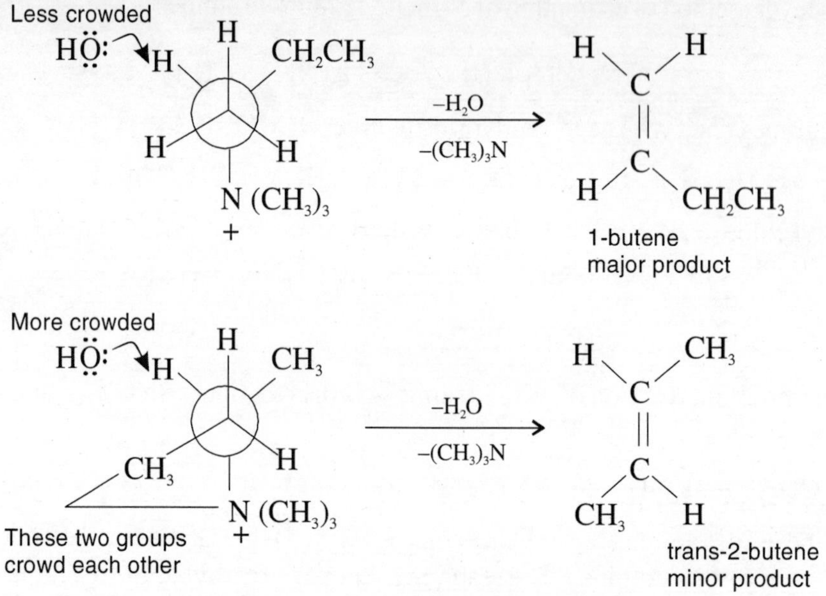

Figure 11.9 Hoffmann elimination

We have already seen in the chapter on alkenes that the more substituted alkene results in high % if alkyl bromides are treated with $NaOCH_3$. The Hofmann elimination is not possible for $(CH_3)_4N^+OH^-$ since it has no beta hydrogen.

DIAZONIUM SALTS

Primary aromatic amines react with nitrous acid generated from $NaNO_2$ and HCl or H_2SO_4 at (5ºC or less) to get stable aromatic diazonium salts.

$$ArNH_2 + HNO_2 + H_2SO_4 \longrightarrow Ar{-}N^+{=}NHSO_4^- + H_2O$$

Nitrous acid reactions of 1°-aryl amines generate relatively stable diazonium species that serve as intermediates for a variety of aromatic substitution reactions. Diazonium cations may be described by resonance contributors, as in Figure 11.10. The left-hand contributor is dominant because it has greater π bonding. Loss of nitrogen is slower than in aliphatic 1°-amines because the C—N bond is stronger, and aryl carbocations are comparatively unstable.

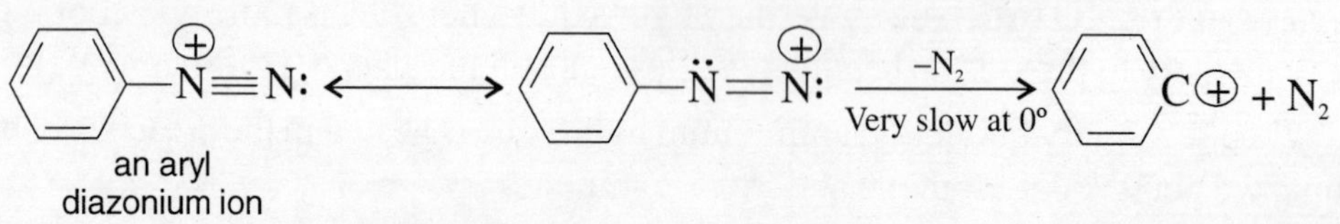

Figure 11.10 Aryldiazonium ion

Aqueous solutions of these diazonium ions have sufficient stability at 0° to 10°C that they may be used as intermediates in a variety of nucleophilic substitution reactions. For example, if water is the only nucleophile available for reaction, phenols are formed in good yield.

$$ArN_2 + HSO_4^- + H_2O \longrightarrow ArOH + N_2 + H_2SO_4$$

The diazonium salt should be used with H_2SO_4, and HCl should not be used because Cl^- competes with OH^-.

Alkyl amines also react with nitrous acid but are unstable and lose nitrogen to give carbocations. For aromatic amines loss of nitrogen to give aryl cation does not take place, since these cations are not stable.

For example 1,1-dimethylpropylamine on treatment with nitrous acid produces carbocations and we get 2-methyl-2-butene, 2-methyl-1-butene and 2-methyl-2-butanol as products. The details are shown in Figure 11.11.

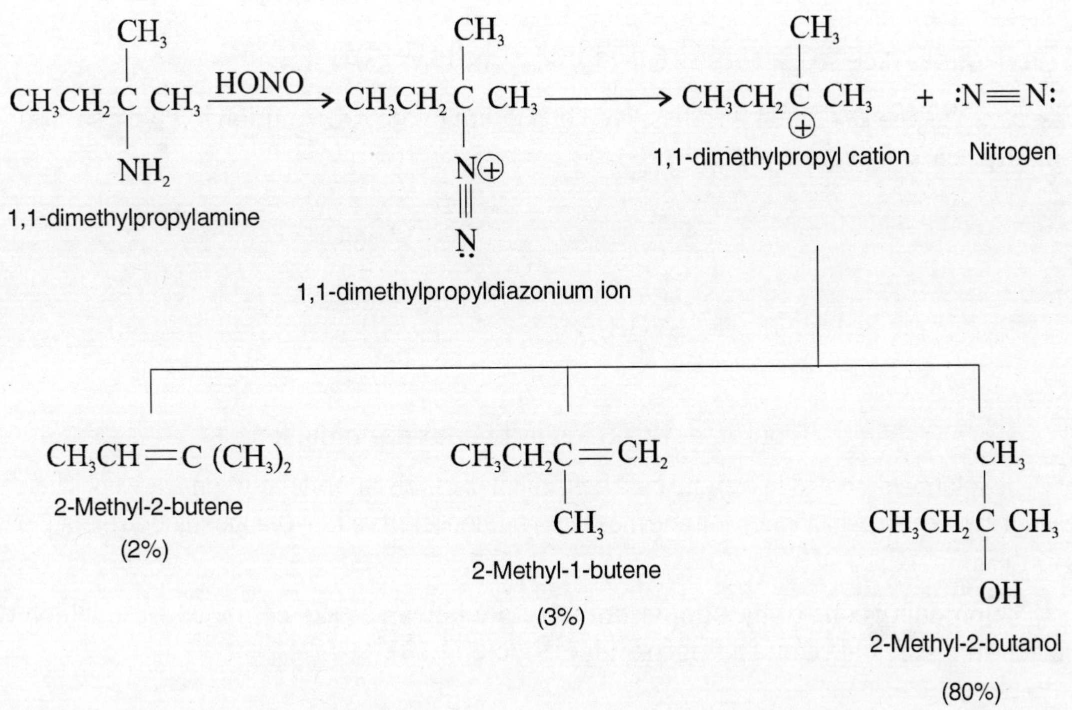

Figure 11.11 Decomposition of aliphatic diazonium compounds

As we had already stated, the aryldiazonium salts are of much synthetic use. This is illustrated in the following reactions.

SYNTHETIC USES OF AROMATIC DIAZONIUM SALTS

1. Sandmayer reaction

   $ArN_2 + HSO_4^- + CuCl(CuBr$ or $CuCN)$ in acid gives ArCl or ArBr or ArCN.
2. The diazonium salt with KI or NaI gives iodobenzene.
3. The diazonium salt on heating with H_3PO_2 gives benzene.
4. When treated with water, phenol is produced.

   Use of diazonium salts in synthesis.

   Let us prepare 3,5-dibromotoluene.

   Toluene + ($HNO_3 + H_2SO_4$) $\longrightarrow$ *p*-nitrotoluene and *o*-nitrotoluene (Step1)

   p-nitrotoluene + Sn/HCl $\longrightarrow$ *p*-aminotoluene + $2Br_2$ $\longrightarrow$ (Step2)

   p-aminotoluene + $2Br_2$ $\longrightarrow$ 3 bromo5bromo-4aminotoluene (Step3)

   3-bromo-5-bromo-4-aminotoluene + HNO_2 $\longrightarrow$ diazonium salt (Step4)

   diazonium salt + H_3PO_2 $\longrightarrow$ 3,5-dibromotoluene (Step5)

COUPLING REACTIONS OF DIAZONIUM SALTS

Reaction with phenol Phenol is dissolved in sodium hydroxide solution to give a solution of sodium phenoxide (Figure 11.12).

$$C_6H_5OH + NaOH \longrightarrow C_6H_5O^-Na^+ + H_2O$$

Sodium phenoxide

Figure 11.12 Sodium phenoxide formation

The solution is cooled in ice, and cold benzenediazonium chloride solution is added. There is a reaction between the diazonium ion and the phenoxide ion and a yellow-orange solution or precipitate is formed.

The product is one of the simplest of what are known as azo compounds, in which two benzene rings are linked by a nitrogen bridge (Figure 11.13).

$$C_6H_5-\overset{+}{N}{\equiv}N + C_6H_5-O^- \longrightarrow C_6H_5-N{=}N-C_6H_4-OH$$

Figure 11.13 Azo dye with phenol

Reaction with naphthalen-2-ol Naphthalen-2-ol is also known as 2-naphthol or beta-naphthol. It contains an —OH group attached to a naphthalene molecule rather than to a simple benzene ring. Naphthalene has two benzene rings fused together.

The reaction is carried out under exactly the same conditions as with phenol. The naphthalen-2-ol is dissolved in sodium hydroxide solution to produce an ion just like the phenol one. This solution is cooled and mixed with the benzenediazonium chloride solution.

An intense orange-red precipitate is formed—another azo compound (Figure 11.14).

Figure 11.14 Azo dye with naphthalen-2-ol

Reaction with phenylamine (aniline) Some liquid phenylamine is added to a cold solution of benzenediazonium chloride, and the mixture is shaken vigorously. A yellow solid is produced (Figure 11.15).

Figure 11.15 Azo dye with aniline

These strongly coloured azo compounds are frequently used as dyes known as azo dyes. The one made from phenylamine (aniline) is known as "aniline yellow." Azo compounds account for more than half of modern dyes.

PROBLEMS

I. "Explain Why" Questions and Small Structure-Based Problems

1. Why is aniline less basic than CH_3NH_2?

Answer

In aniline the lone pair of electrons on nitrogen are delocalized into the benzene ring and they are not available for protonation. Hence aniline has lower basicity.

2. Pyridine is basic but pyrrole is not basic. Why?

Answer

In pyridine, the lone pair on nitrogen is free for protonation but in pyrrole the lone pair along with 4 more pi electrons gives the aromatic sextet and hence pyrrole is not basic.

3. Tetrahydroquinoline is less basic compared to tetrahydroisoquinoline. Why?

Answer

In tetrahydroquinoline the lone pair on nitrogen is next to sp^2 carbon and gets delocalized into the benzene ring, but in case of tetrahydroisoquinoline the lone pair is next to sp^3 carbon and delocalization is not possible. The lone pair is available for protonation and hence it is much more basic than the tetrahydroquinoline.

4. 1-chloro-2-methyl aziridine has 4 isomers but only two boiling fractions. Why?

Answer

The *cis* and *trans* isomers are both optically active giving rise to 4 isomers, but optically active stereoisomers have the same boiling point and hence only two boiling fractions are there.

5. Azacyclopropane or aziridine is less basic than piperidine. Why?

Answer

This is due to internal strain in the 3-membered ring.

6. Cyclohexylmethylamine is more basic than benzyl amine. Why?

Answer

The cyclohexyl carbon is sp^3 but the carbon in phenyl attached to CH_2NH_2 group is sp^2.

7. $(CH_3)_3N^+CH_2CH_2NH_2$ is less basic than $(CH_3)_3CCH_2CH_2NH_2$. Why?

Answer

The lone pair of electrons of nitrogen in NH_2 are pulled by the positive charge on nitrogen attached to the three methyl groups. Also the charge is transmitted through space. Field and inductive effects operate.

8. N-nitrosoamines, $RNHN{=}O$, are insoluble in HCl. Why?

Answer

The electrons of N—O bond are delocalized leading to less basicity and hence they are insoluble in HCl.

9. Aryldiazonium ion is more stable than aliphatic diazonium ion. Why?

Answer

Electron release is there from *ortho* and *para* positions in the aryl diazonium ion.

10. Diazocoupling is prevented in strongly acidic conditions. Why?

Answer

Strong acid converts $ArNH_2$ to ArN_3^+ and the ring gets deactivated to diazocoupling.

II. Multiple Choice Questions (One out of four answers is correct)

1. Arrange in the increasing order of basicity of the following

   I. $C_6H_5NH_2$ II. cyclohexylamine III. pyrrole

   a) $I<II<III$ b) $III<II<I$ c) $III<I<II$ d) $III=II<I$

2. Hoffmann elimination is not possible for

   a) $(C_2H_5)_3N^+CH_3OH^-$ b) $(C_6H_5)_3N^+CH_3OH^-$

   c) $(C_2H_5)_3N^+C_3H_7OH^-$ d) $(C_2H_5)_3N^+C_4H_9OH^-$

3. One of the following methods does not give a primary amine

   a) $RCN + LiAlH_4 \longrightarrow$

   b) $R_2C = NOH + H_2/Rh/C_2H_5OH \longrightarrow$

   c) $RCONH_2 + NaH, CH_3I$ followed by addition of $LAlH_4, H_2O \longrightarrow$

   d) $RMgX + NH_2Cl \longrightarrow$

4. The reagent not used in reduction of compounds to amines is

   a) Ni/H_2 b) $LiAlH_4, H_2O$ c) $NaBH_3CN$ d) Li/NH_3

5. Phthalimide synthesis cannot be done with

   a) $C_6H_5CH_2Br$ b) $C_6H_5NHCH_3$ c) $(CH_3)_3CBr$ d) CH_3Br

6. The compound which is not optically active is

   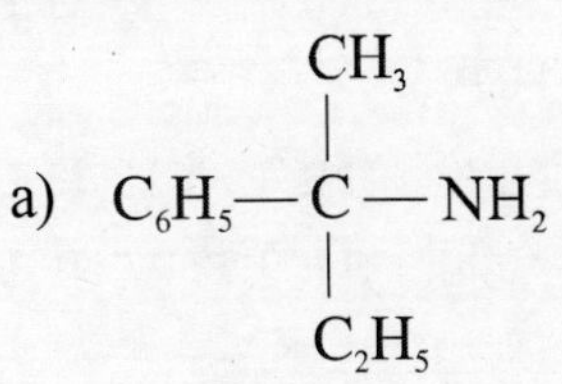

   b) $C_6H_5—N(\to O)(C_2H_5)—CH_3$ (O above N, C_2H_5 below N)

   c) 1-chloro-2-methylaziridine d) $(CH_3)_3N$

7. Aziridine is much less basic than piperidine because

   a) It has internal ring strain b) piperidine has 6-membered ring

   c) piperidine has no nitrogen lone pair d) piperidine is aromatic

8. One of the following has no stereoisomers

   a) 1-chloro-2-methylaziridine b) azobenzene

   c) $(CH_3)_3N$ d) $R_1R_2R_3R_4N^+Cl^-$

9. In aqueous solution secondary amines are much more basic than tertiary amines since
   a) tertiary amine is bulky
   b) hydrogen bonding is more in secondary amines when protonated since more hydrogens are there
   c) lone pair is less available in tertiary amines
   d) +I effect of methyl groups in tertiary amines reduce basicity
10. The ozonolysis products in $(CH_3)_2S$ of the major alkene from Hoffmann elimination of

$$[CH_3-CH_2-\overset{\displaystyle CH_3}{\underset{\displaystyle CH_3}{\overset{|}{\underset{|}{N}}}}-CH_2CH_2CH_3]^+ \ OH^- \text{ will be}$$

   a) 2 moles of HCHO   b) HCHO and CH_3CHO
   c) 2 moles of acetone   d) 2 moles of CH_3CHO
11. In one of the following preparations chain length is unchanged.
   a) $C_6H_{13}COOH$ mixed with $SOCl_2$ followed by addition of NH_3 and reduction by $LiAlH_4$
   b) $C_6H_{13}COOH$ to amine by adding $LiAlH_4$, adding PBr_3, adding KCN, adding $LiAlH_4$
   c) $C_6H_{13}COOH$ to $C_6H_{13}CONH_2$, Br_2/KOH and water addition
   d) $C_6H_{13}COOH$ to $C_6H_{13}COCl$ followed by reduction with Pd, $BaSO_4$, H_2 followed by addition of $C_6H_{13}NH_2$
12. Toluene is converted to *o*-chlorotoluene by the reagent set
   a) I. HNO_3, H_2SO_4 II. Cl_2/Fe III. Sn/HCl, OH^- IV. $NaNO_2$/HCl V. H_3PO_2
   b) I. HNO_3, H_2SO_4 II. Cl_2/Fe III. Sn/HCl, OH^- IV. H_3PO_2
   c) I. HNO_3, H_2SO_4 II. Cl_2/Fe III. Sn/HCl, OH^- IV. $NaNO_2$/HCl V. H_2O
   d) I. HNO_3, H_2SO_4 II. Br_2/Fe III. Sn/HCl, OH^- IV. $NaNO_2$/HCl V. H_3PO_2
13. Aniline is converted to tribromobenzene by
   a) reacting with acetic anhydride and then brominated and hydrolysed
   b) adding bromine water, diazotize and then adding H_3PO_2
   c) adding bromine water and diazotizing
   d) reacting with acetic anhydride and then brominated
14. One of the following reactions leads to products which exhibit geometric isomerism
   a) $ArN_2Cl + OH^-$ followed by addition of NaOH   b) $ArN_2Cl + SnCl_2/HCl$
   c) $ArN_2HSO_4 + H_2O \longrightarrow$ d) $ArN_2Cl + H_3PO_2 \longrightarrow$

15. When a primary aliphatic amine reacts with nitrous acid, the carbonium ion formed leads to
    a) alkenes only b) alcohols only
    c) alkyl halides only d) alkenes, alcohols and alkyl halides
16. Na and alcohol are used to reduce
    a) oximes to amines b) nitriles to amines
    c) amides to amines d) nitro compounds to amines
17. The compound reduced by NH_3, H_2, Ni to benzylamine
    a) answers haloform reaction
    b) answers Cannizarro reaction
    c) undergoes Bayer–Villiger oxidation to give $HCOOC_6H_5$
    d) is an ozonolysis product of phenyl acetylene
18. *m*-dinitrobenzene is reduced to *m*-nitroaniline by
    a) Fe/HCl b) Sn/HCl c) H_2S, NH_3, C_2H_5OH d) H_2, Ni
19. 2,4-dinitrotoluene with H_2S and NH_3 gets reduced to]
    a) 2,4-diaminotoluene b) 2-nitro-4-aminotoluene
    c) 2-amino-4-nitrotoluene d) 2,4-diaminobenzene
20. If we use—— compound in the preparation of nitrogen compounds by reacting with alkyl halides, multiple alkylation does not take place.
    a) NH_3 b) $C_2H_5NH_2$ c) $(CH_3)_2NH$ d) $(CH_3)_3N$

KEY

1. c	2. b	3. c	4. d	5. c	6. d	7. a
8. c	9. b	10. a	11. a	12. a	13. b	14. a
15. d	16. a	17. b	18. c	19. b	20. d	

III. Multiple Choice Questions (More than one answer is correct)

1. Some of these compounds are basic
    a) $C_6H_5NH_2$ b) cyclohexylamine c) pyrrole d) pyridine
2. Hoffmann elimination is possible for
    a) $(C_2H_5)_3N^+CH_3OH^-$ b) $(C_6H_5)_3N^+CH_3OH^-$
    c) $(C_2H_5)_3N^+C_3H_7OH^-$ d) $(C_2H_5)_3N^+C_4H_9OH^-$

3. Some of the following methods give a primary amine
   a) $RCN + LiAlH_4 \longrightarrow$
   b) $R_2C{=}NOH + H_2/Rh/C_2H_5OH \longrightarrow$
   c) $RCONH_2 + NaH, CH_3I$ followed by addition of $LiAlH_4$, $H_2O \longrightarrow$
   d) $RMgX + NH_2Cl \longrightarrow$
4. The following reagents are used in reduction of compounds to amines
   a) Ni/H_2 b) $LiAlH_4$, H_2O c) $NaBH_3CN$ d) Li/NH_3
5. Phthalimide synthesis can be done with
   a) $C_6H_5CH_2Br$ b) $C_6H_5NHCH_3$ c) $(CH_3)_3CBr$ d) CH_3Br
6. The compounds which have two stereoisomers are

   a) $C_6H_5{-}C(CH_3)(C_2H_5){-}NH_2$
   b) $C_6H_5{-}N(\rightarrow O)(C_2H_5){-}CH_3$

   c) 1-chloro-2-methylaziridine d) $(CH_3)_3N$
7. The compounds which have diastereoisomers are
   a) 1-chloro-2-methylaziridine b) azobenzene
   c) $(CH_3)_3N$ d) $R_1R_2R_3R_4N^+Cl^-$
8. The ozonolysis products in $(CH_3)_2S$ of the major alkene from Hoffmann elimination of

   $$[CH_3CH_2{-}N(CH_3)_2{-}CH_2CH_2CH_3]^+OH^-$$

   do not correspond to
   a) 2 moles of HCHO b) HCHO and CH_3CHO
   c) 2 moles of acetone d) 2 moles of CH_3CHO
9. In some of the following preparations, chain length is changed
   a) $C_6H_{13}COOH$ mixed with $SOCl_2$ followed by adding NH_3 and reduced by $LiAlH_4$
   b) $C_6H_{13}COOH$ to amine by adding $LiAlH_4$, adding PBr_3, adding KCN, adding $LiAlH_4$
   c) $C_6H_{13}COOH$ to $C_6H_{13}CONH_2$, Br_2/KOH and water addition
   d) $C_6H_{13}COOH$ to $C_6H_{13}COCl$ followed by reduction with Pd, $BaSO_4$, H_2 followed by addition of $C_6H_{13}NH_2$

10. Some of the statements about the reaction of aniline are true
    a) Aniline with acetic anhydride followed by bromination and hydrolysis gives *p*-bromoaniline
    b) Adding bromine water to aniline gives tribromoaniline
    c) Add bromine water to aniline to get *o*-bromaniine
    d) Diazotize aniline and add H_3PO_2 to get phenol
11. Some of the following reactions lead to products which cannot exhibit geometric isomerism
    a) $ArN_2Cl + OH^-$ followed by addition of NaOH
    b) $ArN_2Cl + SnCl_2/HCl$
    c) $ArN_2HSO_4 + H_2O \longrightarrow$
    d) $ArN_2Cl + H_3PO_2 \longrightarrow$
12. When a primary aliphatic amine reacts with nitrous acid then
    a) carbonium ion is formed and N_2 is evolved
    b) alkenes, alcohols and alkyl halides are formed.
    c) only alcohols are formed
    d) only alkyl halides are formed
13. Na and alcohol cannot be used to reduce
    a) oximes to amines
    b) nitriles to amines
    c) amides to amines
    d) nitro compounds to amines
14. The compound reduced by NH_3, H_2, Ni to benzylamine does not undergo
    a) haloform reaction
    b) Cannizarro reaction
    c) aldol condensation
    d) reaction with Tollen's reagent
15. *m*-dinitrobenzene cannot be reduced to *m*-nitroaniline by
    a) Fe/HCl
    b) Sn/HCl
    c) H_2S, NH_3, C_2H_5OH
    d) H_2, Ni
16. If we use these compounds in preparation of nitrogen compounds by reacting with alkyl halides, multiple alkylation takes place.
    a) NH_3
    b) $C_2H_5NH_2$
    c) $(CH_3)_2NH$
    d) $(CH_3)_3N$

KEY

1. a,b,c	2. a,c,d	3. a,b,d	4. a,b,c	5. a,b,d	6. a,b,c	7. a,b
8. b,c,d	9. b,c,d	10. a,b	11. b,c,d	12. a,b	13. b,c,d	14. a,c
15. a,b,d	16. a,b,c					

IV. Assertion–Reason Questions

Questions 1–10 have an assertion in column 1 and reason in column 2. Use the following key to choose the correct answer

a) If both assertion and reason are correct and reason is the correct explanation of assertion
b) If both assertion and reason are correct but reason is not the correct explanation of assertion
c) If assertion is correct but reason is incorrect
d) If assertion is incorrect but reason is correct

No.	Assertion	Reason
1.	Pyridine is not basic but pyrrole is basic.	In pyridine lone pair on nitrogen is not used for aromatic sextet but in pyrrole they are used for forming aromatic sextet.
2.	Tetrahydroquinoline is much less basic than tetrahydroisoquinoline.	In tetrahydroquinoline the lone pair on nitrogen is next to sp^2 carbon and these electrons are delocalized into benzene ring and protonation is less possible.
3.	Amines cannot be alkylated using tertiary halides.	Tertiary halides are bulky.
4.	Aziridine is much less basic than piperidine.	The three-membered ring in aziridine has internal strain add hence less basic.
5.	1-chloro-2-methylaziridine has two isomers including stereoisomers.	The *cis* and *trans* isomers both have enantiomers and the toal isomers is 4.
6.	In aqueous solution secondary amines are more basic than tertiary amines.	In protonated form, there are two hydrogens which can hydrogen-bond in secondary amines.
7.	n-propylamine has a higher boiling point than1-propanol.	Hydrogen bonding is stronger in alcohols and hence 1-propanol has a higher boiling point.
8.	$(CH_3)_4N^+OH^-$ cannot undergo Hoffmann elimination to give alkenes.	The compound has bulky groups.
9.	Aniline and nitrobenzene can be separated by dissolving in ether and adding HCl.	The nitro compound dissolves in HCl.
10.	Amines are more basic than alcohols.	The nitrogen lone pair is held less tightly than oxygen lone pair and amine nitrogen lone pair is more available for protonation.

KEY

1. d	2. a	3. b	4. a	5. d	6. a	7. d
8. b	9. c	10. a				

V COMPREHENSION PASSAGES

Passage 1

Amines can be prepared by several methods. Gabriel phthalimide synthesis, nitrile reduction, RMgX reacting with $ClNH_2$, RX and NH_3 and so on.

Answer the following questions

1. Primary amine is not a product in one of the following synthesis
   a) phthalimide + KOH, RI, H_2O
   b) RMgX + $ClNH_2$
   c) RNCO + KOH
   d) CaNCN + NaOH + 2RX + H^+
2. One of the following is not a process of conventional reduction in preparing amines
   a) phenylacetonitrile + Raney Ni at 413K
   b) N-methylacetanilide + $LiAlH_4$(ether) + H_2O
   c) $RCONH_2 + Br_2$ + NaOH
   d) RCH═NOH + Na/C_2H_5OH
3. The alkyl halide not used in Gabriel phthalimide synthesis is
   a) RI
   b) $C_6H_5CH_2I$
   c) R_2CHCl
   d) R_3I

KEY

1. d **2.** c **3.** d

Passage 2

Amines can be optically active if NH_2 is connected to C which has three different groups or nitrogen attached to three different groups. The basicity of amines is mainly explained by the resonance effect.

Answer the following questions

1. The compound which is not chiral among the following is

   a) $CH_3N(C_2H_5)(C_3H_7)$ [$CH_3N—C_2H_5$ with C_3H_7 on N]

   b) Phenylethylamine

   c) $C_6H_5CH_2NH_2$

   d) $CH_3N^+(C_3H_7)(C_2H_5)CH_2X^-$ [$CH_3N^+CH_2X^-$ with C_3H_7 above and C_2H_5 below N]

2. Arrange in decreasing order of basicity of I. $C_6H_5CONH_2$ II. $C_6H_5NH_2$ III. CH_3NH_2

a) I>II>II b) III>II>I c) III>I>II d) III=II>I

3. The compound which does not dissolve in HCl is

a) $C_2H_5NH_2$ b) pyridine

c) $C_6H_5CONH_2$ d) $C_6H_5CH_2NH_2$

KEY

1. c 2. b 3. c

Passage 3

Hoffmann elimination is one of the methods of making alkenes which are less substituted. The quaternary hydroxide must have beta hydrogen for elimination.

Answer the following questions

1. The quaternary hydroxide which cannot undergo elimination is

a) $(CH_3)_3N^+C_6H_5OH^-$ b) $(C_2H_5)_4N^+OH^-$

c) $(C_2H_5)_3N^+CH_3OH^-$ d) $C_6H_5N^+(C_2H_5)_3OH^-$

2. The compound which is obtained in maximum yield by decomposition of

$CH_3CH_2CHCH_3$ with $N^+(CH_3)_3OH^-$ on the CH carbon, will be

a) 2-butene b) 1-butene

c) propylene d) 1-pentene

3. When azacyclohexane (piperidine) is treated with CH_3I and AgOH, and heated and the process repeated, it gives the major alkene whose ozonolysis products are

a) 2HCHO + $CHOCH_2CHO$ b) 2HCHO + CHOCHO

c) HCHO + $CH_3COCOCH_3$ d) CH_3CHO + $CH_3COCOCH_3$

KEY

1. a 2. b 3. a

VI. Matrix Matching

Question 1

Column 1	Column 2
a) $CH_3CH_2NH_2$	p) Ethanal + $NH_3/H_2/Ni$
b) $(CH_3)_2NH + CH_3CHO + H_2/Ni$	q) $CH_3CN + LiAlH_4 + H_2O$
c) $RNCO + 2KOH$	r) $CH_3CH_2MgX + NH_2Cl$
d) $CaNCN + NaOH + RX + H^+$	s) $(CH_3)_2NCH_2CH_3$
	t) RNH_2
	u) $R_2NH + CO_2 + NH_3$

Answer

$[a \rightarrow p,q,r], [b \rightarrow s], [c \rightarrow t], [d \rightarrow u]$

Question 2

Column 1	Column 2
a) Iodobenzene	p) $ArN_2HSO_4 + CuBr$
b) bromobenzene	q) $I_2 + CuCl$ + diazonium salt
c) Cyclohexyl $CH_2NO(Me)_2$ on heating	r) $ArN_2HSO_4 + NaI$
d) *p*-nitrobenzoic acid to *p*-cyanobenzoic acid	s) benzene + Fe/Br_2
	t) methylene cyclohexane + Me_2NOH
	u) Nitrogroup is reduced by Zn/HCl and $NaNO_2$ + HCl added and CuCN is added and hydrolysed

Answer

$[a \rightarrow q,r], [b \rightarrow p,s], [c \rightarrow t], [d \rightarrow u]$

Question 3

Column 1	Column 2
a) Pyrrole is nonbasic due to	p) since the protonated form of secondary amines has more hydrogens involved in hydrogen bonding
b) Aniline is separated from nitrobenzene	q) The lone pair electrons on nitrogen used for aromatic sextet
c) Secondary amines in aqueous solution are more basic than tertiary amines	r) by adding HCl
d) Aziridine is less basic than piperidine due to	s) internal ring strain

Answer

$[a \rightarrow q], [b \rightarrow r], [c \rightarrow p], [d \rightarrow s]$

VII. Structure-Based and Numerical Problems

1. Compound A is obtained by decarboxylation of sodium salt of B by NaOH/CaO. B is obtained by oxidation with $KMnO_4$ of $C_6H_5CH_2OH$ or isopropyl benzene or benzyl chloride, etc. A with HNO_3/H_2SO_4 gives C. C with Sn/HCl gives D. D with $NaNO_2/H_2SO_4$ gives E. E with water gives F. E with KI gives G. Find A to G.

   Answer

   A is benzoic acid;

   B is benzene;

   C is nitrobenzene;

   D is aniline;

   E is benzenediazonium hydrogensulphate;

   F is phenol;

   G is iodobenzene.

2. Benzene(A) reacts with CH_3Cl and $AlCl_3$ to give B. B with HNO_3/H_2SO_4 gives C and CC. CC is in large percentage. C with Sn/HCl gives D. D with $NaNO_2$/HBr gives E. E with CuBr gives F. Find A to F.

   Answer

   A is benzene;

   B is toluene;

C is *o*-nitrotoluene;

CC is *p*-nitrotoluene;

D is *o*-aminotoluene;

E is diazonium salt;

F is *o*-bromotoluene.

3. Benzene reacts with HNO_3/H_2SO_4 to give A. A with Sn/HCl gives B.B with $NaNO_2/HCl$ gives C. C with D gives E, F, G and H. H can be mopped up by pyridine. G turns lime water milky. F is always evolved when diazonium salts decompose. E on ozonolysis in reducing conditions gives 2 moles of C_6H_5CHO. D gives effervescence with $NaHCO_3$. Find A to H.

   A is nitrobenzene;

   B is aniline;

   C is benzene diazonium

   D is cinnamic acid;

   E is stilbene;

   F is nitrogen;

   G is CO_2 H$\longrightarrow$HCl

   H is $C_6H_5CH{=}CHCO_2$chloride;

4. Compound A is obtained by treatment of cumene with air along with B as another product. A on complete hydrogenation gives C. C on oxidation gives D. D with NH_2OH gives E. E with H_2, Rh/C in ethanol gives F. Compare the basicity of F with aniline. Find A to F.

   Answer

   A is phenol;

   B is acetone;

   C is cyclohexanol;

   D is cyclohexanone;

   E is cyclohexanoneoxime;

   F is cyclohexyl amine. It is more basic than aniline since nitrogen lone pair of aniline is drawn into benzene ring by delocalization.

5. Nitrobenzene is treated with Sn/HCl to give A. A with $NaNO_2/H_2SO_4$ gives B. B with water gives C. C with HNO_3/H_2SO_4 gives mainly D. D with propylene and H_3PO_4 gives E. E with H_2/Ni gives F. F with $NaNO_2$/HCl followed by addition of H_3PO_2 gives propoful (G) which is an intravenous anaesthetic. G has the structure 1,6-diisopropyl phenol. Find A to F.

Answer

A is aniline;

B is diazonium sulphate of benzene;

C is phenol;

D is *p*-nitrophenol;

E is 1,6-diisopropyl-4-nitrophenol;

F is 1,6-diisopropyl-4-aminophenol.

6. Benzene reacts with CH_3COCl in $AlCl_3$ to give A. A on Clemmenson reduction gives B. B on oxidation with $KMnO_4$ gives C. C with 2 moles of HNO_3/H_2SO_4 mixture gives D. D with Ni/H_2 gives E. E with three moles of ICl gives F. Find A to F.

Answer

A is acetophenone

B is ethylbenzene

C is benzoic acid

D is 3,5-dinitrobenzoic acid

E is 3,5-diaminobenzoic acid

F is 3,5-diamino-2,4,6-triiodobenzoic acid

7. Compound A is intramolecularly hydrogen-bonded and is prepared by carboxylation of B. B on reduction completely gives C. C on dehydration gives D. D on ozonolysis in reducing conditions gives $CHO(CH_2)_4CHO$. A with HNO_3/H_2SO_4 gives E. E with H_2/Ni gives F. F with HNO_3/H_2SO_4 gives G. G on diazotization followed by addition of H_3PO_2 gives H. H with H_2/Ni gives J. J is one of the building blocks in synthesis of caine anaesthetics. Find A to J.

Answer

A is salicylic acid;

B is phenol;

C is cyclohexanol;

D is cyclohexene;

E is 2-carboxyl-4-nitrophenol;

F is 2-carboxyl-4-aminophenol;

G is 2-carboxyl-4-amino-5-nitrophenol;

H is 2-carboxyl-5-nitrophenol;

J is 4-aminosalicylic acid.

8. Compound A is obtained by the oxidation with $KMnO_4$ of $C_6H_5CH_2OH$ or $C_6H_5CH_2Br$ or C_6H_5CHO. A with $LiAlH_4$ gives B. B with PBr_3 gives C. C with KCN gives D. D with $LiAlH_4$ gives E. A with $SOCl_2$ gives F. F with NH_3 gives G. G with $LiAlH_4$ gives H. G with Br_2/KOH followed by addition of water gives J. Find A to J. What is the relationship between E, H and J?

Answer

A is benzoic acid;

B is $C_6H_5CH_2OH$;

C is $C_6H_5CH_2Br$;

D is $C_6H_5CH_2CN$;

E is $C_6H_5CH_2CH_2NH_2$;

F is C_6H_5COCl;

G is $C_6H_5CONH_2$;

H is $C_6H_5CH_2NH_2$;

J is $C_6H_5NH_2$;

H contains one CH_2 group more than J, and E contains one CH_2 group more than H.

12

CARBOHYDRATES, AMINO ACIDS, PEPTIDES AND POLYMERS

POLYMERS

A polymer is a molecular compound distinguished by its high molar mass ranging into thousands and millions of grams, and made up of many repeating units. The properties of these macromolecules differ from small molecules like aldehydes, ketones, hydrocarbons, etc., and special techniques are used to study them. Polymers can be divided into two important types a) Natural and b) Synthetic.

Natural polymers Naturally occurring polymers include proteins, nucleic acids, cellulose (called as polysaccharides) and rubber (polyisoprene).

Synthetic polymers Synthetic polymers are made by joining monomeric units together, one at a time by addition and condensation reactions.

TYPES OF POLYMERIZATION

Addition polymerization A typical example of addition polymerization is the conversion of ethylene into polyethylene (often called polythene). Addition polymers result from the rapid addition of one molecule at a time to a growing polymer chain, usually with reactive intermediates like cations, anions or radicals. Some important addition polymers are listed in Table 12.1.

Table 12.1 Addition polymers

Polymer	Monomer unit	Repeating Unit
Polyethylene	Ethylene	$-(CH_2{=}CH_2)_n-$
Polypropylene	Propylene	$-(CH_3-\underset{\|}{\overset{CH_3}{}}(CH-)_n-$
Polystyrene	Styrene	$-(CH_2-CH(C_6H_5)-)_n-$
Polyvinylchloride	vinyl chloride	$-(CH_2-CH(Cl)-)_n-$
Polytetrafluoroethylene	Tetrafluoroethylene	$-(CF_2{=}CF_2)_n-$

Types of addition polymerization Next we discuss briefly the three types of addition polymerization

a) Free radical polymerization
b) Cationic polymerization
c) Anionic polymerization

Free radical polymerization Free radical polymerization is initiated by radicals like benzoyl peroxide. These radicals generate phenyl radicals and CO_2. The phenyl radicals then react with monomers like styrene to the growing chain of the polymer. Finally polystyrene is formed containing 100 to 10000 monomer units. Ethylene and propylene can also be polymerized using free radical polymerization but the radicals are less stable.

Cationic polymerization In this polymerization carbocations are intermediates. BF_3 is the effective catalyst used. The catalyst protonates the monomer and starts the polymer chain. For example isobutylene is converted to a tertiary butyl carbocation and adds on to another isobutylene molecule. Finally we get polyisobutylene.

Anionic polymerization Here carbanion intermediates are involved. Generally the monomer used has an electron-withdrawing group like cyano or nitro. The polymerization is initiated by a Grignard reagent or organolithium.

Acrylate Polymers

Acrylates (Figure 12.1) are a family of polymers, which are a type of vinyl polymer. Acrylates are of course made from acrylate monomers, Acrylate monomers are the esters which contain vinyl groups, that is, two carbon atoms double-bonded to each other, directly attached to the carbonyl carbon.

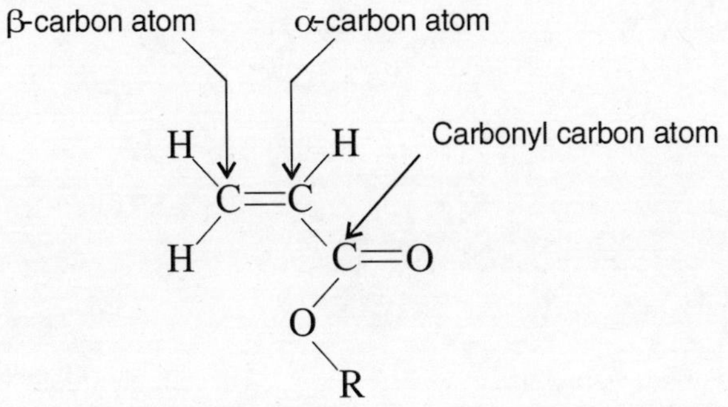

β-carbon atom
α-carbon atom
Carbonyl carbon atom

$$\begin{array}{l} H \quad\quad H \\ \;\; C{=}C \\ H \quad\quad\;\; C{=}O \\ \quad\quad\;\; O \\ \quad\quad\quad\; R \end{array}$$

Figure 12.1 An acrylate

Some acrylates have an extra methyl group attached to the alpha carbon, and these are called *methacrylates*. One of the most common methacrylate polymers is poly(methyl methacrylate).

An example is the polymer formed from methyl acrylate as shown in Figure 12.2.

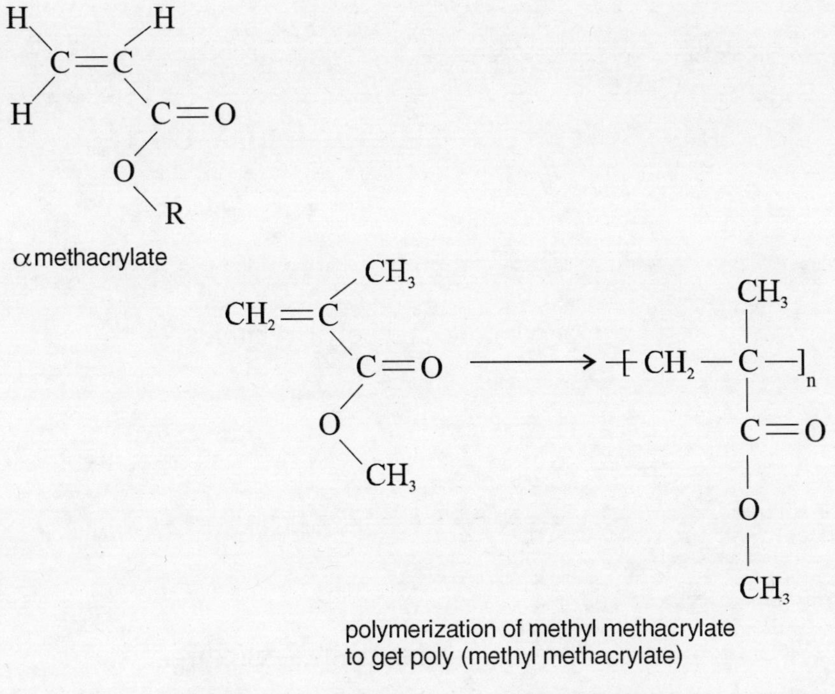

$$\begin{array}{l} H \quad\quad H \\ \;\; C{=}C \\ H \quad\quad\;\; C{=}O \\ \quad\quad\;\; O \\ \quad\quad\quad\; R \end{array}$$

α methacrylate

$$CH_2{=}C(CH_3){-}C({=}O){-}O{-}CH_3 \longrightarrow {+}CH_2{-}C(CH_3)(C({=}O){-}O{-}CH_3){-}]_n$$

polymerization of methyl methacrylate
to get poly (methyl methacrylate)

Figure 12.2 Poly(methyl methacrylate)

Acrylate and methacrylate The little methyl group would make a whole lot of difference in the behaviour and properties of the polymer, and this is true. Poly(methyl acrylate) is a white rubber at room temperature, but poly(methyl methacrylate) is a strong, hard, and clear plastic.

$$-\!\!\left[CH_2-\underset{\underset{\underset{CH_3}{|}}{\underset{O}{|}}}{\underset{C=O}{|}}{CH}\right]\!\!-_n \qquad -\!\!\left[CH_2-\overset{CH_3}{\overset{|}{\underset{\underset{\underset{CH_3}{|}}{\underset{O}{|}}}{\underset{C=O}{|}}{C}}}\right]\!\!-_n$$

This is poly (methyl acrylate).
It is soft and rubbery.

This is poly (methylmethacrylate).
It is a hard plastic.

Polyacrylates containing nitrogen There are several derivatives of polyacrylates which contain nitrogen. Polyacrylamide and polyacrylonitrile are two such polyacrylates and are shown in Figure 12.3. Polyacrylonitrile is used to make acrylic fibers. Polyacrylamide will dissolve in water and is used industrially. Cross-linked polyacrylamides can absorb water. These gels are used to make soft contact lenses. It's the absorbed water in them that makes them soft.

$$H_2C=CH-C(=O)-NH_2 \longrightarrow -\!\!\left[CH_2-\underset{\underset{\underset{NH_2}{|}}{C=O}}{\underset{|}{CH}}\right]\!\!-_n$$

acrylamide and polyacrylamide

$$H_2C=CH-C\equiv N \longrightarrow -\!\!\left[CH_2-\underset{\underset{C\equiv N}{|}}{CH}\right]\!\!-_n$$

acrylonitrile and polyacrylonitrile

Figure 12.3 Polyacrylates containing nitrogen

Condensation Polymerization

These polymers result from the formation of ester or amide linkages between different molecules. Polyamides, polyesters, polycarbonates and polyurethanes are some examples of condensation polymers.

Nylon-6 This is obtained from caprolactam (Figure 12.4).

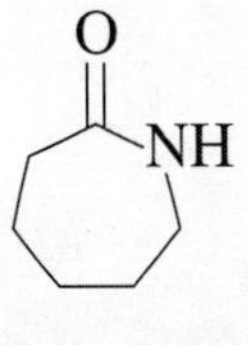

Figure 12.4 Caprolactam

Caprolactam molecule used to synthesize Nylon-6 by ring opening polymerization.

Nylon-6 begins as pure caprolactam. As caprolactam has 6 carbon atoms, it got the name Nylon-6.

When caprolactam is heated at about 533 K in an inert atmosphere of nitrogen for about 4–5 hours, the ring breaks and undergoes polymerization. Then the molten mass is passed through spinnerets to form fibres of Nylon-6.

The formation of condensation polymer Nylon-6 is given in the Figure 12.5 given below.

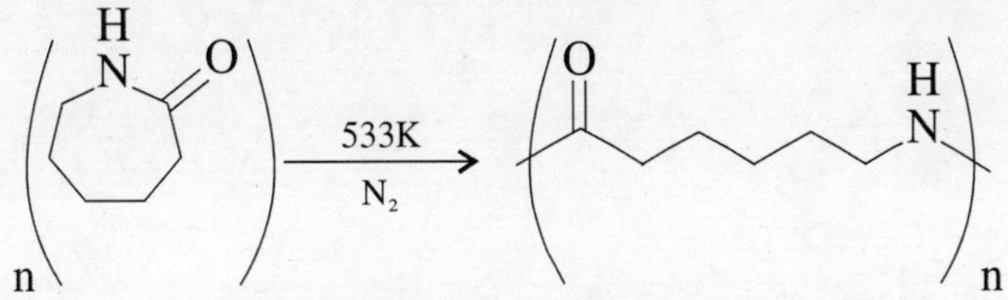

Figure 12.5 Nylon 6

During polymerization, the peptide bond within each caprolactam molecule is broken, with the active groups on each side re-forming two new bonds as the monomer becomes part of the polymer backbone.

The next section discusses specific examples of polymers, their properties and uses.

PROPERTIES AND USES OF SOME IMPORTANT POLYMERS

Natural Rubber Natural rubber is a polymer of isoprene (2-methyl-1,3-butadiene). Pyrolysis degrades natural rubber to isoprene. Ziegler–Natta catalysts are used to polymerize isoprene. Figure 12.6 shows the synthesis of natural and synthetic rubber.

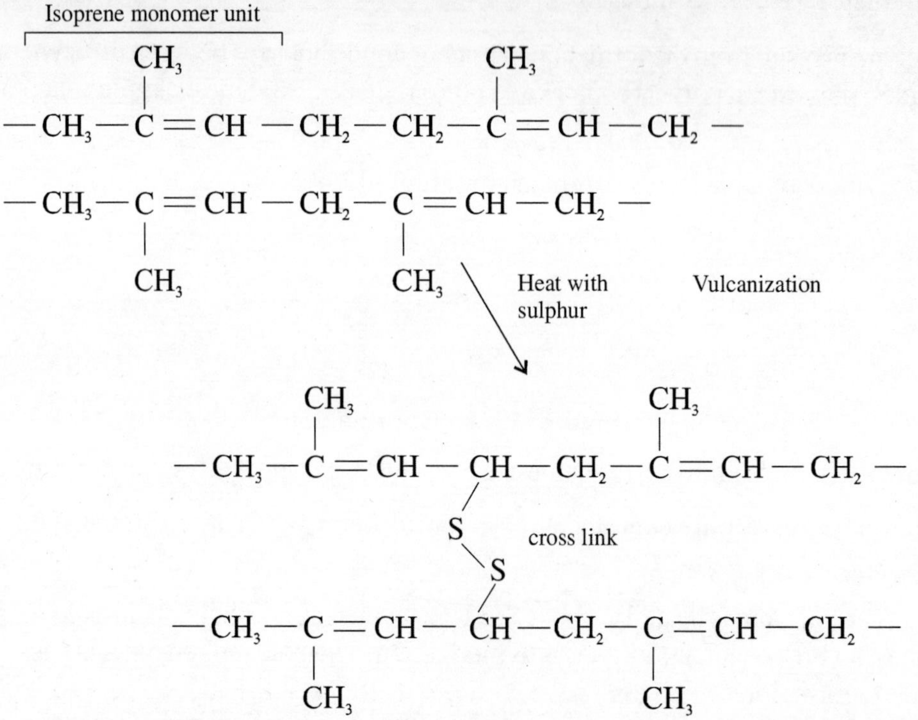

Isoprene monomer unit

$$-CH_2-C(CH_3)=CH-CH_2-CH_2-C(CH_3)=CH-CH_2-$$

$$-CH_2-C(CH_3)=CH-CH_2-C(CH_3)=CH-CH_2-$$

Heat with sulphur

Vulcanization

$$-CH_2-C(CH_3)=CH-CH-CH_2-C(CH_3)=CH-CH_2-$$

S—S cross link

$$-CH_2-C(CH_3)=CH-CH-CH_2-C(CH_3)=CH-CH_2-$$

Natural Rubber

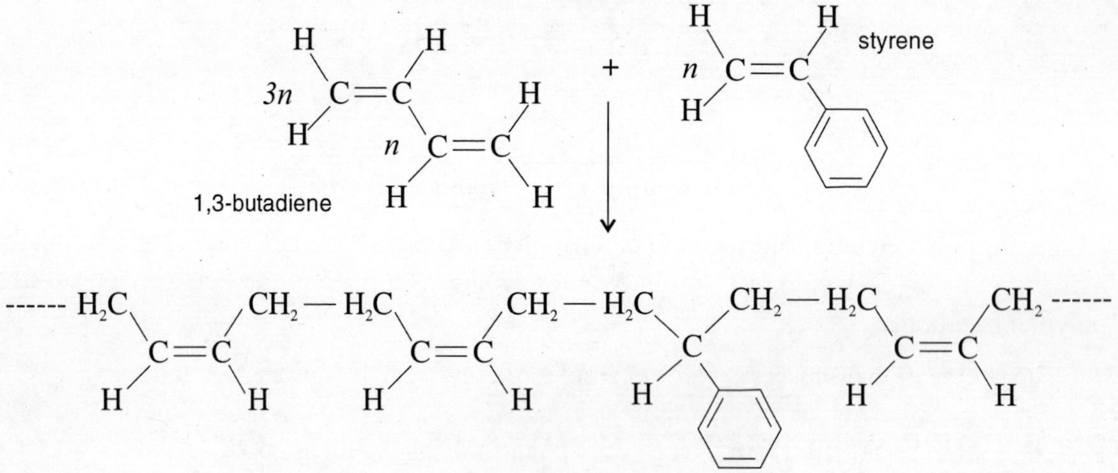

3n $H_2C=CH-CH=CH_2$ (1,3-butadiene) + n $H_2C=CH-C_6H_5$ (styrene)

$$\text{------} H_2C-CH=CH-CH_2-H_2C-CH=CH-CH_2-H_2C-CH(C_6H_5)-CH_2-H_2C-CH=CH-CH_2 \text{------}$$

Styrene-Butadiene Rubber

Figure 12.6 Synthesis of Natural rubber and styrene butadiene rubber

Ozonolysis of natural rubber Ozone is added to natural rubber and treated with H_2O_2 in acetic acid. The product is levulinic acid with structure $CH_3COCH_2CH_2COOH$.

PVC Radical polymerization of vinyl chloride gives PVC. It is used in clothings, seat coverings, pipe-work taps and lab stopclocks. Vinyl chloride can be added to vinylidine chloride to form PVC polymers.

CARBOHYDRATES

Carbohydrates were originally thought of as hydrates of carbon since a carbohydrate like $C_6H_{12}O_6$ can be written as $C_6(H_2O)_6$. This view has been abandoned. It is used to refer to polyhydroxy aldehydes and ketones commonly called sugars. Glucose (dextrose) is pentahydroxyhexanal and has the structure as shown in Figure 12.7.

```
     H—C=O
       |
     H—C—OH
       |
    OH—C—H
       |
     H—C—OH
       |
     H—C—OH
       |
      CH2OH
```

Figure 12.7 Structure of glucose

Carbohydrates are synthesized by green plants during photosynthesis, a complex process in which sunlight provides energy to convert CO_2 into glucose. The reaction is

$$6CO_2 + 6H_2O \xrightarrow{\text{(Sunlight)}} 6O_2 + C_6H_{12}O_6$$

When eaten, glucose can be either metabolized in the body to provide immediate energy or stored in the form of glycogen for later use. Human beings lack enzymes needed for digestion of cellulose and they require starch as their dietary source of carbohydrates. Animals like cows have microorganisms to digest cellulose.

CLASSIFICATION OF CARBOHYDRATES

Carbohydrates are divided into two important categories, simple and complex. Simple sugars or monosaccharides are systems like glucose which cannot be hydrolysed to smaller units. Complex carbohydrates are two or more monosaccharides linked together. Sucrose (table sugar) is a disaccharide made up of one glucose molecule linked to one fructose molecule. Cellulose is a polysaccharide made up of several thousand glucose molecules linked together. Hydrolysis breaks them into the monosaccharide units.

$$\text{Sucrose} + H_2O \longrightarrow \text{Glucose} + \text{fructose}$$
$$\text{Cellulose} + H_2O \longrightarrow \approx 3000 \text{ glucose}$$

Monosaccharides are further classified into *aldoses* and *pentoses*. 'ose' is used to designate a carbohydrate and *aldo* or *keto* indicates the type of carbonyl present. The number of carbons is indicated by tri, tetra, penta, etc. For example, glucose is an aldohexose which is a six-carbon aldehyde sugar. Fructose is a ketohexose and ribose is an aldopentose.

The structures of some monosaccharides are given in Figure 12.8.

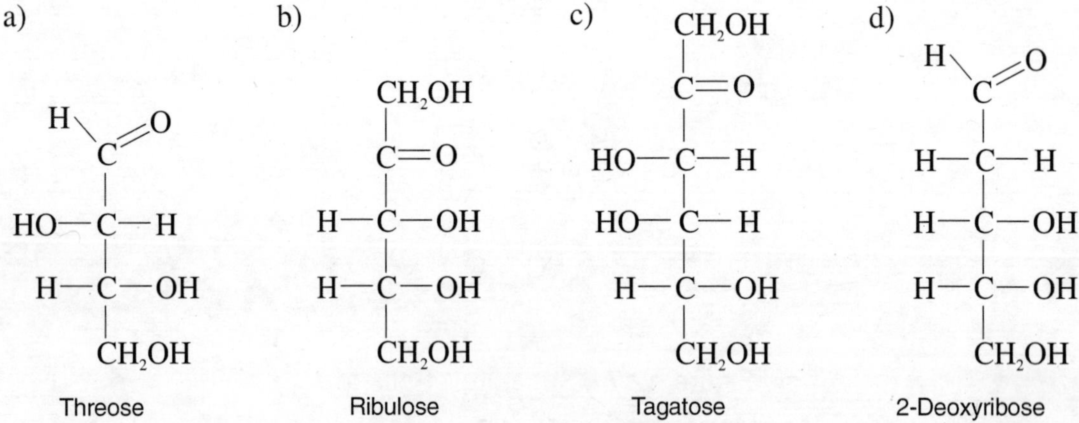

Figure 12.8 Structures of threose, ribulose, tagatose and 2-deoxyribose

Threose is an aldotetrose and ribulose is a ketopentose.

Tagatose is a ketohexose and 2-deoxyribose is an aldopentose.

Fischer Projections of Carbohydrates

All carbohydrates have stereogenic carbons and hence. Fischer projections are formulated to visualize them clearly (*See* chapter 1 for Stereoisomerism). A tetrahedral carbon is represented by two crossed lines. The horizontal lines represent bonds coming out of a page and vertical lines represent bonds going into the page. By convention the carbonyl carbon is placed at or near the top in Fischer projections. Thus R-glyceraldehyde, the simplest monosaccharide is represented by the Fischer projection shown in Figure 12.9.

$$\begin{array}{c} CHO \\ | \\ H-\!\!\!-|-\!\!\!-OH \\ | \\ CH_2OH \end{array}$$

Figure 12.9 Fischer projection of the simplest monosaccharide

Fischer projections can be rotated by 180 degrees but not by 90 or 270 degrees.

Thus, the Fischer projection,

$$\begin{array}{c} CH_2OH \\ | \\ OH-\!\!\!-|-\!\!\!-H \\ | \\ CHO \end{array}$$

is the same as the Fischer projection shown in Figure 12.9

Glyceraldehyde has one stereogenic centre and has two enantiomers. The naturally occurring glyceraldehydes rotate the plane of polarized light in the clockwise direction. It is called (+) glyceraldehydes. At C1 it has R configuration and so R-glyceraldehyde is D glyceraldehyde D for dextrorotatory. Ordinary optically active molecules are represented using d or l, the lower case alphabets, but for sugars we use upper case alphabet (D or L). Almost all naturally occurring monosaccharides have the same stereochemical configuration as D-glyceraldehyde and are called D sugars. The hydroxyl at the lowest stereogenic centre points to the right in Fischer projection. As we already saw in stereoisomerism, absolute configuration has no relation to d and l isomers. Here also, a D sugar may be dextrorotatory or leavorotatory. The D in the monosaccharide indicates that the stereochemistry of the lowest stereogenic centre is to the right in Fischer projection when the molecule has carbonyl at or near the top. The D, L system does not say anything about the stereochemistry of the other stereogenic centres.

Aldotetroses have two stereogenic centres and they are 4-carbon sugars. There will be $2^2 = 4$ isomers. There are two D, L pairs called threose and erythrose.

MONOSACCHARIDES

We have already seen that ketones and aldehydes react with alcohols to form hemiacetals. Similarly the monosaccharides which have C═O group and OH in the same molecule undergo intramolecular nucleophilic addition to form cyclic hemiacetals. Five- and six-membered cyclic hemiacetals are stable and the open-chain form is in equilibrium with cyclic hemiacetal form.

In aqueous solution glucose exists in pyranose form and OH group at C5 adds on C1 carbonyl. Fructose exists 20% in furanose form and 80% pyranose form.

In fructose, OH at C5 adds on to C=O at C2.

Haworth Projections

Pyranose and furanose forms are represented by Haworth projections as shown in Figure 12.10.

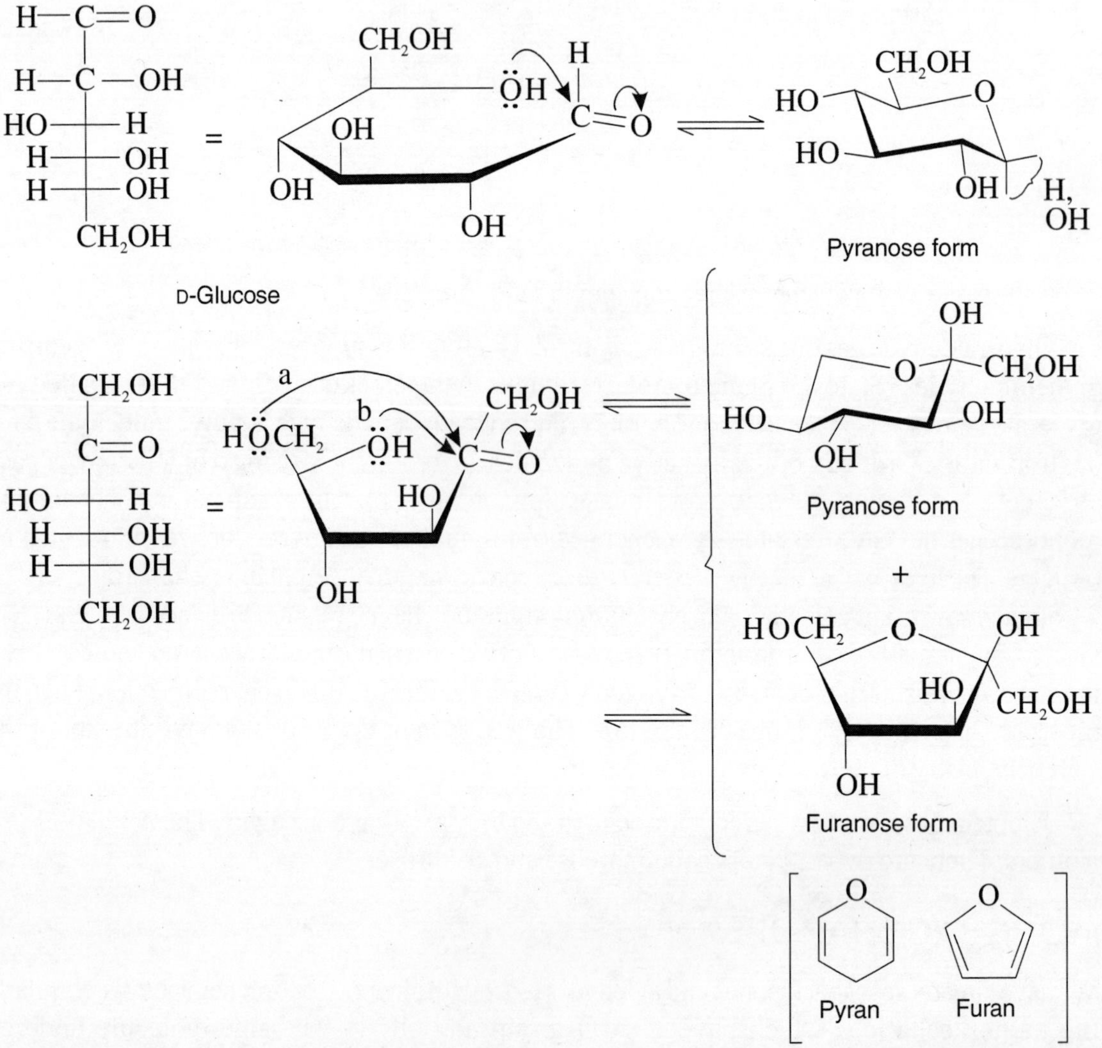

D-Glucose

Pyranose form

Pyranose form

Furanose form

Pyran

Furan

Figure 12.10 Glucose and fructose in their cyclic pyranose and furanose forms

Mutarotation

When an open-chain monosaccharide cyclizes to furanose or pyranose form, a new stereogenic centre is created at the former carbonyl carbon. The two diastereoisomers produced are called anomers and the hemiacetal carbon is called the anomeric centre. For example glucose cyclizes to 36 : 64 mixture of the two anomers. The minor anomer with the OH group at C1 *trans* to CH_2OH substituent at C5 is called α anomer. The complete name is α-D-glucopyranose. The major anomer with OH at C1 *cis* to CH_2OH substituent at C5 is called the β anomer and its name is β-D-glucopyranose (Figure 12.11).

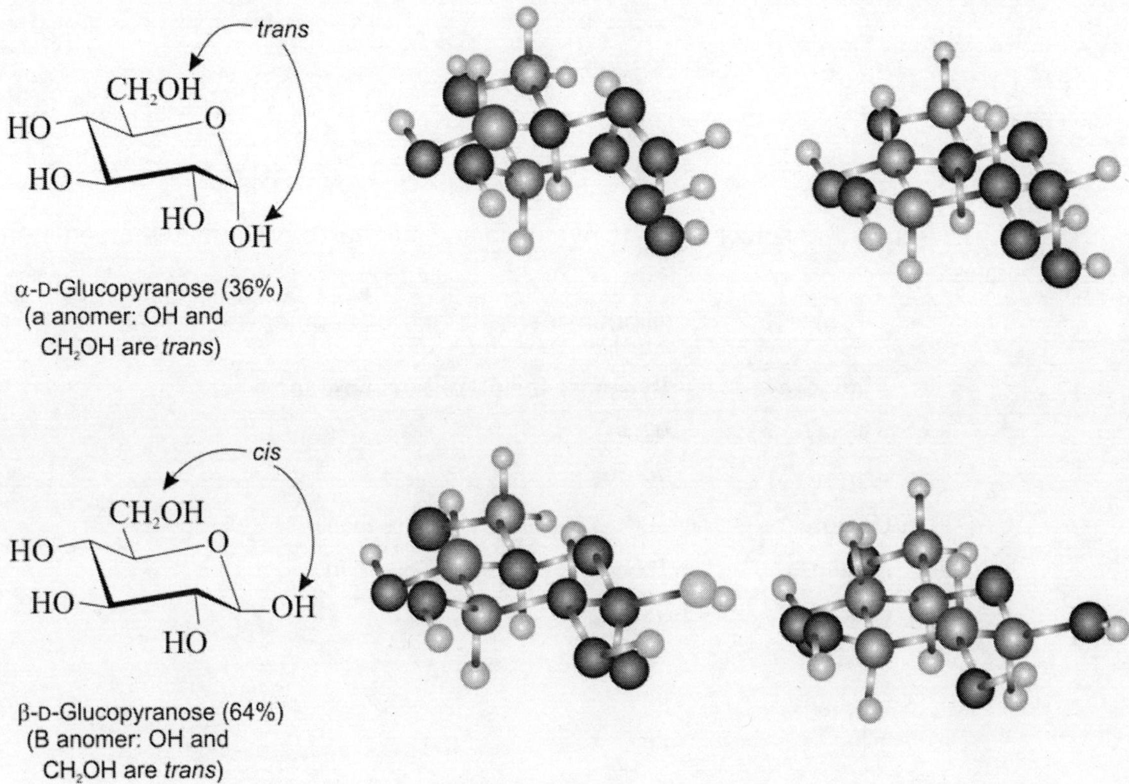

Figure 12.11 α- and β-D-glucopyranose anomers

Both anomers can be crystallized and purified. The minor anomer has melting point equal to 146°C and the major anomer has melting point of 148 to 155°C. The minor anomer has specific rotation equal to +122.2 degrees. The major anomer has specific rotation +18.7 degrees. When a sample of either anomer is dissolved in water, the optical rotation slowly changes and ultimately reaches a constant value of +52.6 degrees. This is the phenomenon of mutarotation due to slow conversion of the two anomers to 36 : 64 equilibrium mixture. The equilibration can be catalysed by acids or bases.

Calculation of equilibrium percentages from optical rotations of anomers.

Solved Example 1

The optical rotation of alpha anomer is 112.2°. The value for beta anomer is 18.7°. The equilibrium value of optical rotation is 52.6°. Find the % of anomers in the equilibrium mixture.

Answer

Let x be fraction of alpha anomer and y that of beta anomer.

$$x + y = 1$$

$$y = 1 - x$$

$$112.2x + 18.7(1 - x) = 52.6$$

Hence $x = 0.362; y = 0.638$. Thus anomer mixture has the % ratio 36 : 64.

Table 12.2 shows the percentages of pyranose and furanose forms in some common carbohydrates.

Table 12.2 Equilibrium percentages of anomers

Aldohexose	Pyranose form	Furanose form
Allose	92	8
Altrose	70	30
Glucose	100	Less than1
Mannose	100	Less than1
Gulose	97	3
Idose	75	25
Galactose	93	7
Talose	69	31

Tests for Reducing Sugars

Reducing sugars exist in hemiacetal form. They give positive answers to Tollens reagent test, Fehling's test and Benedict's reagent test. All aldoses are reducing sugars and some ketoses also are reducing sugars. Fructose answers Tollen's test even though it is not a reducing sugar. Fructose is readily isomerized to aldose in basic solution through many keto–enol tautomerizations. Glycosides are non-reducing since the acetal group cannot open to an aldehyde under basic conditions.

DISACCHARIDES

Disaccharides are carbohydrates which yield two monosaccharides on hydrolysis. They are called glucosides in which the alkoxy group attached to anomeric carbon, is derived from a second sugar molecule, maltose. This is obtained by hydrolysis of starch cellobiose which is obtained by hydrolysis of cellulose. Both these sugars are disaccharides. Lactose is a disaccharide containing 2 to 6 percent of milk and is called milk sugar.

Digestion of lactose is facilitated by the enzyme beta glycosidase lactase. A deficiency of this enzyme makes it difficult to digest lactose and abdominal discomfort results. Sucrose is a disaccharide in which D-glucose and D-fructose are joined at their anomeric carbons by a glycoside bond (Figure 12.2).

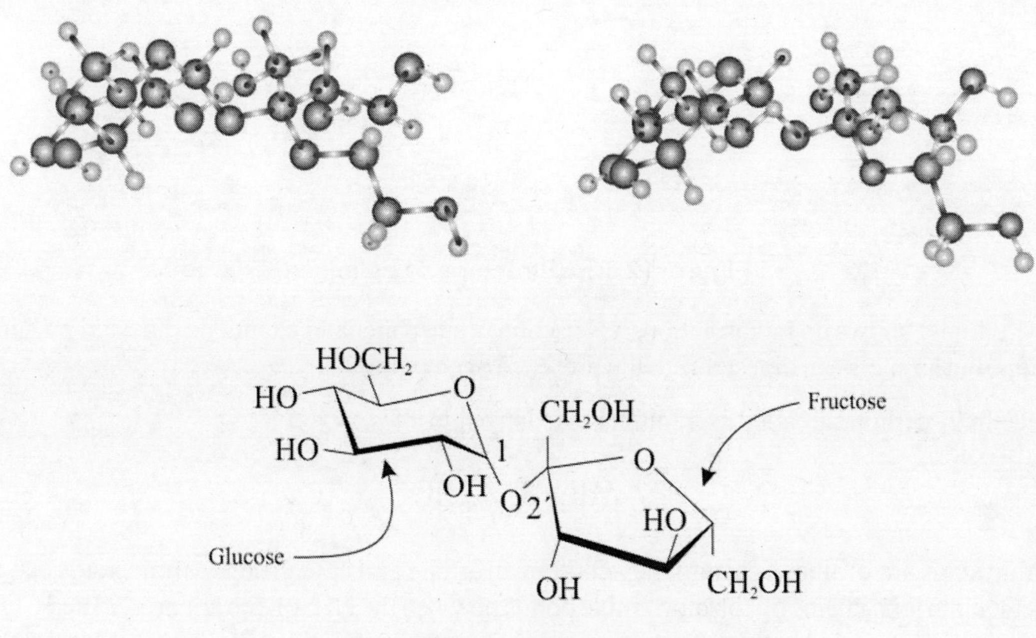

Sucrose, α 1,2′-glycoside
[2-*O*-(α-D-Glucopyranosyl)-β-D-fructofuranoside]

Figure 12.12 Structure of Sucrose

Since sucrose does not have a free anomeric group, it does not undergo mutarotation. Sucrose on hydrolysis gives glucose and fructose. The 1:1 mixture of glucose and fructose, is called invert sugar since sign of optical rotation changes when sucrose hydrolyses (+66.5° to –22°). Insects such as honeybees have enzymes called invertases which catalyse hydrolysis of sucrose to glucose–fructose mixture. Honey itself is a mixture of glucose, fructose and sucrose.

POLYSACCHARIDES

Polysaccharides are those in which thousands of simple sugars are linked together through glycoside bonds. Since they have no free anomeric hydroxyls except for one at the end of the chain, polysaccharides are not reducing sugars and do not show mutarotation. Cellulose and starch are the two most important polysaccharides. Cellulose consists of D-glucose units linked by 1,4′ β-glycoside bonds. Several thousand glucose molecules are linked with hydrogen bonding and Figure 12.13 shows the structure of cellulose.

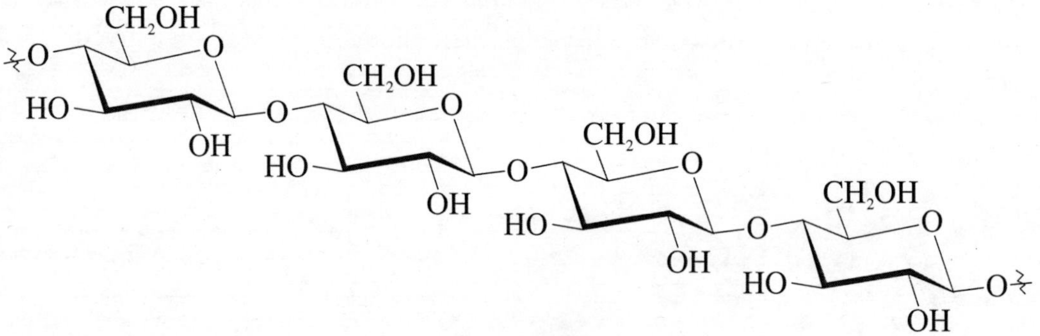

Figure 12.13 Structure of cellulose

Cellulose shows no mutarotation. It is used in tissue paper and cannot be digested by human beings, but animals can digest due to the presence of enzymes.

Cellulose trinitrate called gun cotton is used in explosives.

AMINO ACIDS

Amino acids are bifunctional and they contain an amine and acid group. Amino acids can join together into long chains by forming amide bonds between the NH_2 of one amino acid and COOH of another. Those chains with less than 50 amino acids are called peptides. 20 amino acids commonly found in proteins are listed along with their properties.

All are α amino acids. The amino group in each acid is the substituent in the alpha carbon atom. Except proline all others contain primary amines. Proline is a secondary amine whose nitrogen and carbon form part of a pyrolidine ring.

An amino acid can be referred to in shorthand by a 3-letter code. Foe example *ala* for alanine, *gly* for glycine and so on. In addition a single alphabet indicating the absolute configuration of the optically active amino acid, is present in the name. Except glycine, NH_2CH_2COOH, all others contain stereogenic centres. Only 10 out of 20 can be synthesized by human beings and the remaining come from dietary sources.

Table 12.3 Amino acids and their properties

Amino acid $H_2N-C(CO_2H)(R)-H$ $R-C(H)(NH_2)-CO_2H$

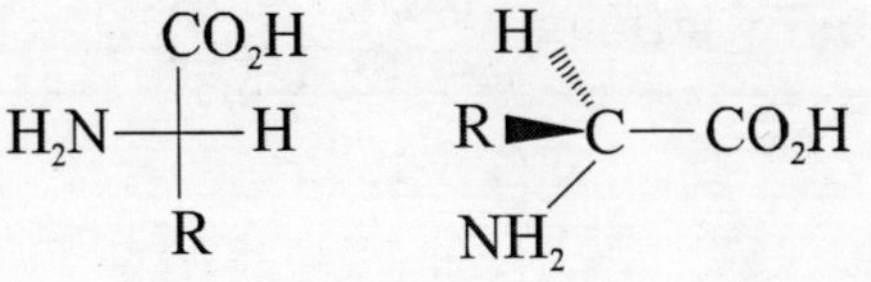

Structure of R	Name	Abbreviations	pK_{a1} α-CO_2H	pK_{a12} α-NH_3^+	pI
Neutral Amino Acids					
$-H$	Glycine	G or Gly	2.3	9.6	6.0
$-CH_3$	Alanine	A or Ala	2.3	9.7	6.0
$-CH_3(CH_3)_2$	Valine	V or Val	2.3	9.6	6.0
$-CH_3CH(CH_3)_2$	Leucine	L or Leu	2.4	9.6	6.0
$-CH_3CH(CH_3)CH_2CH_3$	Isoleucine	I or Ile	2.4	9.7	6.0
$-CH_3-C_6H_5$ 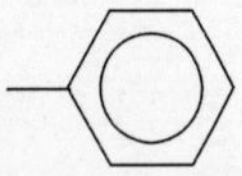	Phenylalanine	F or Phe	1.8	9.1	5.5
$-CH_2CONH_2$	Asparagine	N or Asn	2.0	8.8	5.4
$-CH_2CH_3CONH_2$	Glutamine	Q or Gln	2.2	9.1	5.7
$-CH_3$ 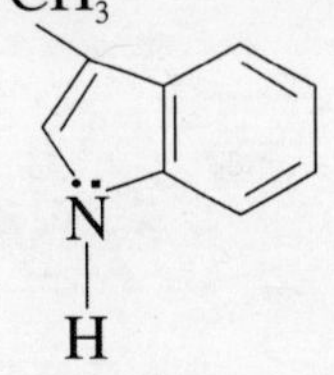 (3-indolyl, N–H)	Tryptophan	W or Trp	2.4	9.4	5.9
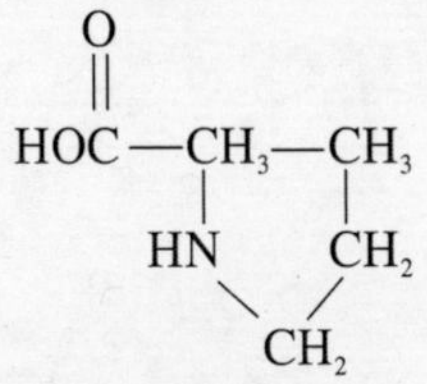 $HOC(=O)-CH_3-CH_3$, HN, CH_2, CH_2 (ring) (Complete structure)	Proline	P or Pro	2.0	10.6	6.3
$-CH_2OH$	Serine	S or Ser	2.2	9.2	5.7

(*Contd.*)

Table 12.3 (Continued)

Structure of R	Name	Abbreviations	pKa_1 α-CO_2H	pKa_{12} α-NH_3^+	pKa_2 R group	pI
Neutral Amino Acids						
$-CH(OH)CH_3$	Threonine	T or Thr	2.6	10.4		6.5
$-CH_2-C_6H_4-OH$	Tyrosine	Y or Tyr	2.2	9.1	10.1	5.7
$HOC(=O)-CH-CH_2$ ring: HN, CH(OH), CH_2 (Complete structure)	Hydroxyproline	Hyp	1.9	9.7		6.3
$-CH_2SH$	Cysteine	C or Cys	1.7	10.8	8.3	5.0
$-CH_2-S-S-CH_2-$	Cystine	Cys-Cys	1.6 2.3	7.9 9.9		5.1
$-CH_2CH_2SCH_3$	Methionine	M or Met	2.3	9.2		5.8
R contains an Acidic (Carboxyl) Group						
$-CH_2CO_2H_2$	Aspartic acid	D or Asp	2.1	9.8	3.9	3.0
$-CH_2CH_2CO_2H$	Glutamic acid	E or Glu	2.2	9.7	4.3	3.2
R contains a Basic Group						
$-CH_2CH_2CH_2CH_2NH_2$	Lysine[e]	K or Lys	2.2	9.0	10.5[b]	9.8
$-CH_2CH_2CH_2NH-C(=NH_2)-NH_2$	Arginine	R or Arg	2.2	9.0	12.5[b]	10.8
$-CH_2-$(imidazolyl, N-H)	Histidine	H or His	1.8	9.2	6.0[b]	7.6

DIPOLAR STRUCTURE OF AMINO ACIDS

Since each amino acid has an acidic and basic group, it undergoes an intramolecular acid–base reaction and exists in the form of a dipolar ion more commonly called as zwitterions (german *switter* = hybrid).

$$H_2N-CHR-COOH \longrightarrow H_3N^+-CHR-COO^-$$

zwitterion

These are internal salts and have properties of salts. They are soluble in water but insoluble in organic solvents like hydrocarbons. They are crystalline substances with high melting points. They are amphoteric and can react with acids or bases depending on circumstances. In aqueous solution a zwitterion accepts a proton to give a cation. In aqueous basic solution, it loses a proton to form an anion.

$$H_3N^+-CHR-COO^- + H_3O^+ \longrightarrow H_3N^+-CHRCOOH + H_2O$$

Isoelectric Point

In acid solution with low pH an amino acid exists as a cation primarily. Similarly in basic solution (high pH) it exists as an anion. Hence at an intermediate pH the amino acid must exist in neutral form (dipolar zwitterions). The pH at this point is called *isoelectric point*. The isoelectric point depends on the amino acid structure. The 15 neutral amino acids have isoelectric point around pH equal to 5 to 6.5. The acidic amino acids have isoelectric point at low pH and the three basic amino acids have isoelectric point at high pH. The difference in isoelectric points of amino acids can be used to separate a mixture of these acids. A technique known as electrophoresis is used. A solution of different amino acids is placed at the centre of a strip of paper or gel. The paper or gel is moistened with an aqueous buffer at a given pH and electrodes are connected at the end of the strip. When an electric potential is applied, the amino acids with negative charges (those are deprotonated because their isoelectric points are lower than the buffer pH) migrate slowly to the positive electrode. Similarly the amino acids with positive charge migrate toward the negative electrode. Different amino acids migrate at different rates depending on isoelectric point and pH of buffer. Thus they can be separated. A solution of ninhydrin is used which gives purple colour with amino acid.

Calculation of % of protonated form, neutral form and deprotonated form is possible using Henderson–Hesselbalch equation (for buffers). Let us calculate these for 1M solution of alanine at pH = 1. The equilibrium between protonated amino acid H_2A^+ and zwitterion HA is considered. The pK_a of protonated form is 2.34.

According to Henderson equation

$$\log[(HA)/(H_2A^+)] = pH - pKa = 1 - 2.34 = -1.34$$

Taking antilog, we get the ratio

$$[HA]/H_2A^+] = 0.046$$

Since

$$[HA] = 1 - [H_2A^+]$$

We have

$$[1 - [H_2A^+]/[HA] = 0.046$$

From this we get $[H_2A^+] = 0.956$ and $[HA] = 0.046$

At pH = 1, 95.6 % exists in protonated form and 4.4% exists in zwitterion form.

SYNTHESIS OF AMINO ACIDS

Strecker synthesis

Aldehydes with NH_3 and HCN followed by acidification and heating give the amino acid.

$$RCHO + NH_3 + HCN \longrightarrow \underset{\displaystyle NH_2}{\underset{|}{RCHCN}} \longrightarrow (H_2O \text{ in acid on heating}) \longrightarrow \underset{\displaystyle NH_3^+}{\underset{|}{RCHCO_2^-}}$$

SOLVED EXAMPLE 2

Synthesize DL tyrosine

Answer

$$HO{-}C_6H_4CH_2CHO + NH_3 + HCN \longrightarrow HO{-}\underset{\displaystyle NH_2}{\underset{|}{C_6H_4CH_2CHCN}}$$

$$\downarrow (H_3O^+ heat, H_2O)$$

$$HOC_6H_4CH_2CHCOO^-$$

POLYPEPTIDES

Enzymes cause alpha amino acids to polymerize, with elimination of water.

$$H_3N^+CHRCOO^- + H_3N^+CHRCOO^- \longrightarrow (-H_2O) \longrightarrow H_3N^+CHRCONHCHRCOO^-$$

(dipeptide)

The CONH bond is the peptide bond. The polymers containing 2 or 3 amino acid residues are called di- or tripeptides respectively. If 3–10 amino acid residues are there, they are called oligopeptides. In a polypeptide

$$H_3N^+—CHR—CO(NH—CHR—CO)_nNHCHRCOO^-$$

The nitrogen end is called N terminal residue and carboxyl end is the C terminal residue. The following is the shorthand notation of glycylvalylphenylalanine (Gly-Val-Phe). The structure is

$$H_3N^+CH_2CO—NH—CH(isopropyl)CONH\underset{\displaystyle CH_2Ar}{\underset{|}{C}}HC\,OO^-$$

We have already indicated that amino acid separation is done based on isoelectric point differences, and ninhydrin gives purple colour. Only proline and hydroxyproline do not answer ninhydrin test since their alpha amino groups are secondary amines and part of a 5-membered ring.

I. "Explain Why" Questions and Small Structure-Based Problems

1. The number of stereoisomers for aldopentoses is 8 but that for aldohexoses is 16. Explain.

   Answer

   Aldopentoses have 3 stereogenic centres and aldohexoses have 4 stereogenic centres. The total number of stereoisomers is given by the formula 2^n where *n* is number of stereogenic centres.

2. Maltose and cellulose answer Tollen's test and Fehling's test. Why?

   Answer

   They are reducing sugars and are in equilibrium with their aldehyde forms.

3. Sucrose does not answer Tollen's test. Why?

   Answer

   It is not a reducing sugar and does not exist in hemiacetal form.

4. Glycine has one distilling fraction and one isomer but glyceraldehydes have one distilling fraction but two isomers. Why?

Answer

Glycine is achiral but glyceraldehydes are chiral.

5. In a 1 M solution of alanine, deprotonated alanine is 17% and zwitterionic alanine is 83%. Explain.

Answer

The pKa values of protonated alanine and zwitterionic alanine are 9.69 and 2.34 and hence % of deprotonated and zwitterionic forms are 17 and 83.

II. Multiple Choice Questions (One out of four answers is correct)

1. One of the following is a disaccharide.
   a) glucose b) starch c) cellulose d) maltose
2. Carbohydrates yielding more than ______ monosaccharides are called polysaccharides.
   a) 2 b) 3 c) 4 d) 10
3. Carbohydrates exist primarily as
   a) aldehydes b) ketones
   c) acids d) acetals or hemiacetals
4. A ketopentose has got
   a) 2 C=O and 3 O—H groups b) one C=O and 4 O—H groups
   c) 3 C=O and 1 OH groups d) 5 OH groups
5. Among the following, one is optically inactive.
   a) sucrose b) maltose
   c) glyceraldehydes d) dihydroxyacetone
6. A pyranose contains monosaccharide which is a ______ membered ring.
   a) 4 b) 5 c) 6 d) 3
7. Mutarotation means
   a) optical rotation becomes negative
   b) optical rotation becomes more positive
   c) change of optical rotation to equilibrium value
   d) optical rotation becomes zero
8. Both aldehydes and ketones answer
   a) haloform reaction b) Liebermann test
   c) $NaHCO_3$ test d) sodium bisulphite test

9. For a carbohydrate to answer Benedict's solution,
   a) it must exist in acetal form
   b) it must exist in hemiacetal form,
   c) it must have keto group
   d) it must have aldehyde group
10. Tollen's test is answered by
   a) all carbohydrates
   b) nonreducing sugars
   c) reducing sugars
   d) polysaccharides
11. Dihydroxyacetone reacts with HIO_4 to give
   a) 2HCHO
   b) 2HCOOH
   c) $2HCHO + CO_2$
   d) $HCHO + 2CO_2$
12. When ozazone is formed from carbohydrates, the number of molar equivalents of phenyl hydrazine used is
   a) 1
   b) 2
   c) 3
   d) 4
13. Ozazone formation leads to
   a) loss of stereocentre at C3
   b) loss of stereocentre at C4
   c) loss of stereocentre at C5
   d) loss of stereocentre at C2
14. One of the following statements is wrong
   a) Sucrose is hydrolysed to glucose and fructose.
   b) Sucrose gives +ve test with Benedict solution.
   c) Sucrose does not form ozazone.
   d) Sucrose does not show mutarotation.
15. One of the following statements is wrong
   a) Maltose is a nonreducing sugar.
   b) Maltose gives +ve test with Fehling's reagent.
   c) Maltose forms ozazone.
   d) Maltose exists in two anomeric forms.
16. On heating starch, it yields
   a) amylose
   b) amylopectin
   c) 10–20% of amylose and 80–90 % of amylopectin
   d) glucose
17. The compound which is optically inactive is
   a) glycine
   b) alanine
   c) sucrose
   d) glyceraldehyde
18. The cation form of alanine has pKa = 2.3 and dipolar form has pKa = 9.7. The isoelectric point is at pH equal to
   a) 12
   b) 4
   c) 3
   d) 6

19. The strongest acid among the following is
    a) cationic form of alanine
    b) propanoic acid
    c) acetic acid
    d) peroxyacetic acid
20. The concentration of dipolar ion is maximum at pH =
    a) 4
    b) 6
    c) 9.7
    d) 2.3
21. The alpha amino acid that does not give purple colour with ninhydrin is
    a) alanine
    b) proline
    c) leucine
    d) valine
22. Hydroxyproline does not answer ninhydrin test because
    a) it is not alpha amino acid
    b) this amino acid is a secondary amine
    c) forms part of 5-membered ring
    d) contains hydroxy group
23. Natural rubber is a polymer of
    a) 1,3-butadiene
    b) 1,2-pentadiene
    c) 2-methyl-1,3-butadiene
    d) 1,3,5-hexatriene
24. The catalyst used for conversion of isoprene to natural rubber is
    a) Ziegler–Natta catalyst
    b) Ni
    c) $AlCl_3$
    d) Fe
25. Natural rubber is a ________ polymer of isoprene
    a) 1,2 addition
    b) 1,3 addition
    c) 1,4 addition
    d) condensation
26. In natural rubber all double bonds are
    a) *trans*
    b) equivalent
    c) *cis*
    d) no suitable answer
27. Vulcanization is possible due to
    a) presence of conjugated double bonds
    b) double bonds providing vinylic hydrogens
    c) double bonds providing allylic hydrogens
    d) double bonds being isolated
28. Paper napkin tissues are made from
    a) fructose
    b) glucose
    c) sucrose
    d) cellulose
29. Rubber on ozonolysis finally produces
    a) a ketone
    b) keto acid called levulinic acid
    c) aldehyde
    d) dialdehyde
30. The carbohydrate which does not show mutarotation is
    a) polysaccharides
    b) sucrose
    c) maltose
    d) α-D-glucopyranose

KEY

1. d	2. a	3. d	4. b	5. a	6. c	7. c
8. d	9. b	10. c	11. c	12. c	13. d	14. b
15. a	16. c	17. a	18. d	19. a	20. b	21. b
22. b	23. c	24. a	25. c	26. c	27. c	28. d
29. b	30. a					

III. Multiple Choice Questions (More than one answer is correct)

1. Some of the following are monosaccharides

a) glucose b) starch c) cellulose d) glyceraldehyde

2. The carbohydrates which answer Tollen's test and Fehling's test are

a) glucose b) sucrose

c) $CH_2OH—(CHOH)_nCHO$ d) maltose

3. Glyceraldehyde by oxidative cleavage with HIO_4 gives

a) HCOOH b) HCHO c) CH_3CHO d) CH_3COCH_3

4. Dihydroxyacetone react with HIO_4 to give

a) HCHO b) CO_2 c) CH_3CHO d) CH_3COCH_3

5. Among the following compounds, optically active compounds are

a) sucrose b) maltose c) glyceraldehydes d) dihydroxyacetone

6. The reagents used to convert $PhCH_2CHO$ to $PhCH_2CHNH_3^+COO^-$ are

a) $CN^- + NH_3$ b) OH^- heat c) H_3O^+ d) Na_2CO_3

7. Adipic acid can be converted to proline by

a) 1 eq. of $SOCl_2 + NH_3$ b) Br_2/KOH

c) Br_2/P followed by intramolecular S_N2 reaction d) $NaHSO_4$

8. Amino acids with more than one chiral centre are

a) glycine b) Isoluecine c) Threonine d) 4-hydroxyproline

9. A protein in water weighing 1 g/litre has osmotic pressure of 4×10^{-3} atm. at 298 K. The molecular weight of protein is

a) 6116 g b) approximately 100 times that of urea

c) 1200 g d) 12000 g

10. An aldohexose and 2-ketohexose
    a) both give positive test by Tollens reagent
    b) Both give the same acid
    c) Both sugars react with aq. Br_2
    d) The keto sugar rearranges to the aldohexose in basic conditions

KEY

1. a,d **2.** a,c,d **3.** a,b **4.** a,b **5.** a,b,c **6.** a,b,c **7.** a,b,c
8. b,c,d **9.** a,b **10.** a,b,d

IV Comprehension Passages

Passage 1

Carbohydrates may be monosaccharides, disaccharides or polysaccharides. Some carbohydrates are reducing sugars and answer Tollen's Test and Fehling's test. Many carbohydrates have more than one stereogenic centre and there are correspondingly many stereoisomers.

Answer the following questions

1. The optically inactive compound among the following is
   a) glyceraldehydes b) glucose
   c) sucrose d) dihydroxyacetone
2. For a carbohydrate to answer Benedict's solution
   a) it must exist in acetal form b) it must exist in hemiacetal form
   c) it must have keto group d) it must have aldehyde group
3. An aldotetrose has ______ chiral centres
   a) 1 b) 2 c) 3 d) 4

KEY

1. d **2.** b **3.** b

Passage 2

Amino acids contain acid group and NH_2 group. Since each amino acid has an acidic and basic group, it undergoes an intramolecular acid–base reaction and exists in the form of a dipolar ion more commonly called as zwitterions. In acid solution with low pH, an amino acid exists as a cation primarily. Similarly in basic solution (high pH) it exists as an anion. Hence at an intermediate pH the amino acid must exist in neutral form (dipolar zwitterions). The pH at this point is called isoelectric point. A knowledge of $pK_a = 1$ and $pK_a = 2$ values can be used to find pH at isoelectric point. Most of the amino acids are optically active.

Answer the following questions

1. The amino acid which is optically inactive is

   a) glycine b) alanine c) threonine d) isoleucine

2. The amino acid with two chiral centres is

   a) alanine b) glycine c) phenylalanine d) isoleucine

KEY

1. a **2.** d

Passage 3

Polymers are large molecules obtained by polymerization of smaller molecules called monomers. The polymerization can happen by

a) Free radical polymerization
b) cationic polymerization
c) anionic polymerization

Answer the following questions

1. Free radical polymerization is initiated by

   a) ethyl radical b) benzoyl peroxide c) allyl radical d) benzyl radical

2. Cationic polymerization is initiated by ________ catalyst

   a) BF_3 b) $AlCl_3$ c) Fe d) Ni/H_2

3. Anionic polymerization is initiated by

   a) Fe b) BF_3
   c) organolithium reagent d) carbonium ions

KEY

1. b **2.** a **3.** c

V. Matrix Matching

Question 1

Column 1	Column 2
a) glycine	p) chiral
b) alanine	q) achiral
c) isoleucine	r) chiral with 2 stereogenic centres
d) fructose	s) gives positive test with Tollen's reagent

Answer

$[a \rightarrow q], [b \rightarrow p], [c \rightarrow r], [d \rightarrow s]$

Question 2

Column 1	Column 2
a) isoelectric point of glycine	p) achiral
b) pKa_2 of alanine	q) 5.97
c) isoelectric point of aspartic acid	r) 9.69
d) dihydroxyacetone	s) 2.77

Answer

$[a \rightarrow q], [b \rightarrow r], [c \rightarrow s], [d \rightarrow p]$

Question 3

Column 1	Column 2
a) glyceraldehydes	p) nonreducing sugar
b) aldotetroses	q) triose
c) dihydroxyacetone	r) achiral
d) sucrose	s) two stereogenic centres

Answer

$[a \rightarrow q], [b \rightarrow s], [c \rightarrow r], [d \rightarrow p]$

VI. STRUCTURE-BASED AND NUMERICAL PROBLEMS

1. Threonine has $pKa_1 = 2.09$ and $pKa_2 = 9.1$. Calculate the percentage of protonated and zwitterions forms at pH = 1.5 and also at pH 10.

   Answer

   (protonated form 79.6% zwitterions form 20.4% at pH =1.5 and at pH =10 11.2 % zwitterions form and 88.8 % for anionic form). The isoelectric point for an amino acid alanine is the average of two pKa values.

2. The molecular weight of a protein is 6116 g. If concentration is 1 g/litre at 298 K, what is its osmotic pressure?

   Answer

   MW = gRT/D

   $D = gRT/MW = 1 \times 0.0821 \times 298/6116 = 4 \times 10^{-3}$ atm

3. Haemoglobin contains 0.35% of Fe. Each haemoglobin has four Fe atoms. What is the molecular weight of haemoglobin?

   Answer

   4 atoms of Fe correspond to 224 g. This is 0.35%.

   Hence molecular weight of haemoglobin is 224/0.0035 = 64000 g.

4. Compound A on hydration in Hg^{2+}/H_2SO_4 gives B. B is a product of ozonolysis of 2-butene in reducing conditions. B reacts with NH_3, HCN to give C. C is treated with acid and heated with water to get D which is an amino acid. Is D optically active?

   Answer

   A is Acetylene;

   B is CH_3CHO;

   C is $CH_3\underset{\displaystyle NH_2}{\underset{|}{C}}HCN$;

   D is $CH_3\underset{\displaystyle N^+H_3}{\underset{|}{C}}HCO_2^-$;

   D is optically active.

5. *o*-xylene on oxidation with $KMnO_4$ gives A. Potassium phthalimide reacts with B to give C. C with KOH/H_2O followed by addition of HCl gives D and A and ethanol. Find A to D

Answer

A is phthalic acid;

B is ethylchloroacetate;

C is phthalimidoethylacetate;

D is glycine.

13

PRACTICAL ORGANIC CHEMISTRY

DETECTION OF ORGANIC ELEMENTS

An organic compound may contain carbon, hydrogen, oxygen, nitrogen, sulphur, halogens and rarely phosphorus.

Ignition Test

Place about 0.1 g of the substance on a metal spatula, heat it gently and finally to dull redness.

Observe whether

1. substance melts and is explosive or inflammable and note nature of flame.
2. gases or vapour evolved and note smell.
3. residue fuses.

Burning with a smoky flame indicates carbon and hydrogen and no metals. A black residue consisting of largely carbon is left. Aromatic compounds burn with sooty flame due to large carbon content relative to hydrogen.

DETECTION OF CARBON AND HYDROGEN

The organic compound is mixed with dry air without moisture or CO_2 along with dry black CuO. The mixture is heated strongly when mixture of oxides are formed.

If gaseous mixture is passed through anhydrous $CuSO_4$ and if it turns blue, hydrogen is confirmed. If lime water turns milky, carbon is confirmed. Generally every organic compound must have carbon and hydrogen and these tests are not very essential.

$$C + 2CuO \longrightarrow CO_2 + 2Cu$$

$$H_2 + CuO \longrightarrow H_2O + Cu$$

$$\underset{\text{White}}{CuSO_4} + 5H_2O \longrightarrow \underset{\text{Blue}}{CuSO_4 \cdot 5H_2O}$$

$$Co_2 + Ca(OH)_2 \longrightarrow \underset{\text{Milky suspension}}{CaCO_3} + H_2O$$

CuO is hygroscopic and thus is always ignited to dryness before use. Both SO_2 and CO_2 turn lime water milky and if sulphur is also present, the mixture is passed through acidified dichromate before passing through lime water which absorbs all SO_2.

$$S + 2CuO \longrightarrow SO_2 + 2Cu$$

$$SO_2 + Ca(OH)_2 \longrightarrow CaSO_3 + H_2O$$

$$3SO_2 + K_2CR_2O_7 + 3H_2SO_4 \longrightarrow K_2SO_4 + CR_2(SO_4)_3 + 3H_2O$$

Heating with Soda lime

Mix about 0.2 g of substance with soda lime. Place the mixture in a pyrex test tube, and close tube with bung and a delivery tube. Incline the test tube so that the liquid formed does not run back to the hot part of the tube. Heat gently and then strongly. Collect any condensate in a test tube with 3 ml of water. Nitrogenous compounds evolve ammonia or vapours alkaline to indicator paper and possessing characteristic odours. Hydroxybenzoic acids yield phenolic formates, acetates yield hydrogen. Carboxylic acids yield hydrocarbons (benzene from benzoic acid). Aromatic amino acids yield amines.

DETECTION OF NITROGEN, SULPHUR AND HALOGENS

Sodium Fusion Test

Organic compounds are generally non-polar and it is necessary to convert them to inorganic salts so that ionic tests of inorganic salts may be applied. The most important method is to use the compound with a piece of sodium (Lassaignes test). In this way NaCN, Na_2S and NaX are formed, if nitrogen, sulphur or halogens are present. It is essential to use excess Na. If N and S are present together, NaCNS will be formed giving red colouration with ferric ions without answering separate tests for nitrogen or sulphur. With excess of Na, NaCN is formed.

$$NaCNS + Na \longrightarrow NaCN + Na_2S$$

Nitrogen is then detected by Prussian blue test. The filtered alkaline solution resulting from action of water upon sodium fusion is treated with freshly prepared ferrous sulphate and thus forms sodium hexacyanoferrate. On boiling the alkaline solution, the Fe(II) gets oxidized by air to Fe(III) and the Fe(II) and Fe(III) hydroxides are dissolved. The hexacyanoferrate reacts with Fe(III) forming Prussian blue (blue colour). The reactions are:

$$FeSO_4 + 6NaCN \longrightarrow Na_4Fe(CN)_6$$

$$Fe^{2+} + OH^- \longrightarrow Fe^{3+}$$

$$3Na_4Fe(CN)_6 + 4Fe^{3+} \longrightarrow 12Na + \underset{\text{Prussian blue}}{Fe_4[Fe(CN)_6]_3}$$

Sometimes the prussian blue appears green due to colloidal state.

The test fails in the case of diazo compounds since nitrogen is lost on heating.

Compounds like NH_2NH_2, NH_2OH do not give this test, (no carbon present). Freshly prepared ferrous sulphate is used because ferrous sulphate on keeping is oxidized to ferric sulphate.

HCl should not be used, since in alkaline solution yellow colour formed due to $FeCl_3$ causes Prussian blue to appear greenish (ferric hydroxide is also green). $FeCl_3$ should also not be added. An alternative test for cyanide ion is the formation of deep purple colouration when solution is treated with (alkaline *p*-nitrobenzaldehyde + dinitrobenzene reagent). Sulphur may be detected by the addition of lead acetate and acetic acid which turns black or purple with sodium nitroprusside (disodium pentacyanonitrosoferrate(III)). Halogens are detected by silver halide formation (curdy white for chlorine, yellow for iodide and light yellow for bromide ions). When nitric acid and silver nitrate are added, the nitrogen and sulphur interfere with the detection of halogens and they must be removed by boiling the solution before silver nitrate addition.

AgCN or Ag_2S formed give white and black precipitates and when we boil with HNO_3 the following reactions take place.

$$NaCN + HNO_3 \longrightarrow NaNO_3 + HCN\downarrow$$

$$Na_2S + 2HNO_3 \longrightarrow 2NaNO_3 + H_2S\downarrow$$

Lassaigne's Test

Sodium metal piece (kept in ether) is added to an ignition tube and heated till it is red-hot, then a small pinch of organic solid or 3 drops of organic liquid are added to the test tube, heated and then plunged into small amount of water kept in a 100 ml beaker. The glass breaks and the solution is filtered and liquid in fusion extract tested in portions for nitrogen, sulphur and halogens. The reactions are

$$\text{Organic compound} + Na \longrightarrow NaCN \text{ or } Na_2S \text{ or } NaX$$

$$Na + C + N \longrightarrow NaCN$$
$$2Na + S \longrightarrow Na_2S$$
$$Na + C + N + S \longrightarrow NaCNS$$
$$2Na + X_2 \longrightarrow 2NaX$$

Sodium must be completely reacted before the substance is plunged into water, otherwise sodium catches fire in water.

Only small amount of water should be used.

Lithium and potassium cannot be used because reaction of lithium is slow and its compounds are covalent. Potassium reacts vigorously and is difficult to handle.

The formation of Prussian blue is already given in equations above.

Reactions for Sulphur Detection

Fusion extract + sodium nitroprusside gives deep violet colour.

$$Na_2S + Na_2[Fe(CN)_5NO] \longrightarrow Na_4[Fe(CN)_5NOS] \text{ violet}$$

$$Na + C + N + S \longrightarrow NaCNS$$

If N and S are both present these is red colouration

$$NaCNS + FeCl_3 \longrightarrow FeCl_2(CNS) + NaCl$$

Halogens

$$\text{Extract} + \text{conc.}HNO_3 \text{ boil and add } AgNO_3$$
$$NaX + AgNO_3 \longrightarrow AgX + NaNO_3$$

Layer test for bromine and iodine Sodium extract + dil. HNO_3 + $CHCl_3$ or CCl_4 + chlorine water $\longrightarrow$ brown layer for bromine, violet layer for iodine.

$$NaBr + Cl_2 \longrightarrow NaCl + Br_2 \text{ (brown layer in chloroform)}$$
$$NaI + Cl_2 \longrightarrow NaCl + I_2 \text{ (violet layer in chloroform)}$$

Beilstein test Halogen-containing organic compounds on heating over a Cu wire loop, impart green colour or blue colour to flame. This is due to the formation of copper halide. Urea, thiourea and pyridine also answer this test and so it is not very useful.

DETECTION OF PHOSPHORUS

Organic compound is treated with Na_2CO_3 and KNO_3 and heated.

Then mixture is dissolved in water and conc. HNO_3 and ammonium molybdate is added. Yellow colour or precipitate indicates phosphorus. The relevant equations are

$$3Na_2CO_3 + 2P + (5/2)O_2 \longrightarrow 2Na_3PO_4 + 3CO_2$$
$$Na_3PO_4 + 3HNO_3 \longrightarrow H_3PO_4 + 3NaNO_3$$
$$H_3PO_4 + 12(NH_4)_2MoO_4 + 21HNO_3 \longrightarrow \underset{\text{Yellow}}{(NH_4)_3PO_4\bullet 12MoO_3} + 21NH_4NO_3 + 12H_2O$$

No separate tests are conducted for oxygen. Heating organic compound in the presence of nitrogen leads to water formation in cooler parts of the test tube and shows the presence of oxygen.

Tests for functional groups like acids, alcohols, aldehydes and amides indicate the presence of oxygen.

Unsaturated compounds decolorize $KMnO_4$ and change bromine water colour.

FUNCTIONAL GROUP IDENTIFICATION

Organic compounds may be acidic (acids), basic (amines), neutral (aldehydes, ketones, nitro compounds), etc. The solubility behaviour of these compounds in water, NaOH and $NaHCO_3$ will be used to classify them as acidic, basic or neutral.

When an organic compound contains more than one functional group, the identification is generally based upon the one easily detected. For example benzoic acid, *p*-chlorobenzoic acid, *p*-nitro- or *p*-methoxybenzoic acid are classified as acids. Elements like nitrogen and chlorine in these compounds may be confirmed by preparing suitable derivatives.

SOLUBILITY BEHAVIOUR OF ORGANIC COMPOUNDS

Carbohydrates like glucose, nitrogen compounds like urea and thiourea dissolve in water due to hydrogen bonding. Also organic salts, monofunctional alcohols, ROR′, amines like NR_3, nitriles are water-soluble. Saturated hydrocarbons, aromatic hydrocarbons, alkyl and aryl halides are water-insoluble.

Carboxylic acids dissolve in $NaHCO_3$ and in NaOH. Sulphonic acids and substituted phenols involving electron-withdrawing groups at *ortho* or *para* position dissolve in $NaHCO_3$. Effervescence is observed when acid is treated with $NaHCO_3$. Phenols dissolve in NaOH due to weak acidity. They are partially soluble in water. Bases like amines dissolve in acids like HCl. Aldehydes, ketones and nitro compounds dissolve in ether. Alkenes and alkynes dissolve in 96% sulphuric acid. Among amino acids, low molecular weight acids are water soluble and high molecular weight amino acids are soluble in acids and bases.

SPECIFIC TESTS FOR FUNCTIONAL GROUPS

Confirmatory Test for Acids

Ester formation Warm a small amount of unknown with 2 parts of absolute ethanol and 1 part of concentrated sulphuric acid for 2 minutes. Cool and pour cautiously into aqueous sodium carbonate solution contained in an evaporating dish and smell immediately. Sweet fruity odour indicates ester. Acids of high molecular weights give odourless esters.

Phenols

Phenols do not liberate CO_2 with $NaHCO_3$ solution. Most of phenols are crystalline solids except probably *m*-cresol and *o*-bromophenol. Monohydric phenols have characteristic odours.

1. *Confirmatory test* Prepare a solution of phenol in water and add one or two drops of neutral 1% $FeCl_3$ solution (add dilute NaOH to $FeCl_3$ (free from HCl) till precipitate of ferric hydroxide is formed. Filter and use solution). Add another drop after 2 minutes. If a transient or permanent purple-blue or green colour appears, it indicates phenol. If colour does not appear, use absolute ethanol or methanol as solvent.

2. *Bromine water test* Many phenols (except those with strong reducing properties) yield crystalline bromo compounds on adding bromine water. Dissolve 0.25 g of compound in 10 ml of dil. HCl and add bromine water dropwise. Colour of bromine is slowly discharged and white precipitate of bromophenol appears. Phenols may also answer $KMnO_4$ (Baeyer's test) but the test is answered by alkenes and alkynes also.

3. *Liebermann's test* Add a crystal of $NaNO_2$ to 2 ml of conc. H_2SO_4 until it dissolves. Add 0.1 g of unknown. A blue colour or change of colour of solution to red when poured into 20 ml of ice-water and return of blue colour after adding 10% NaOH indicates presence of phenol.

4. *Phthalein test* Many phenols answer phthalein test. 0.5 g of phenol is mixed with equal amount of phthalic anhydride and 1 drop of conc. H_2SO_4 is added. The beaker containing the mixture is allowed to stand in paraffin oil bath (previously heated to 160ºC). Remove from bath, cool and add 4 ml of 5% NaOH until fused mass dissolves. Dilute with equal volume of water, filter and examine colour of filtrate against white background. It exhibits fluorescence.

Dicarboxylic acids also answer phthalein fusion test.

Primary and Secondary Aliphatic Nitro Compounds

They dissolve in NaOH to give yellow solution. They can be reduced to hydroxylamine and ammoniacal silver nitrate solution gives silver mirror (Tollens test). To distinguish primary and secondary nitro compounds the following test is conducted.

Dissolve a few drops of the compound in conc. NaOH solution and add excess sodium nitrite solution. Upon careful addition of dil. sulphuric acid adding drop by drop we observe the following effects.

- *Primary nitro compounds*—Intense red colour disappearing on acidification. The colour is due to alkali salt of nitronic acid.

$$RCH_2NO_2 + NO \longrightarrow RCH(NO)\text{-}NO_2 \longrightarrow \underset{\underset{NOH}{\|}}{RC} - NO_2$$

- *Secondary nitro compounds*—dark blue or green colour due to nitroso derivative. The coloured compound is soluble in chloroform.

$$RR'CHNO_2 + NO \longrightarrow RR'CNONO_2$$

Tertiary nitro compounds are neutral and do not give this reaction.

- *Tertiary nitro compounds and aromatic nitro compounds*—reduced by Zn and HCl to hydroxylamines which react with ammoniacal silver nitrate to give silver mirror in cold.

Amines Organic compounds that dissolve in HCl consist of nitrogen compounds, pyrones and anthocyanin pigments are exceptions.

Distinction of Primary, Secondary and Tertiary Amines

Reaction with nitrous acid Dissolve 0.2 g of substance in 5 ml of 2 M HCl and in ice and add 2 ml of ice-cold 10% aq. $NaNO_2$ by means of a dropper and with stirring until after standing for 3 or 4 minutes. Immediate positive test for nitrous acid is obtained with starch iodide paper. A clear solution with continuous evolution of nitrogen indicates primary amines (aliphatic or aryl alkyl). Add one half of the solution to cold solution of 0.4 g of 2-naphthol or 5 ml of 5% NaOH solution. The formation of a coloured (orange-red) azo dye indicates primary aromatic amine. If the other half of the diazotized solution is warmed, we observe evolution of nitrogen and strong phenolic odour is there for the aromatic amines.

Tertiary aliphatic amines With sodium nitrite in ice-cold condition, a colourless solution is obtained and it gives positive test with starch iodide paper on addition of small amount of sodium nitrite indicating that it is a tertiary aliphatic amine.

Secondary amines With ice-cold $NaNO_2$ if orange-yellow oil or low-melting solid separates (N-nitrosoamines) secondary amine is indicated. The nitrosoamine can be confirmed by conducting Liebermann test.

Tertiary aromatic amines Treatment with nitrous acid gives dark orange-red solution or orange crystalline precipitate which results from formation of hydrochloride of C-nitrosoamine. Addition of NaOH or Na_2CO_3 yields bright green nitrosamine base.

Carbylamine test for aromatic primary amines To 1 ml of alcoholic KOH, add 0.05 g of amine and 3 drops of chloroform and heat to boiling. A carbylamine (phenyl isocyanide) is formed and easily identified by extremely nauseating smell.

$$RNH_2 + CHCl_3 + 3KOH \longrightarrow RNC + 3KCl + 3H_2O$$

After reaction is over, add conc. HCl to decompose isocyanide and pour it away after odour stops.

NEUTRAL COMPOUNDS

Aldehydes, ketones, esters, alcohols and ethers are neutral compounds.

Aldehydes Aldehydes and ketones give characteristic solid derivatives with 2,4-dinitrophenyl hydrazine.

1. *Tests for aldehyde alone* 2 drops of the aldehyde and 2 ml of Schiffs reagent are mixed and shaken in cold. A pink colour is reduced. Some aromatic aldehydes do not answer the test.

Schiffs reagent is prepared as follows: add 0.2 *p*-rosaniline hydrochloride in 20 ml of cold freshly prepared solution of sulphur dioxide and allow it to stand for a few hours till it becomes colourless or pale yellow. Dilute to 200 ml and keep in stoppered bottle. Add 2 g of sodium metabisulphite to the solution of rosaniline hydrochloride and 2 ml of HCl in 200 ml of water.

Free alkali or alkali salts of acids redden the reagent like aldehyde.

2. *Tollens reagent* The aldehydes with ammoniacal silver nitrate reduce silver oxide to silver and shining silver mirror is produced when aldehyde and Tollens reagent (ammoniacal $AgNO_3$) are mixed and heated.

After the test, pour contents of test tube into sink and wash with dil. nitric acid. Any silver fulminate present which is explosive when dry, will thus be destroyed.

$$RCHO + 2Ag(NH_3)_2OH \longrightarrow 2Ag(s) + RCOONH_4 + H_2O + 3NH_3$$

The test is conducted for carbohydrates also.

Aromatic amines and some phenols and alpha-dialkylaminoketones also answer this test.

3. *Fehlings solution* Dissolve 17.32 g of hydrated $CuSO_4$ in 200 ml of water and dilute to 250 ml. Dissolve 86.5 g of sodium potassium tartrate in 35 g of NaOH in 100 ml of water and dilute to 250 ml. Mix 2.5 ml of each before use.

0.2 g of substance in 5 ml of water is mixed with 5 ml of Fehlings solution, and heated to boiling. Cool the solution. Precipitation of CuO as red-yellow or yellowish green solid indicates the presence of aldehydes.

$$RCHO + Cu^{2+} \longrightarrow RCOOH + Cu_2O(s)$$

For carbohydrates having the group RCOH (with CHO attached below the C) addition of di-positive Cu leads to RCOCOOH and $Cu_2O(s)$.

Methyl Ketones (iodoform test) Methyl ketones, CH_3CHO, $RCHOHCH_3$ and ethanol answer this test.

$$RCHOHCH_3 + I_2/2NaOH \longrightarrow RCOCH_3 + 2NaI + 2H_2O$$

Ethanol is oxidized to acetaldehyde which has CH_3CO group and answers this test.

The reaction of methyl ketones is given below.

$$RCOCH_3 + 3I_2 + 4NaOH \longrightarrow RCOONa + 3NaI + 3H_2O + CHI_3(s)$$

Alcohols Metallic sodium reacts with alcohols evolving hydrogen.

Lucas test Dissolve 13.6 g of anhydrous $ZnCl_2$ in (0,1mole) of concentrated HCl with cooling. This is Lucas reagent.

To 0.2 ml of sample in a test tube, add 2 ml of Lucas reagent at room temperature. Stopper the tube and shake. Allow the mixture to stand. Note the time required for formation of alkyl chloride, which appears as insoluble emulsion. Note the time required for the reaction to take place and appearance of cloudy second layer of emulsion. If it is immediate or observed in 2 to 3 minutes, it is tertiary alcohol. If turbidity appears in 5 to 10 minutes, it is secondary alcohol. If no reaction takes place, the alcohol is primary.

$$RR'R''OH + HCl \text{ (with } ZnCl_2 \text{ catalyst)} \longrightarrow RR'R''Cl + H_2O$$

The test applies to monofunctional alcohols soluble in reagent (The number of carbons is less than six and some polyhydroxy alcohols also answer this text.)

SEPARATION OF BINARY MIXTURES

Mixtures of different functional groups or mixture of different types of amines, etc. can be separated by various techniques. Particular methods depend on physical and chemical properties of the constituents. Properties of components of the mixture in terms of volatility, solubility characteristics, acidic or basic properties should be looked into. Chromatographic methods and spectroscopy and chemical methods can be employed. Here we will see the chemical methods.

Separation of toluene and aniline If extraction is carried with dilute HCl, aniline passes into aq. layer in the form of the salt, aniline hydrochloride, and can be recovered by neutralization.

Phenol and toluene By treatment with NaOH, phenol goes into aqueous layer as sodium salt and can be recovered by acidification. Phenols are separated from alcohols by the same method.

These are examples due to differences in solubility.

Dibutyl ether and chlorobenzene Concentrated sulphuric acid dissolves the ether and this is recovered by dilution with water. Concentrated sulphuric acid cannot be used with unsaturated compounds because it may lead to their polymerization, sulphonation, etc.

Phenols and acids: *o*-cresol is separated from benzoic acid by a dilute solution of $NaHCO_3$. The weakly acidic phenols are not converted to their salts and may be removed by extraction with ether. The acids convert to sodium salts in aqueous layer and are obtained by acidification.

Separation of amines Hinsberg test: This is based on the fact that the reaction with benzene sulphonyl chloride or *p*-toluene sulphonyl chloride converts primary amines to alkali-soluble sulphonamides, secondary amines into alkali-insoluble sulphonamides and tertiary amines are unaffected.

$$PhSO_2Cl + H_2NR \xrightarrow{(NaOH)} \underset{\text{(Water-soluble)}}{[PhSO_2NR]Na^+} \xrightarrow{HQ} \underset{\text{(Water-insoluble)}}{PhSO_2ClNHR}$$

$$PhSO_2Cl + NHR_2 \xrightarrow{(NaOH)} \underset{\text{(water-insoluble)}}{PhSO_2NR_2}$$

The water insoluble solution with acid has no reaction.

For tertiary amines there is no reaction with benzene sulphonyl chloride in NaOH and the amine gets protonated in HCl to amine hydrochloride which is soluble.

Procedure 2 g of the mixture of amines is mixed with 40 ml of 10% NaOH. Add 4 g (3 ml) of benzene sulphonyl chloride in small portions. Warm on a water bath to complete the reaction. Acidify with dil. HCl when sulphonamides of primary and secondary amines are precipitated. Filter off the solid and wash with cold water. Tertiary amine will be in filtrate since it has not reacted. Any disulphonamide of primary amine may be converted to sulphonamide by boiling the solid with 2 g of sodium dissolved in alcohol for 30 minutes. Dilute and distil off alcohol. Filter precipitate of secondary amine sulphonamide. Acidify filtrate with HCl and precipitate derivative of primary amine. Recrystallize derivatives from ethanol.

Aldehydes and hydrocarbons Benzaldehyde may be separated from liquid hydrocarbons by shaking with a solution of sodium metabisulphite. The aldehyde forms solid bisulphite compound. This is filtered and decomposed with dilute acid or with $NaHCO_3$ to recover aldehyde.

Ketones and neutral compounds The Girard T reagent reacts with carbonyl compounds as

$$RR'CO + [Me_3NCH_2CONHNH_2]Cl \longrightarrow [Me_3NCH_2CONHN{=}CRR']Cl + \text{water}$$

The quaternary ammonium salt is polar and soluble in water. Extraction with ether removes water-insoluble compounds. The ketone is regenerated in aqueous solution by hydrolysis with dilute HCl. Girard T reagent is carbohydrazido methyltrimethyl ammonium chloride and is prepared as follows:

Ethylchloroacetate is reacted with trimethylamine and hydrazine hydrate in alcoholic solution

$$Me_3N + ClCH_2COOEt \longrightarrow Cl[Me_3NCH_2COOEt] \longrightarrow NH_2NH_2$$

$$\downarrow$$

$$Cl[Me_3NCH_2CONHNH_2] + \text{ethanol}$$

Separation is based on volatility differences of components in aq. solution.

Diethylamine and butan-l-ol The mixture is separated by adding sufficient sulphuric acid to neutralize the base and steam distillation removes the alcohol.

The amine is recovered by adding NaOH to residue and repeating distillation.

Diethylketone and acetic acid The mixture is treated with sufficient dilute NaOH to transform acid into acetate and distilling aq. mixture. The ketone will pass over steam and nonvolatile salt remains in flask. Acidification with dil. sulphuric acid liberates acid. This can be extracted or steam-distilled.

PROBLEMS

I. "Explain Why" Questions and Small Structure-Based Problems

1. When a mixture of alcohols and phenols is kept in ether and NaOH added, the phenol separates in aqueous layer. Why?

   Answer

   Phenols are more acidic than alcohols. Hence phenol goes to aq. layer.

2. When a mixture of phenols and acids are kept in ether and $NaHCO_3$ is added, the acid settles in aq. layer. Why?

   Answer

   Acids are more acidic than phenols and hence acid settles in aq. layer.

3. When C_6H_5Cl and $C_6H_5CH_2Cl$ are treated with $AgNO_3$, which will give precipitate with $AgNO_3$?

   Answer

   $C_6H_5CH_2Cl$ will react with $AgNO_3$ to give white precipitate of AgCl. Aryl halides do not undergo substitution reactions of S_N1 or S_N2 type.

4. How will you distinguish allyl chloride and n-propyl chloride?

Answer

Allyl chloride in cold gives white precipitate with $AgNO_3$. n-propyl chloride gives after warming since it reacts slowly.

5. How will you distinguish CH_3CH_2CHO and its isomer acetone?

Answer

Acetone with I_2 and NaOH gives yellow precipitate of CHI_3 and CH_3COONa. This haloform reaction will not be answered by CH_3CH_2CHO since it does not have CH_3CO group.

6. How will you distinguish propylene and propyne?

Answer

Propyne has acidic hydrogen and gives precipitate with ammoniacal $AgNO_3$.

7. How do you distinguish n-propyl alcohol and ethanol?

Answer

Ethanol with NaOH and I_2 answers Iodoform reaction and gives yellow precipitate of CHI_3.

8. How will you distinguish acetic acid from formic acid?

Answer

Formic acid contains CHO group and answers Tollens test by giving silver mirror.

9. How will you distinguish 1-butyne and 2-butyne?

Answer

1-butyne with amm. $AgNO_3$ gives a precipitate since it has acidic hydrogen.

10. How do you distinguish n-propanol from methylethyl ether?

Answer

N-propyl alcohol reacts with metallic Na giving hydrogen gas. Acetyl chloride reacts vigorously with the alcohol giving off HCl. Ethers are unaffected by acetyl chloride.

II. Multiple Choice Questions (One out of Four Answers is Correct)

1. Violet colouration of Lassaigne's test with sodium nitroprusside indicates the presence of

   a) nitrogen b) halogens c) sulphur d) oxygen

2. Nitrogen-containing compounds when heated strongly with conc. H_2SO_4 give

   a) HNO_3 b) $(NH_4)_2SO_4$ c) NH_3 d) NH_4Cl

3. The fusion extract is mixed with HNO_3, and ammonium molybdate is added to give yellow precipitate, the element is

   a) halogen b) sulphur c) nitrogen d) phosphorus

4. Before testing for_______ the sodium fusion extract is heated with HNO_3 to decompose Na_2S and NaCN.

   a) phosphorus b) oxygen c) halogens d) no suitable answer

5. The Lassaigne's test for nitrogen is positive for

   a) urea b) NH_2OH c) NH_2NH_2 d) NH_3

6. In the Lassaigne's test, before halogens are analysed, HNO_3 is added to convert Na_2S to

   a) S b) Na_2SO_3 c) H_2S d) Na_2SO_4

7. One of the following is wrong

   a) HNO_3 converts NaCN to $NaNO_3$ and HCN

   b) HNO_3 converts Na_2S to $NaNO_3$ and H_2S

   c) Freshly prepared $FeSO_4$ is important because $FeSO_4$ on keeping becomes ferric sulphate

   d) Na should not be melted before the substance is added for Lassaigne's test

8. In Lassaigne's test for nitrogen, sodium fusion extract should not be mixed with HCl because

   a) prussian blue colour is changed b) yellow $FeCl_3$ is produced

   c) pH is decreased d) no suitable answer

9. In Lassaigne's test, Na must be heated strongly before introducing the sample since

   a) Na violently reacts with water which will be used for preparing extract

   b) Na is inert to water

   c) Na catches fire in ether

   d) substance decomposes

10. One of the following compounds is soluble in ether and water

    a) alcohols b) hydroxy acids c) amides d) sulphonic acids

11. One of the following is insoluble in ether but soluble in water

    a) alcohols b) aldehydes c) hydroxy acids d) amines

12. One of the following gives CO_2 with $NaHCO_3$

    a) alcohols b) acids c) phenols d) hydrocarbons

13. Lucas test will not be answered by

    a) $CH_3CHOHCH_3$ b) butan-1-ol

    c) be tertiary butyl alcohol f) $C_6H_5CHOHCH_3$

14. For Lucas test the alcohol must_______ in reagent

    a) be forming precipitate b) be soluble

    c) be insoluble d) no suitable answer

15. The ethanolic silver nitrate test is not answered by
   a) alkyl halides b) carboxylic acids c) acyl chlorides d) amines
16. Alkyl chlorides and bromides answer NaI acetone test because
   a) alkyl halides have dipole moment
   b) NaCl and NaBr formed are insoluble in acetone
   c) alkyl halides are polar
   d) no suitable answer
17. The NaI acetone test should be conducted with moderate heat because
   a) alkyl halide decomposes
   b) NaCl formed dissolves at high temperature
   c) excessive heating leads to loss of acetone
   d) enol of acetone is formed
18. The following test is not positive for terminal alkynes.
   a) Baeyer's reagent b) bromine water
   c) sodium detection of active hydrogen d) Fehling's solution
19. Ethanolic silver nitrate test is not answered by
   a) carboxylic acids b) alkyl halides c) acyl halides d) nitro compounds
20. Iodoform test is not answered by
   a) acetaldehyde b) ethanol c) acetone d) benzophenone
21. $CH_3CH=CH_2$ on ozonolysis in reducing conditions gives CH_3CHO and A. A will not answer
   a) Tollens reagent test b) Schiffs reagent test
   c) iodoform test d) Fehling's test
22. Toluene, $C_6H_5CH_2Cl$ and $C_6H_5CH_2CH_3$, get oxidized to same compound A by $KMnO_4$ oxidation. A will answer
   a) haloform reaction b) $NaHCO_3$ test
   c) Fehling's test d) Tollens test
23. 2-butene on ozonolysis in dimethyl sulphide gives 2 moles of A. A will not answer
   a) Tollens test b) Benedict's solution test
   c) haloform test d) $NaHCO_3$ test
24. A is a good S_N2 solvent. It is obtained by the ozonolysis of tetramethylethylene. It will not answer
   a) Tollens reagent test b) sodium bisulphate test
   c) phenyl hydrazine test d) haloform test

25. Compound A very easily undergoes dehydration to give 2-methylpropene. A will answer
    a) haloform test b) $NaHCO_3$ test c) Lucas test d) neutral ferric chloride test
26. This compound A exists 100% in enol form. It is also acidic. It does not answer _______ test.
    a) $NaHCO_3$ b) neutral $FeCl_3$
    c) Liebermann test d) bromine decolorization
27. The carbylamine test is not answered by
    a) CH_3NH_2 b) $C_2H_5NH_2$ c) $C_6H_5NH_2$ d) $(CH_3)_3N$
28. When a mixture of aniline and toluene is extracted with HCl
    a) toluene passes into aq. layer b) aniline passes into aq. layer
    c) NaOH has to be added to separate them d) no suitable answer
29. Hinsberg test leaves_______unaffected
    a) $C_6H_5CH_2NH_2$ b) $(CH_3)_2NH$ c) $(C_2H_5)_3N$ d) CH_3NH_2
30. *Ortho*-nitrophenol can be separated from *p*-nitrophenol by
    a) ether extraction b) distillation
    c) forming azeotrope with benzene d) steam distillation

KEY

1. c	2. b	3. d	4. c	5. a	6. c	7. d
8. b	9. a	10. a	11. c	12. b	13. b	14. b
15. d	16. b	17. c	18. d	19. d	20. d	21. c
22. b	23. d	24. a	25. c	26. a	27. d	28. b
29. c	30. d					

III. Screening Test Questions of Previous IIT-JEE Question Papers

1. Compounds A and B are given below

A is $CH_2{=}\underset{\displaystyle H}{\underset{|}{C}}OCOCH_3$

B is $CH_2{=}\underset{\displaystyle CH_3}{\underset{|}{C}}\text{-}OCOCH_3$

Which of the following can be used to distinguish their hydrolysis products?

a) 2,4-dinitrophenylhydrazine b) Lucas reagent
c) Fehlings solution d) $NaHSO_3$ (2003)

2. Identify a reagent which distinguishes 1-butyne and 2-butyne

a) Bromine in CCl_4 b) H_2 Lindlar catalyst

c) Dilute H_2SO_4, $HgSO_4$ d) Ammoniacal Cu_2Cl_2 (2002)

3. 1-propanol and 2-propanol can be distinguished by December (2000)

a) oxidation by $KMnO_4$ followed by Fehling's solution reaction

b) oxidation by acidic dichromate followed by reaction with Fehling's solution

c) oxidation by Cu followed by reaction with Fehling's solution

d) oxidation with conc. H_2SO_4 followed by reaction with Fehling's solution

4. Propyne and propene can be distinguished by January (2000)

a) Conc. H_2SO_4 b) Br_2 in CCl_4

c) dil. $KMnO_4$ d) ammoniacal silver nitrate

5. A positive carbylamine test is given by

a) N,N-dimethylaniline b) 2,4-dimethylaniline

c) N-methylomethylaniline d) *p*-methyl benzylamine (1999)

6. *p*-chloroaniline and anilinium hydrochloride can be distinguished by

a) Sandmayer reaction b) $NaHCO_3$

c) $AgNO_3$ d) Carbylamine reaction (1998)

KEY

1. c 2. d 3. c 4. d 5. b, d 6. b, c

IV. Multiple Choice Questions (More than one answer is Correct)

1. Violet colouration of Lassaigne's test with sodium nitroprusside indicates the absence of

a) nitrogen b) halogens c) sulphur d) oxygen

2. The compounds which evolve CO_2 on heating to 150 °C are

a) malonic acid b) CH_3CHCH_2COOH

c) CH_3COOH d) succinic acid

3. The Lassaigne's test for nitrogen is positive for

a) urea b) NH_2OH c) SNH_2NH_2 d) thiourea

4. The compounds which answer iodoform reaction are

a) ethanol b) CH_3CHO c) acetone d) $CH_3CH_2COCH_2CH_3$

5. Which of the following statements are correct?
   a) HNO_3 converts NaCN to $NaNO_3$ and HCN.
   b) HNO_3 converts Na_2S to $NaNO_3$ and H_2S.
   c) Freshly prepared $FeSO_4$ is important because $FeSO_4$ on keeping becomes ferric sulphate.
   d) Na should not be melted before the substance is added for Lassaigne's test.
6. Which of the following compounds are insoluble in ether and water?
   a) alcohols b) hydroxy acids c) amides d) sulphonic acids
7. The compounds soluble in HCl are
   a) ketones b) aldehydes c) CH_3NH_2 d) pyridine
8. Some of the following compounds do not give CO_2 with $NaHCO_3$
   a) alcohols b) acids c) phenols d) hydrocarbons
9. The ethanolic silver nitrate test is answered by
   a) alkyl halides b) carboxylic acids c) acyl chlorides d) amines
10. NaI acetone test is answered by
    a) alkyl chlorides b) alkyl bromides c) aryl iodides d) aldehydes
11. The following tests are positive for terminal alkynes.
    a) Baeyer's reagent b) Treatment with $NaHCO_3$ to evolve CO_2
    c) treatment with ammoniacal $AgNO_3$ d) Fehling's solution
12. CH_3CHO answers
    a) Tollens test b) iodoform reaction
    c) Fehling's test d) $NaHCO_3$ test for evolution of CO_2
13. $CH_3CH{=}CH_2$ on ozonolysis in reducing conditions gives CH_3CHO and A. A will answer
    a) Tollens reagent test b) treatment with $NaHCO_3$ to evolve CO_2
    c) iodoform test d) Fehlings' test
14. Toluene, $C_6H_5CH_2Cl$,$C_6H_5CH_2CH_3$ get oxidized to same compound A by $KMnO_4$ oxidation. A will answer
    a) haloform reaction b) $NaHCO_3$ test
    c) ester formation test d) Tollens test
15. A is a good S_N2 solvent. It is obtained by the ozonolysis of tetramethylethylene. It will answer
    a) Tollens reagent test b) sodium bisulphate test
    c) phenyl hydrazine test d) haloform test

KEY

1. a,b,d 2. a,b 3. a,d 4. a,b,c 5. a,b,c 6. b,c,d 7. c,d
8. a,c,d 9. a,b,c 10. a,b 11. a,c,d 12. a,b,c 13. a,d 14. b,c
15. b,c,d

V. Assertion Reason Questions

Questions 1–15 have an assertion in column 1 and reason in column 2. Use the following key to choose the correct answer

a) If both assertion and reason are correct and reason is the correct explanation of assertion
b) If both assertion and reason are correct but reason is not the correct explanation of assertion
c) If assertion is correct but reason is incorrect
d) If assertion is incorrect but reason is correct

No.	Assertion	Reason
1.	Isopropyl alcohol also answers the haloform reaction.	The alcohol is oxidized to methylketone by I_2 + NaOH.
2.	Acetaldehyde does not answer haloform reaction.	It has the CH_3CO group.
3.	*o*-nitrophenol can be separated from *p*-nitrophenol by steam distillation.	*o*-nitrophenol is intramolecularly hydrogen-bonded and has low boiling point.
4.	Terminal alkynes answer test ammoniacal Cu_2Cl_2.	They have acidic hydrogen.
5.	Propyne gives precipitate with ammoniacal silver nitrate but butyne-2 does not give the precipitate.	Propyne has acidic hydrogen.
6.	Phenol decolorizes bromine water.	It is unsaturated.
7.	Tollens test is not answered by HCOOH.	Formic acid has the CHO group.
8.	Acetophenone answers haloform test.	It has the benzene ring.
9.	Lucas test is answered by ethanol.	It is a primary alcohol.
10.	Carbohydrates answer Tollens test.	They contain the aldehyde group.
11.	When toluene and benzylchloride are oxidized with $KMnO_4$, the product answers $NaHCO_3$ test.	All of them give benzoic acid as oxidation product.

(Contd.)

No.	Assertion	Reason
1.	Isopropyl alcohol also answers the haloform reaction.	The alcohol is oxidized to methylketone by I_2 + NaOH.
2.	Acetaldehyde does not answer haloform reaction.	It has the CH_3CO group.
3.	*o*-nitrophenol can be separated from *p*-nitrophenol by steam distillation.	*o*-nitrophenol is intramolecularly hydrogen-bonded and has low boiling point.
4.	Terminal alkynes answer test ammoniacal Cu_2Cl_2.	They have acidic hydrogen.
5.	Propyne gives precipitate with ammoniacal silver nitrate but butyne-2 does not give the precipitate.	Propyne has acidic hydrogen.
6.	Phenol decolorizes bromine water.	It is unsaturated.
7.	Tollens test is not answered by HCOOH.	Formic acid has the CHO group.
8.	Acetophenone answers haloform test.	It has the benzene ring.
9.	Lucas test is answered by ethanol.	It is a primary alcohol.
10.	Carbohydrates answer Tollens test.	They contain the aldehyde group.
11.	When toluene and benzylchloride are oxidized with $KMnO_4$, the product answers $NaHCO_3$ test.	All of them give benzoic acid as oxidation product.
12.	A mixture of aniline and toluene is separated by extraction with dil. HCl.	Aniline forms hydrochloride but toluene will not react.
13.	Phenols are separated from acids by passing into aq. $NaHCO_3$.	Only carboxylic acids are converted to salts.
14.	The haloform reaction is done with I_2 and NaOH only.	$CHCl_3$ and $CHBr_3$ are liquids and hence for qualitative analysis coloured solid will be more suitable. Hence iodination is done.
15.	Conc. sulphuric acid is not chosen as solvent for unsaturated compounds.	The unsaturated compounds polymerize.

KEY

1. a	2. a	3. a	4. a	5. a	6. c	7. d
8. b	9. d	10. a	11. a	12. a	13. a	14. a
15. a						

VI. Comprehension Passages

Passage 1

Certain organic compounds show colour reactions or form characteristic precipitates with reagents which help in identifying them easily.

Answer the following questions

1. Compound A gives precipitate with ammoniacal Cu_2Cl_2. On partial reduction it gives an unsaturated compound which on ozonolysis in reducing conditions gives CH_3CH_2CHO and HCHO. Both A and the unsaturated compound decolorize $KMnO_4$ solution. A is
   a) 2-butyne b) 1-butyne
   c) propyne d) acetylene
2. Compound A reacts with Na evolving hydrogen. A on oxidation gives the solvent B used in S_N2 reactions. B is a product of ozonolysis in oxidizing or reducing conditions of tetramethylethylene. A and B both undergo iodoform reaction. A is
   a) ethanol b) acetaldehyde
   c) isopropyl alcohol d) tertiary butyl alcohol

KEY

1. b **2.** c

Passage 2

Various mixtures of organic compounds can be separated depending on their solubility or acidic nature, etc.

Answer the following questions

1. Diethylamine and 1-butanol can be separated by
   a) adding sufficient H_2SO_4 and performing steam distillation
   b) dissolving in HCl
   c) adding nitrous acid
   d) adding water
2. 1-butyne and 1-butene can be separated by
   a) adding $KMnO_4$ b) adding bromine water
   c) adding HBr d) adding ammoniacal $AgNO_3$

3. CH_3CH_2CHO and acetone can be separated by
   a) adding NaOH and I_2
   b) adding phenyl hydrazine
   c) adding sodium bisulphate
   d) dissolving in acid

KEY

1. a 2. d 3. a

VII. Matrix Matching

Question 1

Column 1	Column 2
a) CH_3CHO	p) answers Tollens test
b) HCOOH	q) answers iodoform test
c) C_2H_5OH	r) answers Fehling's test
d) Acetylene	s) answers $NaHCO_3$ test
	t) gives off H_2 with Na
	u) gives precipitate with a AgNO$_3$.

Answer

$[a \rightarrow p,q,r]$, $[b \rightarrow s,p]$, $[c \rightarrow q,t]$, $[d \rightarrow u]$

Question 2

Column 1	Column 2
a) Malonic acid	p) heating gives CO_2
b) CH_3COOH	q) answers $NaHCO_3$ test
c) Isopropyl alcohol	r) answers iodoform test
d) Lassaigne's test	s) urea shows positive test

Answer

$[a \rightarrow p,q]$, $[b \rightarrow q]$, $[c \rightarrow r]$, $[d \rightarrow s]$

Question 3

a)	Benzene sulphonyl chloride	p)	secondary amines give precipitate.
b)	Phenol is separated from *o*-cresol by	q)	tertiary amines do not react with benzene sulphonyl chloride.
c)	di-butyl ether and chlorobenzene are separated by	r)	by dilute solution of $NaHCO_3$.
d)	Phenol and toluene are separated by	s)	dissolves in conc. H_2SO_4.
		t)	treatment with NaOH.

Answer

$[a \rightarrow p,q], [b \rightarrow r], [c \rightarrow s], [d \rightarrow t]$

VIII. Structure-Based and Numerical Problems

1. An organic compound A gives violet colouration with neutral ferric chloride. It answers Liebermann's nitro test. On complete reduction it gives B. B on oxidation gives C. C does not answer haloform reaction but gives precipitate with 2,4-dinitrophenylhydrazine. C is obtained by ozonolysis of D. The other ozonolysis product of D is E which answers Tollens Test. With NaOH and other aldehydes this always gives HCOONa. Find A to E.

Answer

A is Phenol;

B is cyclohexanol;

C is cyclohexanone;

D is methylene cyclohexane;

E is HCHO.

2. Compound A answers Tollens test and with Fehling's solution gives red solid. A does not undergo aldol condensation. A undergoes Cannizarro reaction with B resulting in C and D. C undergoes oxidation to give E. E gives effervescence with $NaHCO_3$. Acidification of D gives F which also gives effervescence with $NaHCO_3$. F also answers Tollen's test. A is obtained by ozonolysis in Zn/H_2O of G. The other ozonolysis product is B. G answers bromine water test and Bayer's test. G is obtained by partial reduction of phenyl acetylene. Find A to G.

Answer

A is Benzaldehyde;

B is HCHO;

C is benzyl alcohol;

D is HCOONa;

E is C_6H_5COOH;

F is HCOOH;

G is $C_6H_5CH{=}CH_2$;

HCOOH contains HC=O group and hence answers Tollen's test.

3. A is a basic compound. It dissolves in HCl. A reacts with $CHCl_3$ and KOH to give offensive smelling gas. A reacts with $NaNO_2$ and H_2SO_4 to give B. B on hydrolysis gives C which answers neutral $FeCl_3$ test. B with H_3PO_2 gives D. D burns with sooty flame and is a neutral compound. B with CuCl gives E. E answers the test for halogen in Lassaigne's test. When sodium fusion extract is treated with $AgNO_3$ white precipitate forms. E cannot undergo S_N1 and S_N2 reactions. When treated with NaOH under pressure gives C. The formation of C is by benzyne mechanism. Find A to E.

Answer

A is aniline;

B is diazonium salt;

C is phenol;

D is benzene;

E is chlorobenzene.

4. A and B are present in a mixture. On extraction with HCl, B passes into aqueous layer in the form of a salt and is recovered by neutralization. B is prepared by the reaction of C with $NaNH_2$. C does not undergo usual S_N1 S_N2 reactions. The sodium fusion extract of C gives white precipitate with $AgNO_3$ and C burns with sooty flame. The formation of B goes through benzyne intermediate. B reacts with $NaNO_2$ and H_2SO_4 to give D. D with H_3PO_2 gives E. E reacts with CH_3Cl and $AlCl_3$ to give A. A on oxidation with $KMnO_4$ gives F. F shows effervescence with $NaHCO_3$. A reacts with NBS to give G. G reacts with NaOH to give H. H also is oxidized by $KMnO_4$ to give F. H on mild oxidation gives I. I cannot undergo aldol reaction. I undergoes Cannizarro reaction with HCHO to give HCOONa and H. H answers Tollens reagent test. Find A to I.

Answer

A is toluene;

B is aniline;

C is C_6H_5Cl;

D is diazonium salt;

E is benzene;

F is benzoic acid;

G is $C_6H_5CH_2Br$;

H is $C_6H_5CH_2OH$;

I is C_6H_5CHO.

5. A and B are separated by adding sufficient dilute H_2SO_4 to neutralize base, and steam distillation removes B. A reacts with HCl to form salt. It answers Hinsberg test. When benzene sulphonyl chloride is added in NaOH, the compound formed is insoluble and addition of acid does not lead to any reaction. This base is prepared by reaction of 2 moles of C_2H_5I with NH_3 by alkylation. Other compounds are also produced in the reaction. A cannot form stable diazonium salt. B is undergoing S_N2 reactions and does not answer Lucas test. B reacts with HCl in $ZnCl_2$ to give C. C on dehydrohalogenation gives D by E2 mechanism. D with formula C_4H_8 on ozonolysis gives in the presence of Zn and water E and F. Both answer Tollens test. F undergoes Cannizarro reaction and gives HCOONa in that reaction. E undergoes aldol reaction. D answers bromine water and Baeyer's test. Find A to F.

Answer

A is $(C_2H_5)_2NH$;

B is butan-1-ol;

C is butyl chloride;

D is 1-butene;

E is CH_3CH_2CHO;

F is HCHO.

6. An organic compound A with formula $C_6H_{12}O$ forms an oxime but does not answer Tollen's test. A on reduction with sodium amalgam gives B which dehydrates to give C which decolorizes $KMnO_4$. C on ozonolysis produces D and E in dimethyl sulphide. D reduces Tollen's reagent but does not answer haloform reaction. D in enol form has two diastereoisomers. E does not reduce Tollens reagent but answers iodoform reaction. C on further reduction gives F which gives 5 monochloro derivatives. Find A to F.

Answer

A is $CH_3CH_2COCH(CH_3)_2$;

B is $CH_3CH_2CHOHCH(CH_3)_2$;

C is $CH_3CH_2CH{=}C(CH_3)_2$;

D is CH_3CH_2CHO which in enol form has *cis–trans* isomers;

E is acetone;

F is $CH_3CH_2CH_2CH(CH_3)_2$.

7. An aliphatic organic compound A has C, H and N with percentages 61.01, 15.25 and 23.74. A with nitrous acid gives B with percentages 60 and 13.33 for carbon and hydrogen. It does not have nitrogen. A dissolves in HCl. B on oxidation gives C and both B and C answer iodoform reaction. Find A, B and C. Why does A dissolve in HCl?

Answer

Empirical formula of A is C_3H_9N. Empirical formula of B is C_3H_8O.

A is $CH_3\underset{\displaystyle |\atop\displaystyle NH_2}{C}HCH_3$

B is $CH_3\underset{\displaystyle |\atop\displaystyle OH}{C}HCH_3$

C is acetone.

A is an amine which dissolves in HCl forming the hydrochloride.

8. Compound A gives precipitate with ammoniacal $AgNO_3$. On partial reduction it gives B. B on ozonolysis in oxidizing conditions gives C and D. C and D answer test with $NaHCO_3$. D answers Tollen's test. A on hydration in Hg^{2+}/H_2SO_4 gives E. E answers iodoform reaction. B on ozonolysis in reducing conditions gives F and G. F is the isomer of solvent H used in S_N2 reactions. G undergoes Cannizarro reaction and its product in Cannizarro reaction is always HCOONa. Find A to H.

Answer

A is 1-butyne;
B is 1-butene;
C is CH_3CH_2COOH;
D is HCOOH;
E is $CH_3COCH_2CH_3$;
F is CH_3CH_2CHO;
G is HCHO;
H is acetone.

APPENDIX

ADDITIONAL STRUCTURE-BASED PROBLEMS IN ORGANIC CHEMISTRY

1. The % of carbon in a hydrocarbon is 92.3. The molecular weight is 26 g. The compound with $NaNH_2$ gives an anion which reacts with CH_3I to give an alkyne. Find empirical formula and what is the species in reaction of anion with CH_3I.

   Answer

   Empirical formula is CH. The compound is acetylene and the intermediate is a carbanion.

2. Compound A is obtained by reaction between $(CH_3CHCH_3)_2CuLi$ and CH_3CH_2I in diethyl ether. Find the ratio of monosubstituted chlorides obtained for A. What are its other isomers? Hydrogenation of B gives A. How many isomers are there for B. Which exhibit geometric isomerism?

   Answer

   A is $CH_3{-}\overset{\overset{\Large CH_3}{|}}{\underset{\underset{\Large H}{|}}{C}}{-}CH_2CH_3$;

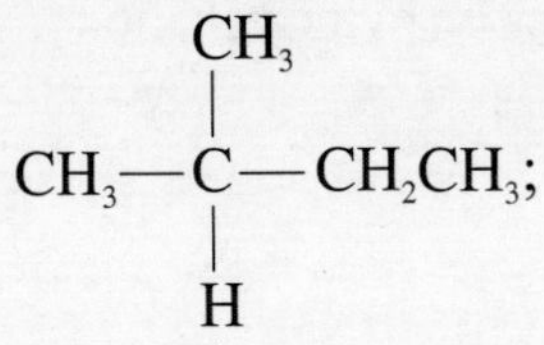

   There are 4 monosubstituted chlorides C, D, E and F.

   C is $ClCH_2{-}\overset{\overset{\Large CH_3}{|}}{\underset{\underset{\Large H}{|}}{C}}{-}CH_2CH_3$;

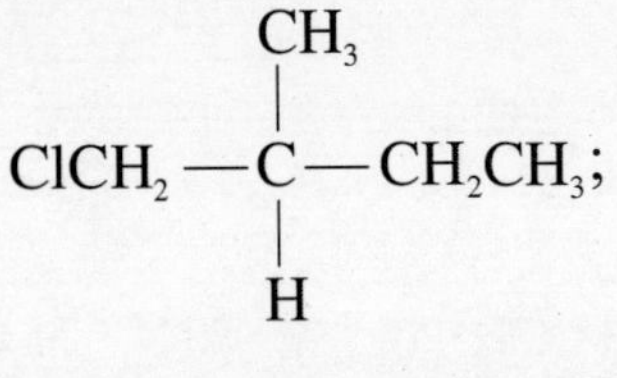

   D is $CH_3{-}\overset{\overset{\Large CH_3}{|}}{\underset{\underset{\Large H}{|}}{C}}{-}CH_2CH_2Cl$;

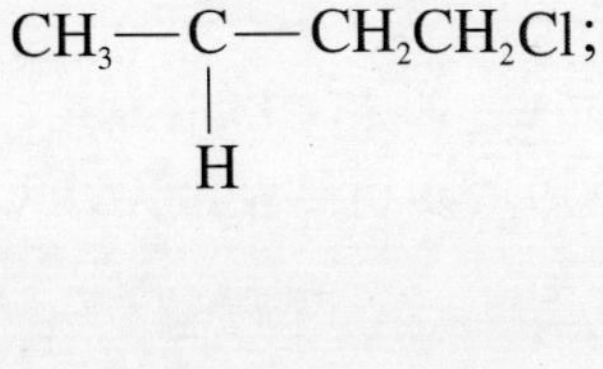

$$\text{E is } CH_3-\underset{\underset{H}{|}}{\overset{\overset{CH_3}{|}}{C}}-CHClCH_3;$$

$$\text{F is } CH_3-\underset{\underset{Cl}{|}}{\overset{\overset{CH_3}{|}}{C}}-CH_2CH_3;$$

They are in ratio 28 : 14 : 35 : 23 percent.

The other isomers of A are $CH_3CH_2CH_2CH_2CH_3$ and $(CH_3)_4C$.

B is $(CH_3)_2C{=}CHCH_3$;

It has the following isomers.

$CH_3CH{=}CHCH_2CH_3$

$CH_2{=}CHCH_2CH_2CH_3$

$CH_2{=}CHCH(CH_3)_2$

Only $CH_3CH{=}CHCH_2CH_3$ exhibits geometric isomerism.

3. One of the three monochlorides of hydrocarbon A is optically active. This chloride has 51% relative to the other 2 chlorides. The molecular weight of the chloride is 106.5 g. Find A. How is A prepared by Corey–House method? Which alkenes on hydrogenation give A. Which of these alkenes exhibit geometric isomerism?

Answer

The molecular weight of chloride indicates structure as

$$CH_3-\underset{\underset{Cl}{|}}{\overset{\overset{H}{|}}{C}}-CH_2CH_2CH_3$$

since it is optically active.

A is prepared from $(CH_3CH_2)_2CuLi$ and $CH_3CH_2CH_2I$ or from $(CH_3CH_2CH_2)_2CuLi$ and CH_3CH_2I. 1-pentene and 2-pentene give rise to A on hydrogenation. 2-pentene exhibits geometric isomerism.

4. A reacts with CH_3CH_2Cl in $AlCl_3$ to give B. B with NBS gives C. C with alcoholic KOH gives D. D is obtained by reduction of E with Lindlar Pd. D with A in HF gives F. E on ozonolysis gives the G, CO_2 and water. G is also obtained by oxidation of B with $KMnO_4$. Find A to G.

Answer

A is Benzene;

B is $C_6H_5CH_2CH_3$;

C is $C_6H_5CHBrCH_3$;

D is $C_6H_5CH=CH_2$;

E is $C_6H_5C\equiv CH$;

F is $(C_6H_5)_2CHCH_3$;

G is C_6H_5COOH.

5. Three compounds A, B and C have formula C_5H_8. All decolorize bromine water, $KMnO_4$ and all are soluble in cold conc. sulphuric acid. A gives precipitate with ammoniacal silver nitrate but B and C have no reaction. A and B on treatment with excess hydrogen give an alkane D which gives three mono substitution products on chlorination. DD, an isomer of D gives only one monosubstitution product on chlorination which cannot undergo E_2 elimination on dehydrohalogenation. C absorbs only one mole of hydrogen to give E. A and B on partial reduction give F and G. F on ozonolysis gives CH_3CHO and CH_3CH_2CHO. G gives by ozonolysis HCHO and $CH_3CH_2CH_2CHO$. E gives only one monosubstitution product on halogenation. C on ozonolysis gives $OCHCH_2CH_2CH_2CHO$. Identify A to G. Among F and G which will show *cis–trans* isomerism. B on oxidation with hot $KMnO_4$ followed by acidification gives acetic acid and propionic acid.

Answer

A is $HC\equiv CCH_2CH_2CH_3$

B is $CH_3C\equiv CCH_2CH_3$

$$CH_3C\equiv CCH_2CH_3 + \text{Oxidation} \longrightarrow CH_3COOH + CH_3CH_2COOH$$

C is Cyclopentene

D is $CH_3CH_2CH_2CH_2CH_3$. 3 monosubstitution products $(CH_3)_4C$ is DD and gives one monochloride only.

E is Cyclopentane which gives only one mono substitution product on halogenation.

F is $CH_2=CHCH_2CH_2CH_3$

G is $CH_3CH=HCCH_2CH_3$

C on ozonolysis gives $OCHCH_2CH_2CH_2CHO$.

F cannot show geometric isomerism but G can show geometric isomerism.

6. Compound A by acid-catalysed dehydration gives B and C. B on ozonolysis gives CH_3CHO, OCHCHO and HCHO. C on ozonolysis gives 2 moles of HCHO and $OCHCH_2CHO$. The heat of hydrogenation of B is lower than that of C. E is another isomer of B or C and has the highest heat of hydrogenation among B, C and E. It gives on ozonolysis HCHO, CO_2 and

CH_3CH_2CHO. B can also be obtained from F by hydrogenation of G with Pt followed by chlorination at 500ºC and treatment with alc. KOH. G is obtained by treating acetylene with $NaNH_2$, NH_3 and compound H. H is obtained as one of the two mono substitution products of I. J on reduction gives I. J on ozonolysis (Zn, acetic acid) gives CH_3CHO and HCHO. Find A to J.

Answer

A is $H_2C{=}CHCH_2CH(OH)CH_3$ (A)
B is $CH_3{-}CH{=}CH{-}CH{=}CH_2$
C is $H_2C{=}CH{-}CH_2{-}CH{=}CH_2$
Since B is conjugated diene its heat of hydrogenation is lower than that of C.
E is $H_2C{=}C{=}CH{-}CH_2CH_3$
Cumulated diene has highest heat of hydrogenation.

$$HC{\equiv}CCH_2CH_2CH_3(G) + H_2 \xrightarrow{pt} H_2C{=}CHCH_2CH_2CH_3(F)$$

$$H_2C{=}CHCH_2CH_2CH_3(F) + Cl_2 \xrightarrow{(500ºC)} H_2C{=}CHCHClCH_2CH_3$$

$$H_2C{=}CHCHClCH_2CH_3 + alc.KOH \rightarrow H_2C{=}CHCH{=}CHCH_3(B)$$

$$HC{\equiv}CH + NaNH_2(NH_3) \rightarrow HC{\equiv}C{-}Na$$

$$HC{\equiv}CNa + ClCH_2CH_2CH_3(H) \rightarrow HC{\equiv}C{-}CH_2CH_2CH_3(G)$$

$$CH_3CH_2CH_3(I) + Cl_2 \rightarrow CH_3CH_2CH_2Cl\ (H) + CH_3CHClCH_3$$

$$CH_3CH{=}CH_2(J) + H_2 - CH_3CH_2CH_3$$

$$CH_3CH{=}CH_2 - Ozone \xrightarrow{(Zn,\ acetic\ acid)} HCHO + CH_3CHO$$

7. You are given a mixture of the following compounds
   1. n-pentane (b.p. = 36ºC)
   2. 2-pentene (b.p. = 36ºC)
   3. 1-chloropropene (b.p. = 37ºC)
   4. Trimethylethylene (b.p. = 39ºC)
   5. 1-pentyne (b.p. = 40ºC)
   6. Methylene chloride (b.p. = 40ºC)
   7. 3,3-dimethyl-1-butene (b.p. = 41ºC)
   8. 1,3-pentadiene (b.p. = 42ºC)

   Give different chemical tests for these and distinguish them.

   Answer

   With $KMnO_4$ we can distinguish the saturated and unsaturated compounds.
   Thus n-pentane, methylene chloride do not answer $KMnO_4$ test. Among n-pentane and

CH_2Cl_2 we can test by doing an S_N2 reaction with CH2Cl2 to liberate Cl^- which can be tested with $AgNO_3$. n-pentane will not undergo S_N2 reaction and will not give AgCl with $AgNO_3$. $CH_2Cl_2 + AgOH \longrightarrow ClCH_2OH + AgCl$

n-pentane + AgOH $\longrightarrow$ no reaction

1-pentyne with amm. $AgNO_3$ gives a precipitate indicating it is a terminal alkyne. 1-chloropropene + $AgNH_2/NH_3$ gives elimination product propyne + AgCl (white precipitate). The remaining contain 3 alkenes and one diene. The number of moles of hydrogen consumed can be estimated for diene and we characterize it. All alkenes will consume only one mole of H_2. Among the three alkenes, two of them give methyl ketones which can be identified by haloform reaction (iodoform–yellow). Only 3,3-dimethyl-1-butene will not yield a C=O compound which answers iodoform reaction. Trimethylethylene gives CH_3CHO and acetone both of which answer haloform reaction. 2-pentene gives CH_3CHO which answers iodoform reaction. The corresponding hydrogenated compounds from 2-pentene and trimethyl ethylene give different number of monochlorides (3 and 4) by halogenation.

8. Compound X has molecular weight 56 g. It decolorizes $KMnO_4$. When HBr is added, it gives Y. Y with AgOH gives an alcohol which can give a very stable carbocation. On hydrogenation it gives a saturated compound Z which gives one monobrominated compound in high yield. Find X to Z.

   Can X exhibit geometric isomerism?

   Answer

   X is $CH_3—C=CH_2$ (with CH_3 on the central C)

$$CH_3-\underset{\displaystyle CH_3}{\underset{|}{C}}=CH_2$$

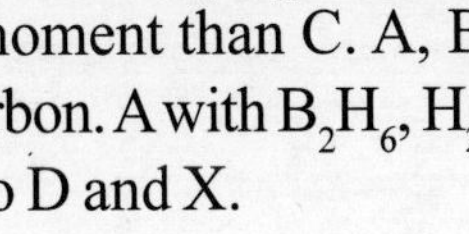

   Y is *t*-butyl chloride. Z is isobutene which selectively gives *t*-butyl bromide in good yield. X cannot exhibit geometric isomerism.

9. X with formula $C_7H_{15}Br$ with strong base gives an alkene mixture A, B and C. B has higher dipole moment than C. A, B and C on hydrogenation give 2-methylhexane as the same hydrocarbon. A with B_2H_6, H_2O_2 and OH^- gives alcohol D. B or C with same reagents give D. Find A to D and X.

   Answer

$$\text{X is } CH_3\underset{\displaystyle CH_3}{\underset{|}{CH}}-\overset{\displaystyle Br}{\overset{|}{CH}}-CH_2CH_2CH_3$$

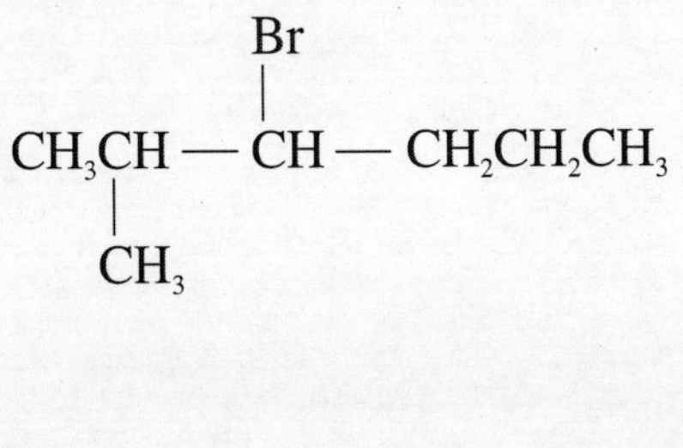

$$\text{A is } \begin{array}{l} CH_3-\underset{\displaystyle CH_3}{\underset{|}{C}}=CHCH_2CH_2CH_3 \end{array}$$

On dehydrohalogenation, alkenes produced are

$$\text{B is } CH_3-\overset{\displaystyle H}{\overset{|}{\underset{\displaystyle CH_3}{\underset{|}{C}}}}-CH=CH-CH_2CH_3 \; \textit{cis}$$

$$\text{C is } CH_3-\overset{\displaystyle H}{\overset{|}{\underset{\displaystyle CH_3}{\underset{|}{C}}}}-\underset{\displaystyle H}{\underset{|}{CH}}=C-CH_2CH_3 \; \textit{trans}$$

(A) *cis* compound *cis*-5-methyl-3-hexene and has high dipole moment C is the *trans* isomer of B with low dipole moment.

From A alcohol formed, according to anti Markovnikov rule is D.

$$\text{D is } CH_3\underset{\displaystyle CH_3}{\underset{|}{CH}}-\overset{\displaystyle OH}{\overset{|}{C}}HCH_2CH_2CH_3$$

B and C are the *cis* alkene and *trans* alkene and by hydroboration oxidation, give D. The distinction between B and C is in terms of dipole moments. C is *trans* compound and B is *cis* compound and has higher dipole moment.

10. Bring out the following conversions.

 i. toluene to *o*-nitrobenzoic acid or *p*-nitrobenzoic acid
 ii. touene to *meta*-nitrobenzoic acid
 iii. toluene to 1-bromo-4-trichloromethylbenzene
 iv. aniline to *p*-nitroaniline
 v. *m*-chloroethylbenzene from benzene
 vi. benzene to *m*-bromobenzoic acid
 vii. benzene to 2-bromo-4-nitrotoluene
 viii. 2-chloro-4-methylphenol
 ix. benzene to *p*-chlorostyrene

Answer

i. Toluene $\xrightarrow{(HNO_3 + H_2SO_4)}$ *p*-nitroalkene $\xrightarrow{(KMnO_4, OH^-, heat, H_3O^+)}$ *p*-nitrobenzoic acid.

ii. Toluene $\xrightarrow{(KMnO_4, OH^-, H_3O^+}$ benzoic acid $\xrightarrow{(HNO_3+H_2SO_4)}$ *m*-nitrobenzoic acid.

iii. Toluene $\xrightarrow{Br_2+Fe}$ *o*-bromotoluene $\xrightarrow{Cl_2(h\nu)}$ 1-bromo-2- trichloromethylbenzene.

vi. Aniline $\xrightarrow{CH_3COCl\ (reflux)}$ $C_6H_5NHCOCH_3$ $\xrightarrow{(HNO_3 + H_2SO_4)}$ *o*-nitroacetanilide + *p*-nitroacetanilide

p-nitroacetanilide $\xrightarrow{H_2O,\ H_2SO_4,\ heat}$ *p*-nitroaniline.

v. Benzene $\xrightarrow{(CH_3COCl + AlCl_3)}$ $C_6H_5COCH_3$ $\xrightarrow{Cl_2 + Fe}$ *m*-chloroacetophenone $\xrightarrow{Zn(Hg)\ HCl}$ *m*.-chloroethylbenzene.

vi. Benzene $\xrightarrow{(CH_3Cl + AlCl_3)}$ toluene $\xrightarrow{(KMnO_4)}$ benzoic acid $\xrightarrow{Br_2 + Fe}$ *m*-bromobenzoic acid.

vii. Benzene $\xrightarrow{(CH_3Cl + AlCl_3)}$ toluene $\xrightarrow{SO_3 + H_2SO_4}$ *p*-toluenesulphonic acid $\xrightarrow{NaOH(H_3O^+)}$ *p*-methylphenol $\xrightarrow{Cl_2 + Fe}$ 2-chloro-4-methylphenol

viii. Benzene $\xrightarrow{(CH_3Cl + AlCl_3)}$ toluene $\longrightarrow$ $SO_3 + H_2SO_4$ $\longrightarrow$ *p*-toluenesulphonic acid $\xrightarrow{NaOH(H_3O^+)}$ *p*-methylphenol $\xrightarrow{Cl_2 + Fe}$ 2-chloro-4-methylphenol

ix. Benzene $\xrightarrow{(CH_3CH_2Cl+AlCl_3)}$ ethylbenzene $\xrightarrow{Cl_2+Fe}$ *p*-chloroethylbenzene (*ortho* isomer) → *p*-chloroethylbenzene $\xrightarrow{NBS}$ *p*-chloro$C_6H_4CHBrCH_3$ $\xrightarrow{Alc.\ KOH}$ *p*-chloro $C_6H_4—CH{=}CH_2$

11. Write the structure of isomeric alkyl bromides of formula $C_5H_{11}Br$

i. which undergoes E_1 elimination at fastest rate.

ii. which cannot produce E_2 product.

iii. which isomers yield a single alkene by E_2 elimination.

iv. which yields complex mixture of alkenes by E_2 elimination.

Answer

Isomers are

A is $CH_3CH_2CH_2CH_2CH_2Br$

$$\begin{array}{c} CH_3 \\ | \\ CH_3—C—CH_2Br \\ | \\ CH_3 \end{array}$$

B is the above structure.

C is $(CH_3)_2CHCH_2CH_2Br$

D is $(CH_3)_2CBrCH_2CH_3$

E is $CH_3CHBrCH_2CH_2CH_3$

F is $CH_3CH_2CHBrCH_2CH_3$

i. D undergoes E_1 elimination at fastest rate since it is tertiary compound.

ii. B cannot produce E_2 product since no beta hydrogen is present.

iii. A, C yield single alkene by elimination since there is only one beta hydrogen and these are $CH_3CH_2CH_2CH{=}CH_2$, $(CH_3)_2CHCH{=}CH_2$. F also yields $CH_3CH_2CH{=}CHCH_3$ as the only alkene either by loss of beta hydrogen on the left or right of Br.

iv. E yields $CH_3CH{=}CH_2CH_2CH_3$ and $CH_2{=}CHCH_2CH_2CH_3$, a mixture of alkenes by E_2 elimination since they have beta hydrogens on both sides of bromine.

12. Write the structure of all isomeric alcohols with formula $C_5H_{12}O$.

i. Which one of these undergoes acid catalysed dehydration most readily?

ii. Write the structure of the most stable carbocation?

iii. Which alkenes may be derived from carbocation in (ii)?

iv. Which alcohols will yield carbocation by hydride shift?

v. Which alcohols will yield carbocation by methyl shift?

vi. What are the different alkenes formed totally?

Answer

$C_5H_{12}O$ is the formula for the following alcohols.

A is $CH_3CH_2CH_2CH_2CH_2OH$

$$B \text{ is } \begin{array}{c} OH \\ | \\ CH_3CCH_2CH_3 \\ | \\ CH_3 \end{array}$$

C is $CH_3CHCH_2CH_2CH_3$ (with OH on the second carbon)

$$\begin{array}{l} CH_3\underset{\displaystyle OH}{\underset{|}{C}}HCH_2CH_2CH_3 \end{array}$$

D is $(CH_3)_2CHCH_2CH_2OH$

$$E \text{ is } CH_3-\underset{\displaystyle CH_3}{\underset{|}{\overset{\displaystyle H}{\overset{|}{C}}}}-\underset{\displaystyle OH}{\underset{|}{\overset{\displaystyle H}{\overset{|}{C}}}}-CH_3$$

$$F \text{ is } CH_3-CH_2-\underset{\displaystyle OH}{\underset{|}{\overset{\displaystyle H}{\overset{|}{C}}}}-CH_2CH_3$$

$$G \text{ is } CH_3-\underset{\displaystyle CH_3}{\underset{|}{\overset{\displaystyle CH_3}{\overset{|}{C}}}}-CH_2OH$$

i. B undergoes dehydration most readily since it gives rise to tertiary carbocation.

ii. $$CH_3\underset{\displaystyle CH_3}{\underset{|}{\overset{+}{C}}}CH_2CH_3$$

iii. The alkenes derived from this cation are

$$(CH_3)_2C{=}CHCH_3 \text{ and } CH_2{=}\underset{\displaystyle CH_3}{\underset{|}{C}}HCH_2CH_3$$

iv. D is $(CH_3)_2CHCH_2CH_2OH$

The cation formed is $(CH_3)_2CHCH_2\overset{+}{C}H_2$

By hydride shift the cation formed is $(CH_3)_2CH\overset{+}{C}HCH_3$. This ion shifts once more leading to cation in (ii).

The alkene formed is $(CH_3)_2C{=}CHCH_3$

F is $CH_3CH_2\underset{\displaystyle OH}{\underset{|}{C}}HCH_2CH_3$ and it gives the cation $CH_3CH_2\overset{+}{C}H-CH_2CH_3$

This by hydride shift gives $CH_3\overset{+}{C}HCH_2CH_2CH_3$

C is $CH_3\underset{|}{\underset{OH}{C}}HCH_2CH_2CH_3$

Gives the cation $CH_3\overset{+}{C}HCH_2CH_2CH_3$

v. $CH_3-\overset{H}{\underset{CH_3}{C}}-\overset{H}{\underset{OH}{C}}-CH_3$

$CH_3-\overset{H}{\underset{CH_3}{C}}-\overset{H}{\underset{+}{C}}-CH_3$

By methyl shift we get $CH_3-\overset{H}{\underset{+}{C}}-\overset{H}{\underset{CH_3}{C}}-CH_3$

Hydride shift leads to tertiary cation

$CH_3-\overset{CH_3}{\underset{+}{C}}-CH_2-CH_3$

vi. The various alkenes formed are

I. $CH_3CH_2CH_2CH=CH_2$ from A.

II. $CH_2=\underset{CH_3}{C}CH_2CH$ from tertiary alcohol B.

And another alkene is $(CH_3)_2C=CHCH_3$.

C is $CH_3\underset{OH}{C}HCH_2CH_2CH_3$

This alcohol gives $CH_3CH{=}CHCH_2CH_3$ in *cis* and *trans* forms.

And also $CH_2{=}CHCH_2CH_2CH_3$

D is $(CH_3)_2CH\,CH_2CH_2OH$ and gives $(CH_3)_2CHCH{=}CH_2$

E is $(CH_3)_2CH\underset{\displaystyle |\atop \displaystyle OH}{C}HCH_3$ and gives $(CH_3)_2C{=}CHCH_3$

F is $CH_3CH_2\overset{\displaystyle OH\atop \displaystyle |}{C}HCH_2CH_3$ and gives $CH_3CH_2CH{=}CHCH_3$ in *cis* and *trans* forms.

13. Compound A has formula $C_{11}H_{16}$ and contains benzene ring. It gives 4 monosubstituted chlorides. Give structure of the monochlorides. One of the monochlorides undergoes dehydrohalogenation by E_2 mechanism to give. $C_6H_5CH{=}CHCH(CH_3)_2$ and $C_6H_5\ CH_2CH{=}C(CH_3)_2$. Which alkene is in higher yield? Find A. Find the ozonolysis products in $(CH_3)_2S$ of the two alkenes.

Answer

A is $C_6H_5CH_2CH_2CH(CH_3)_2$

The 4 monochlorides are $C_6H_5CH_2CHClCH(CH_3)_2$, $C_6H_5CHClCH_2CH(CH_3)_2$,

$$C_6H_5CH_2CH_2\overset{\displaystyle CH_3\atop \displaystyle |}{\underset{\displaystyle |\atop \displaystyle CH_3}{C}}{-}Cl,\quad C_6H_5CH_2CH_2\underset{\displaystyle |\atop \displaystyle CH_3}{C}HCH_2Cl.$$

$C_6H_5CHClCH_2CH(CH_3)_2$ on dehydrohalogenation gives

$C_6H_5CH{=}CHCH(CH_3)_2$ and $C_6H_5CH_2CH{=}C(CH_3)_2$. The first one is conjugated and hence produced in more amount even though it is less substituted. The Hofmann rule is followed. Ozonolysis products are C_6H_5CHO and $CHOCH(CH_3)_2$ for the first alkene. For the second one the products of ozonolysis are $C_6H_5CH_2CHO$ and acetone.

14. Compound A with formula $C_5H_{10}O$ on hydrolysis in dil. H_2SO_4 gives B and C. A does not show test for alcohols. It decolorizes $KMnO_4$. A does not answer Tollens test and does not react with sodium bisulphate. A is not cyclic. B on dehydration gives D. D with Cl_2/H_2O gives E. E with *t*-butyl alcohol gives F. F with sodium acetylide gives G. G on hydrolysis in acid gives 3-butyne-1-ol. Find A to G.

Answer

A is $C_2H_5O{-}\overset{\displaystyle CH_3\atop \displaystyle |}{C}{=}CH_2$;

B is ethanol;

C is acetone;

D is ethylene;

E is $OHCH_2CH_2Cl$;

F is $ClCH_2CH_2$—$O(C(CH_3)_3)$;

G is $(CH_3)_3C-OCH_2CH_2C\equiv CH$. G on hydrolysis gives 3-butyne-1-ol

15. A and B are isomers. A is insoluble in aq. $NaHCO_3$ and is soluble in NaOH. With bromine and $FeBr_3$ A rapidly forms C. C contains three bromine atoms. B does not dissolve in NaOH. B is obtained from E by reaction with CH_2N_2. E is obtained from D by treating with NaOH under high pressure at high temperature. B reacts with Li/NH_3 to give F. F with dil. H_2SO_4 gives G and H. G is the oxidation product of ethylene by ozonolysis in oxidizing conditions. Find A to G.

Answer

A is *m*-cresol;

A and B are isomers;

C is 1,3,5-tribromo-2-methylphenol 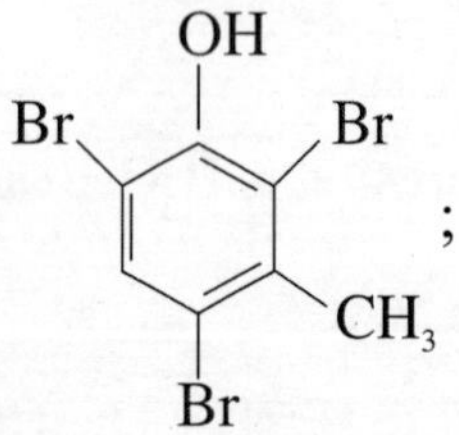 ;

D is C_6H_5Cl;

E is phenol;

F is 1-methoxy-1,4-cyclohexadiene;

G is CH_3OH.

16. Benzene reacts with Br_2/Fe to give A. A with $NaNH_2$ gives B. C is aromatic and has a molecular weight of 28 grams more than A. C does not react with $NaNH_2$ to give substitution product. When C is oxidized with $KMnO_4$ two COOH groups are introduced ortho to the bromine atom and is called D. Benzene reacts with $CH_3COCl/AlCl_3$ to give E. E with Cl_2/Fe gives F. F by Clemmenson reduction gives G. G with $NaNH_2$ gives H, I and J. Find A to J. Why three products are obtained from G with $NaNH_2$.

Answer

A is bromobenzene

B is aniline

C is 2,6-dimethylbromobenzene

D is 2,6-dicarboxyl bromobenzene

E is $C_6H_5COCH_3$

F is *m*-chloroacetophenone

G is *m*-chloroethylbenzene

H is *o*-aminoethylbenzene

I is *m*-aminoethylbenzene

J is *p*-aminoethylbenzene.

Two benzyne intermediates give rise to three products on nucleophilic aromatic substitution.

17. Compound A can give rise to three monochlorides on chlorination in sunlight. The ratio of chlorinated products is 22:52:26 in percentage. One of these chlorinated products (52%) is found to be in S-configuration. Find the number of dichlorides when this is further chlorinated. Find their configurations if they are optically active. Otherwise find which have diastereoisomers, which are achiral, which occur in meso form. If the halogen is replaced by fluorine in this hydrocarbon and elimination is done with CH_3OK, what is the major product? What are the ozonolysis products from this alkene in reducing conditions?

Answer

A is n-pentane. The chloro compound 2-chloropentane is in S-configuration.

The Fischer Projection diagram is

```
        H
        |
Cl ---------- CH3
        |
       C3H7
```

S form of 2-chloropentane

There are 5 dichlorides

```
           H
           |
i.  Cl ---------- CH2Cl
           |
          C3H7
```

R form for the dichloride since priority for CH_2Cl is higher than that for C_3H_7

```
           Cl
           |
ii. Cl ---------- CH3
           |
          C3H7
```

Not chiral since two atoms are same

iii. $CH_3CH_2CHClCHClCH_3$

There are two asymmetric carbons and we have diastereoisomers namely (2S, 3S) and (2S, 3R).

iv. $CH_3CHClCH_2CHClCH_3$

There are two enantiomers (2S, 4R) and (2S, 4S) and occur in a meso form

v. Fischer projection: Cl (top), Cl (left), CH_3 (right), $CH_2CH_2CH_2Cl$ (bottom)

The dehydrohalogenation of 2-fluoropentane gives according to Hoffmann rule 1-pentene as a major product. This by ozonolysis in reducing conditions gives HCHO and $CHOCH_2CH_2CH_3$.

18. Compound A with formula $C_5H_{10}O$ is used in Wittig reaction with the phosphonium ylide obtained from $(C_6H_5)3P$ and C_2H_5Br to obtain B. B on ozonolysis in reducing conditions gives the ketone (A) and C. C is hydration product of acetylene. A by Clemmenson reduction gives D. A gets reduced to $(C_2H_5)_2CHOH$. D gives 3 monochloro products on halogenation.

Find A to D.

Answer

A is diethylketone or 3-pentanone;

B is 3-ethyl-2-pentene;

C is CH_3CHO;

D is n-pentane.

19. A is obtained by Wurtz reaction of CH_3I with Na. A with NBS gives B. B with Mg in ether gives C. C with CH_3CN and acid gives D. D with CH_3MgBr in water gives E. F is obtained by reaction of CHI_3 with Ag. G is partial reduction product of F with Li/NH_3. G on reduction gives A F on hydration in Hg^{2+}/H_2SO_4 gives H. H reacts with CH_3MgBr to give J. Find A to J

Answer

A is C_2H_6;

B is C_2H_5Br;

C is C_2H_5MgBr;

D is $C_2H_5COCH_3$;

$$E \text{ is } C_2H_5-\underset{\underset{CH_3}{|}}{\overset{\overset{CH_3}{|}}{C}}-OH;$$

F is acetylene;

G is ethylene;

H is CH_3CHO;

J is $CH_3CH(OH)CH_3$.

20. Benzene reacts with A in HF at 273K to give 84% of B. B in oxidation in air gives C and acetone. A with NBS gives D. E on partial reduction gives A. E is obtained by hydrolysis of Mg_2C_3. Sodium salt of C with CO_2 followed by acidification gives F. C with CH_3Br in $AlCl_3$ gives G and H. H is formed in minor amounts. G and H with $KMnO_4$ give J and F.F has intramolecular hydrogen bonding. Find A to J.

Answer

A is propylene;

B is isopropyl benzene;

C is phenol;

D is allyl bromide;

E is propyne;

F is salicylic acid;

G is *p*-cresol;

H is *o*-cresol;

J is *p*-hydroxybenzoic acid.

21. A on ozonolysis in oxidizing conditions gives 2 moles of B. Partial reduction of C gives A. C on hydration gives 2-butanone. B with P/Cl_2 gives D. D with KCN gives E. E on hydrolysis gives F which easily decarboxylates on heating to B and CO2. Find A to F.

Answer

A is 2-butene;

B is CH_3COOH;

C is 2-butyne;

D is $ClCH_2COOH$;

E is $CNCH_2COOH$;

F is malonic acid.

22. The disodium salt of A on Kolbe electrolysis gives B. B with Br_2/CCl_4 gives C. C with KCN gives D. D on hydrolysis gives A. A forms anhydride E. E with C_6H_6 gives F in $AlCl_3$. F with Zn amalgam and HCl gives 4-phenylbutanoic acid (G). G with $SOCl_2$ gives H. H in $AlCl_3$ and CS_2 gives α-tetralone. Find A to H.

Answer

A is succinic acid

B is ethylene;

C is ethylene dibromide;

D is $CN(CH_2)_2CN$;

E is succinic acid;

F is succinic anhydride;

G is 3-benzoylpropionic acid;

H is 4-phenylbutanoic acid;

J is 4-phenylbutanoylchloride.

23. Give methods to synthesize B from A using diazotization procedure and other reactions

i. A is o-nitroethylbenzene and B is 2,4-dichloroethylbenzene

ii. A is 4-nitroethylbenzene and B is 2,4-dichloroethylbenzene

iii. A is *p*-nitroethylbenzene and B is 2,6-dichloroethylbenzene

iv. A is 4-nitroethylbenzene and B is 3,4-dichloroethylbenzene

v. A is *p*-aminoethylbenzene and B is 3,5-dichloroethylbenzene

Answer

i. A is treated with Sn/HCl. The product is treated with acetic anhydride. The resulting compound is nitrated with HNO_3/H_2SO_4. Then it is hydrolysed. The hydrolysis product is chlorinated. Then diazotization is done and CuCl added. Then diazotization is done and H_3PO_2 added to get B.

ii. *p*-nitroethylbenzene is chlorinated. This is then treated with Sn/HCl/OH^- and diazotised. Then CuCl is added to get 2,4-dichloroethylbenzene

iii. *p*-nitroethylbenzene is chlorinated with Cl_2/Fe. Then Sn/HCl/OH^- are added, diazotized and H_3PO_2 is added to get 2,6-dichloroethylbenzene.

iv. 4-nitroethylbenzene is treated with Sn/HCl and the product is treated with acetic anhydride. The resulting compound is treated with Cl_2. The product is treated with OH^-, diazotized and CuCl is added to get 3,4-dichloroethylbenzene.

v. *p*-aminoethylbenzene is chlorinated and the product diazotized and H_3PO_2 is added.

24. Benzene reacts with Cl_2/Fe to give A. A with NaOH under pressure at high temperature gives B. B on complete reduction gives C. Con oxidation gives D. D with $KMnO_4$, H^+ and on heating gives E.E with BaO and on heating gives F.F with NH_3, H_2/Pt gives G. Find A to G.

Answer

A is C_6H_5Cl;

B is C_6H_5OH;

C is cyclohexanol;

D is cyclohexanone;

E is $COOH(CH_2)_4COOH$;

F is cyclopentanone;

G is cyclopentylamine.

25. C_6H_{10}(A) on partial reduction gives (B) C_6H_{12} and on complete reduction gives C_6H_{14}(C). C on chlorination gives 3 monochloro derivatives. On ozonolysis in reducing conditions followed by hydrolysis A gives. Two moles of D and one mole of E. On $KMnO_4$ oxidation, B gives an acid of formula $C_3H_6O_2$ (F). F can be obtained by oxidation of G. G answers Tollens test and also undergoes aldol reaction. G in enol form has two diastereoisomers. E answers Cannizarro reaction. What is the product of this Cannizarro reaction? Find A to G.

Answer

C_6H_{10}(A) is $CH_3CH{=}CHCH{=}CHCH_3$;

B C_6H_{12} is $CH_3CH_2CH{=}CHCH_2CH_3$;

C_6H_{14}(C) is n-hexane;

D is CH_3CHO;

E is CHOCHO;

F is CH_3CH_2COOH;

G is CH_3CH_2CHO;

The enol of G has two geometric isomers. These are

$$\underset{\text{(Cis)}}{(CH_3)(H)C{=}C(OH)(H)} \quad \text{and} \quad \underset{\text{(Trans)}}{(CH_3)(H)C{=}C(H)(OH)}$$

The product of Cannizarro reaction of CHOCHO is $CH_2OHCOONa$.

26. On chlorination in CCl_4 an unsaturated compound of A molecular weight 84 g gives B with molecular weight 155 g. B with alc. KOH gives C with molecular weight 82 g. D is an isomer of C which by $KMnO_4$ oxidation gives F and G with formula $C_2H_4O_2$ and $C_2H_2O_4$. A and C on KMnO4 oxidation give E which has formula $C_3H_6O_2$. G can also be obtained by oxidation

of H. H undergoes Cannizarro reaction. Find A to H. Can A exhibit geometric isomerism? Can D undergo Diels–Alder reaction? Can C give precipitate with ammoniacal $AgNO_3$? Give the product of C with Hg^{2+}, H_2SO_4. Can the ketone formed by hydration of C undergo haloform reaction?

Answer

A is EtCH═CHEt;

B is EtCHClCHClEt;

C is EtC≡CEt;

D is MeCH═CHCH═CHMe;

E is EtCOOH;

F is 2MeCOOH;

G is HOOC—COOH;

H is CHOCHO.

A exhibits geometric isomerism. C does not give precipitate with amm. $AgNO_3$ since it has no acidic hydrogen. The ketone $EtCOCH_2Et$ is obtained on hydration of C. Ketone is not a methyl ketone and so does not undergo haloform reaction.

27. With aqueous KOH, a dihalide A with formula $C_3H_6Cl_2$ gives B. A with alcoholic KOH gives C. C_2H_5MgBr reacts with B in water to give D. D forms an iodide E. C on hydroboration oxidation gives B which has 2 enol isomers. Grignard reagent from E reacts with the isomer of B, namely EE to give a tertiary alcohol (F) of formula $C_8H_{18}O$. EE with NaOH/Br_2 gives $CHBr_3$ and salt of acid G. Find A to G.

Answer

A is $C_2H_5CHCl_2$;

B is C_2H_5CHO;

C is $CH_3C{\equiv}CH$;

D is $(C_2H_5)_2CHOH$;

E is $(C_2H_5)_2CHI$;

EE is Acetone;

F is $CH_3—\overset{\overset{\displaystyle OH}{|}}{\underset{\underset{\displaystyle CH_3}{|}}{C}}—CH(C_2H_5)_2$;

G is CH_3COOH;

F cannot undergo oxidation.

28. $C_5H_{10}O$ answers test with $KMnO_4$ for unsaturation and reacts with Na. A on reduction by a suitable catalyst gives $C_5H_{12}O$(B) which does not answer test with $KMnO_4$ but still reacts with Na. B oxidizes first to C and on further oxidation gives D. C is an isomer of A and reacts with MeMgI in H_2O to give E with formula $C_6H_{14}O$ E is oxidized to F and E and F answer Haloform reaction. A on oxidation gives acetone and G with formula $C_2H_2O_4$. H on oxidation gives G and H answers Cannizarro reaction. Find A to H.

Answer

A is $(Me)_2C{=}CHCH_2OH$;

B is $(Me)_2CH{-}CH_2CH_2OH$;

C is isobutylCHO;

D is isobutylCOOH;

E is isobutyl—CHOHMe;

F is isobutyl—COMe;

G is oxalic acid H—CHOCHO.

$$CH_3CH_2-\underset{\underset{CH_2CH_3}{|}}{\overset{\overset{H}{|}}{C}}-MgI + CH_3-\overset{\overset{O}{\|}}{C}-CH_3$$

$$CH_3-\underset{\underset{CH_3}{|}}{\overset{\overset{OH}{|}}{C}}-CH\,(E+)_2$$

29. Compound A with formula $C_4H_8Cl_2$ on heating with Na gives B. B on treatment with Ni/H_2 gives n-butane. A with NH_3 gives C with formula $C_4H_{12}N_2$. On heating the dihydrochloride of C, we get D with formula C_4H_9N. D on exhaustive methylation and pyrolysis of the resulting base gives E with formula $C_5H_{13}N$. The exhaustive methylation is repeated to get F. F undergoes Diels–Alder reactions and on ozonolysis in reducing conditions gives 2HCHO and CHOCHO. Find A to F.

Answer

A is $ClCH_2CH_2CH_2CH_2Cl$;

B is cyclobutane;

C is $NH_2CH_2CH_2CH_2CH_2NH_2$;

D is Tetrahydropyrrole;

E is $H_2C{=}CHCH_2CH_2N(Me)_2$;

F is 1,3-butadiene.

30. Compound A on ozonolysis gives 2 moles of CH_3COOH. A on reduction with Li/NH_3 gives Trans form of B. A with Hg^{2+}/H_2SO_4 gives C. C with I_2 and NaOH gives CHI_3 and sodium salt of acid D. D with $SOCl_2$ gives E. E on reduction with Pd/C in quinoline gives F. F with MeMgI and H_3O^+ gives G. G on oxidation gives C. Find A to G.

Answer

A is 2-butyne;

B is *trans*-2-butene;

C is $CH_3COCH_2CH_3$;

D is CH_3CH_2COOH;

E is CH_3CH_2COCl;

F is CH_3CH_2CHO;

G is $CH_3CH_2CHOHCH_3$.

References

Brown, W. and Poon, T. (2005). *Organic Chemistry*. John Wiley and Sons Inc., New York.

Finar, I.L. (2002). *Organic Chemistry.* Vol.I. Pearson Education.

Maitland Jones Jr. (2000). *Organic Chemistry,* 2nd edn. W.W.Norton Company, New York.

Meislich, E.K., Meislich, H. and Sharefkin, J. (2003). *Schaums 3000 Problems in Organic Chemistry*. Tata McGraw-Hill, New Delhi.

Ongley, P.A. (1959). *Tutorial Questions in Organic Chemistry.* University of London Press Ltd., London.

Solomons, G. and Fryhle, C. (1998). *Organic Chemistry*, 7th edn. John Wiley and Sons Inc., New York.

INDEX

T

W

Z